Dendritic Neurotransmitter Release

Dendritic Neurotransmitter Release

Edited by

Mike Ludwig

Edinburgh University Medical School
Edinburgh, Scotland

 Springer

Library of Congress Cataloging-in-Publication Data

Dendritic neurotransmitter release/edited by Mike Ludwig.
    p.   cm.
   Includes bibliographical references and index.
   ISBN 0-387-22933-7
    1. Neural transmission. 2. Dendrites. 3. Neurotransmitters. I. Ludwig, Mike.

  QP364.5.D46 2004
  612.8′1—dc22

                  2004058904

ISBN 0-387-22933-7

Printed on acid-free paper.

Printed in the United States of America

9 8 7 6 5 4 3 2 1   SPIN 11310020

springeronline.com

Preface

The transmission of the nervous impulse is always from the dendritic branches and the cell body to the axon or functional process. Every neuron, then, possesses a receptor apparatus, the body and the dendritic prolongations, an apparatus of emission, the axon, and the apparatus of distribution, the terminal arborization of the nerve fibers. I designated the foregoing principle: **the theory of dynamic polarization** *(Cajal 1923).*

Ever since the beautiful drawings from Golgi and Cajal, we have been familiar with the organisation of neurones into dendritic, somatic and axonal compartments. Cajal proposed that these cellular compartments were specialised, resulting in his concept of 'dynamic polarisation'. He considered dendrites to be passive elements that simply transferred information from inputs to the soma. Since the discovery that dendrites of many neural populations release neuroactive substances and in doing so, alter neuronal output, it is now apparent that this theory requires qualification.

This book presents recent developments in the neurophysiology of dendritic release of several chemical classes of transmitters in a number of different areas of the mammalian central nervous system. Once released from a neuron, these substances can act as neurotransmitters and/or neuromodulators, to autoregulate the original neuron, its synaptic inputs, and adjacent cells or, by volume transmission, to affect distant cells. In some systems, dendritic transmitter release is part independent of secretion from axon terminal signifying a selective control of the dendritic compartment.

Dendritic transmitter release is a fast moving and rapidly expanding field in neuroscience research, which is fundamentally altering our understanding of the behaviour of the brain. Indeed, modulation of neuronal function by dendritic transmitter release appears to be a common feature rather than a specific phenomenon of individual populations of neurones. The book highlights the diversity and importance of dendritic transmitter release in normal neuronal function and disease.

I thank the participating authors for their excellent contributions to this book. Together, we hope that the book is of interest for scientists and students in neuroscience and physiology and to anyone interested in the functioning of the brain.

Mike Ludwig
Edinburgh, June 2004

CONTENT

3. THE LIFECYCLE OF SECRETORY VESICLES: IMPLICATIONS FOR DENDRITIC NEUROTRANSMITTER RELEASE   35

David K. Apps, Michael A. Cousin, Rory R. Duncan, Ullrich K. Wiegand and Michael J. Shipston.

4. ELECTRICAL PROPERTIES OF DENDRITES RELEVANT TO DENDRITIC TRANSMITTER RELEASE   55

Arnd Roth and Michael Hausser

15. AUTOCRINE MODULATION OF EXCITABILITY BY DENDRITIC PEPTIDE RELEASE FROM MAGNOCELLULAR NEUROSECRETORY CELLS

Colin Brown

16. GALANIN, A NEW CANDIDATE FOR SOMATO-DENDRITIC RELEASE

Marc Landry, Zhi-Qing David Xu, André Calas, and Tomas Hökfelt

ENDOCANNABINOIDS AND GASES

20. SOMATODENDRITIC H_2O_2 FROM MEDIUM SPINY NEURONS INHIBITS AXONAL DOPAMINE RELEASE — 301

Margaret Rice and Marat V. Avshalumov

21. HYDROGEN SULFIDE AS A RETROGRADE SIGNAL — 315

Hideo Kimura

DENDRITIC NEUROTRANSMITTER RELEASE, FROM EARLY DAYS TO TODAY'S CHALLENGES

A. Claudio Cuello[1]

1. INTRODUCTION

When initially proposed the idea that neurotransmitter substances could be released from dendritic processes was somewhat iconoclastic. Nowadays, however, this is widely accepted by neuroscientists. In the following sections I will provide my personal account regarding: i) the nature of dendritic processes, ii) the case of dendritic release of dopamine from neurons of the substantia nigra, iii) the example of dendritic release of GABA from "axonless" neurons, iv) the dendritic release of neuroactive peptides from primary sensory neurons (fulfilling Dale's predictions) and, finally, v) some reflections on the neurochemical complexity of the dendrites as part of the "synaptic/receptive apparatus".

2. SOME FUNDAMENTAL HISTORICAL IDEAS ABOUT THE NATURE OF THE DENDRITIC PROCESSES

The recognition of dendrites as a distinct component of cells of the nervous tissue is attributed to Deiters who described them as "protoplasmatic processes"; they were later renamed as "dendrites" by Wilhem His at the end of XIX century (as quoted by Ramon y Cajal, 1904). Although Camillo Golgi produced the earliest and most accurate representations of dendritic trees for several types of neurons, it was Ramon y Cajal who projected the term with its modern connotation. While Golgi, Gerlach and others entertained the concept that axons and dendrites constitute a continued, diffuse, anatomic and functional network, Ramon y Cajal proposed the notion of individual, interconnected cells through synapses. The unequivocal demonstration of synaptic contacts has been a central aspect of the neuronal theory. Thus, communication between neurons is established at the point of contact. Consequently, individual neurons are the building

---

[1] Departments of Pharmacology and Therapeutics, Anatomy and Cell Biology and Neurology and Neurosurgery, Faculty of Medicine, McGill University, Montreal, PQ, Canada

Dendritic Neurotransmitter Release, edited by M. Ludwig
Springer Science+Business Media, Inc., 2005

blocks of elaborate networks which define the function of the nervous system as opposed to the "reticularism" proposed by Golgi and others.

Furthermore, Sherrington made the sharp observation that there is a delay of the nerve transmission in the reflex arc which he attributed to a valve-like mechanism at the point of neuronal contact, which he termed the "synapse" (Sherrington, 1906). Influenced by Sherrington's electrophysiological observations, Ramon y Cajal proposed in his "neuronal theory" the concept of the dynamic polarization of the cells of the nervous system was elegantly explained in his "*Textura del Sistem Nervioso del Hombre y de los Vertebrados*" (Ramon y Cajal, 1904). This gave a solid framework to modern neurobiology. Since the beginning of the XX century the dogma has been that neuronal electrical activity flows "somatofugally" in axons and "somatopetally" in dendrites, a concept that applies to most neurons (typical neurons) with some exceptions, e.g. dorsal root ganglia sensory neurones and the axonless granule cells of the olfactory bulb. Bodian (1962) resolved the apparent structural and functional contradictions to the neuronal theory found in several neuronal types in the mammalian nervous system by defining the dendritic zone as "the receptor membrane of a neuron, either consisting of a set of tapering cytoplasmatic extensions (dendrites) which receive synaptic endings of other neurons or differentiated to convert environmental stimuli into local-response-generating activity". On the other hand, axons were "cytoplasmatic extensions uniquely differentiated to conduct nervous impulses away from the dendritic zone". This simple, accurate and clear generalization has been missed in many contemporary textbooks.

A great many neuroscientists are probably unaware that not long ago arguments were still exchanged in favour of either the chemical or the electrical nature of the neuronal communication. The realization in the last few decades that the neuronal transmission was a chemical and not an electrical affair (Krnjevic, 1974) added a new dimension to the role of axons and dendrites. Thus, chemical transmitters were expected to be released at axonal terminal sites ("telodendria" in Bodian's terminology, boutons/varicosities of axon terminals in more common terminology) while dendrites were expected to be the main locus of transmitter receptors. This concept was reinforced by the discovery of subcellular units which could very well be the ultimate site of neuronal storage of chemical transmitters, conveniently localized in axon terminals, more specifically in the pre-synaptic side of synapses. In the early days of electron microscopy these units were discovered simultaneously by De Robertis and Bennett (1954) and Palade and Palay (1954) and named "synaptic vesicles". Such convenient vesicular packaging was the anatomical answer to the brilliant concept of quantal release of neurotransmitters which evolved from the laboratory of Bernard Katz, based on electrophysiological observations of miniature potentials in the motor end plate (e.g. Fatt and Katz, 1952; Katz and Miledi, 1963). The actual demonstration that "synaptic vesicles" visualized by electron microscopy were indeed packaging/storing units of neurotransmitters took approximately a decade from its initial ultrastructural discovery. The demonstration of the vesicular storage of transmitters was an emotionally charged and exciting saga. It involved all the efforts of two competing laboratories: De Robertis' lab in Buenos Aires, Argentina, and Whittaker's lab in Cambridge, England. This evidence was provided by the convincing demonstration by these two groups that the archetypical neurotransmitter, acetylcholine, was found clearly enriched in purified subcellular fractions containing isolated synaptic vesicular fractions. These principal investigators wrote an early account of how this was achieved (De Robertis, 1964; Whittaker, 1964).

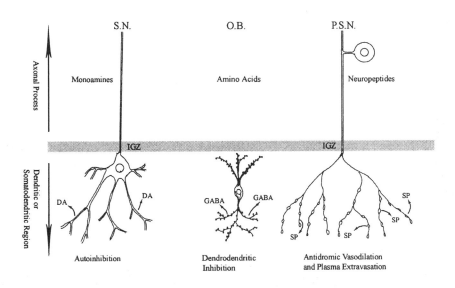

Fig. 1. Schematic illustrations defining the dendritic domains in three very different systems: the substantia nigra (SN) dopaminergic neurons, the olfactory bulb (OB) and primary sensory neurons (PSN) of the spinal cord dorsal root ganglia. These neurons dendritically release monoamines (Dopamine: DA), amino acids (GABA) and neuropeptides (e.g. substance P: SP) respectively . The scheme represents the dendritic and axonal territories following the classical criteria discussed in the chapter. The main function of the dendritically-released substances are indicated below each cell type. A historical account of the evidence supporting dendritic release of transmitters in each neuronal prototype is found in the main text. IGZ indicates impulse generating zone at the origin of the axonal process. Olfactory bulb GABAergic granule cell neurons are considered "axonless".

The basic concepts concerning neurons, their connections, synapses and transmitters were solidly established by the late sixties and early seventies. They are still fundamentally correct and have inspired decades of excellent neurochemical, electrophysiological, neuropharmacological and anatomical studies. The resulting studies have provided us with a reliable framework to understand how the nervous system operates. However, even in the early days of the neuronal theories, there were observations which did not tightly fit with the emerging neuronal principles. A glaring example is that of the sensory axonal reflex. At the beginning of the XX century, Bayliss (1901) and Langley (1923) demonstrated that peripheral terminals (dendrites) of sensory nerves were responsible for the "antidromically"- induced vasodilation. In other words, the dendritic sensory branch which should carry impulses "somatopetally" on occasion carried impulses "somatofugally", against the law of conduction, i.e. antidromically. Later, Lewis and Marvin (1926) proposed that a histamine-like compound released by sensory peripheral branches is responsible for the reported antidromic vasodilation. These unorthodox ideas provoked the imagination of Sir Henry Dale who left us essential principles in physiology and pharmacology. He reasoned that the sensory transmitter which produced antidromic vasodilation in the periphery should be the same transmitter that is released orthodromically at the central ends of the same sensory neurons (Dale, 1935). In other words, he was telling us that if we want to search for sensory neurotransmitters it would be easier to do it in the periphery as opposed to in the CNS where the meshwork of synapses and transmitters would make the task most difficult. This statement was liberally borrowed by many neuroscientists and became the famous "Dale's principle", interpreted wrongly as the "one neuron-one transmitter hypothesis". Dale's reasoning was instead that the same neuron should contain the same transmitters in all its processes which would imply, in modern terms, that dendrites should also contain the same neurotransmitter as axonal terminal processes. The Dale hypothesis ("Dale's principle") does not discuss how many transmitters could be involved.

3. THE CASE FOR DENDRITIC RELEASE OF DOPAMINE FROM SUBSTANTIA NIGRA NEURONS

Although Dale offered a very good lead in the peripheral nervous system to search for putative transmitters released from dendritic processes, it was the central nervous system which offered us the first unequivocal example of such a phenomenon. A great deal of this is due to the cytoarchitecture of the substantia nigra (SN) dopaminergic cell neurons. The majority of their cell bodies are neatly packed in the SN pars compacta, and their axonal processes project dorsally and rostrally toward the neostriatum (caudate and putamen, Ungerstedt, 1971), while their long, branched dendrites radiate profusely through the pars reticulata. This description of the SN dopaminergic neurons corresponds well with Ramon y Cajal's (1904) description: "The Golgi method shows these neurons to have different shapes, predominantly triangular and provided with very long, shaggy and discretely divided dendrites, which expand throughout almost all the nucleus (meaning the area)". The identification of dopamine as the transmitter of the nigro-striatal pathway occupies an interesting and most influencial chapter in the history of modern neuroscience. Arvid Carlsson in the 1950s forcefully promoted dopamine as a transmitter (see Carlsson 1958; and Carlsson's autobiographic account, 1998). The fact that dopamine was found highly concentrated in the basal ganglia (Bertler and Rosengreen, 1964) led Sourkes in Canada and Hornykiewicz in Canada and Austria to demonstrate the dopamine deficits in Parkinson's disease and the consequent proposal for L-DOPA replacement therapy (Sourkes and Poirier, 1965; Hornykiewicz, 1972; and for personal accounts see Hornykiewicz, 1992, and Sourkes, 2000).

The actual anatomical demonstration of the nigral dopaminergic neurons was provided by the revolutionary technique of catecholamine-induced fluorescence (Dahlstrom and Fuxe, 1964). The possibility of some local dendritic synthesis of dopamine was implied in the elegant reports of the occurrence of immunoreactivity to catecholamine synthesizing enzymes in SN dendrites (Pickel et al., 1976; Hokfelt et al., 1973). However, attention to the fate of dopamine in dendritic compartments was provided by the provocative short communication from Bjorklund and Lindvall (1975) which demonstrated glyoxylic acid-induced fluorescence in SN cell bodies and in the long dendrites extending into the rat SN pars reticulata. This report implied a functional role for dendritic dopamine, as the reserpine depletion of the dendritic fluorescence could be re-established by incubating substantia nigra slices *in vitro* with dopamine, and its uptake prevented with desimipramine and benztropine. These observations provoked Laurie Geffen, Thomas Jessell (at the time a Cambridge graduate student), Leslie Iversen

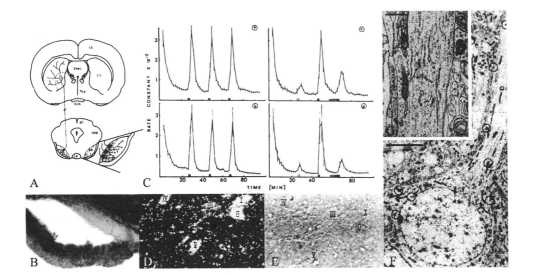

Fig. 2. Composite schematic, illustrating the sites of dopamine incorporation, storage and release from dendritic processes of the rat substantia nigra, A) scheme illustrating the segregation of axonal and dendritic processes in the nigrostriatal pathway. SN, substantia nigra; c, pars compacta; IP, nucleus interpeduncularis; MGB, medial geniculate body; sc, superior colliculus; CS, corpus striatum; Sept, septum; CX, cortex; Hyp, hypothalamus B) illustration of the microdissection of the pars reticulata of the rat substantia nigra from live, unfixed, unstained brain stem tissue slices. The open (dissected out pars reticulata) is observed by transillumination between the crus cerebri (below) and the SN pars compacta (above). On the right the exit of the 3rd cranial nerve can be observed. The darker (myelinated) region in the upper, right corner is part of the medial forebrain bundle C) Illustrates the rate constant outflow of previously incorporated 3H-dopamine, from superfused slices of the rat substantia nigra (top panels, a and c) and corpus striatum (b and d). Panels a and b illustrate repeated K+ stimulated release of dopamine from (microdissected) substantia nigra dendrites, and the marked inhibition of dendritic and axonal release in the absence of Ca^{2+} (open squares in c and d) or presence of high magnesium molarity (reproduced with permission from Geffen et al, 1976). D) Radioautography illustrating the incorporation of 3H-dopamine in a number of substantia nigra neurons (labeled I to V) and long dendritic processes, radiating towards the pars reticulata, dark field micrograph. E) Phase contrast micrograph showing the localization of neurons labeled I to V, as well as the dendritic profiles incorporating radioactive dopamine as displayed with arrows. F) Electron microscopic radioautography illustrating silver grains (circles), depicting subcellular sites of tritiated dopamine in the perisomatic region and the dendritic process of a substantia nigra neuron. Scale bar = 5μm, D-F from (Cuello and Kelly, 1977). G) Electron micrograph, electron dense products (arrow heads) illustrating the subcellular storage sites of incorporated 5-OH-dopamine within tubular profiles resembling the smooth endoplasmic reticulum cisterns (ser) of a substantia nigra dendritic profile (Reproduced with permission from Mercer et al, 1979).

and myself to investigate whether dendrites are also able to release neurotransmitters. The neat organization of the SN dopaminergic cell bodies in the pars compacta and dopaminergic dendrites in the pars reticulata permitted simple microdissection, taking advantage of the fact that fresh, unstained, live tissue slices could be obtained by a procedure developed at the time (for review see Cuello and Carson, 1983) and also of the refinement in the micro superperfusion system developed by Jessell and Iversen. We embarked on the project by comparing the uptake and release of ^{3}H-dopamine in the area of the nerve terminals (caudate putamen) and the dendrites (substantia nigra) in microdissected rat CNS tissue section. In these experiments (Geffen et al., 1976), we provided the first direct evidence for a dopamine uptake and release mechanism in dendrites of the SN neurons. We were able to show that short pulses of high molarity KCl provoked a Ca^{2+}-dependent release of newly incorporated [^{3}H] dopamine from substantia nigra slices (dopaminergic dendrites), analogous to that seen in dopaminergic nerve terminals of the neostriatum. Simultaneously, and to our great surprise, in the same Nature issue Korf et al. (1976) indicated that the antidromic stimulation of the substantia nigra projection to the neostriatum resulted in an elevated amount of dopamine metabolites providing indirect evidence that a release of the amine could occur in the SN somatodendritic area following neuronal stimulation. Soon after these reports the Glowinsky group in Paris provided additional *in vivo* evidence of this mechanism using the push-pull cannula approach (Nieoullon et al., 1977) and later, in 1981, a well quoted review article emphasizing the TTX-insensitive nature of this dendritic monoamine release and its incipient pharmacology (Cheramy et al., 1981).

The rat and human brain contain approximately the same concentrations of dopamine in the substantia nigra, in the order of 0.40 µg/g of wet tissue (Sourkes and Poirier, 1965; Hornykiewicz, 1972; Cuello and Iversen, 1978). At the time we developed a highly sensitive radioenzymatic technique (Cuello et al., 1973) which permitted us to determine monoamine concentrations in minute, separate, microdissected regions of the rat substantia nigra. We observed relatively high concentrations (1.52 µg/g) of dopamine in the SN pars compacta where the dopaminergic elements are densely concentrated in cell bodies and short dendrites, while smaller amounts (0.28 µg/g) were found in the pars reticulata where the long-branched dopaminergic dendrites are present (Cuello and Iversen, 1978). As the weight of the pars reticulata is almost 3 times that of the compacta, it can be concluded that the pars reticulata contributes significantly to the total dopamine content of the substantia nigra.

We were interested in defining the dendritic storage sites of dopamine. We were able to demonstrate by high resolution radioautography that radiolabelled dopamine is incorporated into substantia nigra dendrites (Cuello and Kelly, 1977), however, the electron microscopical analysis of this material did not reveal the expected "synaptic vesicle" localization as the potential site of monoamine storage. The question then was where is dopamine stored in the SN dendritic processes if not in synaptic vesicles? Synaptic vesicles had been described in dendrites of substantia nigra neurons (Hajdu et al., 1973) but were not seen either by Sotelo (1971) using radiolabelled noradrenaline or by us stereotaxically applying ^{3}H-dopamine in the SN (Cuello and Kelly, 1977; Cuello and Iversen, 1978; Cuello, 1982).

We resorted to the use of false transmitters which render an electron dense signal upon incorporation into subcellular compartments. Typically, these were small and large core synaptic vesicles of sympathetic neurons. We stereotaxically injected small amounts of the false transmitter 5-OHD into the rat substantia nigra. This again did not demonstrate amine storage in synaptic vesicles but generated evidence of amine uptake and storage in short cisterns of the smooth endoplasmic reticulum (SER) within dendritic profiles of the substantia nigra (Mercer et al., 1979). Such subcellular compartamentalization would be compatible with what had been previously proposed by Tranzer (1972) as an "immature" catecholaminergic compartment in the axonal shaft of sympathetic nerves. The involvement of these structures was elegantly confirmed by Pickel and collaborators (Nirenberg et al., 1996) using the immunogold technique for the ultrastructural localization of the vesicular monoamine transporter VMAT2. These authors found clear evidence of VMAT2-immunoreactive sites in tubulo-vesicular profiles related to the SER within dendritic processes of ventrotegmental dopaminergic neurons. Such non-synaptic vesicle localization of dendritic dopamine could explain the ionic and electrophysiological characteristics observed for the dopamine release in these neuronal processes (Cheramy et al 1981; Rice et al., 1997) and most importantly, it provided a structural basis for the recent and well supported proposition, that the dendritic dopamine release is the consequence of a functional dopamine transport reversal (Falkenburger et al., 2001). These authors have proposed the interesting possibility of the therapeutic inhibition of the dendritic VMAT2 transporter in early stages of Parkinson disease, preventing dopamine-induced autoinhibition at the nigra level.

4. THE RELEASE OF GABA FROM THE OLFACTORY BULB GRANULE CELL DENDRITES

Early studies on the olfactory bulb by Golgi (1875) demonstrated the existence of small granular neurons with several dendritic branches but deprived of an axonal process. This was confirmed by Ramon y Cajal (1904), who further revealed that the processes ramify largely in the external plexiform layer where they establish contacts with the secondary dendritic processes of the mitral cell neuron. He also observed the presence of "spines" in these dendrites (generally referred to as "gemmules" by modern authors). Ramon y Cajal (1904) also advanced some ideas on the functional aspects of these processes by saying: "The peripheral dendrite would represent dynamically a functional expansion of the neurons conveying nervous flow, cellulifugally as in the genuine axons". In the early sixties, in Oxford, Phillips et al. (1963) gathered electrophysiological data reinforcing this notion. It was Shepherd and collaborators who extensively analyzed this system electrophysiologically and determined that the mitral dendrite was excitatory to the granule cell dendrite while the granule cell dendrite had an inhibitory effect on mitral cell dendrites (for reviews see Shepherd, 1972, 1976). The electron microscopical analysis of these dendritic contacts revealed unusual features for dendritic profiles (Hirata, 1964). Electron microscopic observations and the vast array of electrophysiological data on the physiological properties of these two types of dendrites (mitral and granule cell) permitted Rall et al. (1966) to propose the existence of reciprocal dendro-dendritic synapses. Further studies reinforced this idea (Price and Powell, 1970a, b; Willey, 1973). However, the existence of these reciprocal dendro-dendritic synapses was challenged by Ramon-Moliner (1977), which prompted Jackowski et al. (1978) to re-examine this problem in an extensive electron microscopic study on individual and serial sections and also in material stained by the E-PTA and BIUL methods. These authors concluded that the majority of gemmulofugal and mitrofugal synapses in the external plexiform layer (EPL) of the olfactory bulb can be clearly resolved and in serial sections. These authors confirmed that the synaptic contact occurs in pairs (reciprocal synapses).

The above was the anatomical and physiological background of these curious dendrite to dendrite communications. We decided to look at this problem anew with the perspective of another example of dendritically driven transmitter release system. We looked at the electron microscopy of the region and were able to confirm the existence of the mitral-granule cell reciprocal dendro-dendritic synapses in the EPL of the rat olfactory bulb (Cuello, 1982). However, it was not clear at the time which transmitters were involved in such synapses. Several putative neurotransmitters were attributed to the mitral cell neuron, amongst them acetylcholine, aspartate and glutamate (Bloom et al., 1965; Felix and McLennan, 1971; Nicoll, 1971; Yamamoto and Matsui, 1976; Hunt and Schmidt, 1978) but there was no direct, conclusive evidence for a defined mitral neurotransmitter substance. On the other hand, there was a strong case for the GABAergic nature of the granule cell dendrite. Some early electrophysiological data suggested that GABA could be involved in the inhibitory effects observed secondarily to the antidromic activities of the mitral cell neurons (Felix and McLennan, 1971; McLennan, 1971; Nicoll, 1971).

At the time, Jahr and Nicoll (1980) provided a neat demonstration of the GABAergic nature of the granule cell dendritic inhibition using intracellular recordings of identified mitral cells in tissue slices of the turtle olfactory bulb where antidromically evoked inhibitory postsynaptic potentials could be blocked by bicuculline. The neurochemistry of the region demonstrated that the highest olfactory bulb concentrations of GABA and glutamic acid decarboxylase (GAD) occurred in the EPL (Graham, 1973; Jaffe and Cuello, 1980a) and that pattern of neurotransmitter distribution is not followed by other markers such as acetylcholine or the catecholamines (Jaffe and Cuello, 1980a).

A direct evidence for the GABAergic nature of the granule cell dendrites came from the work of Ribak et al. (1977) who provided immuno-histochemical evidence for the localization of GAD in these processes, both at the light and electron microscopical levels. It was also observed that granule cell dendrites incorporated 3H-GABA both *in vivo* and *in vitro* (Halasz et al., 1979; Jaffe and Cuello, 1980b). The *in vitro* uptake of 3H-GABA in the EPL tissue preparations were shown to be sodium and temperature dependent. We demonstrated that the incorporation of 3H-GABA in the EPL was reduced drastically in the presence of L-2, 4-diaminobutyric acid (DABA) but only marginally affected by p-alanine. DABA and p-alanine were already known to displace GABA preferentially from neuronal or glial sites respectively (Iversen and Kelly, 1975; Kelly and Dick, 1976). The *in vitro* release of previously incorporated 3H-GABA from microdissected slices of the EPL in response to depolarizing stimulation occurs in a manner similar to that observed in GABAergic nerve terminals of the substantia nigra (Jaffe and Cuello, 1980b). High molarity K^+ and fentomolar amounts of veratridine were able to elicit a reproducible Ca^{2+}-dependent release of 3H-GABA, both from the substantia nigra (striatonigral axonal terminations) and EPL slices of the olfactory bulb (granule cell dendrites). As in axon terminals, the veratridine-induced dendritic release of GABA was blocked by tetrodotoxin. These experiments provided neurochemical evidence for yet another example of CNS dendritic release of a classical transmitter substance. This example, however, has to be taken more as an exception than as a rule as these neuronal processes probably combine both the receptive and the impulse-generating zone (IGZ) which, in Bodian's (1962) conceptualization of neuronal processes, characterize dendrites and axons, respectively.

5. THE DENDRITIC RELEASE OF NEUROACTIVE PEPTIDES FROM DENDRITIC BRANCHES OF PRIMARY SENSORY NEURONS

The final historical example of dendritic release of neuroactive substances is that of the peripheral branches of primary sensory neurons whose receptor/transducer terminations correspond to dendritic profiles according to Ramon y Cajal's interpretation

(1904) and Bodian's (1962) more contemporary classification. These branches, as described in any Neuroscience textbook, gather environmental signals which are eventually transformed into nerve impulses. However, as discussed above, these dendritic sensory branches were additionally noted early on to have an "antidromic" (against the direction of the neuronal polarity) effector activity. This effect was classically described as the "sensory antidromic vasodilatatory response". Dale hypothesized that a chemical compound released from these sensory peripheral branches was causing the antidromic vasodilation (Dale, 1935). They did not reflect in their considerations on the dendritic nature of these terminations. Sir Henry Dale hoped that the investigation of these branches could lead to the identification of "the" sensory transmitter in the CNS. The realization of that vision had to wait for the identification of neuroactive peptides as transmitter candidates.

In the thirties, Von Euler and Gaddum (1931) discovered from tissue extracts a chemically unidentified substance with pharmacological effects which included vasodilation. This unknown substance was later named substance P. Lembeck (1953) found that the substance P-like material was enriched in dorsal roots of sensory ganglia and speculated that substance P was indeed "the" sensory transmitter. This concept was introduced by Otsuka and collaborators (1972) in a series of elaborate electrophysiological studies made even prior to the chemical identification of substance P.

The actual nature of the substance P had to wait for the chemical identification and characterization by Leeman and co-workers who described it as a peptide of eleven amino acids of the tachykinin family (Chang et al., 1971). The development of high affinity antibodies against the known substance P peptide allowed Thomas Hokfelt to define the substance P localization sites in the CNS and periphery. Immunoreactivity to substance P was described by Hokfelt et al (1975) in the periphery, in locations compatible with the sensory nature of these processes. With Marina Del Fiacco and George Paxinos, we were able to demonstrate experimentally that the resection of the purely sensory branch of the mental nerve abolished substance P immunoreactivity in peripheral fibres providing evidence for its sensory localization in peripheral dendritic terminals (Cuello et al., 1978). Furthermore, the peptide was observed in the cell bodies of the primary sensory neurons of the spinal cord and trigeminal system (Hokfelt et al., 1975; Del Fiacco and Cuello, 1980) as well as in the CNS in superficial layers of the dorsal horn of the spinal cord localizations which were compatible with the sensory nature of the newly discovered neuroactive peptide (Hokfelt et al., 1975; Cuello et al., 1978). Early after the identification of substance P, a transport mechanism of substance P immunoreactive material towards the periphery was demonstrated for primary sensory neurons (Takahashi and Otsuka, 1975; Hokfelt et al., 1975; Gamse et al., 1979a). Lembeck and Holzer (1979) pointed out that the potent substance P vasodilatory effects are very similar to those resulting from the antidromic stimulation of mixed nerves (Lembeck and Holzer, 1979). More direct evidence supporting the hypothesis that substance P plays a key role in the sensory-elicited antidromic vasodilation came from the application of the first reported substance P antagonists which were capable of inhibiting the antidromic vasodilation both in the saphenous (Lembeck, Donnerer and Bartho, 1982) and in the trigeminal (Couture and Cuello, 1984) sensory peripheral territories. Furthermore, the application of capsaicin (8-methyl-N-vanillyl-6-nonemide) known at the time to be a potent activator of chemo-sensitive peripheral sensory fibers led to desensitization of sensory fibers (Jancso et al., 1967). We demonstrated that this agent was capable of depleting to near totality the immunoreactivity to substance P in presumptive sensory fibers at their point of entry in the spinal cord (Jessell et al., 1978). The neonatal application of capsaicin was shown to elicit a marked impairment of the vasodilatory responses following antidromic stimulation (Lembeck and Holzer, 1979; Gamse et al., 1979b). Nowadays there is abundant evidence for a peripheral effector role of "dendritically" released substance P and a number of neuroactive peptides from peripheral branches of sensory neurons. The current view is that sensory peptides are locally released to facilitate the passage of blood borne substances (plasma extravasation) to respond to peripheral noxia. This mechanism is central to what is known as neurogenic inflammation and might be involved in a number of pathological conditions as varied as migraine and disturbances of the upper airways.

A more complex case of "dendritically" located neuroactive peptides is that of substance P in sensory ganglia. Sympathetic ganglia are richly innervated by substance P-immunoreactive fibers (Hokfelt et al., 1977). We established that in the prevertebral ganglia these substance P-immunoreactive fibers originate in sensory neurons of the dorsal root ganglia by cutting the lumbar splanchnic nerves, while the severance of the hypogastric or colonic nerves produced no obvious changes (Baker et al., 1980; Matthews and Cuello, 1982). These fibers were affected by the application of capsaicin (Gamse et al., 1981; Matthews and Cuello, 1982). Electron microscopy of these peripheral sensory terminals in the prevertebral sympathetic ganglia revealed clear indications of conventional substance P immunoreactive presynaptic sites (Matthews and Cuello, 1982). This synaptic configuration of substance P immunoreactive sensory branch fibres in sympathetic ganglia was consistent with the electrophysiological findings of Otsuka and collaborators (Tsunoo et al., 1982) showing a substance P induced depolarization resembling non-cholinergic slow EPSP in the sympathetic ganglia. The existence of these sensory dendritic synapses in the sympathetic ganglia might be the structural and electrophysiological basis for postulated enteric-automonic reflexes.

6. REFLECTIONS ON OTHER POSSIBLE ROLES OF DENDRITES IN CHEMICAL TRANSMISSION

In recent years there has been renewed interest in the significance of the dendritic release of neurotransmitter substances. The Editors and the contributors to this volume have made significant progress to consolidate and expand the concept of a dendritic mechanism for transmitter signalling. These investigations will play an important role in helping us to understand the complexity of neuron to neuron and neuron to peripheral cell communications. This timely volume should help us in the quest for a deeper understanding of these relationships. Dendrites, besides being the key cellular receptive neuronal component also participate in sending back messages in the form of "non-classical" neuronal communications (Cuello, 1983). This could be via neurotransmitter substances but also through trophic factor-mediated communications. For example, it is likely that dendrites are central to the maintenance of synaptic numbers in the CNS. It has been well illustrated by Thoenen and collaborators (for review see Thoenen, 1995) that the neuronal release of trophic factors is activity dependent. On the other hand, the steady state number of synaptic boutons appears to be dependent on the mature, fully differentiated CNS, by the minute offering of endogenous trophic factors. This is illustrated by the disappearance of pre-existent NGF-sensitive cholinergic synapses

(Debeir et al., 1999) following the immunoneutralization of NGF or the pharmacological blockade of its cognate receptor in the cerebral cortex. A more obvious place for the origin of such trophic stimulation lies in the dendritic processes. A number of *in vitro* investigations do indeed point towards the dendrites as the main site for trophic factor storage and release of neurotrophins (e.g. see Goodman, 1996; Horch and Katz, 2002). These propositions await confirmation from *in vivo* models as well as more direct ultrastructural analysis for which there are currently important limitations.

We presently have a good understanding of the electrophysiological aspects of the conducting properties of dendrites (for reviews see Johnston et al., 1996; Yuste and Tank, 1996). We are also learning day-by-day that dendrites can modulate their receptive capabilities with exquisite precision by modifying their molecular post-synaptic apparatus. Thus, the work of Steward, Schuman and others (for reviews see Steward and Schuman, 2001; Jiang and Schuman, 2002; Glanzer and Eberwine, 2003) teaches us that a great deal of protein synthesis related to the downstream signalling of transmitter receptors occurs locally in dendritic domains and can adapt to the neuron's functional experience.

It is to be expected that in the coming years we will learn a great deal about dendritic-induced neurotransmitter signalling, dendritic trophic interactions and dendritic self regulation of their receptor capabilities. The neuroscience field is ready to generate and accept with great interest these expected developments.

7. ACKNOWLEDGEMENTS

I would like to thank my Cambridge, Oxford and McGill collaborators who have contributed to all aspects of the research discussed in this review, in particular my past graduate students. The investigations quoted here were supported while in the UK by grants from The Medical Research Council, The Wellcome Trust and the E.P. Abraham Cephalosporin Trust. I would like to thank my present research group in Canada at the McGill Department of Pharmacology and Therapeutics for bringing new interests to my current research activities, to the CIHR and the NIH (NIA) for supporting my ongoing research projects and to Sid Parkinson and Mona-Lisa Bolduc for effective editorial assistance.

8. REFERENCES

Baker, S. C., Cuello, A. C., and Matthews, M. R., 1980, Substance P-containing synapses in a sympathetic ganglion, and their possible origin as collaterals from sensory nerve fibers, *J. Physiol.* **308**: 76.
Bayliss, W. M., 1901, On the origin from the spinal cord of the vasodilator fibers of the hind limb, and on the nature of these fibers, *J. Physiol.* **26**: 173.
Bertler, A., and Rosengren, E., 1964, Occurrence and distribution of dopamine in brain and other tissues, *Experientia* **15**: 10.
Bjorklund, A., and Lindvall, O., 1975, Dopamine in dendrites of substantia nigra neurons: suggestions for a role in dendritic terminals, *Brain Res.* **83**: 531.
Bloom, E. F., Costa, E., and Salmoiraghi, G. C., 1965, Analysis of individual rabbit olfactory bulb neuron responses to the microelectrophoresis of acetylcholine, norepinephrine and serotonin synergists and antagonists, *J. Pharmacol. Exp. Ther.* **146**: 16.
Bodian, D., 1962, The generalized vertebrate neuron, *Science* **137**: 323.
Carlsson, A., 1998, Autobiography, in: *The history of neuroscience in autobiography, II.*, L.R. Squire, ed., Academic Press, San Diego, pp. 28-66.

Carlsson, A., Lindqvist, M., Magnusson, T., and Waldeck, B., 1958, On the presence of 3-hydroxytyramine in brain, *Science* **127**: 471.

Chang, M. M., Leeman, S., and Niall, H., 1971, Amino-acid sequence of substance P, *Nat. New Biol.* **232**: 86.

Cheramy, A., Leviel, V., and Glowinski, J., 1981, Dendritic release of dopamine in the substantia nigra, *Nature* **289**: 537.

Couture, R., and Cuello, A. C., 1984, Studies on the trigeminal antidromic vasodilatation and plasma extravasation in the rat, *J. Physiol.* **346**: 273.

Cuello, A. C., 1982, Storage and release of amines, amino acids and peptides from dendrites, *Prog Brain Res.* **55**: 205.

Cuello, A. C., 1983, Nonclassical neuronal communications, *Fed. Proc.* **42**: 2912.

Cuello, A. C., and Carson, S., 1983, Microdissection of fresh rat brain tissue slices, in: *Brain Microdissection Techniques,* A. C. Cuello, ed., John Wiley and Sons, New York, pp. 37-125.

Cuello, A. C., and Iversen, L. L., 1978, Interactions of dopamine with other neurotransmitters in the rat substantia nigra: a possible functional role of dendritic dopamine, in: *International Symposium on Interactions among Putative Neurotransmitters in the Brain,* S. Garatinni, J.F. Pujol and R. Samanin, eds., Raven Press, New York, pp. 127-149.

Cuello, A. C., and Kelly, J. S., 1977, Electron microscopic autoradiographic localisation of ^{3}H-dopamine in the dendrites of the dopaminergic neurones of the rat substantia nigra *in vivo, Brit. J. Pharmacol.* **59**: 527.

Cuello, A. C., Del Fiacco, M., and Paxinos, G., 1978, The central and peripheral ends of the substance P-containing sensory neurons in the rat trigeminal system, *Brain Res.* **152**: 499.

Cuello, A. C., Hiley, R., and Iversen, L. L., 1973, Use of catechol O-methyltransferase for the enzyme radiochemical assay of dopamine, *J. Neurochem.* **21**: 1337.

Dahlstrom, A., and Fuxe, K., 1964, Evidence for the existence of monoamine-containing neurons in the central nervous system. I: Demonstration of monoamines in the cell bodies of brain stem neurons, *Acta Physiol. Scand.* 62, S232: 1.

Dale, H. H., 1935, Pharmacology and nerve endings, *Proc. Roy. Soc. Med.* **28**: 319.

De Robertis, E., 1964, Electron microscope and chemical study of binding sites of brain biogenic amines, *Prog. Brain Res.* **8**: 118.

De Robertis, E., and Bennett, H., 1954, Submicroscopic vesicular components in the synapse, *Fed. Proc.* **13**:35.

Debeir, T., Saragovi, H. U., and Cuello, A. C., 1999, A nerve growth factor mimetic TrkA antagonist causes withdrawal of cortical cholinergic boutons in the adult rat, *PNAS USA* **96**: 4067.

Del Fiacco, M., and Cuello, A. C., 1980, Substance P- and enkephalin-containing neurones in the rat trigeminal system, *Neuroscience* **5**: 803.

Falkenburger, B. H., Barstow, K. L., and Mintz, I. M., 2001, Dendrodendritic inhibition through reversal of dopamine transport, *Science.* **293**: 2465.

Fatt, P., and Katz, B., 1952, The electric activity of the motor end-plate, *Proc. Royal . Soc. Lond. B* **140**: 183.

Felix, D., and McIennan, H., 1971, The effect of bicuculline on the inhibition of mitral cells of the olfactory bulb, *Brain Res.* **25**: 661.

Gamse, R., Lembeck, F., and Cuello, A. C., 1979a, Substance P in the vagus nerve. Immunochemical and immunohistochemical evidence for axoplasmic transport, *Naunyn Schmiedeberg's Arch. Pharmak.* **306**: 37.

Gamse, R., Lembeck, F., and Holzer, P., 1979b, Indirect evidence for presynaptic localisation of opiate receptors on chemosensitive primary sensory neurones, *Naunyn Schmiedeberg's Arch. Pharmak.* **308**: 281.

Gamse, R., Wax, A., Zigmond, R, E., and Leeman, S. E., 1981, Immunoreactive substance P in sympathetic ganglia: distribution and sensitivity towards capsaicin, *Neuroscience* **6**: 437.

Geffen, L. B., Jessell, T., Cuello, A. C., and Iversen, L. L., 1976, Release of dopamine from dendrites in rat substantia nigra, *Nature* **260**: 358.

Glanzer, J. G., and Eberwine, J. H., 2003, Mechanisms of translational control in dendrites, *Neurobiol. Aging* **24**: 1105.

Golgi, C., 1875, Sulla fina struttura dei bulbi olfactory, *Reggio-Guilia* (quoted by Ramon y Cajal, 1904).

Goodman, L. J., Valverde, J., Lim, F., Geschwind, M. D., Federoff, H. J., Geller, A. I., and Hefti, F., 1996, Regulated release and polarized localization of brain-derived neurotrophic factor in hippocampal neurons, *Mol. Cell. Neurosci.* **7**: 222.

Graham, L. T., 1973, Distribution of glutamic acid decarboxylase activity and GABA content in the olfactory bulb, *Life Sci.* **12**: 443.

Hajdu, F., Hassler, R., and Bak, I. J., 1973, Electron microscopic study of substantia nigra and the strio-nigral projection in the rat, *Z. Zellforsch.* **146**: 207.

Halasz, N., Ljungdahl, A., and Hokfelt, T., 1979, Transmitter histochemistry of the rat olfactory bulb, III. Autoradiographic localisation of [^{3}H]GABA, *Brain Res.* **167**: 221.

Hirata, Y., 1964, Some observations on the fine structure of the synapses in the olfactory bulb of the mouse, with particular reference to the atypical synaptic configurations, *Arch. Histol. Jap.* **24**: 293.

Hokfelt, T., Elfvin, L.-G., Schultzberg, M., Goldstein, M., and Nilsson, G., 1977, On the occurrence of substance P-containing fibers in sympathetic ganglia: immunohistochemical evidence, *Brain Res.* **132**: 29.

Hokfelt, T., Fuxe, K., and Goldstein, M., 1973, Immunohistochemical studies on monoamine-containing cell systems, *Brain Res.* **62**: 461.

Hokfelt, T., Kellerth, J.-O., Nilsson, G., and Pernow, B., 1975, Experimental immunohistochemical studies on the localisation and distribution of substance P in cat primary sensory neurons, *Brain Res.* **100**: 235.

Horch, H. W., and Katz, L. C., 2002, BDNF release from single cells elicits local dendritic growth in nearby neurons, *Nat. Neurosci.* **5**: 1177.

Hornykiewicz, O., 1972, Dopamine in the basal ganglia: its role and therapeutic implications (including the clinical use of L-DOPA), *Brit. Med. Bull.* **29**: 172.

Hornykiewicz, O., 1992, From dopamine to Parkinson's disease: a personal research record, in: *The Neurosciences: Paths of Discovery II,* F. Samson, G. Adelman, eds., Birkhauser, Boston, pp. 125-146.

Hunt, S., and Schmidt, J., 1978, Are mitral cells cholinergic? *Neurosci. Abstr.* **3**: 204.

Iversen, L. L., and Kelly, J. S., 1975, Uptake and metabolism of γ-aminobutyric acid by neurons and glial cells, *Biochem. Pharmacol.* **24**: 933.

Jackowski, A., Parnavelas, J. G., and Lieberman, A. R., 1978, The reciprocal synapse in the external plexiform layer of the mammalian olfactory bulb, *Brain Res.* **159**: 17.

Jaffe, E. H., and Cuello, A. C., 1980a, The distribution of catecholamines, glutamate decarboxylase and choline acetyltransferase in layers of the rat olfactory bulb, *Brain Res.* **186**: 232.

Jaffe, E. H., and Cuello, A. C., 1980b, Release of γ-aminobutyrate from the external plexiform layer of the rat olfactory bulb: possible dendritic involvement, *Neuroscience* **5**: 1859.

Jahr, C. E., and Nicoll, R. A., 1980, Dendrodendritic inhibition: demonstration with intracellular recording, *Science* **207**: 1473.

Jancso, N., Jancso-Gabor, A., and Szolesanyi, J., 1967, Direct evidence for neurogenic inflammation and its prevention by denervation and by pretreatment with capsaicin, *Brit. J. Pharmacol.* **31**: 138.

Jessell, T. M., Iversen, L. L., and Cuello, A. C., 1978, Capsaicin-induced depletion of substance P from primary sensory neurones, *Brain Res.* **152**: 183.

Jiang, C., and Schuman, E. M., 2002, Regulation and function of local protein synthesis in neuronal dendrites, *Trends Biochem. Sci.* **27**: 506.

Johnston, D., Magee, J. C., Colbert, C. M., and Cristie, B. R., 1996, Active properties of neuronal dendrites, *Annu. Rev. Neurosci.* **19**: 165.

Katz, B., and Miledi, R., 1963, A study of spontaneous miniature potentials in spinal motoneurones, *J. Physiol.* **168**: 389.

Kelly, J. S., and Dick, F., 1976, Differential labelling of glial cells and GABA inhibitory interneurones and nerve terminals following the microinjection of ³H-β-alanine, ³H-GABA into single folia of the cerebellum, *Cold Spr. Harb. Symp. Quant. Biol.* **40**: 93.

Korf, J., Zieleman, M., and Westerink, B. H. G., 1976, Dopamine release in substantia nigra? *Nature* **260**: 257.

Krnjevic, K., 1974, Chemical nature of synaptic transmission in vertebrates, *Physiol. Rev.* **54**: 418.

Langley, J. N., 1923, Antidromic action, *J. Physiol.* **57**: 428.

Lembeck, F., 1953, Central transmission of afferent impulses. III. Incidence and significance of the substance P in the dorsal roots of the spinal cord, *Naunyn Schmiedebergs Arch. Exp. Pathol. Pharmakol.* **219**: 197.

Lembeck, F., Donnerer, J., and Barthó, L., 1982, Inhibition of neurogenic vasodilation and plasma extravasation by substance P antagonists, somatostatin and (D-Met², Pro⁵) enkephalinamide, *Eur. J. Pharmacol.* **85**: 171.

Lembeck, F., and Holzer, P., 1979, Substance P as a mediator of antidromic vasodilation and neurogenic plasma extravasation, *Naunyn-Schmiedeberg's Arch. Pharmakol.* **310**: 175.

Lewis, T., and Marvin, H. M., 1926, *Herpes zoster* and antidromic impulses, *J. Physiol.* **62**: 19.

Matthews, M. R., and Cuello, A. C., 1982, Substance P-immunoreactive peripheral branches of sensory neurones innervate guinea-pig sympathetic neurones, *PNAS USA* **79**: 1668.

McLennan, H., 1971, The pharmacology of inhibition of mitral cells in the olfactory bulb, *Brain Res.* **29**: 177.

Mercer, L., Del Fiacco, M., and Cuello, A. C., 1979, The smooth endoplasmic reticulum as a possible storage site for dendritic dopamine in substantia nigra neurones, *Experientia* **35**: 101.

Nicoll, R. A., 1971, Pharmacological evidence for GABA as the transmitter in granule cell inhibition in the olfactory bulb, *Brain Res.* **35**: 137.

Nieoullon, A., Cheramy, A., and Glowinski, J., 1977, Release of dopamine *in vivo* from cat substantia nigra, *Nature* **266**: 375.

Nirenberg, M. J., Chan, J., Liu, Y., Edwards, R. H., and Pickel, V. M., 1996, Ultrastructural localization of the vesicular monoamine transporter-2 in midbrain dopaminergic neurons: potential sites for somatodendritic storage and release of dopamine, *J. Neurosci.* **16:** 4135.

Otsuka, M., Konishi, S., and Takahashi, T., 1972, The presence of motoneuron-depolarizing peptide in bovine dorsal roots of spinal nerves, *Proc. Jap. Acad.* **48:** 342.

Palade, G. E., and Palay, S. L., 1954, Electron microscopical observations of interneuronal and neuromuscular synapses, *Anat. Rec.* **118:** 335.

Phillips, C.G., Powell, T. P. S., and Shepherd, G. M., 1963, Responses of mitral cells to stimulation of the lateral olfactory tract in the rabbit, *J. Physiol.* **168:** 65.

Pickel, V. M., Joh, T. H., and Reis, D. J., 1976, Monoamine-synthesizing enzymes in central dopaminergic, noradrenergic and serotonergic neurons: immunocytochemical localisation by light and electron microscopy, *J. Histochem. Cytochem.* **24:** 792.

Price, J. L., and Powell, T., 1970a, The synaptology of the granule cells of the olfactory bulb, *J. Cell Sci.* **7:**125.

Price, J. L., and Powell, T., 1970b, The morphology of the granule cells of the olfactory bulb, *J. Cell Sci.* **7:** 91.

Rall, W., Shepherd, G. M., Reese, I. S., and Brightman, M. W., 1966, Dendrodendritic synaptic pathway for inhibition in the olfactory bulb, *Exp. Neurol.* **14:** 44.

Ramon y Cajal, S., 1904, *Textura del Sistem Nervioso del Hombre y de los Vertebrados,* Libreria de Nicolas Moya, Madrid.

Ramon-Moliner, E., 1977, The reciprocal synapses of the olfactory bulb: questioning the evidence, *Brain Res.* **128:** 1.

Ribak, C. E., Vaughn, T. E., Saito, K., Barber, R., and Roberts, E., 1977, Glutamate decarboxylase localisation in neurons of the olfactory bulb, *Brain Res.* **126:** 1.

Rice, M. E., Cragg, S. J., and Greenfield, S. A., 1997, Characteristics of electrically evoked somatodendritic dopamine release in substantia nigra and ventral tegmental area *in vitro*, *J. Neurophysiol.* **77:** 853.

Shepherd, G. M., 1972, Synaptic organisation of the mammalian olfactory bulb, *Physiol. Rev.* **52:** 864.

Shepherd, G. M., 1976, The olfactory bulb: a simple system in the mammalian brain, in: *Handbook of Physiology. The Nervous System,* pp. 945-968.

Sherrington, C. S., 1906, *The Integrative Action of the Nervous System,* Yale University Press, New Haven.

Sotelo, C., 1971, The fine structural localisation of ^{3}H norepinephrine in the substantia nigra and area postrema of the rat: an autoradiographic study, *J. Ultrastruct. Res.* **36:** 827.

Sourkes, T. L., 2000, How dopamine was recognized as a neurotransmitter: a personal view, *Parkinsonism Relat. Disord.* **6:** 63.

Sourkes, T. L., and Poirier, L., 1965, Influence of the substantia nigra on the concentration of 5-hydroxytryptamine and dopamine of the striatum, *Nature* **207:** 202.

Steward, O., and Schuman, E. M., 2001, Protein synthesis at synaptic sites on dendrites, *Annu. Rev. Neurosci.* **24:** 299.

Takahashi, T., and Otsuka, M., 1975, Regional distribution of substance P in the spinal cord and nerve roots of the cat and the effect of dorsal root section, *Brain Res.* **87:** 1.

Thoenen, H., 1995, Neurotrophins and neuronal plasticity, *Science* **270:** 593.

Tranzer, J.P., 1972, A new amine storing compartment in adrenergic axons, *Nature New Biol.* **237:** 57.

Tsunoo, A., Konishi, S., and Otsuka, M., 1982, Substance P as an excitatory transmitter of primary afferent neurons in guinea-pig sympathetic ganglia, *Neuroscience* **7:** 2025.

Ungerstedt, U., 1971, Stereotaxic mapping of the monoamine pathways in the rat brain, *Acta Physiol. Scand.* **S367:** 1.

Von Euler, U. S., and Gaddum, J. H., 1931, An unidentified depressor substance in certain tissue extracts, *J. Physiol.* **72:** 74.

Whittaker, V. P., 1964, Investigations on the storage sites of biogenic amines in the central nervous system, *Prog. Brain Res.* **8:** 90.

Willey, T. J., 1973, The ultrastructure of the cat olfactory bulb, *J. Comp. Neurol.* **152:** 211.

Yamamoto, C., and Matsui, S., 1976, Effect of stimulation of excitatory nerve tract on release of glutamic acid from olfactory cortex slices *in vitro*, *J. Neurochem.* **26:** 487.

Yuste, R., and Tank, D. W., 1996, Dendritic integration in mammalian neurons, a century after Cajal, *Neuron* **16:** 701.

2

MORPHOLOGICAL STUDIES OF DENDRITES AND DENDRITIC SECRETION

John F. Morris*[1]

1. INTRODUCTION

In September 2000 Andrew Matus and Gordon Shepherd, reviewing a workshop on dendrites for Neuron, entitled their article "The millennium of the dendrite?" Their opening sentence noted that "despite the fact that they constitute as much as 80% of the surface area of most types of neurons, dendrites have not always loomed large in the consciousness of neurobiologists". The article went on to discuss new insights into various aspects of dendrites – development, active properties, calcium handling, plasticity, receptor localization – and although it highlighted developments in protein and RNA targeting, not once was secretion from dendrites mentioned. However, it has been clear for many years that dendrites can secrete peptides and/or amines to affect both their parent neuron and surrounding neurons and glia and even, perhaps, distant targets. Exocytosis is also involved in the insertion of membrane proteins such as synaptic receptors. This volume is therefore a timely correction of that omission.

Ever since the pioneering studies of Ramon y Cajal it has been known that dendritic trees develop in a wide variety of different architectures and that architecture is, to a certain extent, plastic. Proper growth and branching of dendrites are crucial for nervous system function. The variation in architecture is undoubtedly related to the contacts dendrites make with their various afferent projections. It also means that, if dendritic secretion is to modulate that synaptic input (see below), secretion from dendrites must take place at specific sites often very distant from the somata. The active principles secreted may be transported from the cell bodies in vesicles; equally the presence of ribosomes and endoplasmic reticulum in dendrites makes it clear that dendrites can also be sites of the synthesis of proteins to be externalised from the dendrites. The relationship between such protein synthesis and the synaptic input is therefore of considerable importance. This review will briefly discuss very recent contributions to the control of dendritic growth and architecture; the transport of both secretory vesicles and mRNA into the dendrites; the localization of RNA and endoplasmic reticulum in the dendrites; and the evidence for secretion from the dendrites and its function.

---

[1] Department of Human Anatomy and Genetics, University of Oxford, Oxford QX1 3QX, UK

Dendritic Neurotransmitter Release, edited by M. Ludwig
Springer Science+Business Media, Inc., 2005

2. NEURONAL SYSTEMS USED TO STUDY DENDRITES

Every neuroscientist is naturally interested in the dendrites of those neurons which they study. However, a few systems have been extensively investigated: hippocampal and cortical pyramidal neurons and Purkinje cell neurons all have elaborate and well-defined dendritic trees; they are model systems for studying long-term changes in neuronal responsiveness. Slice cultures of hippocampal neurons have been at the forefront of the study of RNA transport into dendrites in the formation, maintenance and plasticity of synapses. However, such systems have scarcely featured at all in the study of secretion from dendrites (except that which inserts receptor proteins into the membrane).

By contrast, it has been peptidergic neurons with much simpler dendritic architecture – in particular the magnocellular hypothalamic neurons which secrete vasopressin and oxytocin – which have been at the forefront of the study of the linked processes of dendritic secretion and dendritic plasticity. It seems almost certain, however, that what has been learned from the magnocellular system is widely applicable, since almost every immunocytochemical study of peptidergic neurons shows that the peptides which are known to be secreted from their axonal terminals are also distributed throughout their dendrites. However, positive proof that there is controlled secretion of peptides from the dendrites of most other peptidergic neurons remains to be demonstrated.

3. REGULATION OF DENDRITIC MORPHOLOGY

A number of recent reviews (McAllister, 2000; McFarlane, 2000; Jan and Jan, 2003) deal with this question in detail and there is still much to be learned, in particular how neurons develop their type-specific dendritic morphology. This is clearly an interactive process which involves both the intrinsic properties of the neuron, substances secreted by surrounding neurons and glia, and the resultant activity of the neurons (Whitford et al., 2002; Miller and Kaplan, 2003). Studies of dendrite-specific expression of green fluorescent protein in living *Drosophila* sensory neurons has shown that dendrites branch either by sprouting of branches from an existing dendrite or by bifurcation of growth cone-like growing dendrite tips (Gao et al., 1999).

Study of the *Drosophila* visual system has shown by mutations of Cdc42, a member of the Rho family of small GTPases, that Cdc42 affects not only dendritic length, girth and branching (Gao et al., 1999) but also dendritic spine density (Scott et al., 2003). The latter is particularly interesting because dendritic spines and their plasticity lie at the heart of the modulation of the responsiveness of many neurons (Segal, 2002). Kalirin, a Rho GDP-GTP exchange factor, causes spine formation when overexpressed and loss of hippocampal pyramidal spines and decreased dendritic complexity when underexpressed (Penzes et al., 2001; Ma et al., 2003). The Rho GTPases regulate actin dynamics and thereby the cytoskeleton of the developing dendrites. They are also important in activity-driven dendritic plasticity; in *Xenopus* tadpoles light-induced visual activity promotes dendritic arborisation and this requires NMDA receptors, decreased RhoA activity and increased Rac and Cdc42 activity (Sin et al., 2002). Importantly, mutations in the signalling pathways to Rho GTPases have been reported in several human neurological diseases (Luo, 2000). The microtubule component of the cytoskeleton is equally important. MAP2 (a major component of microtubule crossbridges) is a well-known marker of dendrites and its mRNA is specifically targeted to dendrites (see below). Other genes involved in cytoskeletal regulation of dendritic morphology including Kakapo, a cytoskeletal protein related to dystrophin (Fuchs and Karakesisoglou, 2001) have been identified by genetic screens. One gene, hamlet, appears to act as a binary genetic switch between single- and multiple-dendrite neuron morphology (Moore et al., 2002). Recent studies of *Drosophila* sensory neurons have also shown that levels of the homeoprotein Cut influence the degree of dendritic branching: high levels are associated with extensive unbranched dendritic terminal protrusions (spikes) and medium levels with complex arbors, whereas neurons with simple dendrites express little or no Cut (Grueber et al., 2003). It will be interesting to determine whether low levels of expression of the mammalian homologue of Cut determine the simple dendritic tree of many peptidergic neurons.

Extrinsic influences on dendritic morphology derive from the synaptic input, from factors secreted by surrounding neurons and glia, and from the hormonal environment of the cells. Matsutani and Yamamoto (2000) showed that mitral cell dendrite development depends on olfactory inputs, and a recent study has shown that dendritic complexity in parietal and visual cortex pyramidal neurons is influenced by the complexity of the cage environment and that the resultant changes vary with sex and age (Kolb et al., 2003). In rat motor neurons, the precise molecular structure of the AMPA receptor GluR1 subunit (but not NMDA receptors) is important; the change of a single amino acid in a critical site which determines single-channel conductance and ion permeability has a profound effect on dendritic architecture during morphogenesis in early postnatal life (Inglis et al., 2002). Various neurotrophic factors, their receptors, and downstream signalling pathways are involved in dendritic growth, linking in to the ras (Alpar et al., 2003), rho, MAPkinase, PI-3 kinase and Jun kinase cascades (Huang and Reichardt 2001). Recent papers explore the role of brain-derived neurotrophic factor (BDNF) on cortical dendrites (Gorski et al., 2003) and hippocampal dendrites (Tolwani et al., 2002). These act through the TrkB receptors. Full length tyrosine kinase receptor B (TrkB) increased proximal dendrite branching in ferret visual cortex, whereas truncated TrkB promoted net elongation of distal dendrites (Yacoubian and Lo, 2000). Differential expression of these receptor subtypes can therefore modify the pattern of dendritic growth, but how the expression is controlled remains to be determined. Bone morphogenetic proteins (BMPs) also influence dendritic arborisation in culture and the sympathetic neuron dendritic growth induced by BMP-7 has recently been shown to require not only the Smad1 signalling pathway but also the involvement of proteasome-mediated degradation events (Guo et al., 2001) though what is degraded has yet to be determined. Coculture studies have shown that BMP actions on dendritic growth are influenced by surrounding glia which modulate the balance between BMPs 5, 6, 7 and the BMP antagonists follistatin and noggin (Lein et al., 2002). The location of the glial cells is also important, because dendrite (but interestingly not axon) outgrowth from corticospinal neurons is promoted by cortical but not by spinal cord astrocytes (Dijkstra et al., 1999). Semaphorin 3A, present at high levels near the pial surface, acts as a chemoattractant for cortical neuron apical dendrites (Polleux et al., 2000). Fibroblast growth factor 2, in contrast, appears to inhibit dendrite formation in hippocampal progenitor cells, decreasing MAP-2 expression and increasing tau expression, indicating a dendrite to axon polarity shift (Tatebayashi et al., 2003)

Appropriate systemic levels of thyroid hormone have long been known to be essential for proper neuronal development especially in the cerebellum. It has now been shown that T3 induces its effect through TRalpha 1 expressed on the Purkinje cells and not on the granule cells; TRbeta isoforms are not involved. The T3 effect is not mediated by neurotrophins as it is unaffected by knockout of BDNF or blockade of TrkB receptors (Heuer and Mason, 2003). Neuroendocrine neuron dendritic architecture is modulated by both the peptides that they produce and by sex steroids. In starlings, the dendritic morphology of GnRH neurons varies dramatically with the reproductive status of the birds (Foster et al., 1987). In magnocellular neurons the relationship between the glia, synapses and dendrites of oxytocin (but not vasopressin) neurons changes dramatically at the end of pregnancy and in lactation (Theodosis et al., 1986; Hatton, 1997). These changes have been shown to require the secretion of oxytocin from the dendrites of the oxytocin neurons (see below). During development, branching of the dendrites of magnocellular neurons requires interplay between the release of peptides from the dendrites and presynaptic terminal activity (Chevaleyre et al., 2000, 2002). One study (Montagnese et al., 1990) suggests that the level of circulating oestrogen is important for this effect, but it does not appear to be essential. Similar changes in the relationship between dendrites and their surrounding glia occur in male homozygous Brattleboro rats in which oxytocin secretion is markedly increased by the osmotic stimulus caused by their failure to secrete vasopressin (Chapman et al., 1986; Ma and Morris, 2002). On the other hand, administration of oestradiol to male rats can induce similar plastic changes, some of which suggest that outgrowths from the cell bodies and dendrites contribute to the plasticity (Morris et al., 1993). The steroids presumably act largely via the estrogen receptor beta which is located in the oxytocin neurons (Alves et al., 1998). More recently, testosterone has been shown to influence the dendritic architecture of male arcuate neuroendocrine neurons identified by retrograde labelling. Castration causes an increase in dendritic outgrowth of these neurons; testosterone replacement inhibited that outgrowth. Interestingly, the changes in dendritic arbor are paralleled by those in LH secretion, suggesting that changes in dendritic morphology are an intrinsic component of testosterone negative feedback (Danzer et al., 2001).

4. MORPHOLOGY OF DENDRITES OF PEPTIDERGIC NEURONS

The dendrites of the hypothalamic supraoptic and paraventricular magnocellular neurons which secrete vasopressin or/and oxytocin are, like the dendrites of most peptidergic neurons, relatively simple structures with few branches and few or no classical dendritic spines, but are the only peptidergic dendrites to have been studied in great detail. The simple morphology has been determined by three different light microscopic procedures: Golgi; dye-filling; and immunocytochemical detection of the peptide contained in the dendrites. Magnocellular neurons have been notoriously difficult to stain with Golgi methods. Armstrong (1995) provides a detailed review of the results achieved with this and other methods in the supraoptic nucleus. The majority show that most of the neurons are bipolar, though some have only one dendrite, and very few are multipolar (particularly in the retrochiasmatic (tuberal) part of the nucleus); branching in all types of dendrite is sparse. In all types, however, one dendrite passes ventrally to the ventral glial lamina, a second may pass dorsally. The dendrites are described as varicose, but few of the cells illustrated show dendritic dilatations of the size seen by electron microscopy (see below). Intracellular fills are most easily made in slice preparations. This technique confirmed and amplified results from Golgi impregnations but, like the Golgi impregnations, did not reveal the large dilations. This may be because such dilatations occur at a relatively long distance from the cell body, often after the ventral dendrites have turned laterally within/just above the ventral glial lamina. Intracellular filling of magnocellular neurons by retrograde uptake of cholera toxin-coupled horseradish peroxidase from the posterior pituitary (Ju et al., 1986) revealed a dense plexus of axon- and dendrite-like processes just beneath the ependyma of the third ventricle. Immunocytochemical staining for vasopressin, oxytocin, or their neurophysins (Fig. 1A) again confirms the rather simple relatively unbranched nature of the dendritic tree. In some studies (e.g. Sofroniew and Glassmann, 1981; Ju et al., 1992) larger dilatations of the dendrites are reported (Fig. 1B). It has also been shown that there appear to be few consistent differences either between the morphology of dendrites of oxytocin and

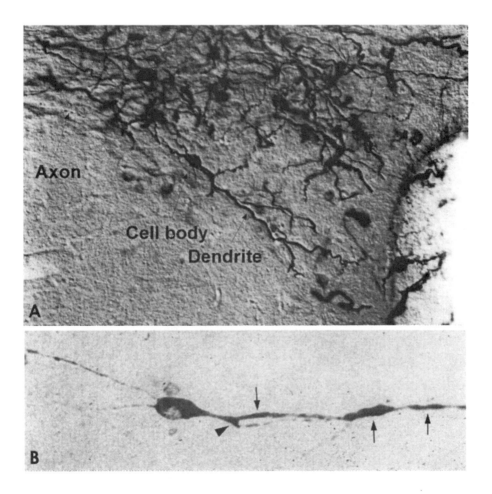

Fig. 1 A: Magnocellular neurons in the paraventricular nucleus immunostained for vasopressin neurophysin. The simple dendrites of the cells pass medially and ventrally toward the third ventricle. The fine beaded axons pass laterally toward the fornix. **B** Isolated multipolar magnocellular neuron immunolabelled for vasopressin neurophysin. The dendrite shows one branch (arrowhead) and a number of smaller and larger dilatations (arrows).

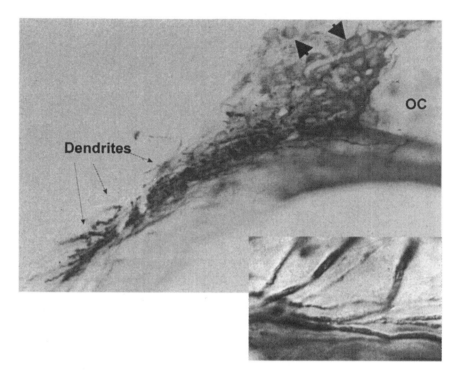

Fig. 2: Rat supraoptic nucleus immunostained for vasopressin neurophyin. The cell bodies (arrowheads) are seen adjacent to the optic chiasm (OC). The dendrites, which are strongly immunostained, pass laterally in a large sheaf along the ventral surface of the hypothalamus (dendrites).   Inset: Rat supraoptic nucleus immunostained for oxytocin neurophysin. Detail of dendrites in the dendrite-rich region adjacent to the ventral glial lamina. Some parts of the dendrites are quite narrow, other parts are more dilated. Immunoreactivity within the dendrite is patchy consistent with the uneven distribution of the hormone-containing dense-cored vesicles.

vasopressin neurons, or between those of neurons in the supraoptic and paraventricular nucleus. The ventral dendrites from most supraoptic neurons pass down into the ventral glial lamina region and then run laterally beneath the surface of the brain, extending in some cases for several hundred microns (Fig. 2). Coronal sections show most dendrites of paraventricular magnocellular neurons running medially and obliquely downward toward the third ventricle (Fig. 1) and apparently ending there. However, horizontal and parasagittal sections reveal that many of these turn abruptly and run for a considerable distance parallel to the ventricular wall. Thus the dendrites of both paraventricular and supraoptic magnocellular neurons project toward and then run along a brain surface in contact with the CSF. The function of this proximity to the CSF has never been resolved. The rapid response of magnocellular neurons to intraventricular administration of osmotic stimuli suggests a sensory role, but equally the presence of vasopressin, oxytocin and their neurophysins in the cerebrospinal fluid is consistent with targeting of secretion to the CSF, though the function of secretion into the CSF remains uncertain.

Electron microscopy is not good for evaluating the overall morphology of peptidergic dendrites, but is very good at showing the fine structure of such dendrites.

Once again, it is the dendrites of the magnocellular neurons that have been most extensively studied. The proximal parts of the dendrites of magnocellular neurons can be followed from the cell body and more distal parts can be recognised when vasopressin- or oxytocin-containing 160nm dense-cored vesicles are present. Electron-microscopic observations are, however, at odds with most light microscopic studies in two respects: electron microscopy reveals large secretory vesicle-filled dilatations of the distal parts of the dendrites particularly in the more lateral parts of the region ventral to the supraoptic nucleus (Fig. 3A); but on the other hand it has failed to reveal classical dendritic spines with afferent boutons and spine apparatus, so the nature of the spine-like processes reported in a number of light microscopic studies remains uncertain. It might be argued that such spines are only present on parts of the dendrites that lack identifying secretory vesicles, but this seems unlikely given the widespread distribution of secretory vesicles in the dendrites.

The other feature that electron microscopy has revealed is that, at or after the end of pregnancy and during lactation (see Russell et al., 2003), the dendrites of oxytocin- but not vasopressin-secreting neurons become bundled together without intervening glia and that many synaptic boutons now contact more than one dendrite or cell body (double or shared synapses); these plastic changes regress after lactation (see Theodosis, 2002) and are caused by oxytocin released from the dendrites (see below).

5. PEPTIDE-CONTAINING SECRETORY VESICLES IN DENDRITES

Both the immunocytochemical and electron-microscopic studies of the magnocellular neurons show that the dendrites contain very substantial amounts of the peptides and that these are present at very considerable distances from the cell body. Electron microscopy shows the peptide to be contained in classical 160nm secretory vesicles. The proximal parts of the dendrites (Fig. 3B) contain relatively few neurosecretory vesicles and in these and many other undilated parts of the dendrites only sparse numbers of vesicles are present, often near or in contact with the central core of microvesicles which presumably transport the vesicles into the dendrites. Although occasional undilated parts of the dendrites are filled with vesicles, the major accumulations of vesicles in dendrites are present in large dilatations of the dendrites (Fig. 3A), which lack most of the other organelles found in dendritic cytoplasm (see below). It is of course possible that some of the peptide is free in the cytosol and not in secretory vesicles, but this seems unlikely in view of the signal on both the oxytocin and the vasopressin gene that directs the nascent prohormones into the rough endoplasmic reticulum and the regulated secretory pathway. Equally, despite the presence of substantial amounts of ribosomes and rough endoplasmic reticulum and of vasopressin mRNA in the dendrites (Mohr et al., 1995), the lack of Golgi stacks and any evidence of dendritic packaging of 160nm secretory vesicles suggests that these vesicles are transported into the dendrites from the cell somata.

There is much less available information on the peptide content of dendrites of other peptidergic neurons and almost no electron microscopic studies of their vesicular packaging. Immunocytochemical studies suggest that, wherever a neuron secretes a peptide from its axonal terminals, that peptide can also be found in the dendrites (e.g. in the suprachiasmatic nucleus; Castel et al., 1996). Within the axonal terminals in the median eminence, the releasing factor peptides are certainly packaged in dense-cored

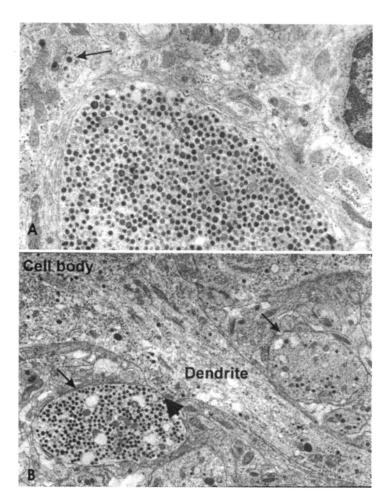

Fig. 3A: Electron micrograph of the dendrite-rich region of the supraoptic nucleus of a normal Long Evans rat to show the marked variation in size and number of dense-cored vesicles. In the top left a cross section of a small dendrite profile contacted by a synaptic bouton (arrowhead) contains a single dense-cored vesicle (arrow). The remainder of the micrograph is dominated by part of an enormously dilated portion of a dendrite containing thousands of dense-cored vesicles.
B: Electron micrograph showing the proximal part of a dendrite (D) as it emerges from the cell body (CB). This part of the dendrite contains relatively few dense-cored vesicles. Adjacent to the proximal dendrite are two other cross-sectioned dendrites (arrows). That on the left contains numerous dense-cored vesicles and makes a dendrite-dendrite contact with the proximal dendrite (arrowhead). That on the right contains many fewer secretory vesicles.

vesicles of about 100nm diameter so there is no reason to believe that the same is not also true of the peptide in the dendrites. However, these parvocellular neurons produce much less peptide and therefore many fewer peptide-containing vesicles so that such vesicles in dendrites may be even more difficult to locate.

One other peptide known to be secreted from dendrites is acetylcholinesterase (AChE) which is secreted with dopamine from dendrites of substantia nigra neurons

(Greenfield, 1985), but in magnocellular neurons AChE is found in dendritic RER and not in dense-cored vesicles (Morris and Ludwig, 2004). Interestingly, a recent experiment shows that AChE promotes dendritic growth synapse formation and expression of AMPA receptors in hippocampal neurons (Olivera et al., 2003).

6. VISUALISATION OF SECRETION OF OXYTOCIN AND VASOPRESSIN FROM DENDRITES

Although earlier studies (Moos et al., 1984; 1989; DiScala Guenot et al., 1987) had indicated the likelihood of dendritic secretion, the visual proof that dendrites are capable of secreting oxytocin and vasopressin came from the application of tannic acid (Buma et al., 1984) to slices of hypothalamus by Pow and Morris (1989) which captures the cores of peptidergic vesicles at the point at which they are exocytosed for analysis by electron microscopy and immunocytochemistry (Fig. 4). Knowledge of the content of single secretory vesicles (Nordmann and Morris, 1984) allowed the calculation that dendritic secretion could account for the peptide released as determined by assay. Stimulation of the slices with transmitters showed, as expected, that activation of the neurons with glutamate, acetylcholine and noradrenaline could all stimulate dendritic exocytosis, as could oxytocin and vasopressin, and also oestradiol by a mechanism which could be blocked by AP5 and therefore presumably involves the NMDA receptors (Morris et al., 2000). The microscopic studies also showed that the soma of magnocellular neurons can exocytose the secretory vesicles but that somatic release makes only a minor contribution to the total. The dendrites do not need to be attached to the cell perikarya as isolated magnocellular dendrosomes can similarly be stimulated to exocytose their peptidergic vesicles (Pow et al., 1990). Some exocytoses occurred in the immediate vicinity of afferent synaptic boutons (Fig. 4B), others at glial-ensheathed parts of the dendrites (Fig. 4C), but whether and how the specific location of release is controlled, remains to be determined.

It is now accepted that the vast majority of the oxytocin and vasopressin that can be located extracellularly in the supraoptic and paraventricular nuclei is secreted by dendrites and microdialysis studies (see Ludwig and Leng, 2004 this volume) now make that assumption. Axon collaterals may contribute a small amount, but are very few in number. Furthermore, intranuclear secretion is often temporally dissociated from secretion from the axons in the posterior pituitary (Moos et al., 1989; Ludwig et al., 1994; Ludwig, 1998). A rise in intradendritic calcium is a prerequisite for exocytosis of peptide-containing vesicles. This can be visualised by the *in vitro* use of calcium dyes in isolated magnocellular neurons (Dayanithi et al., 2000) and isolated dendrosomes (Morris et al., 1993), and derives from an intracellular store since it occurs in cells suspended in a low-calcium medium (Dayanithi et al., 2000). The intracellular calcium store is thapsigargin-sensitive and release of calcium from the store by oxytocin stimulation causes a long-lasting conditional priming of dendritic neuropeptide release which appears to involve movement of secretory vesicles to the periphery of the dendrites (Tobin et al., 2004; Ludwig and Leng, this volume) in much the same way as GnRH priming of gonadotrophin release causes a movement of vesicles to the periphery of pituitary gonadotroph cells (Lewis et al., 1986). The isolated magnocellular neurons used to study intracellular calcium dynamics usually retain only a short portion of the cell's dendritic tree, so the precise localisation of calcium dynamics in the dendrites remains to be

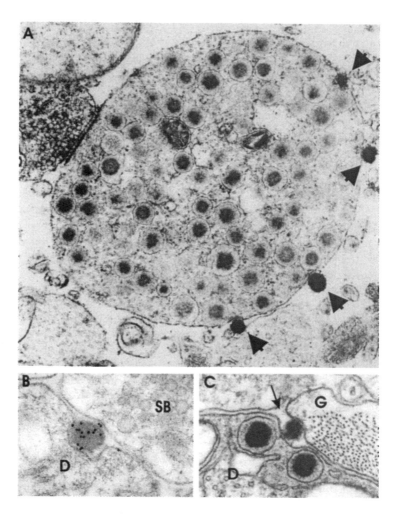

Fig 4A: Electron micrograph showing exocytoses (arrows) from a dendrite in the supraoptic nucleus. A vibratome slice of the supraoptic nucleus region was stimulated in vitro with 10^{-9} M oestradiol for 5 min in the presence of tannic acid to capture the exocytosis. Oestradiol exerts a rapid non-genomic action to cause secretion from the dendrites of vasopressin and oxytocin neurons.
B: High power electron micrograph of the supraoptic nucleus showing exocytosis of a vasopressin neurophyin-immunogold-labelled secretory vesicle from a dendrite (D) the point of contact of a synaptic bouton (SB) identified by the presence of electron-lucent synaptic vesicles.
C: Exocytosis of a dense-cored vesicle from a dendrite (D) next to a glial process (G) identified by its content of glial filaments. Two other dense-cored vesicles are within the dendrite very close to the membrane.

explored. Interestingly, in Purkinje cells, high frequency stimulation of the parallel fibre input causes large calcium transients which last for seconds and are confined to a small part of the dendrite (Kuruma et al., 2003). This raises the possibility that the sensitivity and release of peptides can be locally controlled within the dendrites. In cultured hippocampal neurons release of calcium from internal stores alters the morphology of dendritic spines (Korkotian and Segal, 1999) indicating that it may also locally control dendritic plasticity.

The only other neural system in which tannic acid has been used to visualise dendritic secretion appears to be the suprachiasmatic nucleus (Castel et al., 1996). As in the magnocellular nuclei, exocytosis of large numbers of dense-cored vesicles from dendrites is evoked by depolarisation and stimulation with glutamate.

7. DENDRITIC SECRETION AND CONTROL OF SYNAPTIC INPUT

It has been known for many years that dendrites of all types of neuron contain numerous mRNAs, free ribosomes and rough endoplasmic reticulum (Fig. 5A) and can therefore locally synthesize a variety of proteins (Bodian, 1965; Steward and Levy, 1982). Free polyribosomes translate proteins destined for the cytosol; those which attach to rough endoplasmic reticulum translate proteins destined for exocytosis either as soluble or membrane proteins. However, most dendrites lack any obvious stacks of Golgi membranes and no packaging of dense-cored vesicles in magnocellular dendrites has been reported. This suggests that the oxytocin- and vasopressin-containing dense-cored vesicles in the dendrites are transported there from the cytoplasm. It has also been shown that vasopressin mRNA is present within the dendrites and appears to be transported there specifically along microtubules by a sorting mechanism which involves the last 395 nucleotides of the mRNA and a poly(A)-binding protein (PABP) necessary for mRNA stabilisation and translation initiation (Mohr et al., 1995, 2001). This raises the possibility that some vasopressin might also be secreted from dendrites by the constitutive route in small electron-lucent vesicles.

Although many of the studies on dendritic protein synthesis have used hippocampal neurons (see Steward and Schuman, 2003) the dendrites of magnocellular neurons are also valuable for the study of dendritic synthesis because they can be identified in the neuropil from their content of secretory vesicles and because the activity of the neurons can be altered in a long lasting way either physiologically as in lactation (see above) or by use of homozygous Brattleboro (BB) rats which have life-long hyperactive magnocellular neurons because of the osmotic stimulation which results from their hereditary diabetes insipidus. The BB rats have larger dendrites containing increased numbers of ribosomes, ribosomal RNAs and rough endoplasmic reticulum (Ma and Morris, 2002).

A large number of mRNAs are specifically targeted to dendrites (Eberwine et al., 2002). Indeed, Eberwine et al (2001) report that, from studies using antisense RNA amplification of dendritic mRNA followed by differential display and microarray analysis, about 400 mRNAs can be localized to the dendrites of hippocampal neurons in primary culture. These include the mRNA for the cytoplasmic proteins MAP2 (Fig. 5B) and αCaMKII (Garner et al., 1988, Burgin et al., 1990) both of which are increased in dendrites of BB rats as is that for αSNAP an important component of exocytotic machinery (Ma and Morris 2002). Inactivation of αSNAP shows that it is also important for the growth of dendrites of cultured hippocampal neurons (Grosse et al., 1999) thus linking dendritic secretion and dendritic growth. Interestingly, the cis-acting elements responsible for dendritic targeting of αCaMKII appears to differ markedly from that for MAP2 (Blichenberg et al., 2001), though some proteins involved in mRNA localization appear to be conserved from *Drosophila* to mammals (Roegiers and Jan, 2000, Palacios

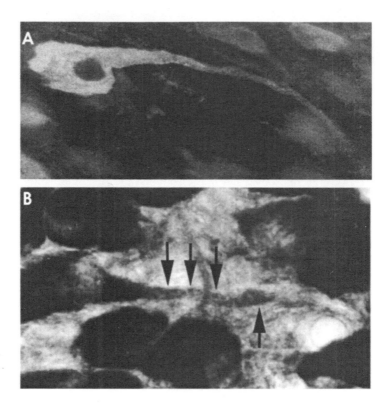

Fig. 5A: Immunolabelling for the rough endoplasmic reticulum marker protein disulphide isomerase in magnocellular neurons of a homozygous Brattleboro rat. The labelling is present in proximal and more distal parts of the dendrites, where some intensely labelled patches appear to be close to the dendritic membrane.
B: High magnification in situ hybridisation labelling for CaMII kinase mRNA in magnocellular neurons of a homozygous Brattleboro rat. The mRNA extends into the dendrite where it is present in a series of patches (arrows), some of which appear to be just beneath the dendritic membrane.

and StJohnston, 2001). BC1 RNA and BC200 RNA are small non-messenger RNAs that are transported into dendrites as ribonucleoprotein particles (Muslimov et al., 2002a, b), which bind to PABP (Muddashetty et al., 2002) and to elF4A and which therefore appear to be involved in protein translation in neuronal dendrites (Wang et al., 2002). BC1 expressing neurons co-express growth-associated protein 43 (GAP-43) but whereas BC1 is detectable in distal dendrites, the GAP43mRNA is largely restricted to the perikarya (Lin et al.,, 2001). However, targeted elimination of the BC1 RNA gene caused no obvious neurological abnormalities and did not interfere with the dendritic distribution of mRNAs for MAP2 or αCaMKII, though it does appear to cause a subtle reduction in exploratory activity in the mutant mice (Skryabin et al., 2003). Post-synaptic densities (PSD) are dynamic structures (Marrs et al., 2001) and their proteins are also likely to be translated on free polysomes. Estrogen, which induces synaptogenesis in CA1

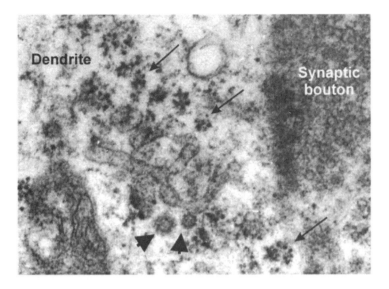

Fig. 6: High power electron micrograph showing part of a dendrite adjacent to synaptic bouton. In the subsynaptic region are many polyribosomes (arrows) some of which are attached to endoplasmic reticulum, and a branched sac of smooth endoplasmic reticulum from which coated vesicles appear to be budding (arrow heads). These could carry receptor proteins to the subsynaptic membrane.

hippocampus, rapidly stimulates local synthesis of PSD95 via the Akt/protein kinase B pathway (Akama and McEwen, 2003).

The possibility that local dendritic synthesis controls the insertion of synaptic receptor proteins contrasts sharply with the classical view that molecules that make up the postsynaptic density are synthesized in the soma then transported to specific synaptic regions (Tang and Schuman, 2002). Most of the translational machinery protein components have been shown to accumulate under synapses and in synaptic spines in various types of neuron (Steward and Schuman, 2001, 2003) and are involved in synaptic plasticity and memory consolidation (Steward, 2002; Steward and Worley, 2002). Interestingly, selective targeting of the mRNA for the activity-regulated cytoskeletal protein Arc to strongly activated synaptic sites requires NMDA receptor activation (Steward and Worley, 2001a, b). There is now conclusive evidence that mRNAs can be translated in dendrites (Eberwine et al., 2001; Villanueva and Steward, 2001) and that, in cultured hippocampal neurons, translation occurs at particular, stable hot-spots characterised by increased ribosome density (Kacharmina et al., 2000; Job and Eberwine 2001).

It is now clear that mRNAs for receptors such as glutamate and GABA are transported into dendrites and that many of the receptor proteins are synthesized locally. Dendritic targeting of the mRNA for NMDA receptor protein NR1, like that for vasopressin, depends on a cis-acting element in the 5'-untranslated region and interaction with RNA-binding proteins (Pal et al., 2003). Targeting of metabotropic glutamate receptors, by contrast appears to be determined by their cytoplasmic carboxy-terminals (Stowell and Craig, 1999). Kacharmina et al (2000) have demonstrated that myc-tagged N-terminal of subunit 2 of the glutamate receptor GluR2 can be synthesized and integrate correctly into the membrane of isolated dendrites. The receptor proteins are presumably inserted into the membrane of the dendrites by the exocytosis of small electron lucent vesicles which bud from the endoplasmic reticulum (Fig. 6), but this has yet to be formally demonstrated. Indeed the lack of an obvious Golgi apparatus in dendrites is a problem, but trans-Golgi network protein 38 can be detected as small points of immunoreactivity in magnocellular dendrites (Ma, 1999). A comparison of the postsynaptic polyribosome in BB and wild-type Long Evans rats shows a markedly greater number in BB rats commensurate with the increased size and activity of the neurons (Ma, 1999; Morris and Ludwig, 2004) and both GABAA and NMDAR1 mRNAs are prominent in magnocellular dendrites (Morris et al., 2000). However, not all receptors are synthesized and inserted locally. Using tagged glycine receptors, Rosenberg et al (2001) have shown that exocytosis of the receptor-bearing vesicles occurred largely at non-synaptic sites in the cell body and initial segment of the dendrites, followed by diffusion in the plasma membrane. Diffusion of glycine receptors has also been elegantly demonstrated by Triller's group (Meier et al., 2001; Dahan et al., 2003). In contrast to both local synthesis/insertion and diffusion, in magnocellular neurons V_{1A} vasopressin receptors are co-packaged with vasopressin in the dense-cored secretory vesicles (Hurbin et al., 2002) so that dendritic secretion of vasopressin also externalises the receptors and presumably inserts them into the membrane. The secreted vasopressin could then act on the parent or adjacent vasopressin neurons (Gouzenes et al., 1999) or on the synaptic input (Ludwig et al., 2000; Yamashita et al., 2001; Ludwig and Pittman, 2003). Oxytocin can also affect afferent excitation (Pittman et al., 2000) but whether oxytocin receptors are similarly co-localized with oxytocin in secretory vesicles remains to be determined.

8. CONCLUSION

This review attempts to highlight the role that morphological studies have played in developing our understanding of the importance of secretion from dendrites. It is now clear that dendritic secretion plays a role in the regulation of dendritic morphology; whether during development or in the morphological plasticity that accompanies both synaptic-induced activity and physiological states such as lactation. Dendrites of peptidergic neurons often contain substantial amounts of vesicle-packaged peptide and, in the magnocellular neurosecretory system and suprachiasmatic nucleus at least, the dense-cored vesicles can be secreted to exert autocrine and paracrine effects on both the magnocellular neurons and their afferent synaptic input. Whether these or other soluble peptides can also be secreted from dendrites via small electron-lucent vesicles and the constitutive route remains to be determined. Finally, dendritic secretion plays a role in the insertion of receptor proteins into the postsynaptic membranes and thus controls the local receptive properties of the neurons. The extent to which all these functions are controlled very locally to produce functional microdomains in the very simple or very elaborately branched dendritic trees is an important question for the future.

9. ACKNOWLEDGEMENTS

Studies in the author's laboratory have been funded by The Wellcome Trust, BBSRC, and the Human Frontiers programme, and conducted in collaboration with a number of students and staff in the laboratory, in particular, Drs David Pow, Andrew Ward, Dan Ma, Hui Wang, and Helen Christian.

10. REFERENCES

Akama, K. T., and McEwen, B., 2003, Estrogen stimulates postsynaptic density-95 rapid protein synthesis via the Akt/protein kinase B pathway, *J. Neurosci.* **23**: 2333.

Alpar, A., Palm, K., Schierwagen, A., Arendt, T., and Gartner, U., 2003, Expression of constitutively active p21H-rasval12 in postmitotic pyramidal neurons results in increased dendritic size and complexity, *J. Comp. Neurol.* **467**: 119.

Alves, S. E., Lopez, V., McEwen, B. S., and Weiland, N. G., 1998, Differential colocalization of estrogen receptor β (ERβ) with oxytocin and vasopressin in the paraventricular and supraoptic nuclei of the female rat brain. An immunocytochemical study, *Neurobiology* **95**: 3281.

Armstrong, W. E., 1995, Morphological and electrophysiological classification of hypothalamic supraoptic neurons, *Prog. Neurobiol.* **47**: 291.

Blichenberg, A., Rehbein, M., Muller, R., Garner, C. C., Richter, D., and Kindler, S., 2001, Identification of a cis-acting dendritic targeting element in the mRNA encoding the alpha subunit of Ca^{2+}/calmodulin-dependent protein kinase II, *Euro. J. Neurosci.* **13**: 1881.

Bodian, D., 1965, A suggestive relationship of nerve cell RNA with specific synaptic sites, *PNAS USA* **53**: 418.

Burma, P., Roubos, E. W., and Buijs, R. M., 1984, Ultrastructural demonstration of exocytosis of neural, neuroendocrine and endocrine secretions with an *in vitro* tannic acid (TARI) method, *Histochemistry* **80**: 247.

Burgin, K. E., Waxham, M. N., Rickling, S., Westgate, S. A., Mobley, W. C., and Kelly, P. T., 1990, *In* situ hybridization histochemistry of Ca^{2+}/calmodulin-dependent protein kinase in developing rat brain, *J. Neurosci.* **10**: 1788.

Castel, M., Morris, J. F., and Belenky, M., 1996, Non-synaptic and dendritic exocytoses from dense-cored vesicles in the suprachiasmatic nucleus, *NeuroReport* **7**: 534.

Chapman, D. B., Theodosis, D. T., Montagnese, C., Poulain, D. A., Morris, J. F., 1986, Osmotic stimulation causes structural plasticity of neurone-glia relationships of the oxytocin- but not the vasopressin-secreting neurones in the hypothalamic supraoptic nucleus, *Neuroscience* **17**: 679.

Chevaleyre, V., Dayanithi, G., Moos, F. C., Desarmenien, M. G., 2000, Developmental regulation of a local positive autocontrol of supraoptic neurons, *J. Neurosci.* **20**: 5813.

Chevaleyre, V., Moos, F. C., Desarmenien, M. G., 2002, Interplay between presynaptic and postsynaptic activities is required for dendritic plasticity and synaptogenesis in the supraoptic nucleus, *J. Neurosci.* **22**: 265.

Dahan, M., Levi, S., Luccardini, C., Rostaing, P., Riveau, B., Triller, A., 2003, Diiffusion dynamics of glycine receptors revealed by single-quantum dot tracking, *Science* **302**: 442.

Danzer, S. C., McMullen, N.T., and Rance, N.E., 2001, Testosterone modulates the dendritic architecture of arcuate neuroendocrine neurons in adult male rats, *Brain Res.* **890**: 78.

Dijkstra, S., Bar, P. R., Gispen, W. H., and Joosten, E. A., 1999, Selective stimulation of dendrite outgrowth from identified corticospinal neurons by homotypic astrocytes, *Neuroscience* **92**: 1331.

DiScala-Guenot, D., Strosser, M. T., Stoeckel, M. E., and Richard, P., 1987, Electrical stimulation of perifused magnocellular nuclei *in vitro* elicits Ca^{2+}-dependent, tetrodotoxin-insensitive release of oxytocin and vasopressin, *Neurosci. Letts.* **76**: 209.

Dayanithi, G., Sabatier, N., and Widmer, H., 2000, Intracellular calcium signalling in magnocellular neurones of the rat supraoptic nucleus: understanding the autoregulatory mechanisms, *Exp. Physiol.* **85S**: 75S.

Eberwine, J., Belt, B., Kacharmina, J. E., and Miyashiro, K., 2002, Analysis of subcellularly localized mRNAs using *in situ* hybridization, mRNA amplification, and expression profiling, *Neurochem. Res.* **27**: 1065.

Eberwine, J., Miyashiro, K., Kacharmina, J. E., and Job, C., 2001, Local translation of classes of mRNAs that are targeted to neuronal dendrites, *PNAS USA* **98**: 7080.

Foster, R. G., Plowman, G., Goldsmith, A. R., and Follett, B. K., 1987, Immunocytochemical demonstration of marked changes in the LHRH system of photosensitive and photo refractory European starlings (*Sturnus vulgaris*), *J. Endocrinol.* **115**: 211.

Fuchs, E., and Karakesisoglou, I., 2001, Bridging cytoskeletal intersections, *Genes Dev.* **15**: 1.

Gao, F. B., Brenman, J. E., Jan, L. Y., and Jan, Y. N., 1999, Genes regulating dendritic outgrowth, branching, and routing in Drosophila, *Genes Dev.* **13**: 2549.

Garner, C.C., Tucker, R. P., and Matus, A., 1988, Selective localization of messenger RNA for cytoskeletal protein MAP2 in dendrites, *Nature* 336: 674.

Gorski, J. A., Zeiler, S. R., Tamowski, S., and Jones, K. R., 2003, Brain-derived neurotrophic factor is required for maintenance of cortical dendrites, *J. Neurosci.* 23: 6856.

Grosse, G., Grosse, J., Tapp, R., Kuchinke, J., Gorsleben, M., Fetter, I., Hohne-Zell, B., Gratzl, M., and Bergmann, M., 1999, SNAP-25 requirement for dendritic growth of hippocampal neurons, *J. Neurosci. Res.* 56: 539.

Gouzènes, L., Sabatier, N., Richard, P., Moos, F. C., and Dayanithi, G., 1999, V1A – and V2-type vasopressin receptors mediate vasopressin-induced Ca^{2+} responses in isolated rat supraoptic neurons, *J. Physiol.* 517: 771.

Greenfield, S. A., 1985, The significance of dendritic release of transmitter and protein in the substantia nigra, *Neurochem. Int.* 7: 887.

Grueber, W. B., Jan, L. Y., and Jan, Y. N., 2003, Different levels of the homeodomain protein Cut regulate distinct dendrite branching patterns of Drosophila multidendritic neurons, *Cell* 112: 805.

Guo, X., Horbinski, C., Drahushuk, K. M., Kim, I. J., Kaplan, P. L., Lein, P., Wang, T., and Higgins, D., 2001, Dendritic growth induced by BMP-7 requires Smad1 and proteasome activity, *J. Neurobiol.* 48: 120.

Hatton, G. I., 1997, Function-related plasticity in hypothalamus, *Annu. Rev. Neurosci.* 36: 203.

Heuer, H., and Mason, C. A., 2003, Thyroid hormone induces cerebellar Purkinje cell dendritic development via the thyroid hormone receptor alpha 1, *J. Neurosci.* 23: 10604.

Huang, E. J., and Reichardt, L. F., 2001, Neurotrophins: roles in neuronal development and function, *Annu. Rev. Neurosci.* 24: 677.

Hurbin, A., Orcel, H., Alonso, G., Moos, F., and Rabie, A., 2002, The vasopressin receptors colocalize with vasopressin in magnocellular neurons of the rat supraoptic nucleus and are modulated by water balance, *Endocrinology* 143: 456.

Inglis, F. M., Crockett, R., Korada, S., Abraham, W. C., Hollmann, M., and Kalb, R. G., 2002, The AMPA receptor subunit GluR1 regulates dendritic architecture of motor neurons, *J. Neurosci.* 22: 8042.

Jan, Y.N., and Jan, L. Y., 2003, The control of dendrite development, *Neuron* 40: 229.

Job, C., and Eberwine, J., 2001, Identification of sites for exponential translation in living dendrites, *PNAS USA* 98: 13037.

Ju, G., Liu, S., and Tao, J., 1986, Projections from the hypothalamus and its adjacent areas to the posterior pituitary in the rat, *Neuroscience* 19: 803.

Ju, G., Ma, D., and Duan, X-Q., 1992, Third ventricular subependymal oxytocin-like immunoreactive neuronal plexus in the rat, *Brain Res. Bull.* 28: 887.

Kacharmina, J. E., Job, C., Crino, P., and Eberwine, J., 2000, Stimulation of glutamate receptor synthesis and membrane insertion within isolated neuronal dendrites, *PNAS USA* 97: 11545.

Kolb, B., Gibb, R., and Gorny, G., 2003, Experience-dependent changes in dendritic arbour and spine density in neocortex vary quantitatively with age and sex, *Neurobiol. Learn. Mem.* 79: 1.

Korkotian, E., and Siegal, M., 1999, Release of calcium from stores alters the morphology of dendritic spines in cultured hippocampal neurons, *PNAS USA* 96: 12068.

Kuruma, A., Inoue, T., Mikoshiba, K., 2003, Dynamics of Ca^{2+} and Na^{+} in the dendrites of mouse cerebellar Purkinje cells evoked by parallel fibre stimulation, *Eur. J. Neurosci.* 18: 2677.

Lein, P. J., Beck, H. N., Chandrasekaran, V., Gallagher, P. J., Chen, H. L, Lin, Y., Guo, X., Kaplan, P. L., Tiedge, H., and Higgins, D., 2002, Glia induced dendritic growth in cultured sympathetic neurons by modulating the balance between bone morphogenetic proteins(BMPs) and BMP antagonists, *J. Neurosci.* 22: 10377.

Lewis, C. E., Morris, J. F., Fink, G., and Johnson, M., 1986, Changes in the granule population of gonadotrophs of hypogonadal (hpg) and normal female mice associated with the priming effect of LH-releasing hormone *in vitro*, *J. Endocrinol.* 109: 35.

Lin, Y., Brosius, J., Tiedge, H., 2001, Neuronal BC1 RNA: co-expression with growth-associated protein-43 messenger RNA, *Neuroscience* 103: 465.

Ludwig, M., 1998, Dendritic release of vasopressin and oxytocin, *J. Neuroendocrinol.* 10: 881.

Ludwig, M., Callahan, M. F., Neumann, I., Landgraf, R., and Morris, M., 1994, Systemic osmotic stimulation increases vasopressin and oxytocin release within the supraoptic nucleus, *J. Neuroendocrinol.* 6: 369.

Ludwig, M., and Leng, G., 2004, Conditional priming of dendritic neuropeptide release. This volume.

Ludwig, M., Onaka, T., and Yagi, K., 2000, Vasopressin regulation of noradrenaline release within the supraoptic nucleus, *J. Neuroendocrinol.* 12: 477.

Ludwig, M., and Pittman, Q. J., 2003, Talking back: dendritic neurotransmitter release, *Trends Neurosci.* 28: 255.

Luo. L, 2000, Rho GTPases in neuronal morphogenesis, *Nat. Rev. Neurosci.* 1: 173.

McAllister, A. K., 2000, Cellular and molecular mechanisms of dendrite growth, *Cereb. Cortex* 10: 693.

McFarlane, S., 2000, Dendritic morphogenesis: building an arbour, *Mol. Neurobiol.* **22**: 1.

Ma, D., 1999, Protein synthetic organelles and mRNAs in the dendrites of hypothalamic magnocellular neurons, *D. Phil . Thesis,* University of Oxford.

Ma, D., and Morris, J. F., 2002, Protein synthetic machinery in the dendrites of the magnocellular neurosecretory neurons of wild-type Long Evans and homozygous Brattleboro rats, *J. Chem. Neuroanat.* **23**: 171.

Ma, X. M., Huang, J., Wang, Y., Eipper, B. A., and Mains, R. E., 2003, Kalirin, a multifunctional Rho guanine nucleotide exchange factor, is necessary for maintenance of hippocampal pyramidal neuron dendrites and dendritic spines, *J. Neurosci.* **23**: 10593.

Marrs, G. S., Green, S. H., and Dailey, M. E., 2001, Rapid formation and remodelling of postsynaptic densities in developing dendrites, *Nat. Neurosci.* **4**: 1006.

Matsutani, S., and Yamamoto, N., 2000, Differentiation of mitral cell dendrites in the developing main olfactory bulbs of normal and naris-occluded rats, *J. Comp. Neurol.* **418**: 402.

Matus, A., and Shepherd, G. M., 2000, The millennium of the dendrite?, *Neuron* **27**: 431.

Meier, J., Vannier, C., Serge, A., Triller, A., and Choquet, D., 2001, Fast and reversible trapping of surface glycine receptors by gephyrin, *Nat. Neurosci.* **4**: 253.

Miller, F. D., and Kaplan, D. R., 2003, Signalling mechanisms underlying dendrite formation, *Curr. Opin. Neurobiol.* **13**: 391.

Mohr, E., Morris, J. F., and Richter, D., 1995, Differential subcellular mRNA targeting: deletion of a single nucleotide prevents the transport to axons but not to dendrites of rat hypothalamic magnocellular neurons, *PNAS USA* **92**: 4377.

Mohr, E., Prakash, N., Vileuf, K., Fuhrmann, C., Buck, F., and Richter, D., 2001, Vasopressin mRNA localization in nerve cells: characterization of cis-acting elements and trans-acting factors, *PNAS USA* **98**: 7072.

Montagnese, C., Poulain, D. A., and Theodosis, D. T., 1990, Influence of ovarian steroids on the ultrastructural plasticity of the adult rat supraoptic nucleus induced by central administration of oxytocin, *J. Neuroendocrinol.* **2**: 225.

Moore, A. W., Jan, L. Y., and Jan, Y. N., 2002, Hamlet, a binary genetic switch between single- and multiple-dendrite neuron morphology, *Science* **297**: 1355.

Moos, F., Freund-Mercier, M. J., Guerné, Y., Guerné, J. M., Stoeckel, M.E., and Richard, P., 1984, Release of oxytocin and vasopressin by magnocellular nuclei *in vitro.* Specific facilitatory action of oxytocin on its own release, *J. Endocrinol.* **102**: 63.

Moos, F., Poulain, D.A., Rodriguez, F., Guerné, Y., Vincent, J. D., and Richard, P., 1989, Release of oxytocin within the supraoptic nucleus during the milk ejection reflex in rats, *Exp. Brain Res.* **76**: 593.

Morris, J. F., Christian, H., Ma, D., and Wang, H., 2000, Dendritic secretion of peptides from hypothalamic magnocellular neurosecretory neurons: a local dynamic control system and its functions, *Exp. Physiol.* **85S**: 131S.

Morris, J. F., and Ludwig, M., 2004, Magnocellular dendrites: prototypic receiver/transmitters, *J. Neuroendocrinol.* **16**: 403.

Morris, J. F., Pow, D. V., Sokol, H. W., and Ward, A., 1993, Dendritic release of peptides from magnocellular neurons in normal rats, Brattleboro rats and mice with hereditary nephrogenic diabetes insipidus, In: *Vasopressin,* P. Gross, D. Richter, G. L. Robertson, eds. John Libbey Eurotext, Paris, pp.171-182.

Muddashetty, R., Khanam, T., Kondrashov, A., Bundman, M., Iacoangeli, A., Kremerskothen, J., Duning, K., Barnekow, A., Huttenhofer, A., Tiedge, H., and Brosius, J., 2002, Poly(A)-binding protein is associated with neuronal BC1 and BC200 ribonucleoprotein particles, *J. Mol. Biol.***321**: 433.

Muslimov, I. A., Lin, Y., Heller, M., Brosius, J., Zakeri, Z., and Tiedge, H., 2002, A small RNA in testis and brain: implications for male germ cell development, *J. Cell. Sci.* **115**: 1243.

Muslimov, I. A., Titmus, M., Koenig, E., and Tiedge, H., 2002, Transport of neuronal BC1 RNA in Mauthner axons, *J. Neurosci.* **22**: 4293.

Nordmann, J. J. and Morris, J. F., 1984, Method for quantitating the molecular content of a subcellular organelle: hormone and neurophysin content of newly formed and aged neurosecretory granules, *PNAS USA* **81**: 180.

Olivera, S., Rodriguez-Ithurralde, D., and Henley, J. M., 2003, Acetylcholinesterase promotes neurite elongation, synapse formation, and surface expression of AMPA receptors in hippocampal neurones, *Mol. Cell. Neurosci.* **23**: 96.

Pal, R., Agbas, A., Bao, X., Hui, D., Leary, C., Hunt, J., Naniwadekar, A., Michaelis, M. L., Kumar, K. N., and Michaelis, E. K., 2003, Selective dendrite targeting of mRNAs of NR1 splice variants without exon 5: identification of cis-acting sequence and isolation of sequence-binding proteins, *Brain Res.* **994**: 1.

Palacios, I. M., and StJohnston, D., 2001, Getting the message across: the intracellular localization of mRNAs in higher eukaryotes, *Ann. Rev. Cell Dev. Biol.* **17**: 569.

Penzes, P., Johnson, R. C., Sattler, R., Zhang, X., Huganir, R. E., Kambampati, V., Mains, R. E., and Eipper, B.A., 2001, The neuronal Rho-GEF Kalirin-7 interacts with PDZ domain-containing proteins and regulates dendritic morphogenesis, *Neuron* **29**: 229.

Pittmann Q. J., Hirasawa, M., Mouginot, D., and Kombian, S. B., 2000, Neurohypophysial peptides as retrograde transmitters in the supraoptic nucleus of the rat, *Exp. Physiol.* **85S**: 139S.

Polleux, F., Morris, T., and Ghosh, A., 2000, Semaphorin 3A is a chemoattractant for cortical apical dendrites, *Nature* **404**: 567.

Pow, D. V., and Morris, J. F., 1989, Dendrites of hypothalamic magnocellular neurons release neurohypophysial peptides by exocytosis, *Neuroscience* **32**: 435.

Pow, D. V., Morris, J. F., and Toescu, E. C., 1990, Dendrosomes, a new preparation of isolated neurosecretory dendrites, *J. Neuroendocrinol.* **2**: 103.

Roegiers, F., and Jan, Y. N., 2000, Staufen: a common component of mRNA transported in oocytes and neurons?, *Trends Cell Biol.* **10**: 220.

Rosenberg, M., Meier, J., Triller, A., and Vannier, C., 2001, Dynamics of glycine receptor insertion in the neuronal plasma membrane, *J. Neurosci.* **21**: 5036.

Russell, J.A., Leng, G., and Douglas, A. J., 2003, The magnocellular oxytocin system, the fount of maternity: adaptations in pregnancy, *Front. Neuroendocrinol.* **24**: 27.

Scott, E. K., Reuter, J. E., and Luo, L., 2003, Small GTPase Cdc42 is required for multiple aspects of dendritic morphogenesis, *J. Neurosci.* **23**: 3118.

Segal,, M., 2002, Changing views of Cajal's neuron: the case of the dendritic spine, *Prog. Brain Res.* **136**: 101.

Sin, W. C., Haas, K., Ruthazer, E. S., and Cline, H. T., 2002, Dendritic growth increased by visual activity requires NMDA receptor and Rho GTPases, *Nature* **419**: 475.

Skryabin, B. V., Sukonina, V., Jordan, U., Lewejohann, L., Sachser, N., Muslimov, I., Tiedge, H., and Brosius, J., 2003, Neuronal untranslated BC1 RNA: targeted gene elimination in mice, *Mol. Cell Biol.* **23**: 6435.

Sofroniew, M. V., and Glassmann, W., 1981, Golgi-like immunoperoxidase staining of hypothalamic magnocellular neurons that contain vasopressin, oxytocin or neurophysin in the rat, *Neuroscience* **6**: 619.

Steward, O., 2002, mRNA at synapses, synaptic plasticity, and memory consolidation, *Neuron* **36**: 338.

Steward, O., and Levy, W. B., 1982, Preferential location of polyribosomes under the base of dendritic spines in granule cells of the dentate gyrus, *J. Neurosci.* **2**: 284.

Steward, O., and Schuman, E. M., 2001, Protein synthesis at synaptic sites on dendrites, *Annu. Rev. Neuroci.* **24**: 299.

Steward, O., and Schuman, E. M., 2003, Compartmentalized synthesis and degradation of proteins in neurons, *Neuron* **40**: 347.

Steward, O., and Worley, P. F., 2001, A cellular mechanism for targeting newly synthesized mRNAs to synaptic sites on dendrites *PNAS USA* **98**: 7062.

Steward, O., and Worley, P. F., 2001, Selective targeting of newly synthesized *Arc* mRNA to active synapses requires NMDA receptor activation, *Neuron* **30**: 227.

Steward O., and Worley, P. F., 2002, Local synthesis of proteins at synaptic sites on dendrites: role in synaptic plasticity and memory, *Neurobiol. Learn. Mem.* **78**: 508.

Stowell, J. N., and Craig, A. M., 1999, Axon/dendrite targeting of metabotropic glutamate receptors by their carboxy-terminal domains, *Neuron* **22**: 525.

Tang, S. J., and Schuman, E. M., 2002, Protein synthesis in the dendrite, *Philos. Trans. Roy. Soc. Lond. B Biol. Sci.* **357**: 521.

Tatebayashi, Y., Lee, M. H., Li, L., Iqbal, K., Grundke-Iqbal, I., 2003, The dentate gyrus neurogenesis: a therapeutic target for Alzheimer's disease, *Acta Neuropathol.* **105**: 225

Theodosis, D. T., Chapman, D. B., Montagnese, C., Poulain, D. A., and Morris, J. F., 1986, Structural plasticity in the hypothalamic supraoptic nucleus at lactation affects oxytocin- but not vasopressin-secreting neurons, *Neuroscience* **17**: 661

Tobin, V. A., Hurst, G., Norrie, L., DalRio, F., Bull, P. M., and Ludwig, M., 2004, Thapsigargin-induced mobilisation of dendritic dense core vesicles in supraoptic neurones, *Eur. J. Neurosci.* **19**: 2909.

Theodosis, D. T., 2002, Oxytocin-secreting neurons. A physiological model of morphological neuronal and glial plasticity in the adult hypothalamus, *Front. Neuroendocrinol.* **23**: 101.

Tolwani, R. J., Buckmaster, P. S., Varma, S., Cosgaya, J. M., Wu, Y., Suri, C., and Shooter, E. M., 2002, bdnf overexpression increases dendrite complexity in hippocampal dentate gyrus, *Neuroscience* **114**: 795.

Villanueva, S., and Steward, O., 2001, Protein synthesis at the synapse: developmental changes, subcellular localization and regional distribution of polypeptides synthesized in isolated dendritic fragments, *Mol. Brain Res.* **91**: 148.

Wang, H., Iacoangeli, A., Popp, S., Muslimov, I. A., Imataka, H., Sonenberg, N., Lomakin, I. B., Tiedge, H., 2002, Dendritic BC1 RNA: functional role in regulation of translation initiation, *J. Neurosci.* **22**: 10232.

Whitford, K. L., Dijkhuizen, P., Polleux, F., and Ghosh, A., 2002, Molecular control of cortical dendrite development, *Annu. Rev. Neurosci.* **25**: 127.

Yacoubian, T. A., and Lo, D. C., 2000, Truncated and full-length TrkB receptors regulate distinct modes of dendritic growth, *Nature Neurosci.* **3**: 342.

Yamashita, T., Liu, X., Onaka, T., Honda, K., Saito, T., and Yagi, K., 2001, Vasopressin differentially modulates noradrenaline release in the rat supraoptic nucleus, *Neuroreport* **12**: 3509.

3

THE LIFECYCLE OF SECRETORY VESICLES: IMPLICATIONS FOR DENDRITIC TRANSMITTER RELEASE

David K. Apps, Michael A. Cousin, Rory R. Duncan, Ulrich K. Wiegand and Michael J. Shipston[*]

1. INTRODUCTION

Intercellular communication is essential to the integrative function of neurones and neuroendocrine cells. The vesicular storage and release of bioactive molecules (transmitter) represents the major pathway for intercellular signals to be transferred from the 'sender' to 'receiver' cell. This process, in which the transmitter-containing vesicles undergo fusion with the plasma membrane and thereby release their contents into the extracellular space, is an adaptation of the constitutive exocytosis that occurs in all eukaryotic cells, fulfilling the 'housekeeping' role of turnover of plasma membrane constituents (Figure 1). Regulated exocytosis occurs only in response to specific extracellular signals that induce the synthesis or release of intracellular second messengers (usually Ca^{2+}); these in turn trigger the membrane-fusion machinery.

As this volume testifies, considerable evidence supports the role of dendritic transmitter release in coordinating neuronal function. However, although evidence is accumulating for regulated exocytosis of 'classical' transmitters and peptides from dendrites, the molecular machinery and vesicle lifecycle in dendrites is poorly understood, in contrast to our understanding of synaptic or endocrine transmitter release.

In an attempt to highlight major questions, challenges and opportunities for investigators analysing dendritic transmitter release, we first overview our understanding of the vesicle lifecycle and molecular events in regulated exocytosis from synaptic and endocrine model systems before exploring our current understanding of dendritic release. In Section 2 we provide a general overview of regulated exocytosis before comparing 'classical' synaptic transmitter release to that of peptide release in endocrine cells in

---

[*] Membrane Biology Group, School of Biomedical and Clinical Laboratory Sciences, University of Edinburgh, EDINBURGH, EH8 9XD, UK

Dendritic Neurotransmitter Release, edited by M. Ludwig
Springer Science+Business Media, Inc., 2005

Section 3. We highlight the key molecular events involved in the vesicle lifecycle (Section 4) and review the evidence for regulated exocytosis in dendrites in Section 5.

2. OVERVIEW OF REGULATED EXOCYTOSIS

2.1. Types of Secretory Vesicle

Non-peptide transmitters are stored in and released from synaptic vesicles (SVs); these are approximately spherical, with a single bounding membrane and a diameter of 50 – 60 nm. There are typically only about 200-250 SVs per synapse but, assuming that there are 1000 synapses per neuron, it can be calculated that the 'concentration' of SVs in the whole brain can still reach about 2.5 µM! As well as SVs, many types of neuron contain large dense-cored vesicles (LDCVs), with a diameter of 150-300 nm and a cargo of peptide hormones and biogenic amines. Exocytosis of SVs and LDCVs is separately controlled, although there is probably a common molecular mechanism, with specificity of release being achieved through differences in sensitivity to Ca^{2+}. In peripheral neurons catecholamines are stored in secretory vesicles of intermediate size (60 – 80 nm) called small dense-cored vesicles (SDCVs), the membranes of which contain proteins characteristic both of SVs and of LDCVs.

2.2. Tools for Studying Secretion at the Level of Secretory Vesicles

The small size of synaptic terminals and of SVs themselves, together with the rapidity of the secretory event in neurons, imposes limitations on the types of experimental approach that can be used to study neuronal exocytosis. However much important information has been obtained through the study of non-neuronal secretory cells, in particular from neuoroendocrine tissues such as the adrenal medulla, the pituitary and the endocrine pancreas (Burgoyne and Morgan, 2003). Important recent developments include the use of patch-clamped cells, permeabilized cells and even cell-free preparations; the introduction of 'caged Ca^{2+} (Neher, 1998), permitting the rapid elevation of intracellular $[Ca^{2+}]$ by a flash of light, together with ratiometric Ca^{2+}-indicator dyes to measure it; the measurement of membrane capacitance to record changes in membrane area during fusion and membrane retrieval; amperometric measurement of catecholamine secretion at the single-vesicle level; and the expression of fluorescently-tagged proteins targeted to LDCV lumen or membranes, so visualizing them for study by confocal or evanescent-wave microscopy. As well as LDCVs, neuroendocrine cells contain synaptic-like microvesicles (SLMVs), small vesicles of low density with typical SV membrane markers, but no dense core of cargo proteins. Their function is poorly understood and the results discussed below refer to LDCVs.

2.3. Secretory Vesicle Pools

Biochemical assays of secretion show that only a relatively small fraction of LDCVs is available for release, the remainder being unaffected by most secretagogues. In contrast to neurons, in neuroendocrine cells access of most LDCVs to the plasma membrane may be limited by the cortical actin network, and dissolution of this by actin-severing proteins may be required for sustained exocytosis. Within the releasable pool (about 2000 vesicles, in chromaffin cells; 10% of the total) about 200 vesicles are rapidly released in response to elevated $[Ca^{2+}]$, while release of the remainder is rate-limited by a much slower process (Voets, 2000). This slow phase of exocytosis, which requires sub-micromolar $[Ca^{2+}]$ and also ATP, is called priming, and probably corresponds to assembly of the *trans*-SNARE complex (see section 4), converting the LDCVs to a fusion-competent state. The rapid, or burst, phase is itself heterogeneous, and takes place from two interconvertible LDCV pools called the slowly releasable and rapidly releasable pools (SRP and RRP). The interconversion of these pools is Ca^{2+}-independent and relatively slow, and its molecular basis is not understood. The fusion process itself is sharply $[Ca^{2+}]$-dependent and involves the binding of at least 3 Ca^{2+} ions with dissociation constants around 10 μM. It results in the very fast formation of a fusion pore, of about 2 nm diameter as inferred from conductance measurements; this rapidly expands, with a corresponding increase in conductance. The seal of vesicles to the plasma membrane is very tight, so that conductance between the cytoplasm and the extracellular space is negligible (<0.1 pS per seal); but fusion pore formation appears to be reversible, at least in its early stages. The molecular structure of the fusion pore is controversial (see section 4) but there is some evidence that the rate of fusion pore expansion can be regulated. Closing of the pore before complete release of the vesicle contents constitutes "kiss-and-run" exocytosis (see section 4); alternatively, complete dilation of the pore leads to complete exocytosis of the vesicle cargo (full fusion) and collapse of the vesicle into the plasma membrane. Full fusion is followed by clathrin-mediated retrieval of LDCV membrane components and their recycling through early endosomes and the Golgi. Kiss-and-run fusion results in the partially empty vesicle pinching-off, after which it could be refilled with small molecules, such as catecholamines, through active transport; however replacement of cargo proteins, if they have been released, would involve recycling through the Golgi (Jahn et al., 2003).

Because chemical fixation is slow, fusion events are rarely revealed in electron micrographs, but LDCVs can be seen close to the plasma membrane, and are then described as "morphologically docked". Although these do not necessarily correspond to the SRP or RRP mentioned above, real-time imaging techniques such as evanescent wave microscopy show that these vesicles are relatively immobile, indicating that they may be tethered to the membrane; and also that exocytosis takes place predominantly from immobile vesicles.

3. COMPARISON OF EXOCYTOSIS OF SVs AND LDCVs

In neurons, both the traffic of peptide-containing LDCVs and the recycling of fast-transmitter containing SVs follow the basic principles outlined in section 2. Since neurotransmission has to be fast and reliable, SV recycling has a number of specialisations that are related to the function and morphology of central neurones, as described below.

3.1. Vesicle Biogenesis and the Synaptic Vesicle Lifecycle

Both LDCVs and SVs are generated from the Golgi apparatus (Figure 1), but with significant differences. The cargo proteins of LDCVs form aggregates, resulting in the condensation of the granule matrix and membrane budding to form immature secretory granules – secretory proteins such as the chromogranins/secretogranins therefore have an active role in LDCV assembly (Tooze et al., 2001). Newly formed LDCVs then mature by fusion with other immature granules while their membranes are remodelled, some proteins being removed in clathrin-coated vesicles. It has been reported that LDCVs containing different secretory proteins are differentially targeted within neurons, suggesting that exocytosis from terminals and dendrites may have different functions (Landry et al., 2003; Langley and Grant, 1997). SVs also bud from the Golgi as immature vesicles with a morphology and protein composition different from those of mature SVs; these reach synapses by axonal transport, and there undergo remodelling.

The greatest differences lie in the fate of these vesicles after fusion with the plasma membrane (Figure 1). After LDCV fusion, vesicle membrane components are retrieved and recycled to the Golgi, where they collect new cargo, thus generating vesicles *de novo* in each cycle. In contrast, SVs can be used many times and are locally recycled up to 1000 – 2000 times before degradation. The major reason for this is that most central nerve terminals are far from the cell body. Thus the *de novo* generation of a SV after each fusion event would not produce enough SVs to maintain rates of neurotransmission that may be as high as 400 Hz in some neurones. Furthermore, transmitters such as GABA are synthesised in the nerve terminal or are already present, as in the amino acid pool (glutamate). Uptake of transmitter by SVs occurs through H^+-coupled transporters in the SV membrane, the driving force for uptake being the protonmotive force generated by a V-type H^+-translocating ATPase that is also present on the SV membrane. The final luminal concentration of transmitter may not reach equilibrium with this protonmotive force, suggesting that control of quantal size may occur at the level of transmitter uptake (Parsons, 2000). The transmitters or their precursors are accumulated into nerve terminals by specific Na^+-coupled symporters in the presynaptic plasma membrane, using the electrochemical gradient of Na^+ created by the Na^+,K^+-ATPase. Under some circumstances this uptake system can run backward, affording a Ca^{2+}-independent release pathway for some transmitters - the significance of this is still debated (Attwell et al., 1993).

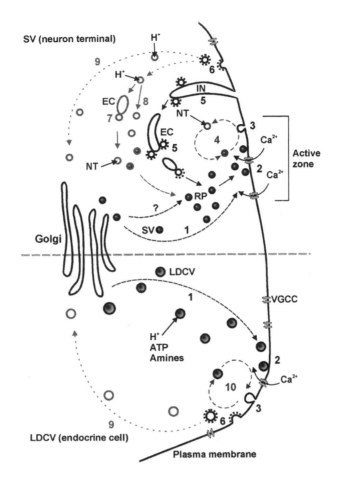

Figure 1. Vesicle lifecycle and membrane trafficking in neuronal and endocrine cells.
Synaptic vesicles (SV) in neurons (top) and large dense cored vesicles (LDCV) in endocrine cells (bottom) bud from the trans-Golgi network of the Golgi apparatus filled with cargo proteins (see text, including peptide transmitters for LDCVs). During their maturation, vesicles are acidified, take up ATP and neurotransmitter (NT, in SVs) or biogenic amines (in LDCVs) and are translocated to the plasma membrane (see step [1]). At the active zone, they dock (see also Figure 2) and are primed [2]. Influx of calcium ions through voltage-gated calcium channels (VGCC) triggers fusion of vesicle and plasma membranes and subsequent release of the vesicle cargo [3]. The vesicle membrane can completely collapse into the plasma membrane and can be retrieved by clathrin-mediated endocytosis [6]. Alternatively, membrane fusion may generate a temporary fusion pore after which the vesicle is quickly retrieved, a mechanism called 'kiss-and-run' exocytosis [4, 10]. In neurons, this mechanism allows quick re-filling of SVs with neurotransmitter (NT) for fast, repetitive exocytosis [4]. Bulk endocytosis provides another mechanism for membrane retrieval in neurons by the invagination of larger membrane areas, which enter endosomal compartments [ECs], from which SVs can bud off [5]. Vesicles retrieved from the plasma membrane by clathrin-mediated endocytosis [6] can either be translocated back to the Golgi for recycling [9] or, in neurons, can fuse to form early endosomal compartments from which new vesicles bud off, ready for re-filling with neurotransmitter [7]. Finally, clathrin-mediated endocytosis can directly deliver vesicles for protonation, refilling and recycling [8].
'Kiss-and-run' exocytosis in neurons provides a mechanism by which SVs can be rapidly retrieved and recycled locally in order to deliver swift, highly repetitive exocytotic response to prolonged stimulation, a long distance away from the cell body. In contrast, because LDCVs in endocrine cells are partially loaded with protein cargo, retrieved vesicles have to return to the Golgi for recycling [9]. However, in endocrine cells LDCVs have also been reported to undergo 'kiss-and-run' exocytosis [10]. This could provide a mechanism to modulate the amount of vesicle cargo released in a single exocytotic event (quantal release).
(A color version of this figure appears in the signature between pp. 256 and 257.)

3.2. Calcium Triggering of Vesicle Fusion

Exocytosis of SVs and LDCVs is stimulated by increases in intracellular free $[Ca^{2+}]$, but the speed of fusion and the required $[Ca^{2+}]$ increase differs between the two vesicle populations. As described above, about 1% of LDCVs are docked at specialised release sites at the plasma membrane and fuse on cell stimulation and subsequent Ca^{2+} influx. Central neurones also have highly specialised release sites within their nerve terminals called, 'active zones' (Murthy and De Camilli, 2003). However at these sites a larger proportion (2-5 %) of the SV pool is present, in keeping with the requirement for a sustained burst of transmitter release on stimulation. Replenishment of the RRP is also much faster in neurones compared to secretory cells (< 10 sec compared to < 60 sec), again reflecting the requirement for high fidelity neurotransmission.

In general SV fusion occurs at least an order of magnitude faster than LDCV fusion, and it requires a greater increase in $[Ca^{2+}]$ (Barg, 2003). Active zones in nerve terminals contain a high density of Ca^{2+} channels and docked vesicles to ensure that (1) Ca^{2+} levels are high enough (~100 μM) to trigger release and (2) there is a minimal delay between Ca^{2+}influx and SV fusion. Such concentrations of free Ca^{2+} are not of course reached uniformly throughout the terminal, and the localized zones of high $[Ca^{2+}]$ are referred to as microdomains (Neher, 1998). The type of Ca^{2+} channel present at release sites also differs between SV and LDCV release sites, with fast-responding P/Q- and N-type Ca^{2+} channels present in active zones, compared to the slower L-type channels in endocrine cells. Ca^{2+} channels have been shown to interact with a number of synaptic proteins, notably synaptotagmin and syntaxin (see section 4).

3.3. Organisation of Vesicle Pools

Both secretory cells and neurons contain reserve and rapidly releasable secretory vesicle pools, but the traffic of vesicles to and from these pools is different. Short-term pulse-labelling experiments suggest that immediately after budding, the membranes of immature secretory vesicles have to be 'remodelled' for the vesicles to become fusion-competent. Imaging of cells expressing fluorescent secretory proteins suggests that LDCVs enter the RRP soon after their biogenesis, but if not fused within a short time they undock and join a reserve pool of vesicles (Duncan et al., 2003). These reserve vesicles can still be used for exocytosis: this is usually by replenishing the RRP during continual stimulation. However, some secretagogues, such as Ba^{2+}, may cause reserve pool vesicles to bypass the RRP, although the physiological relevance of this observation is not understood.

In contrast to LDCVs, little is known about which vesicle pool newly synthesised SVs join, mainly due to a lack of appropriate assays. Much more is known about the routes SVs take once they have joined either the reserve pool or RRP. SVs enter the reserve pool after undergoing clathrin-mediated endocytosis, possibly via a sorting endosome, but they can also join the reserve pool by undocking from the RRP (Murthy and De Camilli, 2003). This is a minor route, compared to secretory cells, because of the rapid secretory rate in neurones.

The main purpose of the reserve pool of SVs is to replenish the RRP on depletion during high-intensity stimulation. However the RRP can also be replenished by a short-circuit pathway, so-called 'reuse'. Reuse is defined as a local recycling of SVs that facilitates replenishment of the RRP without the involvement of endosomal intermediates or the reserve pool. Reuse is a physiological phenomenon and its underlying molecular mechanism could be one of a number of different processes. For example, it could occur by incomplete fusion of SVs and reversal of fusion pore formation, also described as 'kiss-and-run'; by a more rapid form of clathrin-mediated SV endocytosis; or by differential sorting of SVs after clathrin-mediated endocytosis back to the RRP (Rizzoli and Betz, 2004; Harata et al., 2001).

3.4. Retrieval of Vesicles after Fusion

Most vesicle retrieval after fusion of LDCVs and SVs occurs by clathrin-mediated endocytosis, in which a single vesicle is generated *de novo* from the plasma membrane (Murthy and De Camilli, 2003). However, differences between LDCVs and SVs are apparent even at this stage. For example, when a central nerve terminal undergoes prolonged stimulation with depletion of SVs, vesicles can be regenerated by 'bulk endocytosis'. This involves a large invagination of the plasma membrane that allows multiple endocytic budding events to occur simultaneously to facilitate replenishment of SV stocks (Figure 1). This is demonstrated graphically in the Drosophila mutant *shibire,* which has a temperature-sensitive defect in endocytosis. When returned to the permissive temperature huge invaginations are observed by electron microscopy, as the nerve terminal replenishes its SV pools (Koenig and Ikeda, 1989). An alternative mechanism of SV retrieval is to retrieve a vesicle before it completely fuses with the plasma membrane. This process, termed 'kiss-and-run' by Bruno Ceccerelli Fesce (Fesce et al., 1994) is an ideal mechanism for SV retrieval in nerve terminals since it should occur quickly, and with little expenditure of energy. Kiss-and-run has never been conclusively demonstrated in central neurones, mainly because of a lack of appropriate experimental techniques although indirect measurements suggest its existence (Rizzoli and Betz, 2003). Paradoxically, kiss-and-run has been conclusively demonstrated in secretory cells (Alés et al., 1999), perhaps providing the best indication that it occurs in central neurones.

Kiss-and-run is usually regarded as early closure of the fusion pore and detachment of the vesicle before its incorporation into the plasma membrane, and assumed to be a very rapid process. However a different form of kiss-and-run, or more accurately 'intact vesicle retrieval', has been demonstrated for both LDCVs and SVs. This mechanism can retrieve an intact vesicle up to 10 s after fusion in secretory cells (Taraska et al., 2003) and up to 45 sec in central neurones (Gandhi and Stevens, 2003). This implies that a physical restraint must be present to prevent fusion pore expansion during this mechanism, however few proteins have been identified that would fit the requirements of this "expansion clamp".

4. MOLECULAR EVENTS

The molecular mechanisms of vesicle traffic, exocytosis and endocytosis have been extensively characterised in the past 15 years. Much of the molecular machinery identified to date is conserved from yeast to mammals with only small variations in specialised cells such as neurones.

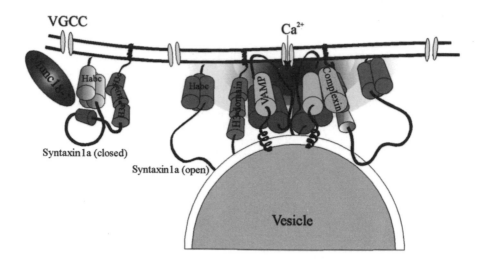

Figure 2. Schematic illustrating the hypothesized conformations of some of the proteins involved in vesicle fusion. The essential t-SNARE protein syntaxin1a exists in the "closed" conformation when bound to the SM protein Munc-18 (also known as nSec-1). In this conformation, the C-terminal SNARE motif of syntaxin1a is occluded by its N-terminal Habc domain thus syntaxin is unavailable to participate with the other SNARE proteins in the formation of the core complex. The subsequent dissociation of Munc-18 results in a large conformational rearrangement of syntaxin1a, "opening" the molecule, and allowing the interaction of the syntaxin1a SNARE motif with the other t-SNARE, SNAP-25, and the v-SNARE, VAMP (also known as synaptobrevin). SNAP-25 is associated with the plasma membrane through a palmitoylated loop, whereas VAMP is an integral vesicle membrane protein. The formation of the tertiary core complex creates a 3-dimensional conformation recognized by complexin, with crystallographic studies revealing its binding to the groove created by syntaxin and VAMP helices, apparently stabilizing the complex. Calcium influx through voltage gated calcium channels (VGCC) creates calcium micro-domains in the vicinity of fusion sites (represented here as purple and blue shading of decreasing intensity to indicate decreasing Ca^{2+} concentration). The precise nature of the "fusion pore" remains undefined, as does the number of core complexes involved. Munc18-mediated sequestering of syntaxin1a has been hypothesized to be an essential modulatory step, preventing ectopic fusion from occurring, however, the regulatory step leading to the dissociation of the two proteins in a cellular environment remains unclear. Furthermore, Munc18 also appears to have other functions important for vesicle docking, as in its absence, docking is reduced.

(A color version of this figure appears in the signature between pp. 256 and 257.)

4.1. The Reserve Vesicle Pool

In neurones, the reserve pool of SVs is maintained by a phosphoprotein called synapsin, which is loosely bound to the outer membrane of the SV. When it is dephosphorylated, synapsin has a high affinity for SVs and also for the actin cytoskeleton, ensuring that SVs are retained within this structure in resting nerve terminals. On nerve terminal depolarisation, synapsin is phosphorylated by various protein kinases, with a consequent reduction in its affinity for both SVs and actin. This releases SVs from the cytoskeleton and also from synapsin, and allows their translocation to the active zone (Hilfiker et al., 1999). This mechanism contrasts with that in neuroendocrine cells, where actin-severing proteins such as scinderin have been implicated (Vitale et al., 1995). Translocation is thought to be an active process mediated by the motor protein myosin, which transports SVs down actin filaments to the active zone. Synapsin is not present in secretory cells therefore the molecular mechanism of maintenance of the reserve pool of LDCVs is still unknown, although there is extensive evidence that depolarization results in disassembly of cortical actin filaments, permitting access of LDCVs to the plasma membrane.

4.2. Tethering and Docking

After translocation to the plasma membrane, secretory vesicles attach to the membrane at release sites. This attachment may occur in several stages, and the initial interaction may be mediated by proteins that are not directly involved in the preparation for fusion. The specificity of the initial vesicle-membrane interaction ensures that vesicles are not sorted to the wrong subcellular compartments. This specificity is encoded in part by rab proteins. The rab family are GTPases that cycle between the cytoplasm and membrane. They have been proposed to act at the earliest stage of vesicle attachment, known as tethering, which is defined as the physical attachment of a transport vesicle to a target membrane before the formation of the fusion complex (Waters and Hughson, 2000). Thus rab proteins would seem to have an important role in conferring specificity to membrane docking. They bind to effectors only at the target membrane in the GTP-bound state and hydrolysis of GTP by their intrinsic activity causes their release. This rab-cycle also involves proteins that activate the GTPase activity of rabs and accelerate their binding to or dissociation from the membrane.

The next stage of the process, docking, involves the SNARE (SNAP Receptor) proteins, which are conserved in almost every subcellular vesicle trafficking event from yeast to mammals, with different homologues residing in different compartments. There are two classes of SNARE protein, called v-SNAREs and t-SNAREs, residing on the vesicle and target membranes respectively (Figure 2). Complex formation between v- and t-SNAREs is thought to provide some specificity in vesicle targeting. In secretory vesicle exocytosis the v-SNARE is synaptobrevin 2 (also known as VAMP - vesicle associated membrane protein), an integral vesicle membrane protein. At the target plasma membrane are syntaxin (an integral membrane protein) and SNAP-25 (a peripheral protein anchored to the membrane by acyl groups). These proteins form a thermodynamically highly stable complex with a 1:1:1 stoichiometry, called the core complex. The stability derives from the interaction of four parallel α-helical SNARE motifs, sequences of 60-70 amino acids containing heptad repeats. VAMP and syntaxin each contribute one helix to the complex, with two contributed by SNAP-25. Within the helix bundle there is a highly conserved layer (called the 'zero-layer', Sutton et al., 1998) of three glutamines and one arginine, which is essential for formation of a functional complex. This has led to the alternative classification of SNARES as R-SNAREs (VAMP) or Q-SNAREs (syntaxin and SNAP-25). In addition to its SNARE motif, syntaxin has a large N-terminal domain of three α-helices, called H_{abc}, which interact with the SNARE motif and prevents its interaction with other SNAREs. In this conformation, syntaxin is said to be in the 'closed' state, and SNARE complex formation can only occur after transition to an 'open' conformation. This conformational transition may provide a means of regulation of exocytosis.

Elucidation of the molecular action of the Clostridial neurotoxins provided important evidence of the essential function of SNARE proteins in neurotransmission (see Burgoyne and Morgan, 2003). Tetanus toxin and the seven types of botulinum toxin are heterodimers, each containing a heavy chain (H) that targets the toxin to a particular neuronal type, and a light chain (L) that is a metalloproteinase. After entry into the cytoplasm, these endoproteinases specifically cleave VAMP (tetanus toxin or botulinum toxins B,D,F and G), SNAP-25 (botulinum toxins A and E) or syntaxin 1 (botulinum toxin C). Even in neuroendocrine cells, which have no receptors for the H-chains, exocytosis is blocked if the toxins are introduced into the cytoplasm, for example by permeabilization of the plasma membrane. In some cell types exocytosis is resistant to Clostridial toxins through the expression of toxin-resistant SNARE isoforms

The formation of the core complex was originally thought to be responsible for tethering of vesicles at release sites, but more recently it has been demonstrated that it participates in the process of membrane fusion itself. Other proteins that have been identified that are involved in the docking process include the SM proteins (sec1 / munc18-like proteins, Gallwitz and Jahn, 2003). These are highly conserved throughout evolution and are thought to link tethering to SNARE assembly. Little is known about the molecular mechanism of this regulation at present. The neuronal SM protein is nsec-1, also known as munc-18. It interacts with the t-SNARE syntaxin to maintain it in the closed conformation and so may restrict its entry into the SNARE complex. Thus munc-18 may regulate docking by controlling the availability of t-SNAREs. However the role of the SM proteins is controversial, and the evidence from munc-18 over-expression or knockout suggests that it may be involved both in SNARE complex assembly and in the fusion process. The syntaxin-binding protein tomosyn has also been proposed as a regulator of SNARE complex formation, as it contains a sequence with homology to the R-SNAREs, and binds to the SNARE motif of syntaxin. (Hatsuzawa et al., 2003). It might therefore inhibit SNARE complex assembly, or even promote it by displacement of munc-18. Physiological studies in secretory cells support the former mechanism, since overexpression of tomosyn reduces the size of the RRP by 50%, without affecting docked vesicles, indicating a putative role in the priming reaction (see below) (Yizhar et al., 2004). Another potential regulatory protein, complexin, forms a tight 1:1 complex with assembled SNARE complexes. It exists in two isoforms, and the double knockout of both isoforms is lethal in mice. It has been proposed that complexin stabilizes particular trans-SNARE intermediates and thereby regulates fusion-pore opening (Archer et al., 2002).

4.3. Priming

Before a docked vesicle can fuse it has to be made competent for release. This step is called priming and seems to be a multi-step process. Priming has requirements for ATP and Ca^{2+}, according to studies using neuroendocrine cells. The ATP requirement has been attributed to a partial disassembly of the SNARE complex by an ATPase called NSF, phosphorylation of nerve terminal proteins, or phosphorylation of plasma membrane lipids so as to generate phosphatidylinositol (4,5) bisphosphate, $PI(4,5)P_2$, (Martin, 2002).

Other proteins have also been implicated in priming. In neuroendocrine cells CAPS (Ca^{2+}-dependent Activator Protein for Secretion) is essential for LDCV priming. CAPS is proposed to mediate this effect through an interaction with recently generated $PI(4,5)P_2$ on the target membrane (Grishanin et al., 2002). In neurones a protein called munc-13 that is specifically localised to active zones is also essential for priming. Munc-13 has been suggested to regulate priming by displacing munc-18 from the t-SNARE syntaxin (Brose and Rosenmund, 2002). This would allow syntaxin to assume its 'open' conformation, resulting in the formation of the SNARE complex. Another munc-13 interacting protein called RIM (rab3-interacting molecule) is also thought to be essential for priming (Martin, 2002). This is because ablation of the RIM gene results in a large reduction in the size of the RRP, but no change in the number of docked SVs, an identical phenotype to the synapses of munc-13-1 knockout mice (Martin, 2002).

4.4. Membrane Fusion

In vitro assays have shown that just three proteins are sufficient for membrane fusion: these are the SNARE proteins (Figure 2), which permit the slow fusion of proteoliposomes containing them (Weber et al., 1998). The 'zippering up' of the helices in the *trans* core complex brings the two membranes into close apposition, a process called a 'loose' to 'tight' transition. The formation of the tight SNARE complex, which is extremely stable, may release enough energy to overcome the thermodynamic barrier to membrane fusion. However, the very slow speed of fusion of SNARE-containing proteoliposomes suggests that other as yet unidentified factors must act to accelerate the fusion process to speeds required for neurotransmission. That the SNARE proteins are essential for fusion but not for docking is indicated by the accumulation of docked vesicles when Clostridial neurotoxins are used to block exocytosis in mammalian cells, although work on some non-mammalian systems has suggested that the SNARE complex may be dissociated before fusion occurs (Peters et al., 2001).

After fusion the SNAREs are now all in the same membrane in a *cis* complex, and have to be separated for reuse. This is performed by the ATPase NSF (N-ethyl maleimide Sensitive Factor), which attaches to the SNARE complex via the small effector protein α-SNAP (Soluble NSF Attachment Protein). ATP hydrolysis by NSF provides the energy to dissociate the *cis* SNARE complex, allowing VAMP to be retrieved by endocytosis and the SNAREs to participate in another round of membrane fusion.

4.5. The Calcium Sensor

Both granule and SV exoyctosis are triggered by an influx of extracellular Ca^{2+}. The protein usually assumed to be the Ca^{2+} sensor in regulated exocytosis is synaptotagmin I, the most widely distributed member of a family that contains at least thirteen members. Synaptotagmin I is a secretory vesicle membrane protein with a short, glycosylated N-terminal domain in the lumen and a large cytoplasmic region containing two C2 domains, sequentially called C2A and C2B, that respectively bind 3 Ca^{2+} and 2 Ca^{2+}. Synaptotagmin I binds to syntaxin and to assembled SNARE complexes, to lipids such as phosphatidylserine and $PI(4,5)P_2$, and it also undergoes Ca^{2+}-dependent oligomerization. A large body of genetic, functional and biochemical evidence supports the idea that synaptotagmin I is a Ca^{2+} sensor in neuronal exocytosis, although it may not be the only one (Yoshihara et al., 2003). Its affinity for Ca^{2+} is consistent with a role in synaptic release (Brose et al 1992). Moreover, mice expressing mutant synaptotagmin I with decreased affinity for Ca^{2+} have a parallel decrease in the apparent Ca^{2+}-affinity of exocytosis (Fernández-Chacón et al., 2001). In *Drosophila* and *C. elegans*, genetic knockout of synaptotagmin I appears to have effects on vesicle docking and SNARE complex assembly, as well as the fusion process itself; this does not seem to be so in the mouse, suggesting that synaptotagmins have multiple functions, some of which can be performed by other proteins, including synaptotagmin isoforms, in mammalian neurons. More acute knockout of synaptotagmin I in *Drosophila* suggests that only the fusion reaction is blocked (Marek and Davis, 2002). Not all synaptotagmins are located on SVs:

for example synaptotagmins III and VII are on the plasma membrane, and it has been suggested (Sugita et al., 2002) that these plasma membrane synaptotagmins are high-affinity Ca^{2+} sensors controlling slow vesicle fusion. However, the mechanism by which synaptotagmins could trigger membrane fusion is still unclear. Ca^{2+} binding is known to regulate interactions of synaptotagmins with both membrane lipids and SNARE proteins, and most significantly both the C2A and C2B domains of synaptotagmin I undergo Ca^{2+}-dependent insertion into lipid bilayers. One current model is that synaptotagmin acts in concert with complexin to stimulate vesicle fusion (Jahn et al., 2003). The assembled *trans*-SNARE complex may be stabilized by complexin in such a way that vesicle and plasma membrane are in a hemi-fusion state. Calcium influx then causes synaptotagmin to penetrate the membrane, destabilising it and allowing fusion to occur.

4.6. The Fusion Pore

Although amperometry and capacitance measurements have yielded a great deal of information about the kinetic characteristics of the fusion pore, its molecular structure is still unresolved. It is clear that in exocytosis, membrane fusion is initiated by proteins, as it is in viral fusion, while the gradual expansion of fusion pores after fusion suggest that at this stage the pore involves lipids. It is not clear whether the pore formed at the onset of fusion is constructed entirely of proteins, and none of the proteins discussed above have been shown to form pores. One radical proposal, derived from work on vacuolar fusion in yeast, is that the fusion pore is formed by the H^+-conducting sector, called V_o, of the vacuolar H^+-ATPase (Bayer et al., 2003). This enzyme also occurs in secretory vesicles, where its function is lumenal acidification, but there has been no demonstration that it has a direct role in exocytosis in neurons or neuroendocrine cells. Recent mutagenesis analysis of syntaxin suggests that the fusion pore may in fact be lined by between 5 to 8 transmembrane domains of syntaxin although whether other proteins are also involved has not been resolved (Han et al., 2004).

4.7. Endocytosis

The majority of proteins that participate in clathrin-mediated endocytosis have been discovered during the past 10 years (Slepnev and De Camilli, 2000). Sites of endocytosis are nucleated by the interactions of the adapter protein AP-2 with synaptotagmin, which resides in the plasma membrane after fusion. AP-2 then recruits clathrin and a protein called AP180 to the plasma membrane. Both AP-2 and AP180 are clathrin assembly molecules and the formation of the clathrin coat is thought to induce a degree of membrane curvature, aiding the invagination process. Invagination is thought to be initiated by the accessory endocytosis protein epsin (Ford et al., 2002), a cytoplasmic protein that is recruited to membranes through its interaction with $PI(4,5)P_2$. On binding $PI(4,5)P_2$, epsin is proposed to undergo a conformational change that allows it to penetrate the membrane, destabilise it and initiate invagination.

Once the nascent endocytic vesicle is fully formed, it is attached to the plasma membrane by a narrow neck. At this stage a GTPase called dynamin is recruited to the vesicle neck by an AP-2-binding protein called amphiphysin. Dynamin is the protein that mediates fission of the vesicle from the plasma membrane. This is a GTP-dependent event and the energy released from the hydrolysis of GTP drives vesicle fission, probably through a conformational change in dynamin structure (Marks et al., 2001). Clathrin-mediated endocytosis is a Ca^{2+}-dependent event in neurones. The Ca^{2+} sensor for endocytosis is not synaptotagmin but the Ca^{2+}-dependent protein phosphatase calcineurin, which on nerve terminal stimulation mediates the dephosphorylation of at least 8 essential endocytosis proteins (the dephosphins) to stimulate the process (Cousin and Robinson, 2001). The mechanism of this stimulation is thought to be due to stimulation of both protein - protein and protein - lipid interactions.

4.8. Kiss-and-Run Exocytosis

The molecular mechanism for either the reuse pathway, kiss-and-run or intact vesicle retrieval is still undetermined, but recent studies have suggested that different isoforms of synaptotagmin may participate in these processes. For example in neuroendocrine cells overexpression of synaptotagmin IV increased the frequency of kiss-and-run events (Wang et al., 2003). In these studies synaptotagmin I favoured full fusion but synaptogamin IV favoured kiss-and-run. Interestingly the regulation of these processes seems to centre on Ca^{2+}, since either full fusion or kiss-and-run was inhibited by mutations in the respective Ca^{2+} binding domains of both synaptotagmin I and IV.

Also different splice variants of a plasma membrane localised synaptotagmin (synaptotagmin VII) may control the speed of SV retrieval in central synapses (Virmani et al., 2003). Overexpression of a truncated form of synaptotagmin VII lacking the C2 Ca^{2+}-binding domains accelerated vesicle retrieval, whereas full length synaptotagmin VII decelerated the process. This suggests that over and above the triggering of vesicle fusion, the synaptotagmin family of proteins can also control the type and speed of vesicle recycling pathways, possibly via their ability to bind Ca^{2+}. The SNARE-binding protein complexin has also been suggested to induce kiss-and-run exocytosis, by stabilizing *trans*-SNARE intermediates that allow rapid fusion pore closure.

5. REGULATED EXOCYTOSIS IN DENDRITES

5.1. Evidence for Vesicular Release

Although retrograde mediators such as endocannabinoids, carbon mononoxide and nitric oxide are not released via a vesicular mechanism (Ludwig and Pittman, 2003), dendritic release of 'classical' transmitters and peptides involves conventional regulated exocytosis (Note that some classical transmitters such as dopamine may also be released via carrier-mediated mechanisms (Attwell et al., 1993; Falkenburger et al., 2001)). Tannic acid fixation of dendrites within the supraoptic nucleus (SON) revealed the classical LDCV morphology with evidence of 'omega' fusion events at the plasma membrane (Pow and Morris, 1989). In several systems, smaller vesicular structures have also been observed that may correspond to SVs (for example see Pow and Morris, 1989). Furthermore, carbon fibre amperometry has also suggested that dopamine release in the striatum is quantal supporting the vesicular hypothesis (Jaffe et al., 1998). However, whether discrete pools (e.g. RRP and reserve pool) of dendritic vesicles exist is essentially unknown although electron micrographic analysis does suggest that a reserve pool of LDCVs can be recruited from within dendrites (Tobin et al., 2004). Whether dendritic vesicles are targeted to 'active sites' at the plasma membrane (for example by docking with Ca^{2+} channels as in nerve terminals), or associate with the cytoskeleton via synapsin-like molecules, remains to be determined.

5.2. Vesicle Targeting

An important question is how secretory vesicles (SVs or LDCVs) are selected for transport to dendritic compartments, rather than the axonal terminal. Although transport of vesicles to dendrites most likely occurs along microtubules (Pow and Morris, 1989), as for axonal transport, direct evidence is still lacking. Several models of dendritic vs axonal targeting have been suggested (see for example Burack et al., 2000) including use of different motor proteins, microtubule polarity and differential retention in the plasma membrane. All such models imply that the vesicle biochemistry of dendritically targeted LDCVs (or SVs) is in some way distinct to those destined for the axon terminals. Indeed, recent evidence suggests that peptidergic vesicles destined for dendrites have a distinct vesicular cargo to that of vesicles destined for the axon terminal (Landry et al., 2003). Whether such differential selection requires distinct motor proteins or recognition by distinct vesicle membrane proteins remains to be determined. Understanding this differential targeting, in particular for peptide secreting neurons, is essential to fully understand how dendritic and axonal terminal peptide transmitter release may be coordinated or independently regulated (Ludwig, 1998; Wotjak et al., 1998), especially under conditions when maintained supply of peptide is required. Increasing evidence also suggests local peptide synthesis in magnocellular dendrites (Ma and Morris, 1998; Ma and Morris, 2002) and endoplasmic reticulum export sites are assembled regularly throughout the entire dendritic tree of hippocampal neurones in culture (Aridor et al., 2004). However whether LDCV biogenesis can also occur from Golgi extensions into dendrites, to support a local dendritic LDCV supply for exocytosis, remains to be shown. Although morphological evidence suggests that SVs and LDCVs may co-exist within the same dendrite definitive evidence for 'classical transmitters' and peptides from the same dendrite, has not been shown. In this regard it is important to note that some non-peptide transmitters (e.g. ATP) can be co-packaged and released from LDCVs.

5.3. Molecules and Mechanisms

What do we know about the molecular mechanisms of vesicular release in dendrites and is it the same as for axonal terminal release? The core complex SNARE proteins SNAP-25 and syntaxin have been identified in dendrites from neonatal hypothalamic neurones (Schwab et al., 2001). Furthermore, botulinum toxin A (which cleaves SNAP-25) inhibits dendritic release of dopamine in the substantia nigra (Bergquist et al., 2002) and tetanus toxin (which cleaves VAMP) disrupts hippocampal dendritic release (Maletic-Savatic and Malinow, 1998). In addition α-SNAP mRNA has been identified in dendrites from hypothalamic neurones (Morris et al., 2000), however, dendritic isoforms of the various core-complex proteins as well as other exocytic proteins such as the postulated Ca^{2+}-sensor, synaptotagmin (Schwab et al., 2001), are essentially unknown. In dopaminergic neurones of the substantia nigra, the Ca^{2+} sensitivity of dopamine release from dendrites is different from its release from axonal terminals (Chen and Rice, 2001), suggesting that distinct isoforms of the Ca^{2+} sensor synaptotagmin may be responsible. Furthermore, botulinum toxin B inhibits dopamine release from axons but not from dendrites in the substantia nigra (Bergquist et al., 2002). This might imply that distinct

VAMP isoforms may exist in the two compartments or that in this system vesicles are already docked. Characterisation of dendritic expression and targeting of core SNARE complex proteins, and of the constellation of core complex regulatory proteins such as munc-18, in conjunction with the application of genetic knockout models and toxin sensitivity, should provide important insights into the biochemistry of dendritic vesicle targeting, tethering and fusion.

5.4. Calcium Dependence

Although increasing evidence supports the obligatory role for Ca^{2+} in mediating dendritic vesicular release (Ludwig, 1998; Moos et al., 1984; Neumann et al., 1993), its source and means of entry may vary with the physiological context as well as the system under investigation. A number of studies suggest that Ca^{2+} entry from the extracellular space occurs through high voltage-activated Ca^{2+} channels (of the L- and N-type). For example, PACAP stimulation of vasopressin release in the SON is blocked by L-type channel blockers (Shibuya et al., 1998) and L- as well as N-type channels have been implicated in dendritic dynorphin release in the dentate gyrus (Simmons et al., 1995). Many studies have indicated a requirement for Ca^{2+} influx (Ludwig, 1998; Moos et al., 1984; Neumann et al., 1993) but in most cases the mode of Ca^{2+} entry is not known. In this context it is important to note that both low- and high- voltage activated Ca^{2+} channels have been reported to be expressed in magnocellular neurones (Fisher and Bourque, 1996; Fisher and Bourque, 2001; Foehring and Armstrong, 1996; Joux et al., 2001) and thus multiple Ca^{2+} entry pathways may exist in different systems. Furthermore, Ca^{2+} influx may be due to activation of dendritic ligand-activated ion channels such as the reported NMDA receptor mediated Ca^{2+} elevation (Bains and Ferguson, 1999) and dendritic peptide release in magnocelluar neurones (Morris et al., 2000) as well as NMDA -mediated dynorphin release in the substantia nigra (Araneda and Bustos, 1989). Preliminary evidence also suggests that Ca^{2+} store operated Ca^{2+} or cationic channels are a source of Ca^{2+} entry in magnocellular neurones (Tobin et al., 2002) although their role in dendritic secretion is not known. It is essentially unknown whether the Ca^{2+} sensitivity of peptide or 'classical transmitter' release from dendrites is distinct and thus whether exocytosis from dendritic SVs or LDCVs is mechanistically more closely related to that in axonal terminals or endocrine cells.

Evidence is also accumulating that release of Ca^{2+} from intracellular stores is an important mode of regulation of dendritic peptide release. Indeed, oxytocin mediated activation of dendritic G-protein linked receptors couples to Ca^{2+} release from intracellular thapsigargin sensitive Ca^{2+} stores to mediate dendritic oxytocin secretion in the SON (Chevaleyre et al., 2000; Lambert et al., 1994; Moos et al., 1984). Thus the source of Ca^{2+}, its mode of entry and whether vesicles are docked in dendritic 'active zones' with Ca^{2+} channels, requires further elucidation.

5.5. Vesicle Recycling and Priming

Intriguing recent work suggests a mode of vesicular priming of oxytocin- containing vesicles by release of intracellular Ca^{2+} from thapsigargin-sensitive stores (Ludwig et al., 2002) and opens the question of how dendritic vesicles are recycled and how recycling is regulated. As discussed above, peptidergic vesicles must recycle through the Golgi to be refilled with peptides. Priming could be a result of recruitment of vesicles from a dendritic reserve pool, tight coupling between vesicle biogenesis in the Golgi and targeting to dendrites or regulation of vesicle docking and priming. In addition, it may indicate that dendritic vesicles undergo a form of kiss-and-run exocytosis with local recycling and re-recruitment to the plasma membrane. Recent EM studies (Tobin et al., 2004) suggest that a reserve pool of vesicles stored within the dendrites several hundred nanometres from the plasma membrane may be recruited to the plasma membrane upon priming. As oxytocin activates a G-protein coupled receptor linked to phospholipase C and the production of IP_3 and diacylglycerol, the IP_3-mediated elevation of intracellular free Ca^{2+} from intracellular stores, in concert with diacylglycerol activation of DAG receptors (protein kinase C or munc-13), may promote translocation of these reserve vesicles to the plasma membrane. PKC and munc-13 play important roles in vesicle trafficking and regulation of the core complex (Brose and Rosenmund, 2002). Elucidation of the fundamental mechanisms underlying such priming events may provide important clues to the molecular mechanisms regulating dendritic release. Analysis of single vesicle targeting, recycling and fusion in real time, although representing a considerable technological challenge, is likely to be amenable to analysis of peptide release from dendritic LDCVs, once suitable assays have been developed.

6. CONCLUSION

In this brief review we have attempted to highlight the core principles and mechanism of regulated exocytosis and highlight some of the key questions and opportunities for analysis of the vesicle lifecycle in dendrites. Many questions remain: how are secretory vesicles targeted to dendrites? Does local vesicle biogenesis or recycling occur in dendrites? Do functionally distinct vesicle pools reside in the dendrite? Are secretory vesicles docked at active zones with distinct Ca^{2+} channels in dendrites? Are SVs and LDCVs differentially targeted and regulated in dendrites? How is vesicle priming and recycling controlled? We are entering a fascinating period in which many key molecular, biochemical, imaging, biophysical and bioinformatics tools can be assembled to allow functional analysis of dendritic release and forge the framework to understand the integrative function of dendritic retrograde signalling in health & disease.

7. REFERENCES

Alés, E., Tabares, L., Poyato, J. M., Valero, V., Lindau, M., and Alvarez de Toledo, G., 1999, High calcium concentrations shift the mode of exocytosis to the kiss-and-run mechanism, *Nat. Cell. Biol.* **1**: 40.

Araneda, R., and Bustos, G., 1989, Modulation of dendritic release of dopamine by N-Methyl-D- Aspartate receptors in rat substantia nigra, *J. Neurochem.* **52**: 962.

Archer, D. A., Graham, M. E., and Burgoyne, R. D., 2002, Complexin regulates the closure of the fusion pore during regulated vesicle exocytosis, *J. Biol. Chem.* **277**: 18249.

Aridor, M., Guzik, A. K., Bielli, A., and Fish, K. N., 2004, Endoplasmic reticulum export site formation and function in dendrites, *J. Neurosci.* **24**:3770.

Attwell, D., Barbour, B., and Szatkowski, M., 1993, Nonvesicular release of transmitter, *Neuron.* **11**: 401.

Bains, J. S., and Ferguson, A. V., 1999, Activation of N-methyl-D-aspartate receptors evokes calcium spikes in the dendrites of rat hypothalamic paraventricular nucleus neurons, *Neuroscience* **90**: 885.

Barg, S., 2003, Mechanisms of exocytosis in insulin-secreting β-cells and glucagon-secreting α-cells, *Pharm. Toxicol.* **92**: 3.

Bayer, M. J., Reese, C., Bühler, S., Peters, C., and Mayer, A., 2003, Vacuole membrane fusion: Vo functions after trans-SNARE pairing and is coupled to the Ca^{2+}-releasing channel, *J. Cell Biol.* **162**: 211.

Bergquist, F., Niazi, H. S., and Nissbrandt, H., 2002, Evidence for different exocytosis pathways in dendritic and terminal dopamine release *in vivo, Brain Res.* **950**: 245.

Brose, N., Petrenko, A. G., Sudhof, T. C., and Jahn, R., 1992, Synaptotagmin: a calcium sensor on the synaptic vesicle surface, *Science* **256**: 1021.

Brose, N., and Rosenmund, C., 2002, Move over, protein kinase C, you've got company: alternative cellular effectors of diacylglycerol and phorbol esters, *J. Cell Sci.* **115**: 4399.

Burack, M. A., Silverman, M. A., and Banker, G., 2000, The role of selective transport in neuronal protein sorting, *Neuron* **26**: 465.

Burgoyne, R. D., and Morgan, A., 2003, Secretory granule exocytosis, *Phys. Rev.* **83**: 581

Chen, B. T., and Rice, M. E., 2001, Novel Ca^{2+} dependence and time course of somatodendritic dopamine release: Substantia nigra versus striatum, *J. Neurosci.* **21**: 7841.

Chevaleyre, V., Dayanithi, G., Moos, F. C., and Desarmenien, M. G., 2000, Developmental regulation of a local positive autocontrol of supraoptic neurons, *J. Neurosci.* **20**: 5813.

Cousin, M. A., and Robinson, P. J., 2001, The dephosphins: dephosphorylation by calcineurin triggers synaptic vesicle endocytosis, *Trends Neurosci.* **24**: 659.

Duncan, R. R., Greaves, J., Wiegand, U. K., Matskevich, I., Bodammer, G., Apps, D. K., Shipston, M. J., and Chow, R. H., 2003, Functional and spatial segregration of secretory vesicle pools according to vesicle age, *Nature* **422**: 176.

Falkenburger, B. H., Barstow, K. L., and Mintz, I. M., 2001, Dendrodendritic inhibition through reversal of dopamine transport, *Science* **293**: 2465.

Fernández-Chacón, R., Königstorfer, A., Gerber, S. H., García, J., Matos, M. F., Stevens, C. F., Brose, N., Rizo, J., Rosenmund, C., and Südhof, T. C., 2001, Synaptotagmin I functions as a calcium regulator of release probability, *Nature* **410**: 41.

Fesce, R., Grohovaz, F., Valtorta, F., and Meldolesi, J., 1994, Transmitter release: Fusion or "kiss-and-run"?, *Trends Cell Biol.* **4**: 1.

Fisher, T. E., and Bourque, C. W., 1996, Calcium-channel subtypes in the somata and axon terminals of magnocellular neurosecretory cells, *Trends Neurosci.* **19**: 440.

Fisher, T. E., and Bourque, C. W., 2001, The function of Ca^{2+} channel subtypes in exocytotic secretion: new perspectives from synaptic and non-synaptic release, *Prog. Biophys. Mol. Biol.* **77**: 269.

Foehring, R. C., and Armstrong, W. E., 1996, Pharmacological dissection of high-voltage-activated Ca^{2+} current types in acutely dissociated rat supraoptic magnocellular neurons, *J. Neurophysiol.* **76**: 977.

Ford, M. G., Mills, I. G., Peter, B. J., Vallis, Y., Praefcke, G. J., Evans, P. R., and McMahon, H. T., 2002, Curvature of clathrin-coated pits driven by epsin, *Nature* **419**: 361.

Gallwitz, D., and Jahn, R., 2003, The riddle of the Sec1/Munc-18 proteins - new twists added to their interactions with SNAREs, *Trends Biochem. Sci.* **28**: 113.

Gandhi, S. P., and Stevens, C. F., 2003, Three modes of synaptic vesicular recycling revealed by single-vesicle imaging, *Nature* **423**: 607.

Grishanin, R. N., Klenchin, V. A., Loyet, K. M., Kowalchyk, J. A., Ann, K., and Martin, T. F., 2002, Membrane association domains in Ca^{2+}-dependent activator protein for secretion mediate plasma membrane and dense-core vesicle binding required for Ca^{2+}-dependent exocytosis, *J. Biol. Chem.* **277**: 22025.

Han, X., Wang, C-T., Bai, J., Chapman, E. R., and Jackson, M . B., 2004, Transmembrane segments of syntaxin line the fusion pore of Ca^{2+}-triggered exocytosis, *Science* **304**: 289.

Harata, N., Pyle, J. L., Aravanis, A. M., Mozhayeva, M., Kavalali, E. T., and Tsien, R. W., 2001, Limited numbers of recycling vesicles in small CNS nerve terminals: implications for neural signaling and vesicular cycling, *Trends Neurosci.* **24**: 637.

Hatsuzawa, K., Lang, T., Fasshauer, D., Bruns, D., and Jahn, R., 2003, The R-SNARE motif of tomosyn forms SNARE complexes with syntaxin 1 and SNAP-25 and down-regulates exocytosis, *J. Biol. Chem.* **278**: 31159.

Hilfiker, S., Pieribone, V. A., Czernik, A. J., Kao, H. T., Augustine, G. J., and Greengard, P., 1999, Synapsins as regulators of transmitter release, *Philos. Trans. Roy. Soc. Lond[B].* **354**: 269.

Jaffe, E. H., Marty, A., Schulte, A., and Chow, R. H., 1998, Extrasynaptic vesicular transmitter release from the somata of substantia nigra neurons in rat midbrain slices, *J. Neurosci.* **18**: 3548.

Jahn, R., Lang, T., and Südhof, T. C., 2003, Membrane fusion, *Cell* **112**: 519.

Joux, N., Chevaleyre, V., Alonso, G., Boissin-Agasse, L., Moos, F. C., Desarmenien, M. G., and Hussy, N., 2001, High voltage-activated Ca^{2+} currents in rat supraotic neurones: Biophysical properties and expression of the various channel alpha 1 subunits, *J. Neuroendocrinol.* **13**: 638.

Koenig, J. H., and Ikeda, K., 1989, Disappearance and reformation of synaptic vesicle membrane upon transmitter release observed under reversible blockage of membrane retrieval, *J. Neurosci.* **9**: 3844.

Lambert, R. C., Dayanithi, G., Moos, F. C., and Richard, P., 1994, A rise in the intracellular Ca^{2+} concentration of isolated rat supraoptic cells in response to oxytocin, *J. Physiol.* **478**: 275.

Landry, M., Vila-Porcile, E., Hökfelt, T., and Calas, A., 2003, Differential routing of coexisting neuropeptides in vasopressin neurons, *Eur. J. Neurosci.* **17**: 579.

Langley, K., and Grant, N. J., 1997, Are exocytosis mechanisms transmitter specific?, *Neurochem. Int.* **31**: 739.

Ludwig, M., 1998, Dendritic release of vasopressin and oxytocin, *J. Neuroendocrinol.* **10**: 881.

Ludwig, M., and Pittman, Q. J., 2003, Talking back: dendritic transmitter release, *Trends Neurosci.* **26**: 255.

Ludwig, M., Sabatier, N., Bull, P. M., Landgraf, R., Dayanithi, G., and Leng, G., 2002, Intracellular calcium stores regulate activity-dependent neuropeptide release from dendrites, *Nature* **418**: 85.

Ma, D., and Morris, J. F., 1998, Local protein synthesis in magnocellular dendrites. Basic elements and their response to hyperosmotic stimuli, *Adv. Exp. Med. Biol.* **449**: 55.

Ma, D., and Morris, J. F., 2002, Protein synthetic machinery in the dendrites of the magnocellular neurosecretory neurons of wild-type Long-Evans and homozygous Brattleboro rats, *J. Chem. Neuroanat.* **23**: 171.

Maletic-Savatic, M., and Malinow, R., 1998, Calcium-evoked dendritic exocytosis in cultured hippocampal neurons. Part I: Trans-Golgi network-derived organelles undergo regulated exocytosis, *J. Neurosci.* **18**: 6803.

Marek, K. W., and Davis, G. W., 2002, Transgenically encoded protein photoinactivation (FlAsH-FALI): acute inactivation of synaptotagmin I, *Neuron* **36**: 805.

Marks, B., Stowell, M. H., Vallis, Y., Mills, I. G., Gibson, A., Hopkins, C. R., and McMahon, H. T., 2001, GTPase activity of dynamin and resulting conformation change are essential for endocytosis, *Nature* **410**: 231.

Martin, T. F., 2002, Prime movers of synaptic vesicle exocytosis, *Neuron.* **34**: 9.

Moos, F., Freundmercier, M. J., Guerne, Y., Guerne, J. M., Stoeckel, M. E., and Richard, P., 1984, Release of oxytocin and vasopressin by magnocellular nuclei *in vitro* - specific facilitatory effect of oxytocin on its own release, *J. Endocrinol.* **102**: 63.

Morris, J. F., Christian, H., Ma, D., and Wang, H., 2000, Dendritic secretion of peptides from hypothalamic magnocellular neurosecretory neurones: a local dynamic control system and its functions, *Exp. Physiol.* **85**: 131S.

Murthy, V. N., and De Camilli, P., 2003, Cell biology of the presynaptic terminal, *Ann. Rev. Neurosci.* **26**: 701.

Neher, E., 1998, Vesicle pools and Ca^{2+} microdomains: new tools for understanding their roles in transmitter release, *Neuron* **20**: 389.

Neumann, I., Russell, J. A., and Landgraf, R., 1993, Oxytocin and vasopressin release within the supraoptic and paraventricular nuclei of pregnant, parturient and lactating rats - a microdialysis study, *Neuroscience* **53**: 65.

Parsons, S. M., 2000, Transport mechanisms in acetylcholine and monoamine storage, *FASEB J.* **14**: 2423.

Peters, C., Bayer, M. J., Bühler, S., Andersen, J. S., Mann, M., and Mayer, A., 2001, Trans-complex formation by proteolipid channels in the terminal phase of membrane fusion, *Nature* **409**: 581.

Pow, D. V., and Morris, J. F., 1989, Dendrites of hypothalamic magnocellular neurons release neurohypophyseal peptides by exocytosis, *Neuroscience* **32**: 435.

Rizzoli, S. O., and Betz, W. J., 2003, Neurobiology: All change at the synapse, *Nature* **423**: 591.

Rizzoli, S. O., and Betz, W. J., 2004, The structural organization of the readily releasable pool of synaptic vesicles, *Science* **303**: 2037.

Schwab, Y., Mouton, J., Chasserot-Golaz, S., Marty, I., Maulet, Y., and Jover, E., 2001, Calcium-dependent translocation of synaptotagmin to the plasma membrane in the dendrites of developing neurones, *Mol. Brain Res.* **96**: 1.

Shibuya, I., Noguchi, J., Tanaka, K., Harayama, N., Inoue, Y., Kabashima, N., Ueta, Y., Hattori, Y., and Yamashita, H., 1998, PACAP increases the cytosolic Ca^{2+} concentration and stimulates somatodendritic vasopressin release in rat supraoptic neurons, *J. Neuroendocrinol.* **10**: 31.

Simmons, M. L., Terman, G. W., Gibbs, S. M., and Chavkin, C., 1995, L-Type calcium channels mediate dynorphin neuropeptide release from dendrites but not axons of hippocampal granule cells, *Neuron* **14**: 1265.

Slepnev, V. I., and De Camilli, P., 2000, Accessory factors in clathrin-dependent synaptic vesicle endocytosis, *Nat. Rev. Neurosci.* **1**: 161.

Sugita, S., Shin, O.-H., Han, W., Lao, Y., and Südhof, T. C., 2002, Synaptotagmins form a heirarchy of exocytic Ca^{2+} sensors with distinct Ca^{2+} affinities, *EMBO J.* **21**: 270.

Sutton, R. B., Fasshauer, D., Jahn, R., and Brunger, A. T., 1998, Crystal structure of a SNARE complex involved in synaptic exocytosis at 2.4Å resolution, *Nature* **395**: 347.

Taraska, J. W., Perrais, D., Ohara-Imaizumi, M., Nagamatsu, S., and Almers, W., 2003, Secretory granules are recaptured largely intact after stimulated exocytosis in cultured endocrine cells, *PNAS USA* **100**: 2070.

Tobin, V. A., Moos, F. C., and Desarmenien, M. G., 2002, ICRAC in magnocellular neurones from adult male rats., *The 5th International Congress of Neuroendocrinology, Bristol, UK P262.*

Tobin, V. A., Hurst, G., Norrie, L., Dal Rio, F., Bull, P. M., and Ludwig, M., 2004, Thapsigargin-induced mobilisation of dendritic dense core vesicles in supraoptic neurones, *Eur. J. Neurosci.* **19:** 2909.

Tooze, S. A., Martens, G. J. M., and Huttner, W. B., 2001, Secretory granule biogenesis: rafting to the SNARE, *Trends Cell Biol.* **11:** 116.

Virmani, T., Han, W., Liu, X., Sudhof, T. C., and Kavalali, E. T., 2003, Synaptotagmin 7 splice variants differentially regulate synaptic vesicle recycling, *EMBO J.* **22:** 5347.

Vitale, M. L., Seward, E. P., and Trifaró, J.-M., 1995, Chromaffin cell cortical actin network dynamics control the size of the release-ready vesicle pool and the initial rate of exocytosis, *Neuron* **14:** 353.

Voets, T., 2000, Dissection of three Ca^{2+}-dependent steps leading to secretion in chromaffin cells from mouse adrenal slices, *Neuron* **28:** 537.

Wang, C. T., Lu, J. C., Bai, J., Chang, P. Y., Martin, T. F., Chapman, E. R., and Jackson, M. B., 2003, Different domains of synaptotagmin control the choice between kiss-and-run and full fusion, *Nature* **424:** 943.

Waters, M. G., and Hughson, F. M., 2000, Membrane tethering and fusion in the secretory and endocytic pathways, *Traffic* **1:** 588.

Weber, T., Zemelman, B. T., McNew, J. A., Westermann, B., Gmachl, M., Parlati, F., Söllner, T. H., and Rothman, J. E., 1998, SNAREpins: minimal machinery for membrane fusion, *Cell* **92:** 759.

Wotjak, C. T., Ganster, J., Kohl, G., Holsboer, F., Landgraf, R., and Engelmann, M., 1998, Dissociated central and peripheral release of vasopressin, but not oxytocin, in response to repeated swim stress: New insights into the secretory capacities of peptidergic neurons, *Neuroscience* **85:**1209.

Yizhar, O., Matti, U., Melamed, R., Hagalili, Y., Bruns, D., Rettig, J., and Ashery, U., 2004, Tomosyn inhibits priming of large dense-core vesicles in a calcium-dependent manner, *PNAS USA* **101:** 2578.

Yoshihara, M., Adolfsen, B., and Littleton, J. T., 2003, Is synaptotagmin the calcium sensor? *Curr. Opin. Neurobiol.* **13:** 315.

4

ELECTRICAL PROPERTIES OF DENDRITES RELEVANT TO DENDRITIC TRANSMITTER RELEASE

Arnd Roth and Michael Häusser[*]

1. INTRODUCTION

Much of the signal processing in the brain takes place in the dendrites of neurons. In the classical view of fast signalling in neurons, chemical signals arriving at the input synapses on dendrites are translated into electrical postsynaptic potentials, which are processed by the cable properties of the dendrites before a binary decision is taken in the axon whether to fire an action potential or not. The action potential then propagates down the axon and is translated back into a chemical signal at the presynaptic terminals, which is passed on to their postsynaptic targets. However, this sequence of chemical and electrical signalling is not the only one that exists in mammalian neurons. For example, the action potential also propagates back into the dendrites of many types of neurons (Stuart et al., 1997b). These "backpropagating action potentials" act as retrograde signals and have implications for synaptic integration and the induction of long-term synaptic plasticity (Magee and Johnston, 1997; Markram et al., 1997), instructing the dendritic synapses that the axon has fired. Furthermore, dendrites can generate local spikes which may not propagate faithfully to the soma (Llinás and Sugimori, 1980; Chen et al., 1997; Stuart et al., 1997a; Golding and Spruston, 1998; Kamondi et al., 1998; Schwindt and Crill, 1998; Martina et al., 2000; Golding et al., 2002). The spread of electrical signals such as backpropagating action potentials and dendritic spikes is governed by the dendritic morphology as well as the passive and active membrane properties (Jack et al., 1983; Segev et al., 1995; Koch, 1999; Stuart et al., 1999; Häusser et al., 2000; Segev and London, 2000; Vetter et al., 2001; Scott, 2002; Häusser and Mel, 2003; Williams and Stuart, 2003). In this chapter we discuss how the electrical properties of dendrites shape the input-output relationship of neurons, focusing on the situation that exists when neuronal output is conveyed not only by the axon but also by the dendrites – via dendritic transmitter release.

---

[*] Arnd Roth and Michael Häusser, Wolfson Institute for Biomedical Research and Dept. of Physiology, University College London, London WC1E 6BT, UK.

Dendritic Neurotransmitter Release, edited by M. Ludwig
Springer Science+Business Media, Inc., 2005

As outlined in other chapters in this volume, a wide range of neuronal types have been shown to exhibit dendritic transmitter release. In these neurons, several mechanisms of coupling between electrical signalling in dendrites and dendritic transmitter release have been distinguished. First, depolarization of the dendritic membrane potential causes activation of voltage-gated Ca^{2+} channels, and the resulting Ca^{2+} influx causes an increase in the intracellular Ca^{2+} concentration, which in turn evokes either Ca^{2+}-dependent release of transmitter substances via exocytosis (Maletic-Savatic and Malinow, 1998) or Ca^{2+}-dependent local synthesis of transmitters such as endocannabinoids (Wilson and Nicoll, 2002; Freund et al., 2003). Second, large increases of Ca^{2+} concentration in the dendritic cytoplasm can be caused by Ca^{2+} release from intracellular stores. For example, Ca^{2+} release from inositol 1,4,5-trisphosphate-sensitive stores can be evoked in the dendrites of hippocampal and neocortical pyramidal neurons (Nakamura et al., 1999; Larkum et al., 2003) by repetitive synaptic activation of metabotropic glutamate receptors which may be paired with Ca^{2+} influx via voltage-gated Ca^{2+} channels activated by electrical depolarization as described in the first mechanism above. Third, Ca^{2+} influx can also be caused directly by synaptic activation of NMDA receptors, which exhibit a large permeability for Ca^{2+}. In granule cells of the olfactory bulb, Ca^{2+} influx through NMDA receptors has been shown to result in dendritic release of GABA (Isaacson and Strowbridge, 1998; Schoppa et al., 1998; Chen et al., 2000; Halabisky et al., 2000; chapter 7, this volume). Finally, Ca^{2+}-independent dendritic transmitter release can occur when depolarized membrane potentials and/or increased intracellular Na^+ concentrations (e.g. Rose and Konnerth, 2001) cause reverse transport by glutamate or GABA transporters (Attwell et al., 1993). Evidence for this voltage-dependent mode of dendritic transmitter release via reverse transport has been provided recently for dopamine release in the substantia nigra (Falkenburger et al., 2001) and for the endocannabinoid anandamide in the striatum (Ronesi et al., 2004). Together, these results suggest that both "classical" and other transmitters can be released from dendrites in a Ca^{2+}-independent fashion by reverse transport.

Following this introduction we will first describe the determinants, phenomenology and possible functional roles of the passive electrical properties of dendrites that govern subthreshold synaptic integration. Next, we will extend this description to include effects due to active voltage-dependent conductances in dendrites. Voltage-dependent conductances such as Na^+ channels, Ca^{2+} channels and NMDA receptors endow dendrites with an electrical positive feedback mechanism that can give rise to both graded boosting of subthreshold synaptic potentials (Stuart and Sakmann, 1995) and all-or-none regenerative events (Yuste et al., 1994; Schiller et al., 1997; Schiller et al., 2000; Ariav et al., 2003; Polsky et al., 2004). Activation of Ca^{2+} channels is particularly important in the context of this chapter as it enables coupling of electrical and Ca^{2+} signalling in dendrites, thus triggering dendritic neurotransmitter release. The chapter concludes with a perspective on the functional consequences of the link between dendritic excitability and dendritic release.

2. THE PASSIVE ELECTRICAL STRUCTURE OF DENDRITES

The dendrites of neurons are cable-like structures, consisting of a conducting core and a surface membrane which can be represented as a capacitance and a resistance in parallel (Jack et al., 1983; Segev et al., 1995; Koch, 1999; Scott, 2002). The core conductor – the intracellular medium – is an electrolyte solution whose electrical conductivity is determined by the concentration of mobile intracellular ions such as K^+ and Cl^- and by the excluded volume occupied by intracellular organelles such as mitochondria. Typical values for the specific resistivity R_i of the intracellular medium in neurons of the mammalian CNS range from 70 to 150 Ω cm (Major et al., 1994; Stuart and Spruston, 1998; Thurbon et al., 1998; Roth and Häusser, 2001; Steuber et al., 2004). The capacitance per unit area of the cell membrane, C_m, is determined by the effective thickness and the effective dielectric constant of the lipid bilayer, both of which are not known exactly since the protein content of the membrane is variable. However, direct measurements of C_m in neurons typically yield values around 1 μF cm^{-2} (Gentet et al., 2000). Among the proteins embedded in the lipid bilayer are various types of ion channels whose density and conductance – which can be voltage-dependent, as described in section 3 below – determine the membrane resistance. Typical values of the specific membrane resistance R_m near the resting membrane potential V_{rest} (about –70 mV; measured as intracellular potential minus extracellular potential) depend mostly on the density of voltage-independent (leak) ion channels in the neuron and typically range from 10 to 100 $k\Omega$ cm^2. As long as R_m is not voltage-dependent, we speak of a "passive" cable; if there is a significant voltage dependence of the specific membrane resistance (see section 3 below), the cable is called "active".

Between dendritic branch points, the geometry of the cable can be approximated by a cylinder. This cylinder is sufficiently long and thin, and the membrane resistance is large compared to the intracellular resistivity, so most of the current inside the dendrite flows parallel to its longitudinal axis. Thus, we do not need to solve for voltage in three dimensions: the problem can be reduced to a description of voltage along a single spatial dimension x. We also assume that the intracellular resistivity is Ohmic since any capacitive effects inside the cytoplasm can be ignored on a millisecond time scale, and inductive effects can be completely neglected (Jack et al., 1983; Koch, 1999). With these assumptions, the membrane potential $V(x,t)$ as a function of spatial position x along the dendrite and time t is governed by the one-dimensional linear cable equation

$$\lambda^2 \frac{\partial^2 V(x,t)}{\partial x^2} = (V(x,t) - V_{rest}) + \tau_m \frac{\partial V(x,t)}{\partial t},$$

where $\lambda = \sqrt{R_m d / (4 R_i)}$ is the steady-state space constant, d is the diameter of the dendrite and $\tau_m = R_m C_m$ is the membrane time constant. Analytical solutions of the linear cable equation have been obtained for a number of situations (Jack et al., 1983; Segev et al., 1995; Koch, 1999; Scott, 2002). Here we give only the solution for an infinite cable in the steady-state, voltage-clamped at V_0 above the resting potential by an electrode located at $x = 0$,

$$V(x) = V_{rest} + V_0 e^{-|x|/\lambda}.$$

The membrane potential drops off exponentially, with space constant λ, on both sides of the location where current enters the dendrite via an electrode or a synapse (see Fig. 1A, solid line). For transient voltage waveforms such as EPSPs and action potentials, the attenuation of their peak amplitude as a function of distance from the origin of the signal is steeper (Fig. 1A, dashed line) since voltage transients need to charge and discharge the membrane capacitance. Thus, synaptic potentials due to transient synaptic conductance are attenuated (and delayed) as they travel away from their origin at the location of the synapse.

In an infinite cable, both steady-state and transient voltage attenuation are symmetric on both sides of the voltage source. In real dendritic trees however, the voltage attenuation is often asymmetric, depending on the dendritic branching pattern (Jack et al., 1983; Segev et al., 1995; Koch, 1999; Scott, 2002). Voltage transients originating in a small distal dendritic branch are more severely attenuated as they travel towards the soma than similar voltage transients travelling from the soma towards the dendritic tips. Thus, voltage transients can be kept isolated more easily in thin terminal dendritic branches than in thick primary dendrites close to the soma. In Purkinje cells, for example, the soma can exert global steady-state voltage control over the rest of the dendritic tree, while voltage control from within a spiny branchlet is much less effective (Segev and London, 2000; Roth and Häusser, 2001).

For passive membrane properties and subthreshold synaptic integration to be directly relevant for dendritic release of neurotransmitter, a mechanism is required to transform subthreshold electrical signals into chemical signals that can trigger dendritic release. One candidate mechanism is activation of low-threshold voltage-gated Ca^{2+} channels in dendrites by synaptic input. Subthreshold synaptic potentials evoke dendritic Ca^{2+} transients in neocortical pyramidal neurons (Markram and Sakmann, 1994; Fig. 2) via low-threshold Ca^{2+} channels. In this scenario, the passive properties of the dendrites regulate the spread of the EPSP and thus the resulting Ca^{2+} transients. Since such signals result from activation of voltage-gated Ca^{2+} channels, they are not strictly passive. However, as long as the activation of these channels stays below the threshold for initiation of regenerative dendritic spikes, their electrical consequences are minor compared to their more important role in activating downstream chemical signalling pathways.

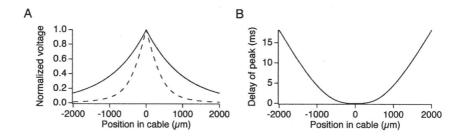

Figure 1. Spread of subthreshold electrical signals in dendrites. A, voltage attenuation in an unbranched passive cable (λ = 1000 µm, τ_m = 40 ms) of infinite length. Solid line, spatial profile of steady-state voltage in response to a voltage clamp at x = 0. Dashed line, spatial profile of the peak amplitude of the transient voltage response to a square pulse voltage clamp command (duration, τ_m/10) at the same location. B, temporal delay of the peak of the transient voltage response in A, plotted as a function of position in the cable.

While it is usually difficult to positively ascribe direct computational functions to the transformations of subthreshold electrical signals performed by neuronal dendrites, there are a few examples where this has been achieved, and some of these examples involve dendritic release of neurotransmitter. The subthreshold electrical signalling in dendrites as governed by the cable equation predestines dendrites to be lowpass filters, both temporally and spatially. A beautiful example of dendrites performing a convolution of retinotopic input with a spatial lowpass filter kernel was recently described in the fly visual system (Cuntz et al., 2003). After spatial lowpass-filtering, the retinotopic input is then passed on via dendrodendritic inhibitory synapses to a downstream neuron which also receives direct excitatory retinotopic input, resulting in a sharpening of the input to the downstream neuron, which has been shown to be involved in figure detection (Egelhaaf, 1985).

Dendrites are also good at picking up particular spatiotemporal sequences of synaptic inputs. The attenuation and delay of synaptic potentials as they travel along a dendrite can be used as a dendritic mechanism of direction selectivity (Rall, 1964). If synaptic inputs along a dendrite are activated sequentially, with the most distal synapses first and the most proximal synapses last, and if the delay in synaptic activation approximately matches the delay of the synaptic potentials as they spread from the site of the synapse to the soma, then the synaptic potentials sum most effectively at the soma, creating a larger compound EPSP amplitude than any other sequence of activation of these synapses. A similar mechanism could underlie direction-selective responses in starburst amacrine cells in the retina (Euler et al., 2002), but with a twist: in these cells, direction-selective Ca^{2+} transients that could underlie dendritic release are observed in the

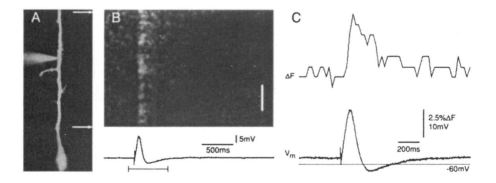

Figure 2. Coupling between subthreshold electrical and chemical (Ca^{2+}) signalling in dendrites. A, confocal image of the soma and proximal apical dendrite of a layer 5 pyramidal neuron loaded with the fluorescent dye Ca^{2+} Green-1 via the dendritic patch pipette. B, dendritic subthreshold EPSPs evoked by synaptic stimulation are associated with dendritic $[Ca^{2+}]_i$ transients observed using Ca^{2+} imaging. Top, spatiotemporal plot of fluorescence change measured with line scans along the dendrite shown in A (length of line scans indicated by arrows in A). Time is represented by the horizontal axis, space by the vertical axis (scale bar, 20 μm). Bottom, dendritic membrane potential measured at the location of the dendritic pipette. C, membrane potential (bottom) and spatially averaged fluorescence change (top) shown on an expanded time scale (time interval indicated by horizontal line in B). Adapted from Markram and Sakmann, 1994.

dendritic tips when sequences of synaptic activation are applied in which synapses near the soma are activated first, and synapses near the dendritic tips last.

3. ACTIVE ELECTRICAL PROPERTIES OF DENDRITES

The passive electrotonic "skeleton" also forms the basis for voltage-dependent, or active electrical signalling in dendrites. This is because any electrical events leading up to a suprathreshold event (involving regenerative activation of voltage-gated conductances) are necessarily subthreshold. Thus, a major determinant of the thresholds for initiation and the conditions for propagation of suprathreshold events in dendrites are the dendritic geometry and passive cable parameters as described in section 2 (Jack, 1983; Segev et al., 1995; Segev and Rall, 1998; Koch, 1999). Of course, whether a dendritic spike is initiated or continues to propagate also depends on the densities and kinetics of voltage-gated channels in the dendrites. The dendrites of different types of neurons contain different sets of channels, and some channel types exhibit distance-dependent gradients in their conductance density (Llinás, 1988; Johnston et al., 1996; Migliore and Shepherd, 2002).

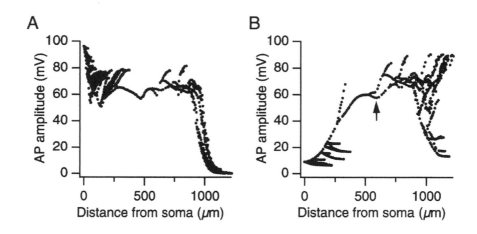

Figure 3. Propagation of suprathreshold electrical signals in dendrites. Simulation of a backpropagating action potential (A) and a dendritic spike (B) in a compartmental model based on the reconstructed morphology of a layer 5 pyramidal neuron. The scatter plots indicate peak voltage reached in each compartment, plotted as a function of distance from the soma. Unlike passive electrical events (Fig. 1), dendritic action potentials can increase in amplitude while propagating away from their site of origin (at the soma in the case of backpropagation, and at the dendritic location indicated by the arrow in B in the case of the dendritic spike). However, when the amplitude falls below a certain threshold determined by the degree of local dendritic excitability, the mode of propagation changes from active to passive and the action potential amplitude decays rapidly. Adapted from Vetter et al., 2001.

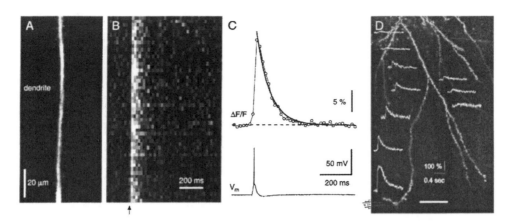

Figure 4. Coupling between suprathreshold electrical events and Ca^{2+} signalling in dendrites. A, confocal reconstruction of a portion of the proximal apical dendrite of a layer 5 pyramidal neuron loaded with 50 μM Ca^{2+} Green-1 via the somatic whole-cell patch pipette. B, line scans along the dendrite shown in A, resulting in a spatiotemporal density plot of relative fluorescence change evoked by Ca^{2+} influx due to a single backpropagating action potential at the time indicated by the arrow. C, spatially averaged relative fluorescence change (top) rises to a peak within a few milliseconds after the somatic action potential (bottom) due to its fast propagation and the fast kinetics of dendritic Ca^{2+} channels. The fluorescence then decays back to baseline (monoexponential fit (thick line), τ = 86 ms) as dendritic Ca^{2+} is removed from the cytoplasm by Ca^{2+} pumps. D, Ca^{2+} transients reflecting a local dendritic spike evoked by synaptic stimulation (at the location of the stimulation pipette indicated in the lower left corner; scale bar, 25 μm). The background image shows basal dendritic branches of a hippocampal CA1 pyramidal neuron filled with Oregon Green 488 Bapta-1 (200 μM) via the somatic patch pipette. Large Ca^{2+} transients are observed only in the stimulated dendritic branch; adjacent branches show much smaller Ca^{2+} transients, suggesting that the electrical spike was localized in a similar manner. Adapted from Markram et al., 1998 (A-C) and Ariav et al., 2003 (D).

To understand how a given set of channels determines the degree of excitability of a dendritic membrane, it is necessary to characterize the biophysical properties of the different channel types, and to measure the electrical signals generated by these channels. Both of these are now possible due to the advent of patch-clamp recordings directly from dendrites (Stuart et al., 1993). Outside-out and cell-attached patch-clamp recordings have provided information on the density, distance-dependent density gradients, and kinetics of Na^+, Ca^{2+}, K^+ and hyperpolarization-activated channels in the dendrites of several types of neurons (Johnston et al., 1996; Migliore and Shepherd, 2002). Recently, methods have been described that allow measurement of dendritic channel densities and kinetics from somatic and dendritic whole-cell recordings (Schaefer et al., 2003a). These methods use simulations of the recorded neuron to correct for distortions of the measured currents due to attenuation of voltage along the dendritic cable. Dendritic whole-cell recordings have also allowed experimentalists to measure directly how voltage-dependent conductances in the dendrites interact to generate regenerative events. Two major types of regenerative events can be distinguished: Na^+ channel driven somatic action potentials that backpropagate into the dendrites (Stuart et al., 1997b; Fig. 3A); and Ca^{2+} and Na^+ channel driven spikes that are initiated in the dendrites of several types of neurons under conditions of intense synaptic stimulation (Llinás and Sugimori, 1980; Chen et al., 1997;

Stuart et al. 1997a; Golding and Spruston, 1998; Kamondi et al., 1998; Schwindt and Crill, 1998; Martina et al., 2000; Golding et al., 2002; Fig. 3B). Backpropagating action potentials represent the propagation of the axonally-generated Na^+ action potential into the dendritic tree. Propagation is associated with a slight broadening and a decrement in amplitude of the action potential in most cell types (Stuart et al., 1997b; Vetter et al., 2001), with the notable exception of substantia nigra dopamine neurons (Häusser et al., 1995) and the primary dendrites of mitral cells in the olfactory bulb (Bischofberger and Jonas, 1997; Chen et al., 1997), both of which support dendritic release of neurotransmitter and exhibit essentially non-decremental backpropagation. Spikes initiated in the dendrites are less stereotyped events whose waveform and spatial spread depends on their exact site of initiation in the dendrites and the relative involvement of Na^+, Ca^{2+} and K^+ channels during their rising, plateau and decay phase. Spikes mediated predominantly by Na^+ channel activation tend to be briefer than Ca^{2+} spikes (Stuart et al., 1997a), but the trial-to-trial and cell-to-cell variability of dendritic spike waveforms is large, even within a given cell type (Schiller et al., 1997; Stuart et al., 1997a; Golding and Spruston, 1998; Larkum et al., 1999; Larkum et al., 2001; Schaefer et al., 2003b).

During network activity *in vivo*, synaptic potentials, backpropagating action potentials and dendritic spikes can be evoked concurrently and interact in different ways (Kamondi et al., 1998; Helmchen et al., 1999; Larkum and Zhu, 2002; Rudolph and Destexhe, 2003; Waters et al., 2003). Two major kinds of interaction occur between backpropagating action potentials and dendritic synaptic potentials, and between dendritic spikes and backpropagating action potentials. Since backpropagation is decremental in most cell types, there usually exists a distance from the soma where the dendritic action potential amplitude falls below the threshold at which sufficient Na^+ channels can be recruited to ensure further active propagation of the action potential (Bernard and Johnston, 2003). Thus, propagation would fail at this point unless an EPSP evoked by a synapse close to this location comes to its rescue (Larkum et al., 2001; Stuart and Häusser, 2001). By a similar mechanism, dendritic IPSPs can push the dendritic action potential amplitude below this threshold and thus prohibit its further propagation.

In layer 5 pyramidal neurons, backpropagating action potentials lower the threshold for evoking a dendritic Ca^{2+} spike, which in turn can depolarize the soma sufficiently to evoke a burst of somatic Na^+ action potentials (Larkum et al., 1999; Schaefer et al., 2003b). This mechanism enables pyramidal neurons to associate deep-layer inputs representing the local receptive field and layer 1 input from different cortical areas. It could also serve as a mechanism underlying particular forms of associative synaptic plasticity (Körding and König, 2000; see below).

Suprathreshold electrical signals in dendrites are coupled to chemical (Ca^{2+}) signalling via dendritic Ca^{2+} channels. While backpropagating action potentials are electrically carried predominantly by activation of dendritic Na^+ channels, they also effectively open dendritic Ca^{2+} channels, resulting in a dendritic Ca^{2+} signal of relatively large amplitude (Fig. 4A-C). Due to the large depolarization they provide, action potentials are able to recruit both low-voltage-activated as well as high-voltage-activated Ca^{2+} channels. Dendritic spikes are associated with Ca^{2+} influx in a similar way. For both types of spikes, the spatial extent of significant Ca^{2+} influx mirrors the spatial extent of the electrical event, which can be large for backpropagating action potentials and fairly small for dendritic spikes restricted to a single dendritic branch (Fig. 4D). Thus, the electrical compartmentalization of dendrites determines the rules for downstream Ca^{2+}-dependent signalling.

4. SUMMARY AND PERSPECTIVES

The dendritic geometry, the passive membrane properties, and the densities, distribution and kinetics of voltage-gated channels together determine the compartmentalization of electrical signals in dendrites, and how these electrical compartments interact. The degree of spatial restriction of electrical signals in dendrites has direct consequences for the input-output relation of dendritic trees. This is true even if neuronal output is regulated only in the classical manner by the action potential propagating down the axon (Mel, 1993; Koch, 1999; Häusser et al., 2000; Segev and London, 2000; Poirazi et al., 2003; Häusser and Mel, 2003; Polsky et al., 2004; Spruston and Kath, 2004), but the role of electrical compartments in determining the neuronal input-output relation is particularly important when neuronal output is mediated by dendritic release of neurotransmitter (Fig. 5). With transmitter release from its dendrites a neuron acquires the potential for multiple outputs, and a single neuron can thus accommodate the function of an entire network of simple units, transforming multiple inputs into multiple outputs (e.g. Euler et al., 2002; Cuntz et al., 2003). This is in contrast

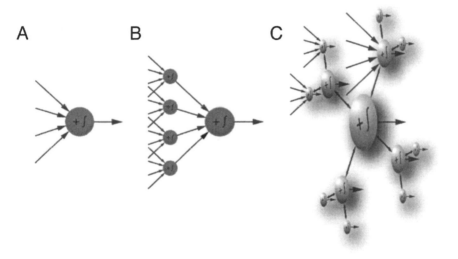

Figure 5. Schematic representations of the input-output relation of neurons. A, "point" neuron. Inputs (left arrows) are summed at the soma, and if a threshold is reached, an output (horizontal arrow, right) is sent down the axon. B, dendritic neuron represented as a two-layer network (Poirazi et al., 2003). Synaptic inputs are grouped onto dendritic subunits (small circles), each of which computes the sum of its local inputs and applies a local thresholding nonlinearity, which could be implemented by a dendritic spike. The results of the local thresholding operations are then summed at the soma (large circle), where the final decision is taken whether to send an output down the axon (horizontal arrow, right). C, neuron with multiple outputs mediated by dendritic release of transmitter. As in B, the neuron is represented by several dendritic subunits (shaded ellipses), each with their own group of inputs (left arrows, only three representative groups are shown for clarity). However, each subunit also has its own output (horizontal arrows) mediated by dendritic transmitter release. As in B, each subunit performs a local thresholding operation, but local thresholds are modifiable via bidirectional electrical interactions between subunits. Subunits can be arranged in a hierarchical manner determined by the morphology as well as the passive and active electrical parameters of the neuron.

to the classical picture with a single output via the axon (unless there are local failures of action potential propagation in the axon: Debanne, 2004, or the action potential waveform in axonal boutons can be modulated: Geiger and Jonas, 2000). The possibility that the backpropagating action potential can trigger and/or enhance dendritic release of transmitter further emphasizes the importance of the retrograde (soma → dendrite) flow of information, which challenges the classical view originally proposed by Cajal.

Beyond the ongoing processing of synaptic inputs, electrical compartmentalization of dendrites also has consequences for long-term synaptic plasticity (Häusser and Mel, 2003; Mehta, 2004). Learning rules based on dendritic spikes (Poirazi and Mel, 2001; Goldberg et al., 2002; Golding et al., 2002; Mehta, 2004) may involve retrograde signalling via dendritic release of transmitters. This has been shown to be the case for the LTD component of spike timing-dependent synaptic plasticity between neocortical layer 5 pyramidal neurons (Sjöström et al., 2003), which is dependent on endocannabinoid release from the postsynaptic dendrites, triggered by the backpropagating action potential.

Hebbian synaptic plasticity is inherently unstable, as it represents a form of positive feedback, this problem is particularly acute when synapses on dendrites are involved (Goldberg et al., 2002; Roth and London, 2004; Rumsey and Abbott, 2004). Various homeostatic plasticity mechanisms (Turrigiano and Nelson, 2004) at different spatial and time scales are probably needed to balance and control the instability due to Hebbian plasticity mechanisms. Homeostatic mechanisms could act via intracellular signalling, particularly via $[Ca^{2+}]_i$ as an indicator of neuronal activity (Liu et al., 1998) or via dendritic release and extracellular spread of a transmitter whose concentration is signalled back to the neuron by dendritic autoreceptors. An intriguing possibility is that dendritic release of transmitter could in turn modify the electrical properties of the dendrites and thus alter subsequent release of transmitter. This would represent a form of intrinsic metaplasticity in which dendritic transmitter release could regulate not only itself but also the potential for future synaptic plasticity.

A prerequisite for a deeper understanding of the contribution of dendritic transmitter release to information processing and learning in neuronal networks is knowledge of the underlying mechanisms. Which experiments are now required to improve our understanding of the link between dendritic excitability and dendritic release? First, we need a more complete characterization of electrical properties of dendrites in neurons exhibiting dendritic release, including the properties and distribution of dendritic voltage-gated ion channels. In only a few model systems, such as mitral cells of the olfactory bulb (chapter 7, this volume), have the electrical properties of dendrites been investigated in detail. Second, a more quantitative understanding of how electrical signals are translated into dendritic Ca^{2+} signals is required, including a complete description of dendritic Ca^{2+} dynamics (Markram et al., 1998). Finally, our understanding of the voltage- and/or Ca^{2+} dependence of dendritic transmitter release and the underlying biophysical mechanisms still lags far behind our understanding of transmitter release at axonal boutons. Together, this information will allow a deeper understanding of how electrical signals in dendrites are translated into dendritic release of transmitters, and provide us with a set of molecular targets which will enable us to better define the role of dendritic transmitter release in relation to behaviour.

5. REFERENCES

Ariav, G., Polsky, A., and Schiller, J., 2003, Submillisecond precision of the input-output transformation function mediated by fast sodium dendritic spikes in basal dendrites of CA1 pyramidal neurons, *J. Neurosci.* **23**: 7750.

Attwell, D., Barbour, B., and Szatkowski, M., 1993, Nonvesicular release of neurotransmitter, *Neuron* **11**: 401.

Bernard, C., and Johnston, D., 2003, Distance-dependent modifiable threshold for action potential back-propagation in hippocampal dendrites, *J. Neurophysiol.* **90**: 1807.

Bischofberger, J., and Jonas, P., 1997, Action potential propagation into the presynaptic dendrites of rat mitral cells, *J. Physiol.* **504**: 359.

Chen, W. R., Midtgaard, J., and Shepherd, G. M., 1997, Forward and backward propagation of dendritic impulses and their synaptic control in mitral cells, *Science* **278**: 463.

Chen, W. R., Xiong, W., and Shepherd, G. M., 2000, Analysis of relations between NMDA receptors and GABA release at olfactory bulb reciprocal synapses, *Neuron* **25**: 625.

Cuntz, H., Haag, J., and Borst, A., 2003, Neural image processing by dendritic networks, *PNAS USA* **100**: 11082.

Debanne, D., 2004, Information processing in the axon, *Nat. Rev. Neurosci.* **5**: 304.

Egelhaaf, M., 1985, On the neuronal basis of figure-ground discrimination by relative motion in the visual system of the fly. II. Figure-detection cells, a new class of visual interneurones, *Biol. Cybern.* **52**: 195.

Euler, T., Detwiler, P. B., and Denk, W., 2002, Directionally selective calcium signals in dendrites of starburst amacrine cells, *Nature* **418**: 845.

Falkenburger, B. H., Barstow, K. L., and Mintz, I. M., 2001, Dendrodendritic inhibition through reversal of dopamine transport, *Science* **293**: 2465.

Freund, T. F., Katona, I., and Piomelli, D., 2003, Role of endogenous cannabinoids in synaptic signalling, *Physiol. Rev.* **83**: 1017.

Geiger, J. R. P., and Jonas, P., 2000, Dynamic control of presynaptic Ca^{2+} inflow by fast-inactivating K^+ channels in hippocampal mossy fiber boutons, *Neuron* **28**: 927.

Gentet, L. J., Stuart, G. J., and Clements, J. D., 2000, Direct measurement of specific membrane capacitance in neurons, *Biophys. J.* **79**: 314.

Goldberg, J., Holthoff, K., and Yuste, R., 2002, A problem with Hebb and local spikes, *Trends Neurosci.* **25**: 433.

Golding, N. L., and Spruston, N., 1998, Dendritic sodium spikes are variable triggers of axonal action potentials in hippocampal CA1 pyramidal neurons, *Neuron* **21**: 1189.

Golding, N. L., Staff, N. P., and Spruston, N., 2002, Dendritic spikes as a mechanism for cooperative long-term potentiation, *Nature* **418**: 326.

Halabisky, B., Friedman, D., Radojicic, M., and Strowbridge, B. W., 2000, Calcium influx through NMDA receptors directly evokes GABA release in olfactory bulb granule cells, *J. Neurosci.* **20**: 5124.

Häusser, M., and Mel, B. W., 2003, Dendrites: bug or feature? *Curr. Opin. Neurobiol.* **13**: 372.

Häusser, M., Spruston, N., and Stuart, G. J., 2000, Diversity and dynamics of dendritic signalling, *Science* **290**: 739.

Häusser, M., Stuart, G., Racca, C., and Sakmann, B., 1995, Axonal initiation and active dendritic propagation of action potentials in substantia nigra neurons, *Neuron* **15**: 637.

Helmchen, F., Svoboda, K., Denk, W., and Tank, D. W., 1999, *In vivo* dendritic calcium dynamics in deep-layer cortical pyramidal neurons, *Nat. Neurosci.* **2**: 989.

Isaacson, J. S., and Strowbridge, B. W., 1998, Olfactory reciprocal synapses: dendritic signalling in the CNS, *Neuron* **20**: 749.

Jack, J. J. B., Noble, D., and Tsien, R. W., 1983, *Electric Current Flow in Excitable Cells,* Oxford Univ. Press, Oxford, UK.

Johnston, D., Magee, J. C., Colbert, C. M., and Cristie, B. R., 1996, Active properties of neuronal dendrites, *Annu. Rev. Neurosci.* **19**: 165.

Kamondi, A., Acsády, L., and Buzsáki, G., 1998, Dendritic spikes are enhanced by cooperative network activity in the intact hippocampus, *J. Neurosci.* **18**: 3919.

Koch, C., 1999, *Biophysics of Computation,* Oxford Univ. Press, Oxford, UK.

Körding, K. P., and König, P., 2000, Learning with two sites of synaptic integration, *Network: Comput. Neural Syst.* **11**: 25.

Larkum, M. E., Watanabe, S., Nakamura, T., Lasser-Ross, N., and Ross, W. N., 2003, Synaptically activated Ca^{2+} waves in layer 2/3 and layer 5 neocortical pyramidal neurons, *J. Physiol.* **549**: 471.

Larkum, M. E., and Zhu, J. J., 2002, Signaling of layer 1 and whisker-evoked Ca^{2+} and Na^+ action potentials in distal and terminal dendrites of rat neocortical pyramidal neurons *in vitro* and *in vivo*, *J. Neurosci.* **22**: 6991.

Larkum, M. E., Zhu, J. J., and Sakmann, B., 1999, A new cellular mechanism for coupling inputs arriving at different cortical layers, *Nature* **398**: 338.

Larkum, M. E., Zhu, J. J., and Sakmann, B., 2001, Dendritic mechanisms underlying the coupling of the dendritic with the axonal action potential initiation zone of adult rat layer 5 pyramidal neurons, *J. Physiol.* **533**: 447.

Liu, Z., Golowasch, J., Marder, E., and Abbott, L. F., 1998, A model neuron with activity-dependent conductances regulated by multiple calcium sensors, *J. Neurosci.* **18**: 2309.

Llinás, R. R., 1988, The intrinsic electrophysiological properties of mammalian neurons: insights into central nervous function, *Science* **242**: 1654.

Llinás, R., and Sugimori, M., 1980, Electrophysiological properties of *in vitro* Purkinje cell dendrites in mammalian cerebellar slices, *J. Physiol.* **305**: 197.

Magee, J. C., and Johnston, D., 1997, A synaptically controlled, associative signal for Hebbian plasticity in hippocampal neurons, *Science* **275**: 209.

Major, G., Larkman, A. U., Jonas, P., Sakmann, B., and Jack, J. J. B., 1994, Detailed passive cable models of whole-cell recorded CA3 pyramidal neurons in rat hippocampal slices, *J. Neurosci.* **14**: 4613.

Maletic-Savatic, M., and Malinow, R., 1998, Calcium-evoked dendritic exocytosis in cultured hippocampal neurons. Part I: trans-Golgi network-derived organelles undergo regulated exocytosis, *J. Neurosci.* **18**: 6803.

Markram, H., Lübke, J., Frotscher, M., and Sakmann, B., 1997, Regulation of synaptic efficacy by coincidence of postsynaptic APs and EPSPs, *Science* **275**: 213.

Markram, H., Roth, A., and Helmchen, F., 1998, Competitive calcium binding: implications for dendritic calcium signalling, *J. Comput. Neurosci.* **5**: 331.

Markram, H., and Sakmann, B., 1994, Calcium transients in dendrites of neocortical neurons evoked by single subthreshold excitatory postsynaptic potentials via low-voltage-activated calcium channels, *PNAS USA* **91**: 5207.

Martina, M., Vida, I., and Jonas, P., 2000, Distal initiation and active propagation of action potentials in interneuron dendrites, *Science* **287**: 295.

Mehta, M. R., 2004, Cooperative LTP can map memory sequences on dendritic branches, *Trends Neurosci.* **27**: 69.

Mel, B. W., 1993, Synaptic integration in an excitable dendritic tree, *J. Neurophysiol.* **70**: 1086.

Migliore, M., and Shepherd, G. M., 2002, Emerging rules for the distributions of active dendritic conductances, *Nat. Rev. Neurosci.* **3**: 362.

Nakamura, T., Barbara, J.-G., Nakamura, K., and Ross, W. N., 1999, Synergistic release of Ca^{2+} from IP_3-sensitive stores evoked by synaptic activation of mGluRs paired with backpropagating action potentials, *Neuron* **24**: 727.

Poirazi, P., Brannon, T., and Mel, B. W., 2003, Pyramidal neuron as a two-layer network, *Neuron* **37**: 989.

Poirazi, P., and Mel, B. W., 2001, Impact of active dendrites and structural plasticity on the memory capacity of neural tissue, *Neuron* **29**: 779.

Polsky, A., Mel, B. W., and Schiller, J., 2004, Computational subunits in thin dendrites of pyramidal cells, *Nat. Neurosci.* **7**: 621.

Rall, W., 1964, Theoretical significance of dendritic trees for neuronal input-output relations, in: *Neural Theory and Modeling*, R. F. Reiss, ed., Stanford Univ. Press, Palo Alto, CA, p. 73.

Ronesi, J., Gerdeman, G. L., and Lovinger, D. M., 2004, Disruption of endocannabinoid release and striatal long-term depression by postsynaptic blockade of endocannabinoid membrane transport, *J. Neurosci.* **24**: 1673.

Rose, C. R., and Konnerth, A., 2001, NMDA receptor-mediated Na^+ signals in spines and dendrites, *J. Neurosci.* **21**: 4207.

Roth, A., and Häusser, M., 2001, Compartmental models of rat cerebellar Purkinje cells based on simultaneous somatic and dendritic patch-clamp recordings, *J. Physiol.* **535**: 445.

Roth, A., and London, M., 2004, Rebuilding dendritic democracy. Focus on "Equalization of synaptic efficacy by activity- and timing-dependent synaptic plasticity", *J. Neurophysiol.* **91**: 1941.

Rudolph, M., and Destexhe, A., 2003, A fast-conducting, stochastic integrative mode for neocortical neurons *in vivo*, *J. Neurosci.* **23**: 2466.

Rumsey, C. C., and Abbott, L. F., 2004, Equalization of synaptic efficacy by activity- and timing-dependent synaptic plasticity, *J. Neurophysiol.* **91**: 2273.

Schaefer, A. T., Helmstaedter, M., Sakmann, B., and Korngreen, A., 2003a, Correction of conductance measurements in non-space-clamped structures: 1. Voltage-gated K^+ channels, *Biophys. J.* **84**: 3508.

Schaefer, A. T., Larkum, M. E., Sakmann, B., and Roth, A., 2003b, Coincidence detection in pyramidal neurons is tuned by their dendritic branching pattern, *J. Neurophysiol.* **89**: 3143.

Schiller, J., Major, G., Koester, H. J., and Schiller, Y., 2000, NMDA spikes in basal dendrites of cortical pyramidal neurons, *Nature* **404**: 285.

Schiller, J., Schiller, Y., Stuart, G., and Sakmann, B., 1997, Calcium action potentials restricted to distal apical dendrites of rat neocortical pyramidal neurons, *J. Physiol.* **505**: 605.

Schoppa, N. E., Kinzie, J. M., Sahara, Y., Segerson, T. P., and Westbrook, G. L., 1998, Dendrodendritic inhibition in the olfactory bulb is driven by NMDA receptors, *J. Neurosci.* **18**: 6790.

Schwindt, P. C., and Crill, W. E., 1998, Synaptically evoked dendritic action potentials in rat neocortical pyramidal neurons, *J. Neurophysiol.* **79**: 2432.

Scott, A., 2002, *Neuroscience: A Mathematical Primer,* Springer-Verlag, New York.

Segev, I., and London, M., 2000, Untangling dendrites with quantitative models, *Science* **290**: 744.

Segev, I., and Rall, W., 1998, Excitable dendrites and spines: earlier theoretical insights elucidate recent direct observations, *Trends Neurosci.* **21**: 453.

Segev, I., Rinzel, J., and Shepherd, G. M., 1995, *The Theoretical Foundation of Dendritic Function,* MIT Press, Cambridge, MA.

Sjöström, P. J., Turrigiano, G. G., and Nelson, S. B., 2003, Neocortical LTD via coincident activation of presynaptic NMDA and cannabinoid receptors, *Neuron* **39**: 641.

Spruston, N., and Kath, W. J., 2004, Dendritic arithmetic, *Nat. Neurosci.* **7**: 567.

Steuber, V., De Schutter, E., and Jaeger, D., 2004, Passive models of neurons in the deep cerebellar nuclei: the effect of reconstruction errors, *Neurocomputing,* in press.

Stuart, G. J., Dodt, H.-U., and Sakmann, B., 1993, Patch-clamp recordings from the soma and dendrites of neurons in brain slices using infrared video microscopy, *Pflügers Arch.* **423**: 511.

Stuart, G. J., and Häusser, M., 2001, Dendritic coincidence detection of EPSPs and action potentials, *Nat. Neurosci.* **4**: 63.

Stuart, G., and Sakmann, B., 1995, Amplification of EPSPs by axosomatic sodium channels in neocortical pyramidal neurons, *Neuron* **15**: 1065.

Stuart, G., Schiller, J., and Sakmann, B., 1997a, Action potential initiation and propagation in rat neocortical pyramidal neurons, *J. Physiol.* **505**: 617.

Stuart, G., and Spruston, N., 1998, Determinants of voltage attenuation in neocortical pyramidal neuron dendrites, *J. Neurosci.* **18**: 3501.

Stuart, G., Spruston, N., and Häusser, M., 1999, *Dendrites,* Oxford Univ. Press, Oxford, UK.

Stuart, G., Spruston, N., Sakmann, B., and Häusser, M., 1997b, Action potential initiation and backpropagation in neurons of the mammalian CNS, *Trends Neurosci.* **20**: 125.

Thurbon, D., Lüscher, H.-R., Hofstetter, T., and Redman, S. J., 1998, Passive electrical properties of ventral horn neurons in rat spinal cord slices, *J. Neurophysiol.* **80**: 2485.

Turrigiano, G. G., and Nelson, S. B., 2004, Homeostatic plasticity in the developing nervous system, *Nat. Rev. Neurosci.* **5**: 97.

Vetter, P., Roth, A., and Häusser, M., 2001, Propagation of action potentials in dendrites depends on dendritic morphology, *J. Neurophysiol.* **85**: 926.

Waters, J., Larkum, M., Sakmann, B., and Helmchen, F., 2003, Supralinear Ca^{2+} influx into dendritic tufts of layer 2/3 neocortical pyramidal neurons *in vitro* and *in vivo, J. Neurosci.* **23**: 8558.

Williams, S. R., and Stuart, G. J., 2003, Role of dendritic synapse location in the control of action potential output, *Trends Neurosci.* **26**: 147.

Wilson, R. I., and Nicoll, R. A., 2002, Endocannabinoid signalling in the brain, *Science* **296**: 678.

Yuste, R., Gutnick, M. J., Saar, D., Delaney, K. R., and Tank, D. W., 1994, Ca^{2+} accumulations in dendrites of neocortical pyramidal neurons: an apical band and evidence for two functional compartments, *Neuron* **13**: 23.

5

SOMATODENDRITIC DOPAMINE RELEASE IN MIDBRAIN

Stephanie J. Cragg and Margaret E. Rice[*]

1. INTRODUCTION

Midbrain dopamine (DA) neurons of the substantia nigra (SN) and adjacent ventral tegmental area (VTA) fall into two main categories of cells, which were originally classified by their anatomical location in the SN and are consequently referred to as dorsal- and ventral-tier neurons (Fallon et al., 1978). These cells can be distinguished by their morphological characteristics, including dendritic arbor, somatic size, major efferent projections, and biochemistry (Fallon et al., 1978; Gerfen et al., 1987a,b). The primary DA cell type in the SN is the ventral-tier cell, which has a large pyramidal cell body (Fig. 1Λ) and extends dendrites laterally along the band of cell bodies in the SN pars compacta (SNc) and ventrally into the pars reticulata (SNr) (Fig. 1B,C). Smaller DA cells are also found in the dorsal tier of the SNc; these dorsal-tier cells are the predominant cell type in the adjacent VTA. Via the median forebrain bundle, DA neurons of the SNc project primarily to the dorsal striatum, whereas those of the VTA project to the nucleus accumbens (ventral striatum), as well as to prefrontal cortex and other mesolimbic structures (Fallon and Moore, 1978; Fallon et al., 1978). The nigrostriatal DA system is essential for motor facilitation by the basal ganglia, whereas the mesolimbic DA system participates in motivation, including reward.

A characteristic of DA neurons in both SN and VTA is the somatodendritic release of DA (Björkland and Lindvall, 1975; Groves et al., 1975; Geffen et al., 1976; Nieoullon et al., 1977); there is evidence for release from somata (Jaffe et al., 1998) and from dendrites (Geffen et al., 1976; Rice et al., 1994). Importantly, release in the SN is exclusively somatodendritic, but that in the VTA is not: the SN receives no identified synaptic DA input or axon collateralization (Juraska et al., 1977; Wassef et al., 1981), whereas the VTA receives DA input from its own axon collaterals, as well as minor input from DA axons from the SNc (Deutch et al., 1988; Bayer and Pickel, 1990).

---

[*] Stephanie J. Cragg, Department of Pharmacology, University of Oxford, OX1 3QT, UK. Margaret E. Rice, Department of Physiology and Neuroscience, NYU School of Medicine, New York, NY 10016, USA.

Dendritic Neurotransmitter Release, edited by M. Ludwig
Springer Science+Business Media, Inc., 2005

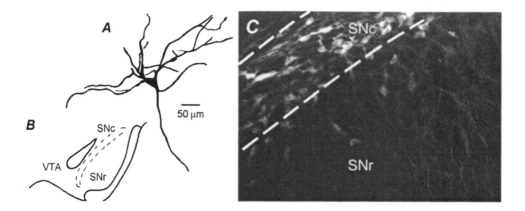

Figure 1. (A) Ventral-tier DA cell of the SNc, with ventral-projecting dendrite extending into the SNr. (B) Coronal view of midbrain with relative locations of SNc, SNr and VTA (midline is to the left of VTA) (from Fallon et al., 1978). (C) Fluorescence micrograph of tyrosine hydroxylase immunoreactivity in SNc and SNr.

This review will focus primarily on studies of somatodendritic DA release in the SNc, with comparison to release in VTA, as well as to release from nigrostriatal axons in dorsal striatum. We review the methods used to study somatodendritic DA release (Section 2), its proposed functions (Section 3), the regulation of extracellular DA concentration by uptake (Section 4), proposed mechanisms of release (Section 5), and receptor regulation of release by synaptic and non-synaptic input (Section 6).

2. EXPERIMENTAL PARADIGMS TO STUDY SOMATODENDRITIC DOPAMINE RELEASE

A wide range of techniques have been used to investigate somatodendritic DA release. Initial studies used detection of ^{3}H-DA to indicate somatodendritic release *in vitro* using midbrain slices (Geffen et al., 1976; Tagerud and Cuello 1979) and *in vivo* using push-pull perfusion (Nieoullon et al., 1977; Cheramy et al., 1981). Development of more sensitive off-line detection methods, especially HPLC with electrochemical detection, permitted monitoring of *endogenous* DA release from midbrain slices *in vitro* (Elverfors et al., 1997). Another advance was seen with the introduction of *in vivo* microdialysis, which permits evaluation of extracellular levels of either exogenous or endogenous DA when coupled with an appropriate off-line analytical method (Elverfors and Nissbrandt, 1991; Robertson et al., 1991; Santiago and Westerink, 1991, 1992; Heeringa and Abercrombie 1995; Bergquist et al., 2002). Microdialysis measurements have been particularly helpful in elucidating factors that influence somatodendritic DA release. Like other *in vivo* methods, microdialysis affords the opportunity to study somatodendritic DA release after systemic drug administration or during behaviour in freely moving animals (*see* Bergquist and Nissbrandt, 2004, this volume). Another advantage of microdialysis is that dialysate analysis is off-line, usually after an HPLC separation step, which permits selective detection of DA as well as the possibility for concurrent monitoring of DA metabolites or other transmitters. These strengths come with caveats, however. Firstly, the spatio-temporal resolution of microdialysis is limited. Secondly, interpretation of *in vivo* studies to address the mechanism of somatodendritic DA release or factors regulating local release are complicated by the unavoidable influence of the overall circuitry governing DA cell activity in the SNc or VTA.

A decade ago, we introduced the use of voltammetric recording using carbon-fibre microelectrodes with fast-scan cyclic voltammetry (FCV) for the study of somatodendritic DA release (Rice et al., 1994). FCV is a high-speed, high spatial resolution detection method that is ideal for monitoring release of DA from discrete brain nuclei, including the SNc and VTA. Indeed, many insights into somatodendritic DA release in midbrain over the last decade have been obtained using FCV or other voltammetric methods. A major advantage of FCV is that it permits real-time monitoring of DA release with sub-second and sub-regional resolution. As with microdialysis, there are also caveats to voltammetric recording. In particular, voltammetric studies of somatodendritic DA release in the SN of some species, including rats (Stamford et al., 1993; Iravani and Kruk, 1997; Bunin and Wightman, 1998) and mice (John et al., 2003), have been hindered by the concomitant or predominant detection of 5-HT. The SN receives one of the highest 5-HT innervation densities in the brain; projections from the raphe nuclei provide direct, asymmetric synaptic 5-HT input to both dopaminergic and non-dopaminergic dendrites of SNc and SNr in primates and rodents (SNr>SNc>VTA) (Steinbusch, 1981; Nedergaard et al., 1988; Lavoie and Parent, 1990; Corvaja et al., 1993; Moukhles et al., 1997). Fortunately, the guinea pig SN receives a less dense 5-HT innervation than that found in rat SN, so that pure somatodendritic DA release can be monitored in guinea-pig SNc *in vitro*, although not in rat SNc (Cragg et al., 1997a). Thus, the guinea pig is the rodent species of choice for the characterisation of somatodendritic DA release in SN and VTA using voltammetry (Cragg et al., 1997a). This is not a concern for microdialysis, because of the separate off-line separation step usually used for DA detection. Interestingly, there are species differences in 5-HT receptor binding profiles, as well, e.g. 5HT$_4$ receptors (Waeber et al., 1994), with the pattern in guinea pig better resembling that in human SN.

3. WHAT IS THE ROLE OF SOMATODENDRITIC DOPAMINE RELEASE?

3.1. Somatodendritic Dopamine is Required for Basal Ganglia-Mediated Movement

The critical role of the nigrostriatal pathway in movement has been convincingly demonstrated by the motor deficits of Parkinson's disease that accompany loss of nigrostriatal DA and which can be ameliorated by the DA precursor, L-DOPA (Wichman and DeLong 1996; Carlsson, 2002). Both somatodendritic and axon-terminal release are required for basal ganglia-mediated movement (Robertson and Robertson, 1989; Timmerman and Abercrombie, 1996; Crocker, 1997; Bergquist et al., 2003). Evidence for this is reviewed in detail in Bergquist and Nissbrandt, 2004. The cellular and receptor targets of somatodendritic dopamine in SN and VTA that underlie these behavioural effects are discussed in the following section.

3.2. Somatodendritic Dopamine Signalling via Volume Transmission

Where does somatodendritically released DA act? Unlike classical synaptic transmission, e.g. subsynaptic receptor activation by glutamate, DA transmission is modulatory and is mediated primarily by extrasynaptic receptors (e.g. Rice, 2000; Pickel, 2000; Cragg and Rice, 2004). Thus, DA in both midbrain and striatum must act via volume transmission (Fuxe and Agnati, 1991; Rice, 2000). As a consequence, understanding DA transmission requires understanding of local diffusion characteristics. Strikingly, the extracellular volume fraction (α) available for DA diffusion in the SN and VTA is 0.30 (Cragg et al., 2001), compared to values of ~0.20 that are typical for forebrain structures, including striatum (Rice and Nicholson, 1991). This means that the extracellular concentration of DA ($[DA]_o$) after release of a given number of molecules will be >30% lower in the SN/VTA than in striatum, in the absence of other regulatory mechanisms. This has obvious implications for concentration-dependent receptor activation in SN/VTA versus striatum, as well as for experimental observations of $[DA]_o$ in these regions. The tortuosity factor, λ, which governs the apparent diffusion coefficient of a diffusing substance, is similar between midbrain and striatum (Cragg et al., 2001).

Somatodendritically released DA acts on D_2 autoreceptors to regulate DA cell activity (Lacey et al., 1988; Chiodo, 1992; Yung et al., 1995; Falkenburger et al., 2001) and subsequent somatodendritic DA release in the SNc (discussed in section 6.1), as well as axonal release in striatum (Santiago and Westerink, 1991; Kalivas and Duffy, 1991). Moreover, somatodendritically released DA in both SN and VTA can act at non-synaptic DA receptors on presynaptic GABAergic and glutamatergic terminals to modulate release of those transmitters (Miyazaki and Lacey, 1998; Radnikow and Misgeld, 1998; Koga and Momiyama, 2000). For example, activation of D_1 receptors on the terminals of striatonigral efferents increases GABA inhibitory transmission to SNr output neurons which decreases the inhibitory SNr output to thalamus (Radnikow and Misgeld, 1998); this would reinforce motor activation by the striatonigral pathway.

4. REGULATION OF SOMATODENDRITIC DOPAMINE BY UPTAKE TRANSPORTERS

Plasma membrane uptake of DA by the DA transporter (DAT) is fundamental to the regulation of $[DA]_o$, such that DAT-mediated uptake, coupled with diffusion, defines the sphere of influence of somatodendritically, as well as synaptically released DA (Cragg et al., 2001; Cragg and Rice 2004). Electrophysiological studies suggest a physiological role for uptake in the modulation of somatodendritic $[DA]_o$ (Lacey et al., 1990). Regulation of $[DA]_o$ by uptake differs between the SNc and VTA, however. Ventral tier neurons of the SNc have greater mRNA and protein levels of DA transporter (and D_2 DA receptor) than dorsal tier cells in dorsal SNc and VTA (Blanchard et al., 1994; Hurd et al., 1994; Sanghera et al., 1994; Ciliax et al., 1995; Freed et al., 1995). It is relevant to note that ventral tier SNc cells are more susceptible to degeneration in Parkinson's disease than dorsal tier cells in either VTA or SNc (Yamada et al., 1990; Fearnley and Lees, 1991; Gibb and Lees, 1991). This pattern of susceptibility is paralleled in the vulnerability to the DA uptake substrate and toxin, 1-methyl-4-phenyl-1,2,3,6-tetrahydropyridine

(MPTP) (German et al., 1988), which can be prevented by DA uptake inhibition. Together, such findings have implicated DA uptake activity or other regionally-specific DA handling mechanisms as possible risk factors in parkinsonian degeneration (Javitch et al., 1985; Sundstrom et al., 1986; Pifl et al., 1993). DA uptake within the midbrain, therefore, is crucial not only for normal DA neuron physiology, but may also contribute to the differential vulnerability of DA cells to pathophysiology.

In vitro experiments indicate that uptake of DA after evoked release or application of exogenous DA in SNc or VTA is primarily via the DAT (Fig. 2) (Cragg et al., 1997b, 2001), consistent with the high density of DAT-expressing cells in these areas (Ciliax et al., 1995; Freed et al., 1995; Nirenberg et al., 1996, 1997). In addition, the DAT is not the only transporter that can transport somatodendritic DA. Uptake of DA by the norepinephrine transporter (NET), albeit modest, is more prominent in VTA than SNc (Cragg et al., 1997b, 2001) and appears to be mediated by a few sparsely packed en passant norepinephrine processes (Cragg et al., 1997b). The role of the DAT on $[DA]_o$ after somatodendritic release may be more marked in SNc than VTA (Cragg et al., 1997b), reflecting differential DAT expression in these regions. Uptake of DA in SNr is much less avid than in either SNc or VTA, enabling DA to diffuse over larger distances without encountering uptake sites (Cragg et al., 2001).

Uptake via the DAT plays an apparently lesser role in the regulation of somatodendritic $[DA]_o$ in the SNc than of axon-terminal in $[DA]_o$ striatum (Cragg et al., 1997b), as shown by a greater increase in $[DA]_o$ during local electrical stimulation in striatum than in SNc when the DAT is blocked. This difference is in keeping with the lower density of the DAT in somatodendritic than axon terminal regions (Donnan et al., 1991; Ciliax et al., 1995; Freed et al., 1995). It can be speculated that a consequence of this apparently less avid regulation of dendritic compared to axon terminal $[DA]_o$, will be a greater sphere of influence of dendritic than for axonal DA (see Cragg et al., 2001; Cragg and Rice, 2004). Nonetheless, uptake via the DAT is an important mechanism of regulating $[DA]_o$ in midbrain (Cragg et al., 1997b, 2001).

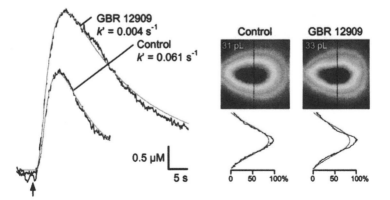

Figure 2. Regulation of diffusing DA in VTA by DATs. Left, example DA diffusion profiles and right, simultaneous images of co-diffused Texas Red-labelled dextran after pressure ejection with and without the DAT inhibitor, GBR-12909 (2 μM). DA diffusion profiles ~100 μm from site of ejection show the decrease in uptake constant (k') and the increase in $[DA]_o$ when DATs are inhibited, despite similar ejection volumes (31-33 pL). Modified from Cragg et al., 2001.

(A color version of this figure appears in the signature between pp. 256 and 257.)

5. WHAT IS THE MECHANISM OF RELEASE?

5.1. Ca^{2+} Dependence of Somatodendritic DA Release

The non-axonal source for somatodendritic DA release has led to much speculation about whether release might be mediated by a novel mechanism. However, few studies have contradicted the original proposal by Geffen et al. (1976) that DA release in the SN is vesicular and mediated by exocytosis, as it is in axon terminals in striatum. Indeed, physiological or pharmacological manipulations that alter $[DA]_o$ in striatum generally affect $[DA]_o$ in SN in a parallel manner (Santiago and Westerink, 1991; Heeringa and Abercrombie, 1995; Cragg and Greenfield, 1997; Cragg et al., 1997b; Hoffman and Gerhardt, 1999; Chen and Rice, 2001), although there is evidence for less similarity between DA release in VTA vs. nucleus accumbens (Iravani et al., 1996).

Consistent with the ionic and pharmacologic characteristics of exocytosis, somatodendritic DA release is depolarization- and Ca^{2+}-dependent (Geffen et al., 1976; Cheramy et al., 1981; Rice et al., 1994; Rice et al., 1997; Chen and Rice, 2001) and sensitive to DA depletion by reserpine (Rice et al., 1994; Heeringa and Abercrombie, 1995). Evidence for Ca^{2+} dependence is often taken as confirmatory of vesicular release, since Ca^{2+} entry is typically required for exocytosis (Douglas and Rubin, 1963; Catterall 1999; but see Parnas et al., 2000), such that the amount of transmitter released can depend on extracellular Ca^{2+} ($[Ca^{2+}]_o$) (Dodge and Rahamimoff, 1967). Consistent with characteristics of classical exocytosis, DA release in striatum requires activation of voltage-dependent Ca^{2+}-channels (Dunlap et al., 1995; Phillips and Stamford, 2000); moreover, both basal and evoked $[DA]_o$ increase when $[Ca^{2+}]_o$ is elevated (Moghaddam and Bunney, 1989; Chen and Rice, 2001).

Somatodendritic DA release in SN also requires Ca^{2+}: incubation in Ca^{2+}-free media with EGTA inhibits evoked DA release by >90% (Rice et al., 1994; Rice et al., 1997). However, in contrast to striatal release, DA release in SNc persists in low-Ca^{2+} media (Rice et al., 1994; Rice et al., 1997; Hoffman and Gerhardt, 1999). Evoked $[DA]_o$ in the SNc in midbrain slices is half-maximal in nominally zero $[Ca^{2+}]_o$ with no EGTA (Fig. 3) (Chen and Rice, 2001). Furthermore, voltage-gated Ca^{2+}-channel blockers have only modest effects on $[DA]_o$ in the SN (Bergquist and Nissbrandt, 2003, 2004). Together, these findings suggest a difference in intracellular Ca^{2+} availability or sensitivity of somatodendritic *versus* axonal release. For example, ready release in low $[Ca^{2+}]_o$ might reflect differential sensitivity of Ca^{2+}-dependent release mechanisms, including fusion proteins, between these compartments; the varying forms of synaptotagmin, a key Ca^{2+}-sensing protein responsible for triggering *synaptic* transmitter release, can differ in Ca^{2+} affinity by 10-20-fold (Südhof and Rizo 1996; Südhof 2002). Differential expression of fusion proteins among cell compartments is not without precedent: Bergquist et al. (2002) recently reported evidence that different synaptobrevin isoforms underlie axonal DA release in striatum *versus* somatodendritic release in the SN.

In further contrast to striatal DA release, evoked $[DA]_o$ in SNc is maximal in 1.5 mM Ca^{2+}, with no increase in higher $[Ca^{2+}]_o$ (Fig. 3) (Chen and Rice 2001). While this might reflect differences in the readily releasable pool of DA in each region, a possible confounding factor is that the train stimulation used to evoke DA release simultaneously elicits release of other transmitters, including GABA and glutamate, which strongly

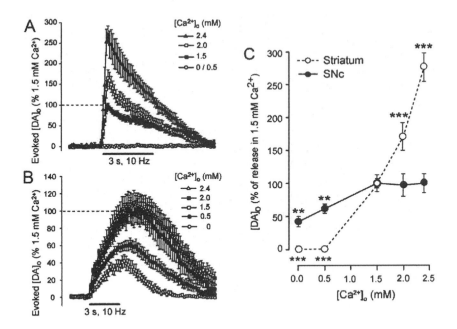

Figure 3. Ca^{2+}-dependence of DA release in (A) dorsal striatum, (B) SNc, and (C) both regions; all data are normalized to peak $[DA]_o$ in 1.5 mM $[Ca^{2+}]_o$ as 100%. DA release was elicited in guinea-pig brain slices by local stimulation (10 Hz, 30 pulses) and monitored with FCV (modified from Chen and Rice, 2001).

regulate somatodendritic DA release in the SNc (Chen and Rice, 2002; *see* section 6.2). This raised the particular concern that the true Ca^{2+}-dependence of somatodendritic release might be masked by concurrent increases in what is presumably Ca^{2+}-dependent GABA release. Such Ca^{2+}-dependent GABA release could decrease DA cell excitability and thereby oppose Ca^{2+}-dependent increases in somatodendritic DA release to cause the apparent "clamping" of evoked $[DA]_o$ for $[Ca^{2+}]_o > 1.5$ mM (Fig. 3C).

We therefore re-examined the Ca^{2+}-dependence of DA release in the SNc (and VTA) using single-pulse stimulation, which elicits DA release that is free from ionotropic GABA or glutamate receptor modulation. With this stimulus, the pattern of Ca^{2+} sensitivity of evoked DA release in the SNc was similar to that seen with pulse-train stimulation, with ready release in low $[Ca^{2+}]_o$ and a plateau in higher $[Ca^{2+}]_o$ (Moran et al., 2003). This suggests that the Ca^{2+} dependent somatodendritic release reflects intrinsic rather than extrinsic factors, with the caveat that as yet undefined tonically active inputs might regulate single-pulse release in a Ca^{2+}-dependent manner. In the VTA, single-pulse DA release was seen readily in low $[Ca^{2+}]_o$, as in the SNc, but then evoked $[DA]_o$ progressively increased with increasing $[Ca^{2+}]_o$ throughout the range tested, as in the striatum (Moran et al., 2003). This pattern might reflect a mixture of somatodendritic and axonal DA release predicted by the anatomy of the VTA (Deutch et al., 1988; Bayer and Pickel, 1990).

5.2. Anatomical Issues

Despite existing evidence for exocytotic release of DA in the SNc, this continues to be questioned on the basis of anatomy: synaptic sites for vesicle fusion are rare. Although dendro-dendritic DA synapses are found in the SNc (Wilson et al., 1977), these are absent in the SNr and make up less than 1% of synapses on DA dendrites overall (Groves and Linder, 1983). Moreover, DA release can be elicited from the SNr in isolation (Geffen et al., 1976; Rice et al., 1994), showing that dendro-dendritic synapses are not required for release. Although vesicular release of catecholamines can occur in the absence of synapses, as in adrenal chromaffin cells (Wightman et al., 1991), vesicles are also rare in DA cells of the SNc. Whereas vesicle density is high in DA terminals in striatum (Nirenberg et al., 1996a; 1997), there are few vesicles in DA somata or dendrites (Wilson et al., 1977; Groves and Linder, 1983; Nirenberg et al., 1996a), implying a limited source for vesicular exocytotic release (Nirenberg et al., 1996a).

Storage of somatodendritic DA has been proposed to be in saccules of smooth endoplasmic reticulum (Mercer et al., 1978; Wassef et al., 1981), as well as in vesicles (Wilson et al., 1977; Groves and Linder, 1983). Consistent with dual storage sites, the vesicular monoamine transporter, VMAT2, is expressed in saccules (so-called tubulovesicles) and, less commonly, in vesicles (Nirenberg et al., 1996b). Whether both storage sites contribute to the releasable pool of DA is not known. Both would be susceptible to DA depletion by reserpine, an irreversible inhibitor of VMAT2, such that reserpine sensitivity alone sheds little light on whether release is vesicular or otherwise.

5.3. Reversal of the DAT

The paucity of vesicles and the abundance of DATs in DA dendrites led several groups to speculate that somatodendritic release might be mediated by vesicle-independent reverse transport (Groves and Linder, 1983; Nirenberg et al., 1996a; Leviel, 2001). Indeed, this is the primary mode of release following certain pharmacological manipulations, including veratridine-induced depolarization (Elverfors et al., 1997) or amphetamine-induced DA displacement from intracellular stores (Sulzer et al., 1995; Jones et al., 1998; Torres et al., 2003). Somatodendritic DA release elicited by veratridine requires voltage-dependent Na^+ channel opening and subsequent Na^+ loading to activate reverse DA transport; interestingly, this release can be blocked by the DAT inhibitor GBR-12909 applied outside the releasing cells (Elverfors et al., 1997). Under normal conditions, cytoplasmic DA concentrations appear to be insufficient to reverse the DAT unless increased by a pharmacological agent, e.g. by amphetamine (Jones et al., 1998).

However, Falkenburger et al. (2001) have reported electrophysiological evidence that stimulation of the subthalamic nucleus can induce dendritic DA release in the SN by a mechanism that can be prevented by low concentrations of GBR-12909. This suggests that under some physiological conditions, somatodendritic release might be mediated by the reversal of the DAT. It is not yet clear how the data of Falkenburger et al. can be reconciled with other *in vitro* and *in vivo* studies of somatodendritic DA release that have typically found an increase in $[DA]_o$ in the presence of an uptake blocker (Elverfors and Nissbrandt, 1992; Santiago and Westerink, 1991,1992; Cragg et al., 1997b; Hoffman et al., 1998; Chen and Rice, 2001).

6. RECEPTOR REGULATION OF RELEASE

6.1. Autoreceptor Regulation of Somatodendritic Dopamine Release

Inhibition of DA release from striatal axon terminals via presynaptic D_2-like autoreceptors plays a powerful role in the regulation of striatal $[DA]_o$ (Limberger et al., 1991; Trout and Kruk, 1992; Cragg and Greenfield, 1997; Benoit-Marand et al., 2001). D_2-receptors are also expressed and located directly on DA dendrites in VTA and SN (Bouthenet et al., 1987; Hurd et al., 1994; Sesack et al., 1994; Yung et al., 1995). In turn, D_2-receptor control of somatodendritic DA release has been documented in SNc indicating that D_2 receptors operate as an autoinhibitory mechanism to regulate somatodendritic release (Cragg and Greenfield, 1997). Nonetheless, the degree of control of somatodendritic $[DA]_o$ by this mechanism is less avid than in striatum. Less effective autoinhibition of release, in conjunction with less avid DA re-uptake via the DAT (Nissbrandt et al., 1991; Cragg et al., 1997b), could result in altogether different regulation of $[DA]_o$ in somatodendritic and axon terminal regions.

The modulation of $[DA]_o$ by D_2 receptors is less marked in VTA than in SNc, consistent with the higher expression and protein levels of the D_2 receptor in ventral tier DA neurons (Hurd et al., 1994). As for differential DAT expression discussed above, regional variation in the management of somatodendritic DA transmission in VTA and SNc by autoreceptor regulation might also contribute to differential susceptibility of these cell groups to degeneration in Parkinson's disease.

6.2. Heteroreceptor Regulation of Somatodendritic Dopamine Release

DA neurons receive significant synaptic input from numerous non-dopaminergic neuron types and concomitantly express somatodendritic receptors for corresponding major classes of neurotransmitters, including GABA, glutamate, 5-HT, and acetylcholine (ACh), as well as co-transmitters like dynorphin and ATP. The role of each of these inputs in the direct heteroreceptor regulation of DA release within the SN and VTA is increasingly being revealed.

Strong inhibitory control of somatodendritic DA release by GABA has been identified, especially in the SNc where both $GABA_A$ and $GABA_B$ receptors control release (Chen and Rice, 2002). The role of GABA in the VTA is less clear. Although measurements *in vivo* using microdialysis show that $GABA_B$ receptors tonically inhibit DA release in VTA (Giorgetti et al., 2002), *in vivo* data can be complicated by concomitant activation of long-loop circuits; more direct *in vitro* methods suggest that $GABA_A$ and $GABA_B$ receptor control of release in VTA is limited and may depend on the strength and timing of other synaptic input (Chen and Rice, 2002).

Glutamate receptor regulation of somatodendritic DA release can take the form of direct excitation by ionotropic receptors (Araneda and Bustos, 1989; Westerink et al., 1992; Gauchy et al., 1994; Rosales et al., 1994; Chen and Rice, 2002), particularly in the VTA (Chen and Rice, 2002), or heterosynaptic inhibition via AMPA-receptor activation on GABA afferents in the SNc (Chen and Rice, 2002). These findings are consistent with the dominance of glutamate input to VTA DA neurons (70% of axodendritic synapses) and the dominance of GABA input to SNc DA cells (70% of axodendritic synapses)

(Bolam and Smith, 1990). Whether local glutamate input enhances or inhibits somatodendritic DA release (via GABA afferents) thus depends on the balance between excitatory and inhibitory inputs at a given moment (Chen and Rice, 2002), and, as such, can vary according to brain region, the pattern of local input circuitry, and experimental conditions.

In keeping with the expression of nicotinic ACh receptors by DA neurons (primarily α4ß2-subunit-containing; Picciotto et al., 1998; Champtiaux et al., 2003), nicotine/ACh facilitatory control of somatodendritic DA release has been identified in VTA using microdialysis with a remote, subcutaneous nicotine administration (Rahman et al., 2003), and more directly in a dendrosomal preparation from SN/VTA using a $[^3H]DA$ release assay (Reuben et al., 2000). Although the effect of nicotinic ACh receptor activation on DA cells is typically excitatory, *inhibitory* effects can also occur by subsequent activation of Ca^{2+}-activated K^+ channels (Fiorillo and Williams, 2000). An additional facilitatory regulation of somatodendritic DA release may be offered by endogenous ATP acting at P2 receptors in VTA, according to experiments using microinjection of P2 receptor antagonists and microdialysis (Krugel et al., 2001). Specific $P2Y_1$-receptor immunoreactivity on neurons in VTA suggests, but does not confirm, that these P2 receptors may be localized directly on DA neurons. Somatodendritic DA release may also be regulated by other classes of molecules, including reactive oxygen species; endogenously generated hydrogen peroxide (H_2O_2) can suppress DA release in the SNc, although not in the VTA (Chen et al., 2002).

Thus, in addition to the role of major synaptic inputs in the direct modulation of DA neuron output activity, these systems can powerfully and variously gate the release of somatodendritic DA. As a consequence, the action of somatodendritic DA release at D_1 and D_2 receptors in VTA and SN will modulate the influence of synaptic input on the final signal integration and the ultimate activity of all dopaminoceptive output neurons of the VTA, SNc and SNr.

7. CONCLUSIONS

Somatodendritic DA release in the SN and VTA serves a number of critical functions in the regulation of animal behaviour from motivation to motion. Released DA in midbrain acts at predominantly extrasynaptic DA receptors to mediate autoreceptor control of DA cell activity and heteroreceptor regulation of synaptic input that is also reciprocal. Somatodendritic DA release is activity- and Ca^{2+}-dependent, although with patterns that differ from those underlying synaptic release in striatum. Although the mechanism of somatodendritic release remains controversial, most evidence points to a conventional exocytotic process. In the 30 years since somatodendritic DA release was proposed, much has been learned about its role and how it is regulated, as summarized in this review. Excitingly, there are many remaining questions to address.

8. ACKNOWLEDGEMENTS

SJC is a Beit Memorial Research Fellow. MER acknowledges support from NIH NS-36362 and NS-45325.

9. REFERENCES

Araneda, R., and Bustos, G., 1989, Modulation of dendritic release of dopamine by N-methyl-D-aspartate receptors in rat substantia nigra, *J. Neurochem.* **52**: 962.

Bayer, V. E., and Pickel, V. M., 1990, Ultrastructural localization of tyrosine hydroxylase in the rat ventral tegmental area: relationship between immunolabelling density and neuronal associations, *J. Neurosci.* **10**: 2996.

Benoit-Marand, M., Borrelli, E., and Gonon, F., 2001, Inhibition of dopamine release via presynaptic D2 receptors: time course and functional characteristics *in vivo, J. Neurosci.* **21**: 9134.

Bergquist, F., and Nissbrandt, H., 2003, Influence of R-type (Cav2.3) and t-type (Cav3.1-3.3) antagonists on nigral somatodendritic dopamine release measured by microdialysis, *Neuroscience* **120**: 757.

Bergquist, F., and Nissbrandt, H., 2004, Dopamine release in the substantia nigra: release mechanisms and physiological function in motor control, in: *Dendritic Transmitter Release*, M. Ludwig, ed., Kluwer Academic/Plenum Press, New York, (in press).

Bergquist, F., Niazi, H. S., and Nissbrandt, H., 2002, Evidence for different exocytosis pathways in dendritic and terminal dopamine release *in vivo, Brain Res.* **950**: 245.

Bergquist, F., Shahabi, H. N., and Nissbrandt, H., 2003, Somatodendritic dopamine release in rat substantia nigra influences motor performance on the accelerating rod, *Brain Res.* **973**: 81.

Björklund, A., and Lindvall, O., 1975, Dopamine in dendrites of substantia nigra neurons: suggestions for a role in dendritic terminals, *Brain Res.* **83**: 531.

Blanchard, V., Raisman-Vozari, R., Vyas, S., Michel, P. P., Javoy-Agid, F., Uhl, G., and Agid, Y., 1994, Differential expression of tyrosine hydroxylase and membrane dopamine transporter genes in subpopulations of dopaminergic neurons of the rat mesencephalon, *Mol. Brain Res.* **22**: 29.

Bolam, J. P., and Smith, Y., 1990, Characterisation of the synaptic inputs to DA-ergic neurones in the rat SN, in: *The Basal Ganglia III.*, G. Bernardini, ed., Plenum Press, New York

Bouthenet, M.-L., Martres, M. P., Sales, N., and Schwartz, J. C., 1987, A detailed mapping of dopamine D2 receptors in rat central nervous system by autoradiography with [^{125}I]iodosulpiride, *Neuroscience* **20**: 117.

Bunin, M. A., and Wightman, R. M., 1998, Quantitative evaluation of 5-hydroxytryptamine (serotonin) neuronal release and uptake: an investigation of extrasynaptic transmission, *J. Neurosci.* **18**: 4854.

Carlsson, A., 2002, Treatment of Parkinson's with L-DOPA. The early discovery phase, and a comment on current problems, *Neural Transm.* **109**: 777.

Catterall, W. A., 1999, Interactions of presynaptic Ca^{2+} channels and snare proteins in neurotransmitter release, *Ann. NY Acad. Sci.* **868**: 144.

Champtiaux, N., Gotti, C., Cordero-Erausquin, M., David, D. J., Przybylski, C., Lena, C., Clementi, F., Moretti, M., Rossi, F. M., Le Novere, N., McIntosh, J. M., Gardier, A. M., and Changeux, J. P., 2003, Subunit composition of functional nicotinic receptors in dopaminergic neurons investigated with knock-out mice, *J. Neurosci.* **23**: 7820.

Chen, B. T., and Rice, M. E., 2001, Novel Ca^{2+} dependence and time course of somatodendritic dopamine release: substantia nigra versus striatum, *J. Neurosci.* **21**: 7841.

Chen, B. T., and Rice, M. E., 2002, Synaptic regulation of somatodendritic dopamine release by glutamate and GABA differs between substantia nigra and ventral tegmental area, *J. Neurochem.* **81**: 158.

Chen, B. T., Avshalumov, M. V., and Rice, M. E., 2002, Modulation of somatodendritic dopamine release by endogenous H$_2$O$_2$: susceptibility in substantia nigra but resistance in VTA, *J. Neurophysiol.* **87**: 1155.

Cheramy, A., Leviel, V., and Glowinski, J., 1981, Dendritic release of dopamine in the substantia nigra, *Nature* **289**: 537.

Chiodo, L. A., 1992, Dopamine autoreceptor signal tranduction in the DA cell body: a "current view", *Neurochem. Int.* **20**: 81S.

Ciliax, B. J., Heilman, C., Demchyshyn, L. L., Pristupa, Z. B., Ince, E., Hersch, S., Niznik, H. B., and Levey, A. I., 1995, The dopamine transporter: immunocytochemical characterization and localization in brain, *J. Neurosci.* **15**: 1714.

Corvaja, N., Doucet, G., and Bolam, J. P., 1993, Ultrastructure and synaptic targets of the raphe-nigral projection in the rat, *Neuroscience* **55**: 417.

Cragg, S. J., and Greenfield, S. A., 1997, Differential autoreceptor control of somatodendritic and axon terminal dopamine release in substantia nigra, ventral tegmental area, and striatum, *J. Neurosci.* **17**: 5738.

Cragg, S. J., and Rice, M. E., 2004, DAncing past the DAT at a DA synapse, *Trends Neurosci.***27**: 270

Cragg, S. J., Hawkey, C. R., and Greenfield, S. A., 1997a, Comparison of serotonin and dopamine release in substantia nigra and ventral tegmental area: region and species differences, *J. Neurochem.* **69**: 2378.

Cragg, S. J., Nicholson, C., Kume-Kick, J., Tao, L., and Rice, M. E., 2001, Dopamine-mediated volume transmission in midbrain is regulated by distinct extracellular geometry and uptake, *J. Neurophysiol.* **85**: 1761.

Cragg, S. J., Rice, M. E., and Greenfield, S. A., 1997b, Heterogeneity of electrically evoked dopamine release and reuptake in substantia nigra, ventral tegmental area, and striatum, *J. Neurophysiol.* **77**: 863.

Crocker, A. D., 1997, The regulation of motor control: an evaluation of the role of dopamine receptors in the substantia nigra, *Rev. Neurosci.* **8**: 55.

Deutch, A. Y., Goldstein, M., Baldino, F. Jr., and Roth, R. H., 1988, Telencephalic projections to the A8 dopamine cell group, *Ann. NY Acad. Sci.* **537**: 27.

Dodge, F. A., Jr., and Rahamimoff, R., 1967, On the relationship between calcium concentration and the amplitude of the end-plate potential. *J. Physiol.* **189**: 90.

Donnan, G. A., Kaczmarczyk, S. J., Paxinos, G., Chilco, P. J., Kalnins, R. M., Woodhouse, D. G., and Mendelsohn, F. A., 1991, Distribution of catecholamine uptake sites in human brain as determined by quantitative [^{3}H]-mazindol autoradiography, *J. Comp. Neurol.* **304**: 419.

Douglas, W. W., and Rubin, R. P., 1963, The mechanism of catecholamine release from the adrenal medulla and the role of calcium in stimulation-secretion coupling, *J. Physiol.* **167**: 288.

Dunlap, K., Luebke, J. I., and Turner, T. J., 1995, Exocytotic Ca^{2+} channels in mammalian central neurons, *Trends Neurosci.* **18**: 89.

Elverfors, A., and Nissbrandt, H., 1991, Reserpine-insensitive dopamine release in the substantia nigra? *Brain Res.* **557**: 5.

Elverfors, A., and Nissbrandt, H., 1992, Effects of *d*-amphetamine on dopaminergic neurotransmission; a comparison between the substantia nigra and the striatum, *Neuropharmacology* **31**: 661.

Elverfors, A., Jonason, J., Jonason, G., and Nissbrandt, H., 1997, Effects of drugs interfering with sodium channels and calcium channels on the release of endogenous dopamine from superfused substantia nigra slices, *Synapse* **26**: 359

Falkenburger, B. H., Barstow, K. L., and Mintz, I. M., 2001. Dendrodendritic inhibition through reversal of dopamine transport, *Science* **293**: 2465.

Fallon, J. H., and Moore, R. Y., 1978, Catecholamine innervation of the basal forebrain. IV. Topography of the dopamine projection to the basal forebrain and neostriatum, *J. Comp. Neurol.* **180**: 545.

Fallon, J. H., Riley, J. N., and Moore, R. Y., 1978, Substantia nigra dopamine neurons: separate populations project to neostriatum and allocortex, *Neurosci. Lett.* **7**: 157.

Fearnley, J. M., Lees, A. J., 1991, Ageing and Parkinson's disease: substantia nigra regional selectivity, *Brain* **114**: 2283.

Fiorillo, C. D., and Williams, J. T., 2000, Cholinergic inhibition of ventral midbrain dopamine neurons, *J. Neurosci.* **20**: 7855.

Freed, C., Revay, R., Vaughan, R. A., Kriek, E., Grant, S., Uhl, G. R., and Kuhar, M. J., 1995, Dopamine transporter immunoreactivity in rat brain, *J. Comp. Neurol.* **359**: 340.

Fuxe, K., and Agnati, L. F., eds., 1991, *Volume Transmission in the Brain: Novel Mechanisms for Neural Transmission*, Raven Press, New York.

Gauchy, C., Desban, M., Glowinski, J., and Kemel, M. L., 1994, NMDA regulation of dopamine release from proximal and distal dendrites in the cat substantia nigra, *Brain Res* **635**: 249.

Geffen, L. B., Jessell, T. M., Cuello, A. C., and Iversen, L. L., 1976, Release of dopamine from dendrites in rat substantia nigra, *Nature* **260**: 258.

Gerfen, C. R., Herkenham, M., and Thibault, J., 1987a, The neostriatal mosaic. II. Patch- and matrix-directed mesostriatal dopaminergic and non-dopaminergic systems, *J. Neurosci.* **7**: 3915.

Gerfen, C. R., Baimbridge, K. G., and Thibault, J., 1987b, The neostriatal mosaic. III. Biochemical and developmental dissociation of patch-matrix mesostriatal systems, *J. Neurosci.* **7**: 3935.

German, D. C., Dubach, M., Askari, S., Speciale, S. G., and Bowden, D. M., 1988, MPTP-induced parkisonian syndrome in *Macaca fascularis*: which midbrain dopaminergic neurons are lost? *Neuroscience* **24**: 161.

Gibb, W. R. G., Lees, A. J., 1991, Anatomy, pigmentation, ventral and dorsal subpopulations of the substantia nigra, and differential cell death in Parkinson's disease, *J. Neurol. Neurosurg. Psych.* **54**: 388.

Giorgetti, M., Hotsenpiller, G., Froestl, W., and Wolf, M. E., 2002, *In vivo* modulation of ventral tegmental area dopamine and glutamate efflux by local GABA(B) receptors is altered after repeated amphetamine treatment, *Neuroscience* **109**: 585.

Groves, P. M., and Linder, J. C., 1983, Dendro-dendritic synapses in substantia nigra: descriptions based on analysis of serial sections, *Exp. Brain Res.* **49**: 209.

Groves, P. M., Wilson, C. J., Young, S. J., and Rebec, G. V., 1975, Self-inhibition by dopaminergic neurons, *Science* **190**: 522.

Heeringa, M. J., and Abercrombie, E. D., 1995, Biochemistry of somatodendritic dopamine release in the substantia nigra: an *in vivo* comparison with striatal dopamine release, *J. Neurochem.* **65**: 192.

Hoffman, A. F., and Gerhardt, G. A., 1999, Differences in pharmacological properties of dopamine release between the substantia nigra and striatum: an *in vivo* electrochemical study, *J. Pharmacol. Exp. Ther.* **289**: 455.

Hoffman, A. F., Lupica, C. R., and Gerhardt, G. A., 1998, Dopamine transporter activity in the substantia nigra and striatum assessed by high-speed chronoamperometric recordings in brain slices, *J. Pharmacol. Exp. Ther.* **287**: 487.

Hurd, Y. L., Pristupa, Z. B., Herman, M. M., Niznik, H. B., and Kleinman, J. E., 1994, The DA transporter and DA D2 receptor mRNAs are differentially expressed in limbic- and motor-related subpopulations of human mesencephalic neurons, *Neuroscience* **63**: 357.

Iravani, M. M., and Kruk, Z. L., 1997, Real-time measurement of stimulated 5-hydroxytryptamine release in rat substantia nigra pars reticulata brain slices, *Synapse* **25**: 93.

Iravani, M. M., Muscat, R., and Kruk, Z. L., 1996, Comparison of somatodendritic and axon terminal dopamine release in the ventral tegmental area and the nucleus accumbens, *Neuroscience* **70**: 1025.

Javitch, J. A., D'Amato, R. J., Strittmatter, S. M., and Snyder, S. H., 1985, Parkinsonism-inducing neurotoxin, MPTP: uptake of the metabolite N-methly-4-phenly pyridine by dopamine neurons explains selective toxicity, *PNAS USA* **82**: 2173.

Jaffe, E. H., Marty, A., Schulte, A., and Chow, R. H., 1998, Extrasynaptic vesicular transmitter release from the somata of substantia nigra neurons in rat midbrain slices, *J. Neurosci.* **18**: 3548.

John, C. E., Budygin, E. A., Mateo, Y., and Jones, S. R., 2003, Voltammetric characterization of monoamines in mouse midbrain, in: *Monitoring Molecules in Neuroscience 2003*, J. Kehr, K. Fuxe, U. Ungerstedt, T. H. Svensson, eds., Karolinska University Press, Stockholm, pp. 490-491.

Jones, S. R., Gainetdinov, R. R., Wightman, R. M., and Caron, M. G., 1998, Mechanisms of amphetamine action revealed in mice lacking the dopamine transporter, *J. Neurosci.* **18**:1979.

Juraska, J. M., Wilson, C. J., and Groves, P. M., 1977, The substantia nigra of the rat: a Golgi study. *J. Comp. Neurol.* **172**: 585.

Kalivas, P. W., and Duffy, P., 1991, A comparison of axonal and somatodendritic DA release using *in vivo* dialysis, *J. Neurochem.* **56**: 961.

Koga, E., and Momiyama, T., 2000, Presynaptic dopamine D2-like receptors inhibit excitatory transmission onto rat ventral tegmental dopaminergic neurones, *J. Physiol.* **523**: 163.

Krugel, U., Kittner, H., Franke, H., and Illes, P., 2001, Stimulation of P2 receptors in the ventral tegmental area enhances dopaminergic mechanisms *in vivo*, *Neuropharmacology* **40**: 1084.

Lacey, M. G., Mercuri, N. B., and North, R. N., 1988, On the potassium conductance increase activated by GABA_B and dopamine D2 receptors to increase potassium conductance in rat substantia nigra neurones of the zona compacta, *J. Physiol.* **401**: 437.

Lacey, M. G., Mercuri, N. B., and North, R. N., 1990, Actions of cocaine on rat dopaminergic neurones *in vitro*, *Br. J. Pharmacol.* **99**: 731.

Lavoie, B., and Parent, A., 1990, Immunohistochemical study of the serotonergic innervation of the basal ganglia in the squirrel monkey, *J. Comp. Neurol.* **299**: 1.

Leviel, V., 2001, The reverse transport of DA, what physiological significance?, *Neurochem Int.* **38**: 83.

Limberger, N., Trout, S. J., Kruk, Z. L., and Starke, K., 1991, "Real-time" measurement of endogenous DA release during short trains of pulses in slices of rat neostriatum and nucleus accumbens: role of autoinhibition, *Naunyn Schmiedeberg's Arch. Pharmacol.* **344**: 623.

Mercer, L., del Fiacco, M., and Cuello, A. C., 1978, The smooth endoplasmic reticulum as a possible storage site for dendritic dopamine in substantia nigra neurones, *Experientia* **35**: 101.

Miyazaki, T., and Lacey, M. G., 1998, Presynaptic inhibition by dopamine of a discrete component of GABA release in rat substantia nigra pars reticulata, *J. Physiol.* **513**: 805.

Moghaddam, B., and Bunney, B. S., 1989, Ionic composition of microdialysis perfusing solution alters the pharmacological responsiveness and basal outflow of striatal dopamine, *J. Neurochem.* **53**: 652.

Moran, K. A., Chen, B. T., Tao, L., and Rice. M. E., 2003, Calcium dependence of dopamine release in the VTA reveals both somatodendritic and synaptic characteristics, Program number 461.16, *2003 Abstract Viewer/Itinerary Planner*, Soc. Neurosci. Washington, DC.

Moukhles, H., Bosler, O., Bolam, J. P., Vallee, A., Umbriaco, D., Geffard, M., and Doucet, G., 1997, Quantitative and morphometric data indicate precise cellular interactions between serotonin terminals and postsynaptic targets in rat substantia nigra, *Neuroscience* **76**: 1159.

Nedergaard, S., Bolam, J. P., and Greenfield, S. A., 1988, Facilitation of a dendritic calcium conductance by 5-HT in the substantia nigra, *Nature* **333**: 174.

Nieoullon, A., Cheramy, A., and Glowinski, J., 1977, Release of dopamine *in vivo* from cat substantia nigra, *Nature* **266**: 375.

Nirenberg, M. J., Vaughan, R. A., Uhl, G. R., Kuhar, M. J., and Pickel, V. M., 1996a, The dopamine transporter is localized to dendritic and axonal plasma membranes of nigrostriatal dopaminergic neurons, *J. Neurosci.* **16**: 436.

Nirenberg, M. J., Chan, J., Liu, Y., Edwards, R. H., and Pickel, V. M., 1996b, Ultrastructural localization of the vesicular monoamine transporter-2 in midbrain dopaminergic neurons: potential sites for somatodendritic storage and release of dopamine, *J. Neurosci.* **16**: 4135.

Nirenberg, M. J., Chan, J., Pohorille, A., Vaughan, R. A., Uhl, G. R., Kuhar, M. J., and Pickel, V. M., 1997, The dopamine transporter: comparative ultrastructure of dopaminergic axons in limbic and motor compartments of the nucleus accumbens, *J. Neurosci.* **17**: 6899.

Nissbrandt, H., Engberg, G., and Pileblad., E., 1991, The effects of GBR 12909, a dopamine re-uptake inhibitor, on monoaminergic neurotransmission in rat striatum, limbic forebrain, cortical hemspheres and substantia nigra, *Naunyn Schmiedeberg's Arch. Pharmacol.* **344**: 16.

Parnas, H., Segel, L., Dudel, J., and Parnas, I., 2000, Autoreceptors, membrane potential and the regulation of transmitter release, *Trends Neurosci.* **23**: 60.

Phillips, P. E. M., and Stamford, J. A., 2000, Differential recruitment of N-, P- and Q-type voltage-operated calcium channels in striatal dopamine release evoked by 'regular' and 'burst' firing, *Brain Res.* **884**: 139.

Picciotto, M. R., Zoli, M., Rimondini, R., Lena, C., Marubio, L. M., Pich, E. M., Fuxe, K., and Changeux, J. P., 1998, Acetylcholine receptors containing the beta2 subunit are involved in the reinforcing properties of nicotine, *Nature* **391**: 173.

Pickel, V. M., 2000, Extrasynaptic distribution of monoamine transporters and receptors, *Prog Brain Res.* **125**: 267.

Pifl, C., Giros, B., and Caron, M. G., 1993, Dopamine transporter expression confers cytotoxicity to low does of the parkinsonism-inducing neurotoxin MPTP, *J. Neurosci.* **13**: 4246.

Radnikow, G., and Misgeld, U., 1998, Dopamine D_1 receptors facilitate $GABA_A$ synaptic currents in the rat substantia nigra pars reticulata, *J. Neurosci.* **18**: 2009.

Rahman, S., Zhang, J., and Corrigall, W. A., 2003, Effects of acute and chronic nicotine on somatodendritic dopamine release of the rat ventral tegmental area: *in vivo* microdialysis study, *Neurosci. Lett.* **348**: 61.

Reuben, M., Boye, S., and Clarke, P. B., 2000, Nicotinic receptors modulating somatodendritic and terminal dopamine release differ pharmacologically, *Eur. J. Pharmacol* **393**: 39.

Rice, M. E., 2000, Distinct regional differences in dopamine-mediated volume transmission, *Prog. Brain Res.* **125**: 275.

Rice, M. E., and Nicholson, C., 1991, Diffusion characteristics and extracellular volume fraction during normoxia and hypoxia in slices of rat neostriatum, *J. Neurophysiol.* **65**: 264.

Rice, M. E., Richards, C. D., Nedergaard, S., Hounsgaard, J., Nicholson, C., and Greenfield, S. A., 1994, Direct monitoring of dopamine and 5-HT release in substantia nigra and ventral tegmental area *in vitro*, *Exp. Brain Res.* **100**: 395.

Rice, M. E., Cragg, S. J., and Greenfield, S. A., 1997, Characteristics of electrically evoked somatodendritic dopamine release in substantia nigra and ventral tegmental area *in vitro*, *J. Neurophysiol.* **77**: 853.

Robertson, G. S., and Robertson, H. A., 1989, Evidence that L-DOPA-induced rotational behavior is dependent on both striatal and nigral mechanisms, *J. Neurosci.* **9**: 3326.

Robertson, G. S., Damsma, G., and Fibiger, H. C., 1991, Characterization of dopamine release in the substantia nigra by *in vivo* microdialysis in freely moving rats, *J. Neurosci.* **11**: 2209.

Rosales, M. G., Flores, G., Hernandez, S., Martinez-Fong, D., and Aceves, J., 1994, Activation of subthalamic neurons produces NMDA receptor-mediated dendritic dopamine release in substantia nigra pars reticulata: a microdialysis study in the rat, *Brain Res.* **645**: 335.

Sanghera, M. K., Manaye, K. F., Liang, C.-L., Iacopino, A. M., Bannon, M. J., and German, D. C., 1994, Low dopamine transporter mRNA levels in midbrain regions containing calbindin, *Neuroreport* **5**: 1641.

Santiago, M., and Westerink, B. H. C., 1991, Characterization and pharmacological responsiveness of dopamine release recorded by microdialysis in the substantia nigra of concious rats. *J. Neurochem.* **57**: 738.

Santiago, M., and Westerink, B. H. C., 1992, Simultaneous recording of the release of nigral and striatal dopamine in the awake rat, *Neurochem. Int.* **20**: 107S.

Sesack, S. R., Aoki, C., and Pickel, V. M., 1994, Ultrastructural localization of D2-receptor-like immunoreactivity in midbrain dopamine neurons and their striatal targets, *J. Neurosci.* **14**: 88.

Stamford, J. A., Palij, P., Davidson, C., and Millar, J., 1993, Simultaneous "real-time" electrochemical and electrophysiological recording in brain slices with a single carbon-fibre microelectrode, *J. Neurosci. Meth.* **50**: 279.

Steinbusch, H. W. M., 1981, Distribution of serotonin-immunoreactivity in the central nervous system of the rat- cell bodies and terminals, *Neuroscience* **6**: 557.

Südhof, T. C., 2002, Synaptotagmins: Why so many? *J. Biol Chem.* **277**: 7629

Südhof, T. C., and Rizo, J., 1996, Synaptotagmins: C2-domain proteins that regulate membrane traffic, *Neuron* **17**: 379.

Sulzer, D., Chen, T. K., Lau, Y. Y., Kristensen, H., Rayport, S., and Ewing, A. (1995) Amphetamine redistributes dopamine from synaptic vesicles to the cytosol and promotes reverse transport, *J. Neurosci.* **15**:4102.

Sundstrom, E., Goldstein, M., and Jonsson, G., 1986, Uptake inhibition protects nigro-striatal dopamine neurons from the neurotoxicity of 1-methyl-4-phenylpyridine (MPP+) in mice, *Eur. J. Pharmacol.* **131**: 289.

Tagerud, S. E. O., and Cuello, A. C., 1979, Dopamine release from the rat substantia nigra *in vitro*. Effect of raphe lesions and veratrine stimulation, *Neuroscience* **4**: 2021.

Timmerman, W., and Abercrombie, E. D., 1996, Amphetamine-induced release of dendritic dopamine in substantia nigra pars reticulata: D1-mediated behavioral and electrophysiological effects, *Synapse* **23**: 280.

Torres, G. E., Gainetdinov, R. R., and Caron, M. G., 2003, Plasma membrane monoamine transporters: structure, regulation and function, *Nat. Rev. Neurosci.* **4**:13.

Trout, S. J., and Kruk, Z. L., 1992, Differences in evoked dopamine efflux in rat caudate putamen, nucleus accumbens and tuberculum olfactorium in the absence of uptake inhibition: influence of autoreceptors, *Br. J. Pharmacol.* **106**: 452.

Waeber, C., Sebben, M., Nieoullon, A., Bockaert, J., and Dumuis, A., 1994, Regional distribution and ontogeny of 5-HT4 binding sites in rodent brain, *Neuropharmacology* **33**: 527.

Wassef, M., Berod, A., and Sotelo, C., 1981, Dopaminergic dendrites in the pars reticulata of the rat substantia nigra and their striatal input. Combined immunocytochemical localization of tyrosine hydroxylase and anterograde degeneration, *Neuroscience* **6**: 2125.

Westerink, B. H., Santiago, M., and Vries, J. B., 1992, The release of dopamine from nerve terminals and dendrites of nigrostriatal neurons induced by excitatory amino acids in the conscious rat, *Naunyn Schmiedeberg's Arch. Pharmacol.* **345**: 523.

Wichmann, T., and DeLong, M. R., 1996, Functional and pathophysiological models of the basal ganglia, *Curr. Opin. Neurobiol.* **6**: 751.

Wightman, R. M., Jankowski, J. A., Kennedy, R. T., Kawagoe, K. T., Schroeder, T. J., Leszczyszyn, D. L., Near, J. A., Diliberto, E. J., Jr., and Viveros, O. H., 1991, Temporally resolved catecholamine spikes correspond to single vesicle release from individual chromaffin cells, *PNAS USA* **88**: 10754.

Wilson, C. J., Groves, P. M., and Fifkova, E., 1977, Monoaminergic synapses, including dendro-dendritic synapses in the rat substantia nigra, *Exp. Brain Res.* **30**: 161.

Yamada, T., McGeer, P. L., Baimbridge, K. G., and McGeer, E. G., 1990, Relative sparing in Parkinson's disease of substantia nigra dopamine neurons containing calbindin-D28k, *Brain Res* **526**: 303.

Yung, K. K. L., Bolam, J. P., Smith, A. D., Hersch, S. M., Ciliax, B. J., and Levey, A. I., 1995, Immunocytochemical localization of D1 and D2 dopamine receptors in the basal ganglia of the rat: light and electron microscopy, *Neuroscience* **65**: 709.

6

DOPAMINE RELEASE IN SUBSTANTIA NIGRA: RELEASE MECHANISMS AND PHYSIOLOGICAL FUNCTION IN MOTOR CONTROL

Filip Bergquist and Hans Nissbrandt[*]

1. INTRODUCTION

The anatomic organisation of the nigro-striatal dopaminergic projection neurones provides a biological system where terminal release sites are distinctly separated from the somatodendritic region. Together with the fact that dopaminergic neurones are not dispersed throughout the central nervous system, this makes the nigro-striatal pathway well suited for comparisons of terminal and somatodendritic release mechanisms.

Moreover, the importance of the loss of nigro-striatal dopaminergic neurones for the development of Parkinson's disease draws interest to a possible physiological relevance of somatodendritic dopamine release. In terms of basal ganglia anatomy, an important target area of nigral somatodendritic dopamine release, the substantia nigra pars reticulata, holds a privileged position where much of the output from the basal ganglia converges. Furthermore, nigral dopamine released in substantia nigra pars compacta can influence the activity of the dopaminergic nigrostriatal pathway via activation of autoreceptors (Aghajanian and Bunney, 1973; Aghajanian and Bunney, 1977; Santiago and Westerink, 1991b). For these reasons, it is possible that at least some symptoms of Parkinson's disease relate in part to the loss of somatodendritic dopamine release in substantia nigra. In line with this, there are several indications that dopamine in substantia nigra pars reticulata can affect motor behaviour and motor performance in rats as well as in primates. Until recently, however, this assumption has been based on indirect evidence (Crocker, 1997; Gerhardt et al., 2002) or on data from non-physiological conditions (Yurek and Hipkens, 1993, 1994; Timmerman and Abercrombie, 1996).

---

[*]Institute of Physiology and Pharmacology, Department of Pharmacology, Göteborg University, Box 431, SE405 30 Göteborg, Sweden.

Dendritic Neurotransmitter Release, edited by M. Ludwig
Springer Science+Business Media, Inc., 2005

2. RELEASE MECHANISMS

When somatodendritic dopamine release was first suggested, the idea was based on the finding that reserpine depleted dopamine from soma and dendrites in substantia nigra (Björklund and Lindvall, 1975). This similarity with the previously described reserpine sensitive dopamine storage in terminals was a strong argument for the existence of somatodendritic dopamine release. Studies, which demonstrated that somatodendritic dopamine release had other characteristics in common with classical neurotransmitter release, soon followed. For example, there were early reports on calcium dependency (Cuello and Iversen, 1978; Westerink and De Vries, 1988; Kalivas and Duffy, 1991; Westerink et al., 1994) and tetrodotoxin sensitivity (Kalivas and Duffy, 1991; Westerink et al., 1994) as well as reserpine sensitivity (Heeringa and Abercrombie, 1995) of dopamine release in nigral preparations.

However, it also became evident that under some experimental conditions, calcium dependency or tetrodotoxin sensitivity is less apparent or even absent (for example Nieoullon et al., 1977a; Chéramy et al., 1981; Elverfors et al., 1997; Hoffman and Gerhardt, 1999; Chen and Rice, 2001). Different explanations for these discrepancies have been put forward, most notably the suggestion that dopamine can be released from dendrites by reversal of the dopamine carrier (Adam-Vizi, 1992; Elverfors et al., 1997; Falkenburger et al., 2001) rather than by conventional exocytosis.

Whereas the mechanisms of release by exocytosis from nerve terminals have been unravelled in detail in the last decade, the understanding of somatodendritic dopamine release mechanisms is incomplete. Consequently there is an ongoing debate about the release mechanism of somatodendritic dopamine. Recent findings, which advocate either release by exocytosis, or release mechanisms independent of exocytosis, will be discussed in the first part of this chapter.

2.1. Complex and Delicate Structures Means Complex and Delicate Problems

A common problem with studies of somatodendritic release in contrast to terminal release is the close proximity to the neuron somata, both anatomically and functionally. Somata can be expected to contain a more complicated set-up of ion channels than terminals, and they compromise the firing pattern generator of the neuron. Because the firing patterns are generated both by intrinsic ion channel activity and by afferent synaptic signals in the entire somatodendritic field, any local intervention that changes intrinsic ion transport and/or afferent synaptic activity, can also affect the firing pattern of the studied neuron. If firing patterns are important, not only for terminal, but also for somatodendritic release, the influence of dendritic activity on somatic firing introduces a factor of uncertainty in almost every study of mechanisms of somatodendritic release. If for example extracellular calcium is depleted, does a change in release depend on inhibition of the dendritic release machinery, or on a change in firing pattern or dendritic post-synaptic activity secondary to decreased synaptic input? To fully resolve these questions it would be necessary to be able to combine electrophysiological clamping of dendrites and techniques that permit release detection from single dendrites. Closer investigation of the relationship between back-propagating action potentials and somatodendritic release are also warranted.

In spite of this general complication, it is feasible to address some of the fundamental questions about dendritic release mechanisms with available methods, albeit not as unambiguously as in the case of terminal release, where the relationship between action potentials and release is strong, and the electrical activity of the structure is more easily controlled.

2.2. The Calcium Dependency of Somatodendritic Dopamine Release

Recent studies have reported that somatodendritic dopamine release is less dependent on extracellular calcium than is terminal dopamine release (Bergquist et al., 1998; Chen and Rice, 2001). The first study found that 30% of spontaneous nigral dopamine release in alert rats remained during calcium depleted perfusion, and the second study showed that approximately 40% of electrically evoked nigral release remained in calcium depleted medium. There are however also some conflicting results. In one study using KCl to evoke somatodendritic dopamine release from rat slices (Hoffman and Gerhardt, 1999), the evoked nigral dopamine release was calcium independent, whereas a similar experiment with KCl-evoked release from guinea pig reported complete calcium dependency (Elverfors et al., 1997). In the same preparation, however, veratridine-induced dopamine release persisted in calcium-depleted superfusion. Another study, that stimulated nigral dopamine release by inducing endogenous glutamate release, also reported calcium independent release (Falkenburger et al., 2001). There is no obvious reason for these contradictory results on the calcium dependency of somatodendritic dopamine release, but a possible explanation is that there is more than one mechanism of release, and that the different conditions used in the various studies favours mechanisms with more or less calcium dependency. A less likely explanation is that there may be different release mechanisms in different species (see also Nieoullon et al., 1977a). The overall impression is nevertheless that nigral dopamine release is calcium dependent under most conditions (Kalivas and Duffy, 1991; Robertson et al., 1991; Santiago and Westerink, 1991a; Westerink et al., 1994; Elverfors et al., 1997), but often to a lesser extent than striatal dopamine release (Rice et al., 1994; Rice et al., 1997; Bergquist et al., 1998; Chen and Rice, 2001).

The difference in calcium dependency between terminal and somatodendritic dopamine release can be explained if somatodendritic dopamine release is controlled to some extent by mobilisation of intracellular calcium stores, similar to what has been described for dendritic release of oxytocin (Ludwig et al., 2002). However, this possibility remains to be explored.

Like other terminal neurotransmitter release in CNS, striatal dopamine release is sensitive to inhibition of P/Q-type and N-type voltage sensitive calcium channels (VSCC) (Dooley et al., 1987; Herdon and Nahorski, 1989; Kato et al., 1992; Bergquist et al., 1998; Okada et al., 1998; Haubrich et al., 2000; El Ayadi et al., 2001), so there is reason to believe that these channels are capable of regulating release in dopaminergic neurons. The presence of N-type, P/Q-type, L-type, T-type and R-type calcium currents have been described in dopaminergic neurones in mesencephalon (Cardozo and Bean, 1995), and any of these could be important for the regulation of nigral dopamine release.

Some attempts have been made to identify the calcium channels that mediate the calcium dependent portion of nigral dopamine release (Elverfors et al., 1997; Bergquist et al., 1998; Bergquist and Nissbrandt, 2003). The results of these studies are summarized in Table 1.

In short, the findings are that, in guinea-pig slices, KCl-evoked dopamine release is blocked by neomycin or ω-agatoxin IVA, whereas in rat substantia nigra, spontaneous dopamine release *in vivo* is blocked by neomycin, all concentrations of SNX-482, or high concentrations of mibefradil, but not by ω-agatoxin IVA. The L-type blockers nifedipine and nimodipine were ineffective in both cases, whereas ω-conotoxin GVIA increased dopamine release *in vivo*. The most likely explanation to these findings is that neither L-, N-, T-, nor P/Q-type VSCC are essential for somatodendritic dopamine release *in vivo* in the rat, but that P/Q-type VSCC are important for KCl-evoked release in guinea pig. The effects of SNX 482 and mibefradil indicate a role for $Ca_v2.3$, a R-type VSCC, in spontaneous somatodendritic dopamine release *in vivo*.

The experiments in Table 1 involve the application of drugs to a tissue that contains several mutually dependent cellular elements, which can all be affected by the drugs. Is it then possible to know whether the observed effect is mediated directly in the dendrites or via indirect action on afferent terminals? Theoretically, the answer is yes, but it assumes not only that we know all afferent transmitters, but also that they can be measured, or that the effects of them can be eliminated in some way. Because this cannot be readily done, conclusions must be made with some caution. The results this far indicate that $Ca_v2.3$ (R-type VSCC) is important for the observed calcium dependency of somatodendritic dopamine release *in vivo*, but the issue of whether the calcium dependency is directly mediated by dendritic calcium channels or not, remains to be settled.

The effects of selective inhibition of different VSCC raise another question. For some time it has been known that in particullar L-type VSCC, or at least VSCC sensitive to dihydropyridines, have an important influence on the regulation of firing rate, and firing pattern in dopaminergic neurones (Nedergaard et al., 1993; Mercuri et al., 1994; Durante et al., 2004; see however Fujimura and Matsuda, 1989; Kang and Kitai, 1993; Wolfart

Table 1: The response of striatal and nigral extracellular dopamine to different VSCC blockers

	VSCC-type blocked					Response in extracellular DA		
	L	P/Q	N	R	T	SN[a]	SN[b]	STR[b]
Nifedipine	X					+/-		
Nimodipine	X				?		+/-	+/-
ωATX		X				↓	↑	↓
ωCTX			X			+/-	↑	↓
ωATX + ωCTX		X	X				↑	
SNX L				X			↓	+/-
SNX H	X	X	X	X	?		↓	↓
Mibefradil L					X		↑	+/-
Mibefradil H	X	X	X	X	X		↓	↓
Nickel					X	+/-	↑	↑
Neomycin	X	X	X	?	?	↓	↓	↓

Blocked channel types are indicated by X. Constellations of channel block that lead to decreased extracellular dopamine in substantia nigra in any of the systems are shadowed, and resulting decreases are framed for clarity. Up and down-arrows indicate significant changes in extracellular dopamine concentrations and +/- indicates no significant change.

[a] KCl-evoked dopamine (DA) release from guinea-pig substantia nigra slices (Elverfors et al., 1997)

[b] Spontaneous dopamine release, microdialysis in alert rats (Bergquist et al., 1998, Bergquist and Nissbrandt 2003).

Abbreviations: ω-ATX ω-agatoxin IVA, ω-CTX ω-conotoxin GVIA, DA dopamine, H high concentration, L low concentration, SN substantia nigra, SNX SNX482, STR striatum, VSCC voltage sensitive calcium channels.

and Roeper, 2002). The observation that drugs that block L-type VSCC do not change somatodendritic dopamine release (Bergquist et al., 1998) could be important, because it suggests indirectly that there is no clear correlation between somatodendritic dopamine release and the generation of action potentials in the soma. This challenges the interpretation of tetrodotoxin sensitivity, which is often taken as a support for action potential regulated release. Perhaps the regulation of somatodendritic release should be regarded as a local phenomenon, driven by ion channel activity intrinsic to the dendrite and by afferent activity on a local scale. In that case, tetrodotoxin sensitivity would not necessarily be the result of decrease in action potentials in the cell body, but could emanate from inhibition of excitatory afferents or from the inhibition of dendritic depolarisation by sodium currents, or even by inhibition of local dendritic action potentials (Hausser et al., 1995) and not necessarily from a decrease in action potentials in the cell body.

2.3. Is Somatodendritic Dopamine Released by Exocytosis or by Carrier Reversal?

Early histological studies of false transmitter (5-hydroxydopamine) storage in substantia nigra indicated that the main storage site for dopamine in dendrites is not vesicular structures, as in terminals, but a compartment of the smooth endoplasmatic reticulum (Wilson et al., 1977; Hattori et al., 1979; Mercer et al., 1979; Groves and Linder, 1983). Later, those findings were confirmed by an immunohistochemical study of the distribution of VMAT2, a vesicular monoamine transporter (Nirenberg et al., 1996). Due to this difference, it cannot be presupposed that terminal and dendritic release use the same release mechanisms, and there is indeed a plethora of evidence for important differences.

In recent years, studies supporting either exocytosis or release by carrier reversal have been published. In 1998 Jaffe and co-workers (Jaffe et al., 1998) described quantal dopamine release from somata of dopamine neurones in substantia nigra, indicating release by exocytosis, but in 2001 Mintz's group found that dendrodendritic autoinhibition induced by endogenous glutamate release in substantia nigra involves dopamine released by reversal of the dopamine carrier/transporter DAT (Falkenburger et al., 2001). Interestingly the DAT can have an excitatory influence on dopaminergic neurones via anion currents (Ingram et al., 2002) so theoretically it is possible that the decrease in dopamine release observed by Falkenburger and co-workers after DAT inhibition is due to loss of DAT-mediated excitation, rather than to inhibited release via DAT. In the study by Elverfors et al (1997), veratridine-induced nigral dopamine release was abolished by DAT-inhibition, indicating that this stimulus leads to release that depends selectively on the DAT. However, in many other preparations DAT-inhibition leads to increased dopamine concentrations in line with the conventional view of the DAT as mainly mediating re-uptake (Santiago and Westerink, 1992; Cragg et al., 1997; Chen and Rice, 2001).

If somatodendritic dopamine is released by exocytosis, it can be expected to be inhibited by SNARE-cleaving toxins. In one study, the acute effect of nigral or striatal injections with clostridial toxins was determined (Bergquist et al., 2002). It was found that both spontaneous and KCl-evoked striatal and nigral dopamine release is highly sensitive to local injections of botulinum toxin A, which cleaves the membrane associated SNARE-protein SNAP-25. However, injections with botulinum toxin B, which cleaves the vesicle associated SNARE-proteins VAMP2 and cellubrevin only blocked dopamine release in the striatum. These findings illustrate another fundamental difference between terminal and somatodendritic dopamine release, and it can be speculated that the insensitivity of nigral dopamine release to botulinum toxin B indicates that another VAMP-isoform, for example VAMP1 or TI-VAMP, is involved in the process of somatodendritic dopamine release.

It should be remembered that although these findings lend support to the exocytosis hypothesis, it could be expected that an effect of botulinum toxin A injected in substantia nigra is to block the release from excitatory afferents. Thus, the inhibitory effect on nigral

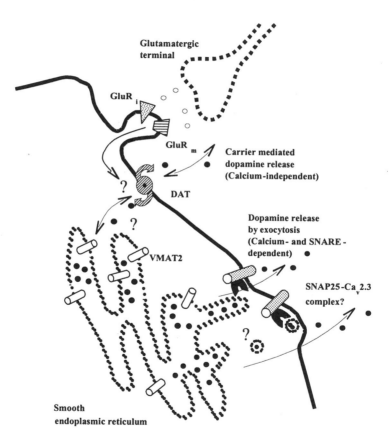

Figure 1. Suggested mechanisms of somatodendritic dopamine release.
Filled circles indicate dopamine, and open glutamate. It is unclear whether exocytosis and carrier reversal exist and work in parallel, but there is evidence for both of them in different preparations. A mechanism by which dopamine can be shuttled from storage compartment to carrier mediated release remains to be established. Although exocytotic release has been suggested to occur from tubulovesicular structures, it is also possible that dopamine is released from regular vesicles budding off from the storage compartment, SER. The existence of a SNAP25-Ca_v2.3 complex in dopaminergic dendrites is a speculative suggestion.
Abbreviations: $GluR_i$ ionotropic glutamate receptor, $GluR_m$ metabotropic glutamate receptor, DAT, dopamine transporter, VMAT2, vesicular monoamine transporter type 2

release could be secondary. Observations contrary to this explanation are that inhibition of the best described excitatory pathway to substantia nigra with AMPA- and NMDA-receptor antagonists does not inhibit nigral dopamine release (Chen and Rice, 2002; Cobb and Abercrombie, 2002), that the inhibitory effect of botulinum toxin A is so pronounced that it is only surmounted by the effects of tetrodotoxin, and that the inhibitory effect also remains with a strong depolarising stimulus (Bergquist et al., 2002).

Nevertheless, the support for carrier-mediated release under physiological conditions is strong (Falkenburger et al., 2001), and possibly both release mechanisms exist in parallel as suggested in Fig. 1.

3. THE PHYSIOLOGICAL ROLE OF SOMATODENDRITIC DOPAMINE RELEASE IN MOTOR CONTROL

As indicated in the introduction of this chapter, a reason for the interest in somatodendritic dopamine release is the possibility of finding new remedies for Parkinson's disease, and exploring the aetiology of this second most common neurodegenerative disorder.

The clinical characteristics of Parkinson's disease can be diverse, but always include at least two of the following symptoms: hypokinesia/bradykinesia, tremor, rigidity and postural imbalance with a semiflected stance. With progression of the disease, there is also a risk of autonomic failure and an increased risk of depression and dementia, symptoms that develop as complications to a more generalised neurodegeneration.

The classical Parkinsonian symptoms can be related to an imbalance in basal ganglia activity, mainly due to the loss of dopaminergic neurones in substantia nigra. Since the early findings by Ehringer and Hornykiewicz (1960), the focus of interest has been on the resulting loss of terminal dopamine release in striatum (caudate nucleus and putamen). If we consider present models of basal ganglia circuitry (Parent and Hazrati, 1995), somatodendritic dopamine transmission may also be important because it takes place in substantia nigra pars reticulata, one of the two major output nuclei of the basal ganglia. In spite of this anatomical arrangement, it is not evident that somatodendritic dopamine release is extensive enough to have physiological effects. Most measurements of nigral dopamine release indicate that extracellular dopamine concentrations are not more than 1/10th of what is found in the striatum (Robertson et al., 1991; Westerink et al., 1992) (Rice et al., 1997). This was recently verified with the "no-net-flux" method (Bergquist et al., 2002). With that technique, the extracellular dopamine concentration was determined to be 2.6 nM in substantia nigra, and 25 nM in the striatum. The concentration in substantia nigra is sufficient to influence dopaminergic receptors (see for example Kebabian et al., 1997), and because all the measurements cited above average the dopamine concentration over space and time, considerably higher concentrations can be expected near the release site (Cragg et al., 2001).

Thus, in spite of considerably lower average extracellular dopamine concentrations in substantia nigra than in striatum, it is not surprising that several experiments indicate an influence of somatodendritic dopamine release on motor control. The problem however, with many of these studies is that the conditions are non-physiological. Studies of turning behaviour in rats after unilateral 6-OHDA lesions (LaHoste and Marshall, 1990; Robertson, 1992; Yurek and Hipkens, 1993, 1994) showed that inhibition of D1-receptors in substantia nigra can modulate rotational behaviour following agonist treatment. This model is analogous to treated (or over treated) Parkinson patients, and indicates a possible role of nigral dopamine receptors in dyskinetic conditions, but does not really tell us if physiologically released nigral dopamine influences motor control. This is also the case with a number of studies reporting rotational motor activity in response to nigral application of dopamine agonists in intact rats (Jackson and Kelly, 1984; Starr and Starr, 1989; Trevitt et al., 2002). There is however also evidence for effects of endogenously released nigral dopamine on motor behaviour (Jackson and Kelly, 1983a, b; Timmerman and Abercrombie, 1996). The latter of these papers showed that local infusion with amphetamine into substantia nigra caused a short-lasting increase in motor activity, which c ould b e b locked w ith a D 1-antagonist. H owever, i t i s likely t hat a mphetamine infusion causes a much more pronounced dopamine release than what could be expected to occur physiologically. Similar methods were used in earlier studies (Jackson and Kelly, 1983a,b) where nigral or systemic amphetamine treatment was combined with nigral application of dopamine antagonists. This means that although these studies clearly

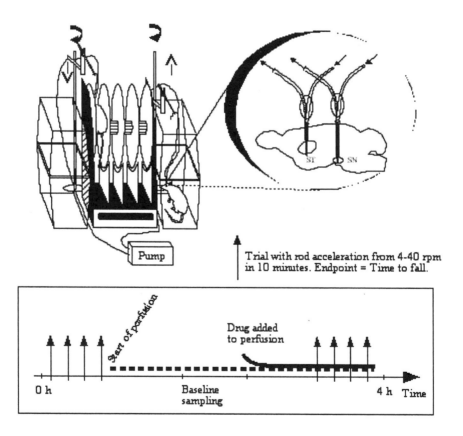

Figure 2. Comprehensive illustration of arrangement with Rotarod equipment and microdialysis cages. Left rat is performing the accelerating rod task and right rat is resting. Rats were tested in four trials before perfusion started, and then in four trials during continuous treatment via the nigral microdialysis probe (schematic diagram). After the first hour of perfusion, samples from both probes were collected and analysed every 10 minutes.

show a potential role of nigral dopamine release for motor control, they do not tell us if this occurs physiologically.

Interestingly, it seems that some of the symptoms of dopaminergic degeneration in animal models can be alleviated by transplantation of mesencephalic tissue into substantia nigra (Nikkhah et al., 1994; Yurek, 1997; Collier et al., 2002). Although this can be taken as further support for a motor control function of nigral dopamine release, it is possible that such effects relate to the introduction of other neuronal or glial elements, since only a minority of the cells in mesencephalic transplants are dopaminergic. In fact, similar or better effects are obtained with GABA-ergic transplants (Winkler et al., 1999).

The most compelling evidence for the involvement of somatodendritic dopamine in motor control comes from a number of studies with local application of dopamine antagonists in substantia nigra (Double and Crocker, 1995; Hemsley and Crocker, 1998, 2001; Trevitt et al., 2001; Trevitt et al., 2002; Bergquist et al., 2003). Trevitt and co-workers evaluated motor activity measured by lever-pressing or open field activity before and after injections of the D1-antagonist SCH 23390 into substantia nigra reticulata or other brain areas. They found that the most potent inhibition of lever-pressing developed after injection into substantia nigra reticulata, and that nigral D1-inhibition was equipotent to local D1-inhibition in striatum or nucleus accumbens in the open field activity test. The conclusions are hampered somewhat by the use of rather large doses SCH 23390 (0.25 µg) which makes the D1-selectivity uncertain. In the later study, however, they found that a D1-agonist (SKF 82958) injected into substantia nigra reticulata induced an increase in nigral GABA-release together with increased locomotor activity, and that these effects could be prevented with SCH 23390.

Consistent with this, EMG-measurements of muscle tone in alert rats show that inactivation of nigral dopamine receptors with the irreversible antagonist EEDQ (N-ethoxycarbonyl-2-ethoxy-1,2-dihydroquinoline) induces an increase in muscle tone in rat (Double and Crocker, 1995; Hemsley and Crocker, 1998). Increased muscle tone will furthermore follow from inhibition of either D1 or D2 receptors in striatum or from D1-receptor inhibition in substantia nigra (Hemsley and Crocker, 2001). This obviously has a potential relevance for motor control, but some uncertainty remains as to whether the increase in muscle tone is enough to affect the relevant physiological endpoint, motor control.

D1-receptor mediated effects of nigral dopamine release make sense for several reasons. Histological evidence show that substantia nigra reticulata is one of the brain regions with the greatest density of D1, or D1-like receptors (Dubois et al., 1986; Filloux et al., 1987; Wamsley et al., 1989), and numerous investigators have found that activation of D1-receptors can presynaptically modulate the release of GABA from striato-nigral afferents (Starr, 1987; Timmerman and Westerink, 1995; Trevitt et al., 2002) as well as postsynaptically modulate the activity of nigrothalamic GABA-ergic output neurones (Ruffieux and Schultz, 1980; Matthews and German, 1986; Waszczak, 1990).

To evaluate the role of somatodendritic dopamine release during physiological conditions, a method was developed that combines real time microdialysis measurements with a commonly used motor performance task, the accelerating rod test (Bergquist et al., 2003). This method has the advantage of providing a relatively sustained spontaneous motor activity on the spot, as well as simultaneous quantitative measures of performance and extracellular neurotransmitter concentrations in one or more brain areas (Fig. 2).

Microdialysis was performed both in substantia nigra and the ipsilateral striatum to find out if the animals' activity on the rod was associated with changes in extracellular dopamine concentrations, indicating a change in terminal and somatodendritic release. To determine if physiological dopamine release in substantia nigra has any influence on motor functions, dopamine antagonists were applied in the nigral probe by reverse dialysis. In another experiment, apomorphine was applied by reverse dialysis in the substantia nigra of 6-OHDA hemilesioned rats in an attempt to restore nigral, but not striatal, dopaminergic activity.

When the animal was tested on the accelerating rod, an increase in nigral and striatal dopamine concentrations could be detected (Bergquist et al., 2003). The increase in striatum was in absolute terms larger than in substantia nigra, but the percentage increase in substantia nigra was higher and reached approximately 30%. The range of this increase is almost identical to what Nieoullon and co-workers observed when stimulating sensory afferents during push-pull recording of nigral dopamine concentrations in cat (Nieoullon et al., 1977b), and it is therefore likely that physiological variations in average extracellular dopamine concentrations in substantia nigra are of approximately this magnitude. Similar to the findings by Trevitt and co-workers (Trevitt et al., 2001) we found that SCH 23390 dose-dependently decreased the time on rod, indicating decreased motor performance when D1-receptors were blocked (Fig. 3). However, it was also clear that SCH 23390 induced pronounced increases in nigral, but not striatal dopamine and 5-HT. Like the study by Trevitt (2001), the interpretation is obstructed to some extent by the use of high drug concentrations at expense of drug selectivity. Explicitly, it is likely that SCH 23390 at the concentrations used is no longer D1-selective. The increase in nigral dopamine could also indicate that the worsened performance is not only an antagonist effect.

An unexpected discovery was made when the D2-antagonist raclopride was added to the nigral probe. It turned out that very low concentrations potently impaired rod performance, but that this effect was partially reversed when the drug concentration was increased (Fig. 3). With increasing raclopride concentrations, we also observed increased striatal dopamine concentrations, presumably due to inhibition of nigral D2 auto-

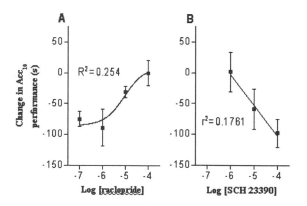

Figure 3. Dose-response of local perfusion with dopamine antagonists in substantia nigra. Outcome is measured as change in time on rod during an acceleration test. Concentrations are inlet concentrations, and highest tissue concentration can be expected to be lower, possibly by a factor of 10 (From Bergquist et al., 2003).

receptors. A possible explanation to the restored rod performance is thus that the increase in striatal dopamine concentrations compensates the loss of transmission in substantia nigra. Because the motor impairments observed with very low raclopride concentrations occurred without any recognisable alterations of striatal dopamine transmission, it is likely that the impairments are mediated by local effects in substantia nigra, and probably via inhibition of raclopride sensitive receptors other than D2 autoreceptors.

An attempt to substitute for a loss of somatodendritic dopamine release was made in 6-OHDA hemilesioned rats that developed a 30% or more decrease in rod performance after lesion. In these rats, acute nigral treatment with apomorphine resulted in a dose-dependent improvement in rod performance from $54\pm6\%$ to $84\pm8\%$ of pre-lesion performance, indicating that restoring nigral dopaminergic tone is sufficient to restore some of the symptoms of dopamine depletion. In some of the animals, it was also evident that the post-lesion asymmetric movement pattern was partly normalised when the highest apomorphine concentration was used (Bergquist et al., 2003).

Altogether, there is plenty of evidence that somatodendritic dopamine release has an important influence on motor control in rats. The question remains, can these results be replicated in other species? Models of Parkinson's disease in rodents are valuable tools in research, but they are not equivalent to the disease, and there is therefore a need to verify the results from animal models in clinical settings.

Although it would not be possible to repeat the experiments described here in human patients, there is a study that supports a motor control mechanism of somatodendritic dopamine release in ageing rhesus monkeys (Gerhardt et al., 2002). This study reveals interesting age-related decreases in baseline dopamine metabolites, and in amphetamine evoked dopamine release in dialysates from substantia nigra, whereas these indices were not significantly changed with age in striatum. In parallel with these changes, movement dysfunction ratings indicated impairments in motor abilities with increasing age. It should however, be noted that the observations by Gerhardt and co-workers do not establish a causal relationship between dendritic dopamine release and motor functions, only co-variation. Also, since there was a similar trend towards decreased dopamine indices in striatum, an alternative explanation to these findings could be that age-dependent changes in striatum are buffered more than the changes in substantia nigra because of the higher density of release sites.

The experimental trials with transplantation of mesencephalic tissue in Parkinson patients have this far concentrated on restoring striatal dopamine transmission. Due to the evidence for nigral somatodendritic dopamine transmission and the far from satisfactory outcome of striatal implants, the alternative of dual target transplantation is now being tested in patients (Mendez et al., 2002) with the hope that a more complete restoration of basal ganglia circuits will lead to improved clinical outcome. To our knowledge, this is the first example of bringing the physiological importance of dendritic release into clinical practice.

4. FUTURE PROSPECTS

4.1. Release Mechanisms

As suggested in the first part of this chapter, several issues remain to be settled regarding the mechanisms of somatodendritic dopamine release. How can the variation in calcium dependency be explained? Is intracellular calcium mobilization involved in any way? Does release occur by carrier reversal or by exocytosis, or are there several release mechanisms operating in parallel? There is also the issue of the co-release of other transmitters, in the case of dopaminergic neurones especially the issue of glutamate co-release (Sulzer and Rayport, 2000). Does this occur also in dopaminergic dendrites?

Some of these questions will have to be addressed with new techniques. Recent advances in evanescent field microscopy have made it possible to visualize exocytosis on a single vesicle level (Oheim, 2001). Possibly this could also be applied in research on dendritic release. The development of specific electrodes is also a promising prospect, which may allow the monitoring of neurotransmitters with a selectivity approaching that of microdialysis combined with the high time and space resolution of voltammetry.

Histological studies of the distribution of SNARE-proteins could advance the knowledge on dendritic release considerably, if they can be done in sufficient detail.

A more direct way to observe carrier mediated release would be fascinating, because the existence of such a mechanism could herald a more general change in the way we look on synaptic transmission, and on the pharmacodynamic action of some drugs interacting with uptake processes.

4.2. Somatodendritic Dopamine Release and Motor Control

The idea that nigral dopamine release has a role in motor control is now accepted by many neuroscientists, but to which degree it influences the clinical picture in Parkinson's disease is still unknown, and it has, in fact, been difficult to compare the relative importance of striatal and nigral dopamine release even in animal models. There is further work to be done in this area, and it would be most interesting to determine its' role in common movement disorders. For example, do nigral dopamine receptors contribute to dyskinesia? The developments in genetic techniques have established a variety of different dopamine receptors, which makes the older division in D1-like and D2-like receptors very crude. Pharmacological development is now beginning to catch up developing more selective substances, and hopefully some of them will turn out to be useful for exploring the role of somatodendritic dopamine release in substantia nigra in detail, as well as for finding new strategies to treat movement disorders.

5. REFERENCES

Adam-Vizi, V., 1992, External Ca^{2+}-independent release of neurotransmitters, *J. Neurochem.* **58:** 395.
Aghajanian, G., and Bunney, B., 1973, Central dopaminergic neurons: neuro-physiological identification and responses to drugs, in: *Frontiers in catecholamine research,* E. Usdin, S. Snyder, eds., Pergamon Press, New York, pp 643-648.
Aghajanian, G. K., and Bunney, B. S., 1977, Dopamine"autoreceptors": pharmacological characterization by microiontophoretic single cell recording studies, *Naunyn Schmiedebergs Arch. Pharmacol.* **297:** 17.
Bergquist, F., and Nissbrandt, H., 2003, Influence of R-type (Cav2.3) and T-type (Cav3.1-3.3) antagonists on nigral somatodendritic dopamine release measured by microdialysis, *Neuroscience* **120:** 757.
Bergquist, F., Niazi, H. S., and Nissbrandt, H., 2002, Evidence for different exocytosis pathways in dendritic and terminal dopamine release *in vivo, Brain Res.* **950:** 245.
Bergquist, F., Shahabi, H. N., and Nissbrandt, H., 2003, Somatodendritic dopamine release in rat substantia nigra influences motor performance on the accelerating rod, *Brain Res.* **973:** 81.
Bergquist, F., Jonason, J., Pileblad, E., and Nissbrandt, H., 1998, Effects of local administration of L-, N-, and P/Q-Type Calcium Channel Blockers on spontaneous dopamine release in the striatum and the substantia nigra: A microdialysis study in rat, *J. Neurochem.* **70:** 1532.
Björklund, A., and Lindvall, O., 1975, Dopamine in dendrites of substantia nigra neurons: suggestions for a role in dendritic terminals, *Brain Res.* **83:** 531.
Cardozo, D. L., and Bean, B. P., 1995, Voltage-dependent calcium channels in rat midbrain dopamine neurons: modulation by dopamine and $GABA_B$ receptors, *J. Neurophysiol.* **74:** 1137.

Chen, B. T., and Rice, M. E., 2001, Novel Ca^{2+} dependence and time course of somatodendritic dopamine release: substantia nigra versus striatum, *J. Neurosci.* **21**: 7841.

Chen, B. T., and Rice, M. E., 2002, Synaptic regulation of somatodendritic dopamine release by glutamate and GABA differs between substantia nigra and ventral tegmental area, *J. Neurochem.* **81**: 158.

Chéramy, A., Leviel, V., and Glowinski, J., 1981, Dendritic release of dopamine in the substantia nigra, *Nature* **289**: 537.

Cobb, W. S., and Abercrombie, E. D., 2002, Distinct roles for nigral GABA and glutamate receptors in the regulation of dendritic dopamine release under normal conditions and in response to systemic haloperidol, *J. Neurosci.* **22**: 1407.

Collier, T. J, Sortwell, C. E., Elsworth, J. D., Taylor, J. R., Roth, R. H., Sladek, J. R., and Redmond, D.E., 2002, Embryonic ventral mesencephalic grafts to the substantia nigra of MPTP-treated monkeys: Feasibility relevant to multiple-target grafting as a therapy for Parkinson's disease, *J. Comp. Neurol.* **442**: 320.

Cragg, S. J., Nicholson, C., Kume-Kick, J., Tao, L., and Rice, M. E., 2001, Dopamine-mediated volume transmission in midbrain is regulated by distinct extracellular geometry and uptake, *J. Neurophysiol.* **85**: 1761.

Cragg, S. J., Rice, M. E., and Greenfield, S. A., 1997, Heterogeneity of electrically evoked dopamine release and reuptake in substantia nigra, ventral tegmental area, and striatum, *J. Neurophysiol.* **77**: 863.

Crocker, A. D., 1997, The regulation of motor control: an evaluation of the role of dopamine receptors in the substantia nigra, *Rev. Neurosci.* **8**: 55.

Cuello, A. C., and Iversen, L. L., 1978, *Interactions of Dopamine with other Neurotransmitters in the Rat Substantia Nigra: A Possible Functional Role of Dendritic Dopamine*, Raven Press, New York, pp 127-149.

Dooley, D. J., Lupp, A., and Hertting, G., 1987, Inhibition of central neurotransmitter release by omega-conotoxin GVIA, a peptide modulator of the N-type voltage-sensitive calcium channel, *Naunyn-Schmiedebergs Arch. Pharmacol.* **336**: 467-470.

Double, K. L, and Crocker, A. D., 1995, Dopamine receptors in the substantia nigra are involved in the regulation of muscle tone, *PNAS USA* **92**: 1669.

Dubois, A., Savasta, M., Curet, O., and Scatton, B., 1986, Autoradiographic distribution of the D1 agonist [3H]SKF 38393, in the rat brain and spinal cord. Comparison with the distribution of D2 dopamine receptors, *Neuroscience* **19**: 125.

Durante, P., Cardenas, C. G., Whittaker, J. A., Kitai, S. T., and Scroggs, R. S., 2004, Low-threshold L-type calcium channels in rat dopamine neurons, *J. Neurophysiol.* **91**: 1450.

Ehringer, H., and Hornykiewicz, O., 1960, Verteilung von Noradrenalin und Dopamin (3-Hydroxytyramin) im Gehirn des Menschen und Ihr Verhalten bei Erkrankungen des extrapyramidalen Systems, *Klinische Wochenschrift* **38**: 1236.

El Ayadi, A., Afailal, I., and Errami, M., 2001, Effects of voltage-sensitive calcium channel blockers on extracellular dopamine levels in rat striatum, *Metab. Brain Disease* **16**: 121.

Elverfors, A., Jonason, J., Jonason, G., and Nissbrandt, H., 1997, Effects of drugs interfering with sodium channels and calcium channels, *Synapse* **26**: 359.

Falkenburger, B. H., Barstow, K. L., and Mintz, I. M., 2001, Dendrodendritic inhibition through reversal of dopamine transport, *Science* **293**: 2465.

Filloux, F. M., Wamsley, J. K., and Dawson, T. M., 1987, Presynaptic and postsynaptic D1 dopamine receptors in the nigrostriatal system of the rat brain: a quantitative autoradiographic study using the selective D1 antagonist [3H]SCH 23390, *Brain Res.* **408**: 205.

Fujimura, K., and Matsuda, Y., 1989, Autogenous oscillatory potentials in neurons of the guinea pig substantia nigra pars compacta *in vitro*, *Neurosci. Lett.* **104**: 53.

Gerhardt, G. A., Cass, W. A., Yi, A., Zhang, Z., and Gash, D. M., 2002, Changes in somatodendritic but not terminal dopamine regulation in aged rhesus monkeys, *J. Neurochem.* **80**: 168.

Groves, P. M., and Linder, J. C., 1983, Dendro-dendritic synapses in substantia nigra: descriptions based on analysis of serial sections, *Exp. Brain Res.* **49**: 209.

Hattori, T., McGeer, P. L., and McGeer, E. G., 1979, Dendro-axonic neurotransmission. II. Morphological sites for the synthesis, binding and release of neurotransmitters in dopaminergic dendrites in the substantia nigra and cholinergic dendrites in the neostriatum, *Brain Res.* **170**: 71.

Haubrich, C., Frielingsdorf, V., Herzig, S., Schroder, H., Schwarting, R., Sturm, V., and Voges, J., 2000, N-type calcium channel blockers - tools for modulation of cerebral functional units?, *Brain Res.* **855**: 225.

Hausser, M., Stuart, G., Racca, C., Sakmann, B., 1995, Axonal initiation and active dendritic propagation of action potentials in substantia nigra neurons, *Neuron* **15**: 637.

Heeringa, M. J., and Abercrombie, E. D., 1995, Biochemistry of somatodendritic dopamine release in substantia nigra: an *in vivo* comparison with striatal dopamine release, *J. Neurochem.* **65**: 192.

Hemsley, K. M., and Crocker, A. D., 1998, The effects of an irreversible dopamine receptor antagonist, N-ethoxycarbonyl-2-ethoxy-1,2-dihydroquinoline (EEDQ), on the regulation of muscle tone in the rat: the role of the substantia nigra, *Neurosci. Lett.* **251**: 77.

Hemsley, K. M., and Crocker, A. D., 2001, Changes in muscle tone are regulated by D1 and D2 dopamine receptors in the ventral striatum and D1 receptors in the substantia nigra, *Neuropsychopharmacology* **25**: 514.

Herdon, H., and Nahorski, S. R., 1989, Investigations of the roles of dihydropyridine and omega-conotoxin-sensitive calcium channels in mediating depolarisation-evoked endogenous dopamine release from striatal slices, *Naunyn Schmiedebergs Arch. Pharmacol.* **340**: 36.

Hoffman, A. F., and Gerhardt, G. A., 1999, Differences in pharmacological properties of dopamine release between the substantia nigra and striatum: An *in vivo* electrochemical study, *J. Pharmacol. Exp. Therap.* **289**: 455.

Ingram, S. L., Prasad, B. M., and Amara, S. G., 2002, Dopamine transporter-mediated conductances increase excitability of midbrain dopamine neurons, *Nat. Neurosci.* **5**: 971.

Jackson, E. A., and Kelly, P. H., 1983a, Nigral dopaminergic mechanisms in drug-induced circling, *Brain Res. Bull.* **11**: 605.

Jackson, E. A., and Kelly, P. H., 1983b, Role of nigral dopamine in amphetamine-induced locomotor activity, *Brain Res.* **278**: 366.

Jackson, E. A., and Kelly, P. H., 1984, Effects of intranigral injections of dopamine agonists and antagonists, glycine, muscimol and N-methyl-D,L-aspartate on locomotor activity, *Brain Res. Bull.* **13**: 309.

Jaffe, E. H., Marty, A., Schulte, A., and Chow, R. H., 1998, Extrasynaptic vesicular transmitter release from the somata of substantia nigra neurons in rat midbrain slices, *J. Neurosci.* **18**: 3548.

Kalivas, P. W., and Duffy, P., 1991, A comparison of axonal and somatodendritic dopamine release using *in vivo* dialysis, *J. Neurochem.* **56**: 961.

Kang, Y., and Kitai, S. T., 1993, A whole cell patch-clamp study on the pacemaker potential in dopaminergic neurons of rat substantia nigra compacta, *Neurosci. Res.* **18**: 209.

Kato, T., Otsu, Y., Furune, Y., and Yamamoto, T., 1992, Different effects of L-, N- and T-type calcium channel blockers on striatal dopamine release measured by microdialysis in freely moving rats, *Neurochem. Internat.* **21**: 99.

Kebabian, J. W., Tarazi, F. I., Kula, N. S., and Baldessarini, R. J., 1997, Compounds selective for dopamine receptor subtypes, *Drug Discovery Today* **2**: 333.

LaHoste, G. J., and Marshall, J. F., 1990, Nigral D1 and striatal D2 receptors mediate the behavioral effects of dopamine agonists, *Behav. Brain Res.* **38**: 233.

Ludwig, M., Sabatier, N., Bull, P. M., Landgraf, R., Dayanithi, G., and Leng, G., 2002, Intracellular calcium stores regulate activity-dependent neuropeptide release from dendrites, *Nature* **418**: 85.

Matthews, R. T., and German, D. C., 1986, Evidence for a functional role of dopamine type-1 (D-1) receptors in the substantia nigra of rats, *Eur. J. Pharmacol.* **120**: 87.

Mendez, I., Dagher, A., Hong, M., Gaudet, P., Weerasinghe, S., McAlister, V., King, D., Desrosiers, J., Darvesh, S., Acorn, T., and Robertson, H., 2002, Simultaneous intrastriatal and intranigral fetal dopaminergic grafts in patients with Parkinson disease: a pilot study. Report of three cases, *J. Neurosurg.* **96**: 589.

Mercer, L., del Fiacco, M., and Cuello, A. C., 1979, The smooth endoplasmic reticulum as a possible storage site for dendritic dopamine in substantia nigra neurones, *Experientia* **35**: 101.

Mercuri, N. B., Bonci, A., Calabresi, P., Stratta, F., Stefani, A., and Bernardi, G., 1994, Effects of dihydropyridine calcium antagonists on rat midbrain dopaminergic neurones, *Br. J. Pharmacol.* **113**: 831.

Nedergaard, S., Flatman, J. A., and Engberg, I., 1993, Nifedipine- and omega-conotoxin-sensitive Ca^{2+} conductances in guinea-pig substantia nigra pars compacta neurones, *J. Physiol.* **466**: 727.

Nieoullon, A., Cheramy, A., and Glowinski, J., 1977a, Release of dopamine *in vivo* from cat substantia nigra, *Nature* **266**: 375.

Nieoullon, A., Cheramy, A., and Glowinski, J., 1977b, Nigral and striatal dopamine release under sensory stimuli, *Nature* **269**: 340.

Nikkhah, G., Bentlage, C., Cunningham, M. G., and Bjorklund, A., 1994, Intranigral fetal dopamine grafts induce behavioral compensation in the rat Parkinson model, *J. Neurosci.* **14**: 3449.

Nirenberg, M. J., Chan, J., Liu, Y., Edwards, R. H., and Pickel, V. M., 1996, Ultrastructural localization of the vesicular monoamine transporter-2 in midbrain dopaminergic neurons: potential sites for somatodendritic storage and release of dopamine, *J. Neurosci.* **16**: 4135.

Oheim, M., 2001, Imaging transmitter release. I. Peeking at the steps preceding membrane fusion, *Lasers Med. Sci.* **16**: 149.

Okada, M., Wada, K., Kiryu, K., Kawata, Y., Mizuno, K., Kondo, T., Tasaki, H., and Kaneko, S., 1998, Effects of Ca^{2+} channel antagonists on striatal dopamine and DOPA release, studied by *in vivo* microdialysis, *Br. J. Pharmacol.* **123**: 805.

Parent, A., and Hazrati, L. N., 1995, Functional anatomy of the basal ganglia. I. The cortico-basal ganglia-thalamo-cortical loop, *Brain Res. Rev.* **20**: 91.

Rice, M. E., Cragg, S. J., and Greenfield, S. A., 1997, Characteristics of electrically evoked somatodendritic dopamine release in substantia nigra and ventral tegmental area *in vitro*, *J. Neurophysiol.* **77**: 853.

Rice, M. E., Richards, C. D., Nedergaard, S., Hounsgaard, J., Nicholson, C., and Greenfield, S. A., 1994, Direct monitoring of dopamine and 5-HT release in substantia nigra and ventral tegmental area *in vitro*, *Exp. Brain Res.* **100**: 395.

Robertson, G. S., Damsma, G., and Fibiger, H. C., 1991, Characterization of dopamine release in the substantia nigra by *in vivo* microdialysis in freely moving rats, *J. Neurosci.* **11**: 2209.

Robertson, H. A., 1992, Dopamine receptor interactions: some implications for the treatment of Parkinson's disease, *Trends Neurosci.* **15**: 201.

Ruffieux, A., and Schultz, W., 1980, Dopaminergic activation of reticulata neurones in the substantia nigra, *Nature* **285**: 240.

Santiago, M., and Westerink, B. H., 1991a, Characterization and pharmacological responsiveness of dopamine release recorded by microdialysis in the substantia nigra of conscious rats, *J. Neurochem.* **57**: 738.

Santiago, M., and Westerink, B. H., 1991b, The regulation of dopamine release from nigrostriatal neurons in conscious rats: the role of somatodendritic autoreceptors, *Eur. J. Pharmacol.* **204**: 79.

Santiago, M., and Westerink, B. H., 1992, Simultaneous recording of the release of nigral and striatal dopamine in the awake rat, *Neurochem. Int.* **20**: S107.

Starr, M. S., 1987, Opposing roles of dopamine D1 and D2 receptors in nigral gamma- [3H]aminobutyric acid release?, *J. Neurochem.* **49**: 1042.

Starr, M. S., and Starr, B. S., 1989, Circling evoked by intranigral SKF 38393: a GABA-mediated D-1 response?, *Pharmacol. Biochem. Behav.* **32**: 849.

Sulzer, D., and Rayport, S., 2000, Dale's principle and glutamate corelease from ventral midbrain dopamine neurons, *Amino Acids* **19**: 45.

Timmerman, W., and Westerink, B. H., 1995, Extracellular gamma-aminobutyric acid in the substantia nigra reticulata measured by microdialysis in awake rats: effects of various stimulants, *Neurosci. Lett.* **197**: 21.

Timmerman, W., and Abercrombie, E. D., 1996, Amphetamine-induced release of dendritic dopamine in substantia nigra pars reticulata: D1-mediated behavioral and electrophysiological effects, *Synapse* **23**: 280.

Trevitt, J. T., Carlson, B. B., Nowend, K., and Salamone, J. D., 2001, Substantia nigra pars reticulata is a highly potent site of action for the behavioral effects of the D1 antagonist SCH 23390 in the rat, *Psychopharmacology* **156**: 32.

Trevitt, J. T., Carlson, B. B., Correa, M., Keene, A., Morales, M., and Salamone, J. D., 2002, Interactions between dopamine D1 receptors and gamma-aminobutyric acid mechanisms in substantia nigra pars reticulata of the rat: neurochemical and behavioral studies, *Psychopharmacology* **159**: 229.

Wamsley, J. K., Gehlert, D. R., Filloux, F. M., and Dawson, T. M., 1989, Comparison of the distribution of D-1 and D-2 dopamine receptors in the rat brain, *J. Chem. Neuroanat.* **2**: 119.

Waszczak, B. L., 1990, Differential effects of D1 and D2 dopamine receptor agonists on substantia nigra pars reticulata neurons, *Brain Res.* **513**: 125.

Westerink, B. H., and De Vries, J. B., 1988, Characterization of *in vivo* dopamine release as determined by brain microdialysis after acute and subchronic implantations: methodological aspects, *J. Neurochem.* **51**: 683.

Westerink, B. H., de Boer, P., Santiago, M., and De Vries, J.B., 1994, Do nerve terminals and cell bodies of nigrostriatal dopaminergic neurons of the rat contain similar receptors?, *Neurosci. Lett.* **167**: 109.

Westerink, B. H., Santiago, M., and De Vries, J. B., 1992, *In vivo* evidence for a concordant response of terminal and dendritic dopamine release during intranigral infusion of drugs, *Naunyn Schmiedebergs Arch. Pharmacol.* **346**: 637.

Wilson, C. J., Groves, P. M., and Fifkova, E., 1977, Monoaminergic synapses, including dendro-dendritic synapses in the rat substantia nigra, *Exp. Brain Res.* **30**: 161.

Winkler, C., Bentlage, C., Nikkhah, G., Samii, M., and Bjorklund, A., 1999, Intranigral transplants of GABA-rich striatal tissue induce behavioral recovery in the rat Parkinson model and promote the effects obtained by intrastriatal dopaminergic transplants, *Exp. Neurol.* **155**: 165.

Wolfart, J., and Roeper, J., 2002, Selective coupling of T-type calcium channels to SK potassium channels prevents intrinsic bursting in dopaminergic midbrain neurons, *J. Neurosci.* **22**: 3404.

Yurek, D. M., 1997, Intranigral transplants of fetal ventral mesencephalic tissue attenuate D1-agonist-induced rotational behavior, *Exp. Neurol.* **143**: 1.

Yurek, D. M., and Hipkens, S. B., 1993, Intranigral injections of SCH 23390 inhibit amphetamine-induced rotational behavior, *Brain Res.* **623**: 56.

Yurek, D. M., and Hipkens, S. B., 1994, Intranigral injections of SCH 23390 inhibit SKF 82958-induced rotational behavior, *Brain Res.* **639**: 329.

7

NEUROTRANSMITTER MECHANISMS AT DENDRODENDRITIC SYNAPSES IN THE OLFACTORY BULB

Nathan E. Schoppa[*]

1. INTRODUCTION

The olfactory bulb is remarkable in that, with few exceptions, most of its classes of neuronal contacts are dendrodendritic rather than conventional axo-dendritic or axo-somatic synapses. At these synapses, bulb neurons release the amino acids glutamate and GABA, which are typically involved in rapid synaptic transmission, as well as dopamine. In the past several years, an explosion of work, mainly done in the *in vitro* bulb slice preparation, has examined the mechanisms of release of these neurotransmitters as well as their synaptic actions. In parallel with these studies, there have been significant advances in the understanding of the functional anatomical organization of the olfactory bulb, which has greatly facilitated interpretation of the mechanistic data. In this chapter, I will discuss what is known about dendritic neurotransmitters in the bulb, as well as how they function more broadly, in mechanisms such as signal-sharpening and decoding during olfactory information processing. I will focus almost entirely on dendritic glutamate and GABA, as considerably less is known about dopamine.

2. OLFACTORY BULB ANATOMY

Figure 1 illustrates a simplified version of the circuitry of the mammalian olfactory bulb, along with its major types of dendrodendritic synapses (Mori et al., 1999, Shepherd et al., 2004). The main excitatory neurons are mitral cells (M) and tufted cells (T), which release glutamate from primary and secondary dendrites. M/T cells perform all of the information-transfer function of the bulb, acting both to receive input from incoming axons of olfactory receptor neurons (ORNs) within glomeruli, and also to pass bulb output onto olfactory cortical centers. The activity of M/T cells is modulated by two different types of dendrodendritic synapses, formed between the dendrites of

[*] Nathan E. Schoppa, University of Colorado Health Sciences Center, Denver, Colorado, 80211.

Dendritic Neurotransmitter Release, edited by M. Ludwig
Springer Science+Business Media, Inc., 2005

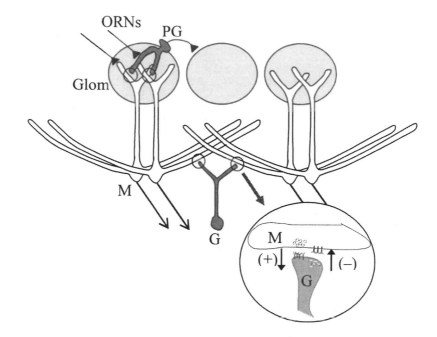

Figure 1. A simplified olfactory bulb circuit. Mitral cells (M) receive afferent input from olfactory receptor neurons (ORNs) at their primary dendrites in glomeruli (Glom) and make dendrodendritic synapses (circled) with periglomerular (PG) cells at primary dendrites and with granule cells (G) at secondary dendrites. PG cells also have axons that terminate on neighboring glomeruli. Enlarged (bottom right) is a reciprocal synapse between the shaft of a mitral cell secondary dendrite and a dendritic spine of a granule cell. Note that the M-to-G glutamatergic synapse (+) is directly adjacent to the G-to-M GABAergic synapse (−). Not shown are other cell-types in the bulb with less well-defined functions, including tufted cells and short-axon cells.

periglomerular (PG) cells and M/T cell primary dendrites or between granule cells and M/T cell secondary dendrites. PG cells can be either GABAergic or dopaminergic, whereas most available data indicate that granule cells are exclusively GABAergic (but see Didier et al., 2001). Interestingly, many of the dendrodendritic synapses are reciprocal contacts, in which the site of the M/T-to-interneuron glutamatergic synapse is adjacent to the site of GABA release on the interneuron spines. This anatomy suggests that bulb interneurons can mediate actions that are highly localized, at the level of a single synapse, as well as actions that require depolarization in large regions of their dendritic arborization. In addition to dendritic transmitters, some bulb neurons can release transmitters from axons, for example from axons of PG cells and the less-common short-axon cells. However, the actions of these transmitters are only just beginning to be understood (Aungst et al., 2003), and it is probably safe to assume that most of the work of the bulb circuit is done by dendritic transmitters.

Beyond this simple classification of neurons and synapse-types, perhaps the most striking property of the mammalian bulb is its high degree of order. M/T cells are organized into discrete networks, each of which are anatomically defined to be the group of neurons (~25 mitral cells and ~50 tufted cells) that send their primary dendrites to one of ~2000 glomeruli. Because each glomerulus receives input from ORNs that express only one type of odorant receptor (OR), each network is OR-specific. This organization of M/T cells into functionally-defined networks is analogous to columns in the visual cortex and barrels in the somatosensory cortex. The bulb, like these other structures, also appears to have anatomical features that promote the mutual excitation or synchronization of neurons within a network. Recent experiments in bulb slices from rats have provided evidence that the primary dendrites of different mitral cells make gap junctional connections within glomeruli (Schoppa and Westbrook, 2002). Additionally, glomeruli are ensheathed and subcompartmentalized by extensive glial processes (Chao et al., 1997, Kasowski et al., 1999), which could lead to local accumulations of glutamate released from mitral cell primary dendrites and "spill-over" of excitatory effects between mitral cells (Isaacson, 1999).

3. MECHANISMS OF DENDRITIC NEUROTRANSMITTER RELEASE

The first clues into the mechanisms of glutamate and GABA release at dendrodendritic synapses in the bulb were provided by the electron micrographs of early anatomical studies (Rall et al., 1966; Price and Powell, 1970; Pinching and Powell, 1971). These images showed evidence for presynaptic elements in M/T, PG, and granule cells that included a complement of small 10-20 nm vesicles, strongly suggestive of vesicular release. More recent physiological studies done in bulb slices have also supported the idea of vesicular release. For example, recordings of miniature excitatory post-synaptic currents (mEPSCs) in granule cells (Fig. 2), include a dominant AMPA receptor-mediated component with rise-time and decay kinetics that are <2 milliseconds, along with a slower component mediated by NMDA receptors. The rapid kinetics of the AMPA receptor-mediated component, which are similar to what has been seen at conventional glutamatergic synapses, are consistent with glutamate being released from mitral cell secondary dendrites in brief "quantal" bursts following vesicle fusion and exocytosis. Rapid kinetics have also been observed for evoked "unitary" EPSCs in PG cells (Schoppa and Westbrook, 2001) and unitary inhibitory post-synaptic currents (IPSCs) in mitral cells (Isaacson and Strowbridge, 1998; Schoppa et al., 1998). Evoked synaptic currents can also be abolished by Ca^{2+} channel blockers that are known to block vesicular release at other synapses (see below). That dendritic neurotransmitter release in the bulb would be vesicular is not surprising, given that dendrodendritic synapses likely subserve functions that are similar to those of axodendritic/somatic synapses in other circuits where there are demands for temporal fidelity in cell-to-cell signaling.

Considerable debate has surrounded the question of the source of Ca^{2+} that mediates vesicular release. The focus thus far has been on events leading to GABA release from granule cell spines, following glutamate release from mitral cells. The reciprocal morphology of the M/T-granule cell synapses (Fig. 1) has led to the intriguing hypothesis that GABA release could be driven directly by Ca^{2+} permeating NMDA receptors located on spines. This mechanism would be in contrast to more conventional mechanisms involving voltage-gated Ca^{2+} channels. Experimental evidence in favor of this hypothesis is the presence of an IPSC in mitral cells that is sensitive to NMDA receptor-blockers but insensitive to blockade of Ca^{2+} channels by cadmium and nickel (Chen et al., 2000;

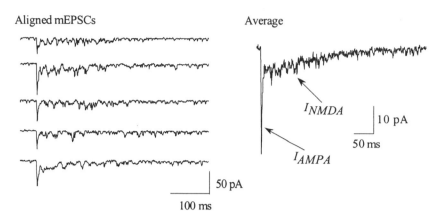

Figure 2. Whole-cell patch recordings of mEPSCs in granule cells illustrate the quantal nature of glutamate release from mitral cell secondary dendrites. The individual mEPSCs (left) and average trace (right) show a fast component mediated by AMPA receptors, along with a slow NMDA receptor-mediated component. Measurements were made in a magnesium-free bathing solution containing tetrodotoxin. The holding potential (V_h) was −60 mV. Displayed data were taken with permission from Schoppa et al., 1998 (Copyright 1998 by the Society for Neuroscience).

Halabisky et al., 2000). These data, however, do not firmly establish a functional role for NMDA receptor-mediated Ca^{2+} influx, since cadmium-insensitive IPSCs are seen only under conditions that promote the activity of NMDA receptors (see also Isaacson, 2001). An additional open question is how the amount of Ca^{2+} influx through NMDA receptors compares to influx through high voltage-gated Ca^{2+} channels. The activation of such channels can provide a potent drive for GABA release (Isaacson, 2001), and, moreover, granule cells fire large-amplitude action potentials (Wellis and Scott, 1990; Luo and Katz, 2001; Cang and Isaacson, 2003) required for opening these channels. The Ca^{2+} contribution of NMDA receptors versus Ca^{2+} channels for GABA release *in vivo* may depend on the strength of an odor stimulus (whether it drives action potentials in granule cells), as well as the location of GABA release (whether it is the same spine activated by glutamate or a different spine; see Section 4.3).

In addition to the Ca^{2+}-source issue, there remain many unresolved questions concerning the mechanisms of dendrodendritic vesicular release. Among these include the Ca^{2+}-dependence of release, the nature of the vesicular release machinery in dendrites, and factors that regulate release (see Davila et al., 2003; Isaacson and Vitten, 2003). Answers to many of these questions will require pair-recordings between M/T cells and PG or granule cells, reports of which, to this point, have remained limited.

4. SYNAPTIC ACTIONS OF DENDRITIC NEUROTRANSMITTER RELEASE

Studies examining the synaptic actions of dendritic neurotransmitters in the bulb have identified several major effects. These include direct excitatory actions on mitral cell activity due to glutamate release, along with inhibitory actions through activation of interneuron-mediated local circuits. Much more is known about the actions of transmitters on mitral cells, as compared to tufted cells, and so the majority of the discussion below will be limited to mitral cells.

4.1. Self-Excitation of Mitral Cells

Self-excitation in mitral cells, which results from glutamate release from their own dendrites, was first seen as a prominent after-depolarization in recordings made in the turtle olfactory bulb (Nicoll and Jahr, 1982). More recent patch-clamp studies in rodent bulb slices (Aroniadou-Anderjaska et al., 1999; Isaacson, 1999; Friedman and Strowbridge, 2000; Salin et al., 2001; Schoppa and Westbrook, 2001) have shown that both AMPA and NMDA glutamate autoreceptors can mediate self-excitation, and, moreover, that such responses occur at both primary and secondary dendrites. Most available data, including the insensitivity of the responses to tetrodotoxin, suggest that self-excitation results from direct activation of presynaptic dendritic autoreceptors on mitral cells rather than axon collaterals or a polysynaptic path. Such presynaptic autoreceptors have been localized anatomically on mitral cells near dendrodendritic synapses (Montague and Greer, 1999), although their exact location with respect to the glutamate release site is not known (Sassoe-Pognetto and Ottersen, 2000). The effect of self-excitation on mitral cell activity may depend on the stimulus conditions. As will be discussed in the following section, AMPA autoreceptors may be involved in lateral excitatory signaling between mitral cells that could be active under a range of conditions. Under conditions that favor NMDA autoreceptors, self-excitation can drive an after-discharge in mitral cells that lasts for hundreds of milliseconds (Nicoll and Jahr, 1982; Friedman and Strowbridge, 2000). Slow kinetics of self-excitation may be required for each mitral cell to remain synchronized to the rest of the cells within its glomerulus-specific network, since mechanisms exist that drive long-lasting "lateral" excitation within a network (see Section 4.2).

4.2. Lateral Excitation of Mitral Cells

In addition to self-excitation, there is accumulating evidence from bulb slice experiments that glutamate release from mitral cell dendrites can cause lateral excitation of other mitral cells. Thus far, two different such effects have been characterized, which can be distinguished by their kinetics and receptor-dependence. The first effect has been seen in whole-cell patch recordings from pairs of mitral cells (Fig. 3a; Schoppa and Westbrook, 2002; Urban and Sakmann, 2002). Upon stimulation of one of the two cells, depolarizations appear in the other cell that are ~1 mV in amplitude and that have a half-width of ~20 milliseconds. The depolarizations (called D_{AMPA}) are blocked by selective blockers of AMPA receptors (NBQX and GYKI-53655), indicating that they are driven by AMPA receptors activated by glutamate release from mitral cells; moreover, they appear to originate at mitral cell primary dendrites. The exact mechanism of generation has not been established. At the moment, the most plausible hypothesis incorporates evidence that mitral cells make gap junctional connections at primary dendrites, with D_{AMPA} being generated through electrical coupling of an AMPA autoreceptor potential that originates there (Fig. 3b). While the absence of selective pharmacological blockers for gap junctions has precluded direct testing of this model, it is strongly supported by computational studies (Schoppa and Westbrook, 2002). Additionally, an alternate model, in which lateral excitation is caused by spill-over-mediated effects of glutamate between mitral cells appears unlikely, based on the low affinity of AMPA receptors for glutamate, as well as the results of voltage-clamp experiments (Schoppa and Westbrook, 2002). The role of D_{AMPA} appears to be the generation of synchronized spike activity in mitral cells that project to the same glomerulus (Fig. 3c,d).

A second type of lateral excitatory response in mitral cells that originates within glomeruli is an NMDA receptor-driven depolarization that persists for several hundred milliseconds (Carlson et al., 2000; Puopolo and Belluzzi, 2001; Schoppa and Westbrook, 2001). These depolarizations, which can occur as a single event or in repeated, oscillatory

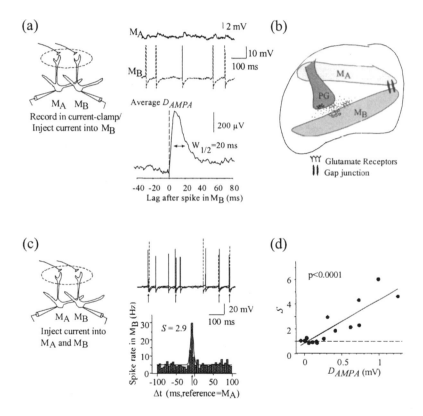

Figure 3. Rapid, AMPA receptor-mediated lateral signaling between glomerulus-specific mitral cells. (a) During whole-cell recordings in a pair of mitral cells, spiking in one cell M_B elicited small depolarizations (D_{AMPA}) in M_A, apparent in both the raw data traces (top) as well as the average trace. (b) Components in hypothesized mechanism for D_{AMPA}. In the model, M_B fires an action potential and releases glutamate from its primary dendrite, which activates glutamate receptors on the PG cell and presynaptic AMPA autoreceptors. D_{AMPA} is generated in M_A when the autoreceptor potential is coupled between M_B and M_A by gap junctional connections. (c,d) D_{AMPA} underlies synchronized spiking in glomerulus-specific mitral cells. Synchrony is seen in part (c) in the raw data traces from one mitral cell pair (vertical arrows), as well as in a prominent peak in the spike-lag histogram from this experiment. The plot in part (d) shows that the magnitude of synchrony across many pairs (S, derived from the lag histograms) is strongly correlated with D_{AMPA}. Displayed data were taken with permission from Schoppa and Westbrook, 2002 (Copyright 2002 by the Nature Publishing Group).

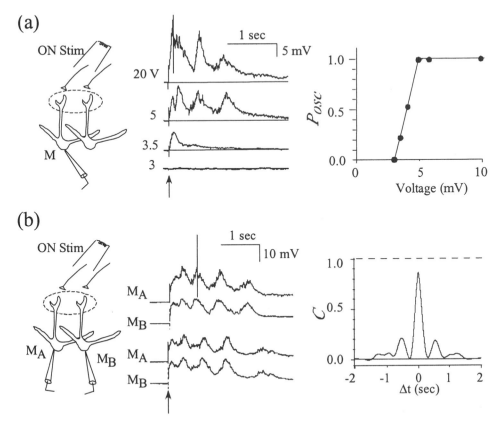

Figure 4. Slow lateral excitatory signals in mitral cells. (a) In a whole-cell current-clamp recording in a single mitral cell, stimulation of the olfactory nerve (ON) at high intensities (≥5 V) drove long-lasting oscillatory responses that included several depolarizing peaks. The plot (right) shows that the probability of evoking oscillations (P_{osc}) was steeply dependent on the magnitude of the stimulus. (b) The oscillations were highly synchronized in a pair of glomerulus-specific mitral cells. Synchrony is apparent in both the raw data traces (left) as well as in the cross-correlogram (right). Displayed data were taken with permission from Schoppa and Westbrook, 2001 (Copyright 2001 by Elsevier Ltd).

bouts over several seconds (Fig. 4a), have most often been characterized following strong olfactory nerve (ON) stimulation. However, focal stimulation of mitral cell bodies is also effective in driving the depolarizations (N. Schoppa, personal communication), which indicates that the key event in their generation is glutamate release from mitral cell dendrites rather than ON terminals (since stimulation of mitral cell somata does not activate ON terminals). Additionally, the excitatory responses can be evoked even when a test mitral cell is voltage-clamped (at –100 mV), arguing that most of it results from glutamate release from mitral cells other than the test cell; hence, it is primarily a "lateral" excitatory response. The most striking aspects of the slow depolarizations are their large amplitude (~10 mV), as well as the fact that, when pair mitral cell recordings are made, they are highly synchronized for cells that project to the same glomerulus (Fig.

4b). Because synchrony has been seen in virtually all glomerulus-specific mitral cell pair recordings (8 out of 9; Carlson et al., 2000; Schoppa and Westbrook, 2001), it is likely that most if not all ~25 mitral cells that comprise a glomerulus-specific network engage in the synchronized depolarizations at the same time. The mechanisms driving synchronization are not clear. The high affinity of NMDA receptors for glutamate suggests that synchronization could be driven by spill-over of glutamate between mitral cells within glomeruli, but it could also be driven by electrical coupling between glomerulus-specific mitral cells.

In comparing the two types of lateral excitatory signals, one major difference that should be emphasized regards the number of mitral cells that must be activated to see a response: the NMDA, but not the AMPA, receptor-mediated signal requires strong stimulus conditions that activate multiple mitral cells. The dependence for multiple-cell activation likely reflects the sensitivity of NMDA receptors to voltage-dependent blockade by magnesium, with magnesium-unblock requiring a large depolarization generated cooperatively across a network of mitral cells. An additional consequence of voltage-dependent magnesium blockade of NMDA receptors is that the slow depolarizations occur in an "all-or-none" fashion (Schoppa and Westbrook, 2001), reflecting the fact that magnesium-unblock is a positive-feedback process. I will argue in Section 5 that the all-or-none feature of the depolarizations is significant in their function in olfactory processing.

4.3. Granule Cell Excitation/Mitral Cell Inhibition

In discussing the mechanisms and actions of the granule cell-mediated inhibitory local circuit, it is first useful to separate inhibition into two components. Following the release of glutamate from a mitral cell, GABA release from granule cells can inhibit that same mitral cell (causing "self"-inhibition) or inhibit other mitral cells that share connections with those granule cells (causing "lateral" inhibition). These two forms of inhibition may differ significantly in their mechanisms of generation. Self-inhibition can be driven completely by local signaling within single spines of granule cells, by either the NMDA receptor-mediated Ca^{2+} signal (Section 3) or by a local depolarization that opens voltage-gated Ca^{2+} channels. In contrast, lateral inhibition depends on the spread of depolarization between spines of the granule cell. Such spread could occur through entirely local passive or active mechanisms (Woolf et al., 1991), although spread across the dendritic arborization of a granule cell will be dramatically augmented by action potential-firing. Self-inhibition has been experimentally measured from the feedback IPSC or inhibitory post-synaptic potential (IPSP) that results from stimulation of a mitral or tufted cell (Jahr and Nicoll, 1982; Isaacson and Strowbridge, 1998; Schoppa et al., 1998; Christie et al., 2001; Halabisky and Strowbridge, 2003). Lateral inhibition has been assessed following stimulation of one other mitral cell in pair recordings (Isaacson and Strowbridge, 1998; Urban and Sakmann, 2002), stimulation of networks of mitral cells (Schoppa and Westbrook, 1999), and during *in vivo* recordings (Margrie et al., 2001). Most of the inhibitory response is generally assumed to be derived from granule cells rather than PG cells, because granule-to-mitral cell synapses vastly outnumber PG-to-mitral cell synapses.

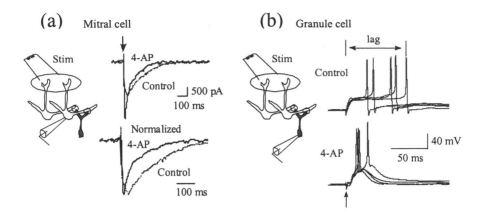

Figure 5. Evidence that the slow kinetics of mitral cell inhibition are due to I_A channels in granule cells. Inhibitory currents in a mitral cell (a) and excitatory responses in a granule cell (b) were prolonged under control conditions, but much-accelerated when I_A was blocked by 4-aminopyridine (4-AP). Studies done in parallel to these (Schoppa and Westbrook, 1999) indicated that 4-AP unmasked a previously-attenuated AMPA receptor-mediated excitatory response in granule cells. Displayed synaptic responses were evoked by electrical stimulation of a glomerulus, which mainly results in a lateral inhibitory response in mitral cells. Voltage-clamp recordings of inhibitory currents in part (a) were made with a KCl-containing patch electrode ($V_h=-70$ mV). Displayed data were taken with permission from Schoppa and Westbrook, 1999 (Copyright 1999 by the Nature Publishing Group).

The most striking characteristic of the inhibitory responses, regardless of the type of inhibition, is their slow kinetics, as they typically last for hundreds of milliseconds (see also Mori and Takagi, 1978; Nowycky et al., 1981). An example of such a slow response is shown in Fig. 5a ("control" trace). The slow kinetics for mitral cell inhibition compare with a ~20 millisecond-decay for unitary IPSCs in mitral cells, as well as with ≤20 millisecond-decays for IPSCs at most other inhibitory synapses in the brain. The slow kinetics of self-inhibition, which depends on NMDA receptor activation on granule cells (Isaacson and Strowbridge, 1998; Schoppa et al., 1998), likely reflect the slow deactivation kinetics of NMDA receptors, resulting in a prolonged Ca^{2+} signal or local depolarization in a granule cell spine. For lateral inhibition, the slow kinetics are likely to be largely due to a novel mechanism of regulating spike-generation in granule cells that involves the interaction between their AMPA and NMDA receptor-mediated synaptic inputs and intrinsic voltage-gated ion channels (Fig. 5; Schoppa and Westbrook, 1999). Specifically, granule cells have a potent, transient A-type K^+ conductance (I_A) in their dendrites that can attenuate the early depolarizing response mediated by their AMPA receptors, preventing early spiking, but is outlasted by the slower NMDA receptor-mediated depolarization. The resulting delayed spiking leads to long-lasting GABA release across a network of granule cells. It should be noted that this prominent role for I_A in regulating granule cell activity and lateral inhibition does not exclude actions of other intrinsic channels in granule cells. For example, experiments in the frog bulb indicate that a Ca^{2+}-activated non-selective cation current can boost the NMDA receptor-mediated depolarization in granule cells that underlies delayed spiking (Hall and Delaney, 2002). Additionally, T-type Ca^{2+} channels in granule cells can drive an after-depolarization

(Egger et al., 2003), which could prolong spiking; these channels may also provide an independent source of Ca^{2+} for GABA release.

While the granule cell-mediated local circuit undoubtedly has a prominent role in olfactory processing (see Section 5), it has been difficult to demonstrate convincing effects on mitral cell activity in cellular studies in slices. The self- and lateral inhibitory potentials following activation of a single mitral cell have typically been very small (~0.3 mV; Urban and Sakmann, 2002) or unmeasurable, when experiments have been done in physiological ionic conditions. Many existing studies of granule cell function have examined the effects of IPSPs evoked by focal stimulation of granule cells or local application of GABA. Such inhibition can alter the timing of action potentials in mitral cells (Desmaisons et al., 1999) or block action potential propagation in mitral cell secondary dendrites (Lowe, 2002; Xiong and Chen, 2002). Presumably, granule cells can also regulate the probability of action potential-firing in mitral cells, through either self- or lateral inhibition. Future studies of granule cell actions in slices could employ stronger stimuli that better mimick a natural stimulus. Because inhibition is NMDA receptor-dependent, the resulting network depolarizations are expected to produce much larger inhibitory responses. In one recent slice study (Friedman and Strowbridge, 2003), strong glomerular stimulation drove a prolonged barrage of inhibitory currents at the gamma frequency. These currents were speculated to be involved in synchronizing mitral cells.

4.4. Effects of Transmitter Release from PG Cell Dendrites

Of the different components of the bulb circuit, the effects of the dendritic transmitters of PG cells (GABA and dopamine) are amongst the least well-understood. Much of the available information about these transmitters has to do with potential modulatory effects on transmitter release. Exogenous application of agonists of $GABA_B$ or dopamine receptors can reduce glutamate release from ON terminals (Wachowiak and Cohen 1999; Aroniadou-Anderjaska et al., 2000; Berkowicz and Trombley, 2000; Ennis et al., 2001) or mitral cells (Davila et al., 2003), suggesting that GABA and dopamine from PG cells could have negative feedback effects on incoming input and mitral cell excitation. The $GABA_A$ receptor-mediated actions of GABA from PG cells are only just beginning to be understood. One recent study showed that GABA release can lead to self-inhibition of PG cells (Smith and Jahr, 2002). No study has successfully isolated a $GABA_A$ receptor-mediated current or potential in mitral cells due to PG cells, though such a response is generally assumed to be inhibitory (but see Shepherd, 1963; Martinez and Freeman, 1984). Inhibition from PG cells would produce fundamentally different effects on bulb circuit function as compared to granule cell inhibition. Because PG cell dendritic contacts onto mitral cells are restricted to one glomerulus (Fig. 1), they would exclusively mediate "intra-glomerular" lateral inhibitory interactions between mitral cells. In contrast, granule cell connections can mediate both intra-glomerular as well as between-network, or "inter-glomerular", interactions.

5. FUNCTIONS IN OLFACTORY PROCESSING

Numerous functional studies have shown that the first effect of odor in the olfactory bulb is to elicit a pattern of activated glomeruli (Bozza and Mombaerts, 2001; Korsching, 2002), reflecting the different classes of odorant receptors activated in the olfactory epithelium. These experiments have led to the idea that the odor "code" begins in the bulb as a spatial map of glomerular activation in the glomerular layer. The well-ordered arrangement of mitral cells into OR-specific networks (see Section 2) leads to very specific hypotheses about the effects of the different types of bulb transmitters on the odor code, as it is reflected at its next level of mitral cell activity. The most commonly-proposed functional mechanism applies the idea of lateral inhibition in cellular studies to the activity of networks of mitral cells (Meredith, 1986; Wilson and Leon, 1987; Yokoi et al., 1995; Luo and Katz, 2001). By causing mitral cells affiliated with strongly-activated ORs to inhibit cells affiliated with weakly-activated ORs, lateral inhibition could sharpen the odor code. Such an effect, which is likely to be driven primarily by granule cells, could enhance both odor response-reproducibility and discrimination. While the general concept of lateral inhibition is likely to be correct, I favor a modified formulation that also incorporates the slow, glomerulus-specific lateral excitatory responses that have been characterized in slice studies (Section 4.2). According to this idea, the probability that any one mitral cell fires an action potential will be dictated by the probability that its entire OR-specific network undergoes a slow depolarization, which in turn will be a function of the strength of the ORN inputs and lateral antagonism between networks. Because the slow depolarizations occur in an all-or-none fashion, such a mechanism will result in a significant reformatting of the odor code that goes beyond that which could be accomplished by lateral inhibition alone, with the new odor code being a much-reduced map of mitral cell activity comprised of a small number of highly and uniformly-active networks. Such dramatic sharpening of the code has been observed between ORNs and glomerular output neurons during odor responses in the insect olfactory system (Sachse and Galizia, 2002).

In addition to sharpening the odorant code, it is often suggested that the olfactory bulb circuit synchronizes the activity of mitral cells (Mori et al., 1999; Laurent et al., 2001). Synchronization of spiking in mitral cells of the same OR-specificity could amplify an OR-specific signal, by increasing the probability that synaptic inputs summate on post-synaptic neurons in the olfactory cortex, whereas synchrony amongst mitral cells of differing OR-specificities could allow the cortex to integrate or "decode" information about different ORs. The synchrony-hypothesis is based on the long-standing experimental observation that odor evokes rapid gamma frequency oscillations in bulb field potentials (Adrian, 1950; Rall and Shepherd, 1968; Freeman, 1972), as well as data from the insect olfactory system that such oscillations (in the structure analogous to the bulb) are correlated with odor discrimination (Stopfer et al., 1997). Odor can also elicit synchronized activity in pair extracellular-unit recordings in the bulb (Kashiwadani et al., 1999). Odor-evoked synchrony is generally hypothesized to be due to the granule cell-mediated local circuit (Rall and Shepherd, 1968; Laurent et al., 2001). Experiments presented here (Fig. 3) also suggest that OR-specific mitral cells could be synchronized by an alternate mechanism involving AMPA receptor-mediated lateral excitatory interactions. The ~1 mV D_{AMPA} signals could work in conjunction with the much larger slow synchronized depolarizations to drive synchronized spikes, with the ~10 millivolt slow depolarizations acting to bring the mitral cell from their resting potential (~-55 mV; Cang and Isaacson, 2003) to near spike threshold (~-45 mV).

6. CONCLUSIONS

A common feature of many sensory systems is the anatomical organization of neurons into functionally-defined units. The mutual excitation of neurons within a functional unit, together with lateral inhibition between units, is thought to sharpen sensory signals, leading to phenomenon such as contrast-enhancement in vision. The studies outlined in this chapter show that the olfactory bulb, like brain regions in other sensory systems, has its excitatory M/T cells organized into functionally-defined networks (defined by their OR-specificity), while the bulb also has synaptic mechanisms that promote excitation within and inhibition between networks. The bulb mechanisms however differ from those in other sensory structures in a number of important ways. First, excitation/inhibition occurs almost entirely through dendrodendritic synaptic interactions rather than at conventional axo-dendritic or axo-somatic synapses. These dendritic actions are mediated by the neurotransmitters glutamate and GABA, and, in one example, by a mechanism that appears to involve the concerted activity of glutamate and gap junctional connections within glomeruli. At least for excitation, dendritic mechanisms may be required because M/T cells, unlike functionally-defined neurons in other systems, are not segregated into physically-defined structures. The glomerulus is the one place in the bulb where M/T cells of the same OR-specificity are exclusively in close proximity with each other, and thus any specific contacts must be at dendrites located there. Another unusual aspect of much of the excitatory and inhibitory activity in the bulb is its long-lasting kinetics, reflecting a prominent role for NMDA receptors. The slow kinetics may be functionally important, allowing the bulb mechanisms to be well-matched to the slow, respiration-dependent afferent input.

A major objective in the future will be to link the mechanistic information that has been obtained mainly in bulb slices with *in vivo* studies of M/T cell activity. Studies are needed, for example, to test whether the synchronized spiking in glomerulus-specific mitral cells seen in slices occurs during natural odorant responses. Some evidence already exists that odor can evoke slow depolarizations in mitral cells similar to the slow excitatory responses in slices (Margrie and Schaefer, 2003; Cang and Isaacson, 2003). The function of the slow depolarizations can be tested from the mitral cell population-response to odor. For example, a role in sharpening odorant signals (see Section 5) predicts that specific blockers of the depolarizations will cause a reduction in the peak mitral cell response, increased variability in responses of different mitral cells, and a loss of the center-surround structure of the response correlation map (Luo and Katz, 2001). There are also many unresolved issues at the single-synapse or cellular level that require additional studies in slices. These include the mechanisms of dendritic transmitter release, the roles of local versus action potential-dependent mechanisms of activating interneurons, and mechanisms of synaptic plasticity, in particular as they relate to cortical feedback circuits. Special focus should be on PG cells and tufted cells, about which very little is known.

7. REFERENCES

Adrian, E. D., 1950, The electrical activity of the mammalian olfactory bulb, *Electroencephalogr. Clin. Neurophysiol.* **2**: 377.

Aroniadou-Anderjaska, V., Ennis, M., and Shipley, M. T., 1999, Dendrodendritic recurrent excitation in mitral cells of the rat olfactory bulb, *J. Neurophysiol.* **82**: 489.

Aroniadou-Anderjaska, V., Zhou, F. M., Priest, C. A., Ennis, M., and Shipley, M. T., 2000, Tonic and synaptically evoked presynaptic inhibition of sensory input to the rat olfactory bulb via GABA(B) heteroreceptors, *J. Neurophysiol.* **84**: 1194.

Aungst, J. L., Heyward, P. M., Puche, A. C., Karnup, S. V., Hayar, A., Szabo, G., and Shipley, M. T., 2003, Centre-surround inhibition among olfactory bulb glomeruli, *Nature* **426**:623.

Berkowicz, D. A., and Trombley, P. Q., 2000, Dopaminergic modulation at the olfactory nerve synapse, *Brain Res.* **855**: 90.

Bozza, T. C., and Mombaerts, P., 2001, Olfactory coding: revealing intrinsic representations of odors, *Curr. Biol.* **11**: R687.

Cang, J., and Isaacson, J. S., 2003, *In vivo* whole-cell recording of odor-evoked synaptic transmission in the rat olfactory bulb, *J. Neurosci.* **23**: 4108.

Carlson, G. C., Shipley, M. T., and Keller, A., 2000, Long-lasting depolarizations in mitral cells of the rat olfactory bulb, *J. Neurosci.* **20**: 2011.

Chao, T. I., Kasa, P., and Wolff, J. R., 1997, Distribution of astroglia in glomeruli of the rat main olfactory bulb: exclusion from the sensory subcompartment of neuropil, *J. Comp. Neurol.* **388**:191.

Chen, W. R., Xiong, W., and Shepherd, G. M., 2000, Analysis of relations between NMDA receptors and GABA release at olfactory bulb reciprocal synapses, *Neuron* **25**: 625.

Christie, J. M., Schoppa, N. E., and Westbrook, G. L., 2001, Tufted cell dendrodendritic inhibition in the olfactory bulb is dependent on NMDA receptor activity, *J. Neurophysiol.* **85**: 169.

Davila, N. G., Blakemore, L. J., and Trombley, P. Q., 2003, Dopamine modulates synaptic transmission between rat olfactory bulb neurons in culture, *J. Neurophysiol.* **90**: 395.

Desmaisons, D., Vincent, J. D., and Lledo, P. M., 1999, Control of action potential timing by intrinsic subthreshold oscillations in olfactory bulb output neurons, *J. Neurosci.* **19**: 10727.

Didier, A., Carleton, A., Bjaalie, J. G., Vincent, J. D., Ottersen, O. P., Storm-Mathisen, J., and Lledo, P. M., 2001, A dendrodendritic reciprocal synapse provides a recurrent excitatory connection in the olfactory bulb, *Proc. Natl. Acad. Sci. U. S. A.* **98**: 6441.

Egger, V., Svoboda, K., and Mainen, Z. F., 2003, Mechanisms of lateral inhibition in the olfactory bulb: efficiency and modulation of spike-evoked calcium influx into granule cells, *J. Neurosci.* **23**: 7551.

Ennis, M., Zhou, F. M., Ciombor, K. J., Aroniadou-Anderjaska, V., Hayar, A., Borrelli, E., Zimmer, L. A., Margolis, F., and Shipley, M. T., 2001, Dopamine D(2) receptor-mediated presynaptic inhibition of olfactory nerve terminals, *J. Neurophysiol.* **86**: 2986.

Freeman, W. J., 1972, Measurement of oscillatory responses to electrical stimulation in olfactory bulb of cat, *J. Neurophysiol.* **35**: 762.

Friedman, D., and Strowbridge, B. W., 2000, Functional role of NMDA autoreceptors in olfactory mitral cells, *J. Neurophysiol.* **84**: 39.

Friedman, D., and Strowbridge, B. W., 2003, Both electrical and chemical synapses mediate fast network oscillations in the olfactory bulb, *J. Neurophysiol.* **89**: 2601.

Halabisky, B., Friedman, D., Radojicic, M., and Strowbridge, B.W., 2000, Calcium influx through NMDA receptors directly evokes GABA release in olfactory bulb granule cells, *J. Neurosci.* **20**: 5124.

Halabisky, B., and Strowbridge, B. W., 2003, Gamma-frequency excitatory input to granule cells facilitates dendrodendritic inhibition in the rat olfactory bulb, *J. Neurophysiol.* **90**: 644.

Hall B. J., and Delaney, K. R., 2002, Contribution of a calcium-activated non-specific conductance to NMDA receptor-mediated synaptic potentials in granule cells of the frog olfactory bulb, *J. Physiol.* **543**: 819.

Isaacson, J. S., 1999, Glutamate spillover mediates excitatory transmission in the rat olfactory bulb, *Neuron* **23**: 377.

Isaacson, J. S., 2001, Mechanisms governing dendritic gamma-aminobutyric acid (GABA) release in the rat olfactory bulb, *Proc. Natl. Acad. Sci. U S A* **98**: 337.

Isaacson, J. S., and Strowbridge, B. W., 1998, Olfactory reciprocal synapses: dendritic signaling in the CNS, *Neuron* **20**: 749.

Isaacson, J. S., and Vitten, H., 2003, GABA(B) receptors inhibit dendrodendritic transmission in the rat olfactory bulb, *J. Neurosci.* **23**: 2032.

Jahr, C. E., and Nicoll, R.A., 1982, An intracellular analysis of dendrodendritic inhibition in the turtle *in vitro* olfactory bulb, *J. Physiol.* **326**: 213.

Kashiwadani, H., Sasaki, Y. F., Uchida, N., and Mori, K., 1999, Synchronized oscillatory discharges of mitral/tufted cells with different molecular receptive ranges in the rabbit olfactory bulb, *J. Neurophysiol.* **82**: 1786.

Kasowski, H. J., Kim, H., and Greer, C. A., 1999, Compartmental organization of the olfactory bulb glomerulus, *J. Comp. Neurol.* **407**: 261.

Korsching, S., 2002, Olfactory maps and odor images, *Curr. Opin. Neurobiol.* **12**: 387.

Laurent, G., Stopfer, M., Friedrich, R. W., Rabinovich, M. I., Volkovskii, A., and Abarbanel, H. D., 2001, Odor encoding as an active, dynamical process: experiments, computation, and theory, *Annu. Rev. Neurosci.* **24**: 263.

Lowe, G., 2002, Inhibition of backpropagating action potentials in mitral cell secondary dendrites, *J. Neurophysiol.* **88**: 64.

Luo, M., and Katz, L. C., 2001, Response correlation maps of neurons in the mammalian olfactory bulb, *Neuron* **32**: 1165.

Margrie, T. W., Sakmann, B., and Urban, N. N., 2001, Action potential propagation in mitral cell lateral dendrites is decremental and controls recurrent and lateral inhibition in the mammalian olfactory bulb, *Proc. Natl. Acad. Sci.* **98**: 319.

Margrie, T. W., and Schaefer, A. T., 2003, Theta oscillation coupled spike latencies yield computational vigour in a mammalian sensory system, *J. Physiol.* **546**: 363.

Martinez, D. P., and Freeman, W. J., 1984, Periglomerular cell action on mitral cells in olfactory bulb shown by current source density analysis, *Brain Res.* **308**: 223.

Meredith, M., 1986, Patterned response to odor in mammalian olfactory bulb: the influence of intensity, *J. Neurophysiol.* **56**: 572.

Montague, A. A., and Greer, C. A., 1999, Differential distribution of ionotropic glutamate receptor subunits in the rat olfactory bulb, *J. Comp. Neurol.* **405**: 233.

Mori, K., Nagao, H., and Yoshihara, Y., 1999, The olfactory bulb: coding and processing of odor molecule information, *Science* **286**: 711.

Mori, K., and Takagi, S. F., 1978, An intracellular study of dendrodendritic inhibitory synapses on mitral cells in the rabbit olfactory bulb, *J. Physiol.* **279**: 569.

Nicoll, R. A., and Jahr, C. E., 1982, Self-excitation of olfactory bulb neurons, *Nature* **296**: 441.

Nowycky, M. C., Mori, K., and Shepherd, G. M., 1981, GABAergic mechanisms of dendrodendritic synapses in isolated turtle olfactory bulb, *J. Neurophysiol.* **46**: 639.

Pinching, A. J., and Powell, T. P., 1971, The neuropil of the glomeruli of the olfactory bulb, *J. Cell Sci.* **9**: 347.

Price, J. L., and Powell, T. P., 1970, The synaptology of the granule cells of the olfactory bulb, *J. Cell Sci.* **7**: 125.

Puopolo, M., and Belluzzi, O., 2001, NMDA-dependent, network-driven oscillatory activity induced by bicuculline or removal of Mg^{2+} in rat olfactory bulb neurons, *Eur. J. Neurosci.* **13**: 92.

Rall, W., and Shepherd, G. M., 1968, Theoretical reconstruction of field potentials and dendrodendritic synaptic interactions in the olfactory bulb, *J. Neurophysiol.* **31**: 884.

Rall, W., Shepherd, G. M., Reese, T. S., and Brightman, M. W., 1966, Dendrodendritic synaptic pathway for inhibition in the olfactory bulb, *Exp. Neurol.* **14**: 44.

Sachse, S., and Galizia, C. G., 2002, Role of inhibition for temporal and spatial odor representation in olfactory output neurons: a calcium imaging study, *J. Neurophysiol.* **87**: 1106.

Salin, P. A., Lledo, P. M., Vincent, J. D., and Charpak, S., 2001, Dendritic glutamate autoreceptors modulate signal processing in rat mitral cells, *J. Neurophysiol.* **85**: 1275.

Sassoe-Pognetto, M., and Ottersen, O. P., 2000, Organization of ionotropic glutamate receptors at dendrodendritic synapses in the rat olfactory bulb, *J. Neurosci.* **20**: 2192.

Schoppa, N. E., Kinzie, J. M., Sahara, Y., Segerson, T. P., and Westbrook, G. L., 1998, Dendrodendritic inhibition in the olfactory bulb is driven by NMDA receptors, *J. Neurosci.* **18**: 6790.

Schoppa, N. E., and Westbrook, G. L., 1999, Regulation of synaptic timing in the olfactory bulb by an A-type potassium current, *Nat. Neurosci.* **2**: 1106.

Schoppa, N. E., and Westbrook, G. L., 2001, Glomerulus-specific synchronization of mitral cells in the olfactory bulb, *Neuron* **31**: 639.

Schoppa, N. E., and Westbrook, G. L., 2002, AMPA autoreceptors drive correlated spiking in olfactory bulb glomeruli, *Nat. Neurosci.* **5**:1194.

Shepherd, G. M., 1963, Neuronal systems controlling mitral cell excitability, *J. Physiol. (Lond.)* **168**: 101.

Shepherd, G. M., Chen, W. R., and Greer, C. A., 2004, Olfactory bulb, in: *The Synaptic Organization of the Brain*, G. M. Shepherd, ed., Oxford UP, New York, pp. 165-216.

Smith, T. C., and Jahr, C. E., 2002, Self-inhibition of olfactory bulb neurons, *Nat. Neurosci.* **5**: 760.

Stopfer, M., Bhagavan, S., Smith, B. H., and Laurent, G., 1997, Impaired odour discrimination on desynchronization of odour-encoding neural assemblies, *Nature* **390**: 70.

Urban, N. N., and Sakmann, B., 2002, Reciprocal intraglomerular excitation and intra- and interglomerular lateral inhibition between mouse olfactory bulb mitral cells, *J. Physiol.* **542**: 355.

Wachowiak, M., and Cohen, L. B., 1999, Presynaptic inhibition of primary olfactory afferents mediated by different mechanisms in lobster and turtle, *J. Neurosci.* **19**: 8808.

Wellis, D. P., and Scott, J. W., 1990, Intracellular responses of identified rat olfactory bulb interneurons to electrical and odor stimulation, *J. Neurophysiol.* **64**: 932.

Wilson, D. A., and Leon, M., 1987, Evidence of lateral synaptic interactions in olfactory bulb output cell responses to odors, *Brain Res.* **417**: 175.

Woolf, T. B., Shepherd, G. M., and Greer, C. A., 1991, Serial reconstructions of granule cell spines in the mammalian olfactory bulb, *Synapse* **7**: 181.

Xiong, W., and Chen, W. R., 2002, Dynamic gating of spike propagation in the mitral cell lateral dendrites, *Neuron* **34**: 115.

Yokoi, M., Mori, K., and Nakanishi, S., 1995, Refinement of odor molecule tuning by dendrodendritic synaptic inhibition in the olfactory bulb, *Proc. Natl. Acad. Sci. U.S.A.* **92**: 3371.

CLASSICAL NEUROTRANSMITTERS AS RETROGRADE MESSENGERS IN LAYER 2/3 OF THE NEOCORTEX: EMPHASIS ON GLUTAMATE AND GABA

Yuri Zilberter[1] and Tibor Harkany[2]

1. INTRODUCTION

The contribution of retrograde signalling to information processing in the brain was contemplated for a long time, with particular respect to central nervous system development and long-term synaptic plasticity (for review see Fitzsimonds and Poo, 1998). In the past several years however, the concept of retrograde signalling has been significantly expanding, and also included short-term modifications of synaptic efficacy. The classical view on synaptic transmission postulated a unidirectional information transfer from an active presynaptic to a corresponding postsynaptic site. This hypothesis was questioned by multiple experimental studies on neurons in different brain regions suggesting that a fast retrograde signal, which provides negative feedback, is present in active synaptic contacts.

Llano et al. initially reported that postsynaptic membrane depolarisation induced significant reduction of $GABA_A$ receptor-mediated spontaneous activity in cerebellar Purkinje cells (Llano et al., 1991). Shortly thereafter, a similar effect was observed in hippocampal CA1 pyramidal cells (Pitler and Alger, 1992). Depression of inhibitory activity in both cell types was due to a transient (<1 min) reduction in the frequency of spontaneous synaptic events and thus had a presynaptic origin. Importantly, this synaptic depression was initiated by postsynaptic depolarisation, suggesting the release of retrograde messenger(s) affecting the presynaptic terminal, and hence termed depolarisation-induced suppression of inhibition (DSI).

DSI is triggered by an increase in the dendritic Ca^{2+} concentration as loading the postsynaptic cells with a Ca^{2+} chelator, e.g., ethylenedioxybis(o-phenylenenitrilo)-tetraacetic acid (BAPTA), can abolish it (Llano et al., 1991; Pitler and Alger, 1992).

---

[1] [1]Department of Neuroscience, Laboratory of Neuronal Network, Berzelius väg 8:B2-2 and [2]Laboratory of Molecular Neurobiology, Department of Medical Biochemistry and Biophysics, Scheeles väg 1:A1, Karolinska Institutet, S-17177 Stockholm, Sweden

Moreover, DSI becomes significantly reduced when applying metabotropic glutamate receptor (mGluR) antagonists (Glitsch et al., 1996; Morishita et al., 1998). Taking these results together, glutamate, released from a subsynaptic dendrite, was suggested as the candidate retrograde messenger.

In contrast, novel findings showed that endogenous ligands with functional similarity to Δ^9-tetrahydrocannabinol, endocannabinoids, can serve as retrograde messengers in hippocampal CA1 pyramidal neurons (Wilson and Nicoll, 2001), cerebellar Purkinje cells (Kreitzer and Regehr, 2001a; 2001b) and neocortical pyramidal cells (Sjöström et al., 2003; Trettel and Levine, 2003; Trettel et al., 2004) acting on type 1 cannabinoid (CB$_1$) receptors located in presynaptic terminals. G-protein coupled CB$_1$ receptors were found expressed throughout the rodent brain (Herkenham et al., 1990), while endogenous lipids, such as anandamide and 2-arachidonylglycerol, were identified as potent endogenous agonists on the CB$_1$ receptor (Devane et al., 1992; Stella et al., 1997). Importantly, endocannabinoids are membrane-diffusible molecules and thus, their release does not require morphologically defined active zones (Alger, 2002), or cell-surface exposure of the corresponding receptors in presynaptic terminals (Hsieh et al., 1999). Accordingly, endocannabinoid effects are not confined to specific synapses and once synthesized, they can affect populations of neurons (Ohno-Shosaku et al., 2001; Wilson and Nicoll, 2001). Endocannabinoid synthesis is controlled by mGluR activation (Maejima et al., 2001; Varma et al., 2001), and triggered by an increase in the dendritic Ca^{2+} concentration (Di Marzo et al., 1994; Di Marzo et al., 1998). In conclusion, endocannabinoid-mediated fast retrograde signalling can modulate the excitability of complex neuronal networks during synaptic activity.

Retrograde signalling has also been described in both excitatory (Zilberter et al., 1999) and inhibitory (Zilberter, 2000) synaptic connections between pyramidal cells and interneurons in layer 2/3 of the neocortex. Inhibitory GABAergic interneurons, showing an amazing morphological and functional diversity (Kawaguchi, 1995; Kawaguchi and Kubota, 1997; Gupta et al., 2000), constitute a relatively small fraction (~10-12%) of all neocortical neurons (Jones and Peters, 1984). Interneurons can effectively control cortical excitability by entraining pyramidal cells. Pyramidal cells and interneurons frequently form reciprocal synaptic connections (Buhl et al., 1997; Reyes et al., 1998; Holmgren et al., 2003) thereby creating elementary neuronal microcircuits. Recent results show (Zilberter et al., 1999; Zilberter, 2000; Harkany et al., 2004) that the mechanism(s) of retrograde signalling in synapses between pyramidal cells and particular subtypes of interneurons in neocortex may be qualitatively different from those reposted in hippocampus and cerebellum.

2. RETROGRADE SIGNALLING AT INHIBITORY NEOCORTICAL SYNAPSES: DENDRITIC GLUTAMATE RELEASE.

Fast-spiking (FS) cells comprise a neurochemically and electrophysiologically heterogeneous population of interneurons, while invariably showing an abrupt start of non-adapting repetitive discharges at a threshold frequency ranging from 40 to 150 Hz. Conditioned depolarisation by constant currents makes FS cells ready to fire continuously in response to further depolarising current pulses. FS cells frequently contain the Ca^{2+}-binding protein parvalbumin (Kawaguchi and Kondo, 2002), are surrounded by poly-anionic chondroitin sulphate-rich perineuronal nets that may serve as rapid local buffers

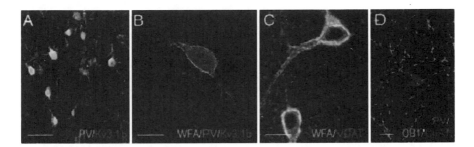

Figure 1. Some distinctive neurochemical hallmarks of cortical fast-spiking (FS) interneurons. (A) FS cells frequently contain the Ca^{2+}-binding protein parvalbumin (PV) and express unique combinations of ion channels that provide anatomical substrates for the maintenance of high-frequency firing. For instance, cortical FS cells show selective expression of the potassium channel 3.1b (Kv3.1b) subunit. (B) In addition, FS cells are often surrounded by perineuronal nets, as revealed by *Wisteria floribunda* agglutinin (WFA) labelling, that likely function as polyanionic buffers for rapid imbalances in cation concentrations during electrical activity (Hartig et al., 1999). (C) FS cell terminals contain high concentrations of vesicular GABA transporter (VGAT) and are embedded in a dense GABAergic interneuron network. (D) FS cells are devoid of CB_1 receptor immunoreactivity but receive modulatory input from various sources including CB_1 receptor-positive and cholinergic (choline-acetyltransferase (ChAT)-immunoreactive) origin. *Scale bars* = 85 μm (A), 20 μm (B,C), 16 μm (D).

(A color version of this figure appears in the signature between pp. 256 and 257.)

of excess cation changes in the extracellular space (Hartig et al., 1999), and selectively express potassium channel (Kv) subunits (e.g. Kv3.1b; Weiser et al., 1995) facilitating the fast repolarisation of their membranes (Fig. 1). FS cells form axo-somatic and/or axo-dendritic contacts on pyramidal cells (Kawaguchi and Kubota, 1997; Gupta et al., 2000; Kawaguchi and Kondo, 2002; Wang et al., 2002; Angulo et al., 2003). Importantly, FS cells contain low levels, if any, of CB_1 receptor mRNA transcripts and lack CB_1 receptor protein shown by *in situ* hybridisation (Marsicano and Lutz, 1999) and immunocytochemistry (Katona et al., 1999; Harkany et al., 2003), respectively (Fig. 1). One FS cell subtype in layer 2/3 of the neocortex, fast-spiking non-accommodating (FSN) interneurons, was chosen for studying endocannabinoid-independent retrograde signalling at inhibitory connections onto pyramidal cells.

In unitary connections between FSN interneurons and pyramidal cells, the elevation of postsynaptic dendritic Ca^{2+} concentration by backpropagating action potentials (bAPs) (Fig. 2A) induces release of retrograde messenger(s). This, in turn, leads to the inhibition of GABA release probability in interneuron axon terminals and consequent short-term synaptic depression (Fig. 2B) (Zilberter, 2000; Harkany et al., 2004) as was indicated by the decreased mean IPSP amplitude during conditioning (51% of control). The on- and off-rates of IPSP depression were relatively fast, less than one second, consigning this type of synaptic plasticity as short-term.

Importantly, the number of APs in the conditioning train determines the magnitude of IPSP depression, suggesting a direct relationship with the extent of dendritic Ca^+ rise (Zilberter, 2000). Moreover, the IPSP depression can be completely prevented by the Ca^{2+} chelators BAPTA and ethylene-bis(oxyethylenenitrilo)tetraacetic acid (EGTA) when applied intracellularly in the postsynaptic cell. These results implicate dendritic Ca^{2+} as a trigger of synaptic depression at FSN-pyramidal cell synapses. A significant

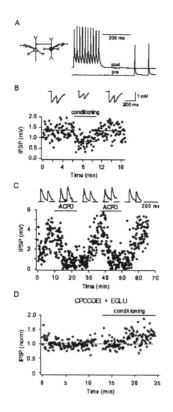

Figure 2. Synaptic depression induced by retrograde signalling in inhibitory connections between FSN interneurons and pyramidal cells. (A) Stimulation protocol used for initiation of the retrograde messenger release from the pyramidal cell dendrites. A train of bAPs (10 or less with a frequency of 50 Hz) was initiated in a pyramidal cell by somatic current injection. The train provided an increase in dendritic Ca^{2+} concentration due to activation of voltage-dependent Ca^{2+} channels. Then, 250 ms after the train, two APs were induced in an interneuron. The two sequential synaptic responses provided a paired-pulse ratio (the IPSP amplitude in response to the second AP divided by the IPSP amplitude in response to the first AP) that is a measure of synaptic behaviour: either depressing (ratio < 1) or facilitating (ratio > 1). A depressing mode is characteristic for synapses formed by FSN interneuron (see B). During conditioning by AP trains the amplitude of first IPSP was measured and plotted in the graph as shown in (B). (B) IPSP depression induced by conditioning with bAPs. Horizontal lines show the mean IPSP amplitude within indicated times. These IPSPs are also shown on the top. (C) ACPD, a group 1 and 2 mGluR agonist, reversibly blocked synaptic transmission at FSN-pyramidal cell connections. (D) A mixture of the antagonists of group 1 and 2 mGluRs, CPCCOEt and EGLU, prevented retrograde signalling at FSN- pyramidal cell synapses. Each open circle represents the average of normalized IPSPs at equivalent times in different experiments. Closed circles represent the average within each minute of the experimental protocol.

increase in the paired-pulse ratio during synaptic depression suggests a presynaptic locus of this phenomenon implying that a retrograde messenger should be released from a postsynaptic dendrite affecting some type of detecting, likely non-endocannabinoid, receptor in FSN axon terminals. Fig. 2B shows that the group I/II mGluR agonist aminocyclopentane-1,3-dicarboxylic acid (ACPD) strongly inhibited synaptic efficacy at FSN-pyramidal cell connections. Importantly, a synaptic mode of behaviour estimated by the IPSP paired-pulse ratio changed from depressing in control, to facilitating under ACPD exposure. Moreover, a number of synaptic failures occurred under ACPD exposure, while absent under control conditions. These data suggest that mGluRs may be present in axon terminals of FSN interneurons. Further experiments revealed that a mixture of the selective mGluR group I/II antagonists 7-(Hydroxyimino)-cyclopropa[b]chromen-1a-carboxylate ethyl ester (CPCCOEt)/(2S)-α-Ethylglutamic acid (EGLU) completely prevented the initiation of IPSP depression by bAP trains (Fig. 2C).

These findings suggest that glutamate, released from a pyramidal cell dendrite, could be a likely candidate for serving the role of a retrograde messenger. This hypothesis was however challenged by the observations that endocannabinoid signalling can also be enhanced by an elevation in dendritic Ca^{2+} levels or activation of mGluRs (Di Marzo et al., 1994; Di Marzo et al., 1998; Maejima et al., 2001; Varma et al., 2001).

Therefore, it is important to consider the potential involvement of endocannabinoids in the retrograde control of synaptic efficacy at FSN-pyramidal cell synapses.

The question whether endocannabinoids play a role at FSN-pyramidal cell synapses can first be addressed by examining the pattern of CB_1 receptor mRNA levels (Marsicano and Lutz, 1999) and immunoreactivity in layer 2/3 of the neocortex (Tsou et al., 1998; Egertova et al., 2003; Morozov and Freund, 2003) with a particular interest in the presence, or absence, of CB_1 receptors in the axons of FSN interneurons (Fig. 3A). In general terms, CB_1 receptor immunoreactivity reveals a dense fibre network in conjunction with perisomatic CB_1 receptor labelling in cholecystokinin-containing GABAergic interneurons in the neocortex. Application of intracellular labelling methods with CB_1 receptor immunocytochemistry confirmed a lack of CB_1 receptors in pre-synaptic terminals of FSN interneurons (Fig. 3B, B'). Electrophysiological recordings verified the lack of endocannabinoid-mediated retrograde signalling in FSN-pyramidal cell synapses inasmuch as the selective cannabinoid receptor agonist WIN 55,212-2 did not affect the efficacy of synaptic transmission, while synaptic depression still could be induced during conditioning in the presence of CB_1 receptor antagonist AM 251 (Harkany et al., 2004). Therefore, it seems plausible to argue that FSN terminals do not contain functionally relevant amounts of CB_1 receptors and endocannabinoids do not contribute to retrograde signalling at inhibitory synapses between FSN interneurons and pyramidal cells.

Multiple findings indicate that glutamate can act as a retrograde messenger at synapses with endocannabinoid-independent retrograde signalling. For example, this type of retrograde signalling can be prevented by botulinum toxin D-induced blockade of exocytosis of synaptic vesicles (Zilberter, 2000). Hence, a mechanism involving vesicular exocytosis of glutamate from pyramidal cell dendrites could be proposed. Dendritic exocytosis of neurotransmitters is a broadly accepted regulatory mechanism for tuning synaptic efficacy in various brain regions (for review see Ludwig and Pittman, 2003). In addition, Ca^{2+}-dependent exocytosis of dendritic organelles was directly observed in cultured hippocampal pyramidal cells using the fluorescent indicator FM1-43 (Maletic-Savatic and Malinow, 1998). Although electrophysiological studies highlight a role for the classical neurotransmitter glutamate as retrograde messenger, identification of the release machinery is still awaited.

Quantal neurotransmitter release requires accumulation of neurotransmitters in synaptic vesicles. Transport of classical neurotransmitters into storage vesicles relies on a proton electrochemical gradient generated by the vacuolar H^+-dependent adenosine triphosphatase (Forgac, 2000) and involves the exchange of protons for the cytoplasmic neurotransmitter. Recent studies (Bellocchio et al., 2000; Takamori et al., 2000; Fremeau et al., 2001; Takamori et al., 2001) demonstrated that glutamate uptake into synaptic vesicles in nerve endings is mediated by vesicular glutamate transporters 1 and 2 (VGLUT1 and VGLUT2), which are expressed in largely non-overlapping subpopulations of glutamatergic neurons throughout the central nervous system (Fremeau et al., 2001; Fujiyama et al., 2001; Kaneko et al., 2002; Li et al., 2003). In contrast, a third subtype of vesicular glutamate transporters, VGLUT3, exhibiting high structural and functional similarity to VGLUT1 and VGLUT2 (Fremeau et al., 2002; Gras et al., 2002; Schäfer et al., 2002; Takamori et al., 2002), was additionally found in various non-glutamatergic nerve cells including cholinergic, serotonergic and GABAergic neurons (Fremeau et al., 2002; Gras et al., 2002; Schäfer et al., 2002; Takamori et al., 2002; Harkany et al., 2003). In addition to its presence in nerve terminals of GABAergic

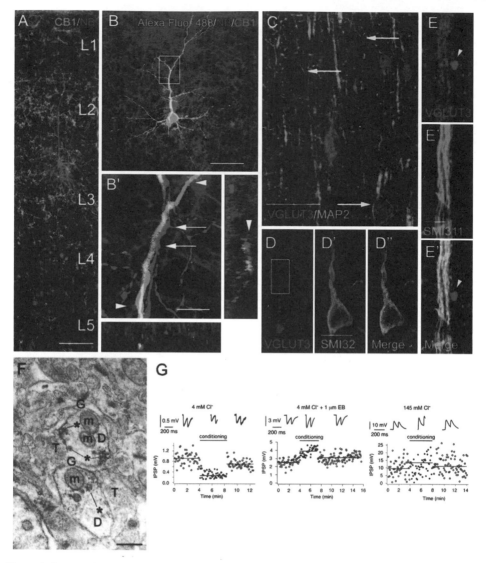

Figure 3. Retrograde signalling at FSN-pyramidal cell synapses is mediated via dendritic vesicular release of glutamate. (A) CB_1 receptor immunocytochemistry revealed highest density of CB_1 receptor-immunoreactive fibres in layer 2/3 of the neocortex, whereas CB_1 receptor-positive cells were predominantly found in deep cortical layers. A neurobiotin (NB)-labelled FSN interneuron-pyramidal cell pair appears in red, and lack CB_1 receptor immunolabelling. L1 – L5 denote cortical layers. (B) Co-localization of CB_1 receptors with identified neuron pairs demonstrated a lack of CB_1 immunoreactivity at inhibitory connections (*open square*). (B') High-power reconstruction of particular inhibitory contacts along the z-axis revealed CB_1 receptor-negative contacts between FSN interneurons and pyramidal cells. Arrows denote CB_1 receptor-negative synaptic contacts between identified cells, whereas arrowheads indicate separate CB_1 receptor-immunoreactive synapses of as yet unidentified origin. AF-488: Alexa Fluor 488. (C-F) VGLUT3-like immunoreactivity is present in dendrites of pyramidal cells in L2/3 of neocortex. (C) Laser-scanning microscopy demonstrated the presence of VGLUT3-like immunoreactivity in populations of putative pyramidal cells immunoreactive for microtubule-associated protein 2 (MAP2, *arrows*). (D-D'') High-power image of a pyramidal cell in layer 2/3 of the neocortex with perisomatic and dendritic VGLUT3-like immunoreactivity. Open square indicates the general localization of (E-E''). (E-E'') VGLUT3-like immunoreactivity in the apical dendritic shaft of a pyramidal cell. Arrowhead denotes VGLUT3-like immunoreactive terminal with a presumed origin in interneurons. (F) Immuno-electron microscopy revealed VGLUT3-like immunoreactivity in vesicle-like structures (*) of pyramidal cell dendrites in rat neocortex. VGLUT3-immunoreactive vesicle-like structures frequently appeared post-synaptically in the vicinity of terminals. *Abbreviations*: D, dendrite; G, glia; m, mitochondria; T, nerve terminal. (G) Evans blue (EB) and high chloride concentrations inhibit glutamate uptake by VGLUT3 also prevent retrograde signalling during conditioning by bAPs. *Scale bars* = 120 µm (A), 65 µm (B), 12 µm (B'), 28 µm (C), 15 µm (D'), 3 µm (E'), 500 nm (F). **(A color version of this figure appears in the signature between pp. 256 and 257.)**

interneurons, VGLUT3 was also localized to membranous structures in dendrites of striatal and hippocampal neurons with interneuron-like morphology (Fremeau et al., 2002). Although the functional significance of VGLUT3 in dendritic compartments of various non-glutamatergic neuron populations is unclear, its possible participation in retrograde signalling has been proposed (Fremeau et al., 2002; 2004).

Although the widespread presence of a VGLUT family member in non-glutamatergic neuron populations and astroglial cells (Montana et al., 2004) is functionally puzzling, its expression in glutamatergic pyramidal cell dendrites could subserve a function for controlling retrograde glutamate release. Recent data in our laboratories using high-resolution confocal laser-scanning microscopy demonstrated the presence of VGLUT3-like immunolabelling in the somata and dendritic arbour of layer 2/3 pyramidal cells (Fig. 3B-D). VGLUT3-like immunoreactivity was localized to proximal, primary and secondary compartments of apical and basal dendrites (Fig. 3C,D) in pyramidal cells. VGLUT3-like labelling was evident in dendritic shafts, the primary site of inhibitory synapses (Freund and Gulyas, 1991), but not in dendritic spines known to receive predominantly excitatory synaptic contacts (Baude et al., 1995). Occasionally, VGLUT3-immunoreactive terminals with a presumed origin in interneurons were in close apposition to VGLUT3-containing dendrites (Fig. 3E-E''). On the electron microscopic level, VGLUT3-like immunoreactivity was present in dendrites of pyramidal cells and vesicle-like structures labelled for this transporter were associated with post-synaptic densities in sub-synaptic dendrites (Fig. 3F). In accordance with the results obtained by laser-scanning microscopy, VGLUT3-like labelling was present in dendritic shafts (Fig. 3F). Additional tests showed that VGLUT3 is the only known VGLUT present in pyramidal cell dendrites (Harkany et al., 2004).

If VGLUT3 participates in dendritic vesicular release of glutamate, modulation of the transporter efficacy should directly affect retrograde signalling. Dyes structurally related to glutamate, such as a biphenyl derivative of 1,3-naphthalene disulfonic acid, known as Evans Blue (EB), are potent, competitive VGLUT antagonists in the nanomolar range (EB: IC_{50} = 87 nM) (Roseth et al., 1995; Bellocchio et al., 2000; Schäfer et al., 2002). In addition, VGLUT subtypes are sensitive to changes in the intracellular Cl^- concentration ($[Cl^-]_i$; potentiation: 1 – 4 mM, maximum inhibition: >140 mM)

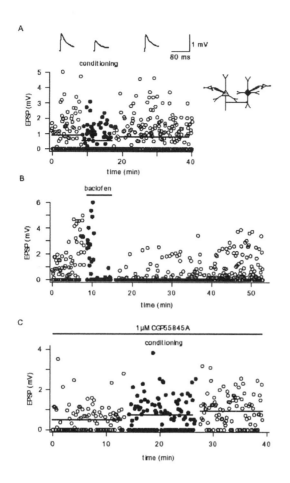

Figure 4. Synaptic depression induced by dendritic GABA release at excitatory synapses between pyramidal cells and bitufted interneurons. (A) EPSP depression induced by conditioning with bAPs. Horizontal lines show the mean EPSP amplitude within indicated times. These EPSPs are also shown on the top. (B) Baclofen, a selective agonist of GABA_B receptors, reversibly blocked synaptic transmission at pyramidal-bitufted cell connections. (C) A selective antagonist of GABA_B receptors, CGP55845A, prevented retrograde signaling during conditioning by bAPs.

(Bellocchio et al., 2000; Schäfer et al., 2002). Intracellular application of EB (0.5-1 μM) in pyramidal cells abolished synaptic depression during conditioning, inducing IPSP potentiation instead (Fig. 3G). In addition, retrograde signalling was abolished by an increase of $[Cl^-]_i$ to 145 mM (Fig. 3G). As in the case of EB application, conditioning induced potentiation of IPSPs. Our findings provide evidence that accumulation of glutamate in storage vesicles, likely controlled by VGLUT3 or related, as yet unidentified VGLUTs, and subsequent dendritic glutamate release, mediate retrograde signalling between FSN interneurons and pyramidal cells.

3. RETROGRADE SIGNALLING IN PYRAMIDAL CELL-BITUFTED INTERNEURON EXCITATORY SYNAPSES: SPOTLIGHT ON GABA

Retrograde signalling was also observed at excitatory synapses formed by pyramidal cells onto dendrites of bitufted interneurons in layer 2/3 (Zilberter et al., 1999). Bitufted interneurons contain somatostatin and, opposite to FSN interneurons, show a strongly facilitating behaviour of synaptic transmission (Reyes et al., 1998). Conditioning by trains of bAPs induced depression of EPSPs in pyramidal-to-bitufted cell synapses (Fig. 4A). The on- and off- rates of EPSP depression were fast, similar to those in FSN-pyramidal cell synapses. Intracellular application of Ca^{2+} buffers, such as BAPTA and EGTA, prevented synaptic depression. Both the paired-pulse ratio and the number of synaptic failures significantly changed during conditioning, indicating that the depression was expressed at the presynaptic site. As in case of inhibitory synapses, presynaptic metabotropic receptors seemed to be involved in mediating retrograde signalling. Fig. 4B demonstrates that baclofen, a selective agonist of $GABA_B$ receptors, reversibly inhibited the efficacy of pyramidal-bitufted cell synapses. Meanwhile, the selective $GABA_B$ receptor antagonist CGP55845A completely prevented the EPSP depression (Fig. 4C). Thus, dendritic release of GABA was suggested as a mechanism underlying synaptic depression in pyramidal-bitufted cell connections. In order to exclude the possibility of endocannabinoid involvement into this process, immunocytochemical and electrophysiological analyses were carried out showing the lack of CB_1 expression in the pyramidal cell axon terminals (*unpublished observations*) in conjunction with the inability of CB_1 receptor agonists and antagonists to affect synaptic efficacy between pyramidal cells and bitufted interneurons.

4. CAN RETROGRADE SIGNALLING BE LOCALIZED TO ACTIVE SYNAPTIC CONTACTS?

Another important consideration was whether retrograde signalling can be localized to the individual synaptic contacts. This question was recently addressed by Kaiser et. al. (2004) where individual synapses between pyramidal cells and bitufted interneurons were identified and synaptic Ca^{2+} transients were recorded. Reconstruction of synaptically connected cell pairs and electron microscopy revealed that pyramidal cells formed on average 3.9 ± 2.8 (mean±SD) excitatory synapses, which were located on the shaft of proximal dendrite segments of bitufted interneuron (33 ± 18 μm; distance from soma). Fig. 5A illustrates one of the Ca^{2+} imaging experiments. An active synaptic contact was identified at the region of intersection of a pyramidal cell axon with a bitufted interneuron dendrite (region 1). APs initiated in the pyramidal cell induced sub-threshold EPSPs in the interneuron and the corresponding Ca^{2+} signal was recorded in the region 1 (upper trace in Fig. 5A). In the dendritic regions located nearby (as labelled on the same dendrite, Fig. 5A), Ca^{2+} signals were much smaller and disappeared 10 μm away from the active synaptic contact. Utilizing the facilitating mode of synaptic behaviour it was possible to initiate an AP in the interneuron by repeated stimulation of the pyramidal cell. In this case, when synaptic stimulation coincided with a bAP, the Ca^{2+} signal in the active synaptic contact was strongly supralinearly amplified (lower trace in Fig. 5A). The amplitude of such Ca^{2+} transients reached 200 nM which is high enough to reliably induce synaptic retrograde signalling, as follows from the graph shown in Fig. 5B. The graph demonstrates the dependence of EPSP depression on an increase in dendritic Ca^{2+} concentration (Zilberter et al., 1999). In summary, reduction of glutamate release by dendritic electrical activity is mediated by a global rise of dendritic $[Ca^{2+}]_i$ with a threshold of 50 nM and a half-maximal effect at 264 nM. Thus, even a single bAP induced by synaptic stimulation can initiate retrograde signalling at individual synaptic contacts indicating that this type of retrograde signalling can be highly compartmentalized.

5. FUNCTIONAL IMPORTANCE OF RETROGRADE SIGNALLING IN LOCAL NEOCORTICAL NETWORKS

FSN, as well as bitufted interneurons, communicate with the majority of local pyramidal cells (Reyes et al., 1998; Holmgren et al., 2003). In reciprocally connected pyramidal cells and interneurons, retrograde signalling regulates synaptic efficacy at both excitatory and inhibitory connections (Fig. 6A). What may be the functional significance of short-term synaptic depression in the pyramidal-interneuron microcircuit, assuming that the main role of the interneuron is to control the excitability of the innervated pyramidal cell? In the absence of synaptic depression induced by retrograde messengers, backward inhibition by the interneuron creates a deep negative feedback for the pyramidal cell; the more the pyramidal cell becomes excited, the more it will be inhibited by interneuron(s). In both types of synapses, however, retrograde messengers decrease the regulatory effects of the interneuron by inhibiting its excitatory input and inhibitory output. Thus, short-term synaptic depression mediated by retrograde messengers in this microcircuit supports the excitation of pyramidal cells. The balance between impacts of axonal and dendritic parts of transmitter release on pyramidal cell excitation presumably modulates the temporal pattern of pyramidal cell output activity.

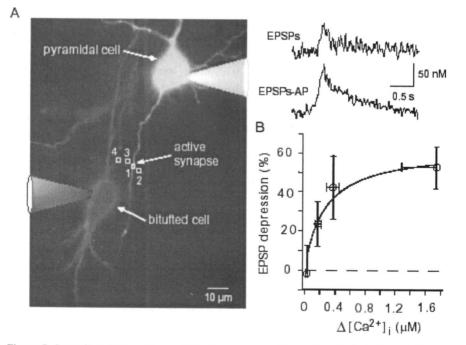

Figure 5. Synaptic activation of an individual synapse coincident with a single bAP can initiate retrograde signaling. (A) Fluorescence imaging of a pyramidal cell (*yellow*) synaptically connected with a bitufted interneuron (*blue*). Ca^{2+} transients at the active synapse during synaptic activation only, and synaptic activation coincident with a bAP are shown on the right. (B) Dependence of EPSP depression induced by retrograde GABA release on the Ca^{2+} concentration.

A

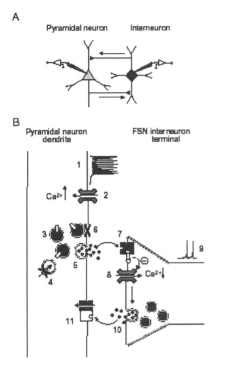

Figure 6. Regulation of the efficacy of inhibitory transmission by dendritic vesicular glutamate release. (A) Pyramidal cells in L2/3 of neocortex are frequently interconnected with inhibitory interneurons. (B) Different steps of retrograde glutamate signaling in the FSN-pyramidal cell synapse. In a postsynaptic pyramidal neuron, bAPs (*1*) induce Ca^{2+} influx into a dendrite *via* voltage-dependent Ca^{2+} channels (*2*). VGLUT3 mediates accumulation of glutamate in storage vesicles (*3*), while glutamate uptake can be terminated by VGLUT inhibition (*4*). Ready for release vesicles exocytose glutamate into the synaptic cleft (*5*). Exocytosis can be prevented by selective blockers (*6*). In the axon terminal of an interneuron, glutamate activates mGluRs (*7*) that results in a down-regulation of Ca^{2+} influx into the terminal *via* voltage-dependent Ca^{2+} channels (*8*) activated by axonal APs (*9*) with a consequent decrease in release probability of GABA (*10*). Finally, this results in a reduced activation of $GABA_A$ receptors (*11*), i.e., in synaptic depression.

Retrograde signalling mediated by dendritic glutamate release is not limited to the neocortical layer 2/3. In layer 5 of neocortex, Ali et al. found that depolarisation of postsynaptic pyramidal cells caused the dendritic release of an endogenous messenger, most likely glutamate, which activated kainate receptors on fast spiking interneuron axon terminals and induced depression of unitary IPSCs (Ali et al., 2001). This synaptic feedback might use a glutamate release mechanism similar to that in layer 2/3 pyramidal cells, although the presynaptic mechanism is different. It is noteworthy that the same pyramidal cell in layer 2/3 can presumably utilize functionally different mechanisms of retrograde signalling. Pyramidal cells receive inhibitory input from CB_1 receptor-positive nerve fibres (Fig. 3B) most likely belonging to cholecystokinin-positive local interneurons. Synaptic efficacy of corresponding perisomatic connections is modulated by dendritic endocannabinoid release (Trettel et al., 2004). Thus, hypothetically, pyramidal cells can simultaneously utilize distinct mechanisms of synaptic feedback to control the efficacy of input synapses along their dendritic arbours.

6. CONCLUSIONS

The results of retrograde signalling studies in FSN- pyramidal cell and pyramidal cell-bitufted interneuron connections show that (*i*) retrograde signalling is mediated *via* activation of presynaptic metabotropic receptors; (*ii*) retrograde signalling strongly depends on postsynaptic Ca^{2+} and is prevented by antagonists of vesicular exocytosis; (*iii*) a single bAP coincident with synaptic activity may initiate retrograde signalling; (*iv*) retrograde signalling is restricted to particular active synapses, and (*v*) vesicular neurotransmitter transporters in dendritic compartments may underlie such mechanisms.

Thus, the axonal and dendritic release machineries provide a dual role for a neurotransmitter in neuronal communication. Presynaptically localized metabotropic receptors serve as high affinity sensors of retrograde messengers. Importantly, neurotransmitter release from cell dendrites follows the classical scenario of quantal exocytosis, suggesting that neurotransmitter-mediated retrograde synaptic communication, in contrast to endocannabinoid signalling, affects only particular synapses in a 'targeted' fashion.

7. SOME UNRESOLVED QUESTIONS

Although available results strongly suggest vesicular exocytosis as a mechanism of retrograde messenger release, detailed studies exploring particular facets of the molecular machinery controlling dendritic neurotransmitter release are lacking. It is unclear whether such processes are similar, if at all, to neurotransmitter release from axon terminals. Both mechanisms are sensitive to SNARE antagonists, however their Ca^{2+} sensitivity seems to be different. In addition, the density of synaptic vesicles in dendritic specializations is lower than that in axon terminals.

Since vesicular glutamate release is likely to be involved in retrograde signalling in FSN-pyramidal cell connections, it is tempting to speculate that this type of dendritic transmitter release may be a phenomenon broadly present at neocortical synapses. However, additional data must be gathered in other local circuitries before making any such conclusions.

A critical unsolved question related to the functional role of inhibitory interneurons in local neocortical networks is how the activity of the network during information processing influences the pattern of firing in the output of a pyramidal neuron considering the limited synaptic signalling received by each pyramidal cell from other network members. According to our recent hypothesis (Holmgren et al., 2003), modulation of the firing rate could be achieved by pyramidal-interneuron-pyramidal communications. In this context, communication implies a modulation of interspike intervals, which can be achieved either *via* excitatory or inhibitory inputs. Thus, the number of interacting pyramidal cells may be considerably increased by suggesting that pyramidal cells communicate both directly and *via* 'relaying' local interneurons. In this case the number of communicating pyramidal cells would appear at least ten fold higher.

In layer 2/3 of the neocortex, at least four different types of interneurons have been found that form reciprocal connections with pyramidal cells (Buhl et al., 1997; Markram et al., 1998; Reyes et al., 1998; Zilberter, 2000; Rozov et al., 2001; Wang et al., 2002). It was also shown that a pyramidal cell might simultaneously be connected with more than one interneuron subtype (Markram et al., 1998; Reyes et al., 1998). Importantly, the properties of pyramidal-interneuron synaptic transmission differ considerably depending on the interneuron type (Markram et al., 1998; Reyes et al., 1998). Thus, pyramidal cells in local networks may form functionally distinct microcircuits with various types of local interneurons, differentially affecting each others' firing patterns. Hence, detailed, realistic computational models are required to appreciate the role of retrograde signalling, originating from different interneurons, within local neuronal networks and their output in modulating the firing patterns of pyramidal cells.

8. REFERENCES

Alger, B. E., 2002, Retrograde signalling in the regulation of synaptic transmission: focus on endocannabinoids, *Progr. Neurobiol.* **68**: 247.

Ali, A. B., Rossier, J., Staiger, J. F., and Audinat, E., 2001, Kainate receptors regulate unitary IPSPs elicited in pyramidal cells by fast-spiking interneurons in the neocortex, *J. Neurosci.* **21**: 2992.

Angulo, M. C., Staiger, J. F., Rossier, J., and Audinat, E., 2003, Distinct local circuits between neocortical pyramidal cells and fast-spiking interneurons in young adult rats, *J. Neurophysiol.* **89**: 943.

Baude, A., Nusser, Z., Molnár, E., McIlhinney, R. A., and Somogyi, P., 1995, High-resolution immunogold localization of AMPA type glutamate receptor subunits at synaptic and non-synaptic sites in rat hippocampus, *Neuroscience* **69**: 1031.

Bellocchio, E. E., Reimer, R. J., Fremeau, R. T. Jr., and Edwards, R. H., 2000, Uptake of glutamate into synaptic vesicles by an inorganic phosphate transporter, *Science* **289**: 957.

Buhl, E. H., Tamas, G., Szilagyi, T., Stricker, C., Paulsen, O., and Somogyi, P., 1997, Effect, number and location of synapses made by single pyramidal cells onto aspiny interneurons of cat visual cortex, *J. Physiol.* **500**: 689.

Devane, W. A., Hanus, L., Breuer, A., Pertwee, R. G., Stevenson, L. A., Griffin, G., Gibson, D., Mandelbaum, A., Etinger, A., and Mechoulam, R., 1992, Isolation and structure of a brain constituent that binds to the cannabinoid receptor, *Science* **258**: 1946.

Di Marzo, V., Melck, D., Bisogno, T., and De Petrocellis, L., 1998, Endocannabinoids: endogenous cannabinoid receptor ligands with neuromodulatory action, *Trends Neurosci.* **21**: 521.

Di Marzo, V., Fontana, A., Cadas, H., Schinelli, S., Cimino, G., Schwartz, J. C., and Piomelli, D., 1994, Formation and inactivation of endogenous cannabinoid anandamide in central neurons, *Nature* **372**: 686.

Egertova, M., Cravatt, B. F., and Elphick, M. R., 2003, Comparative analysis of fatty acid amide hydrolase and cb(1) cannabinoid receptor expression in the mouse brain: evidence of a widespread role for fatty acid amide hydrolase in regulation of endocannabinoid signalling, *Neuroscience* **119**: 481.

Fitzsimonds, R. M., and Poo, M.-M., 1998, Retrograde signalling in the development and modification of synapses, *Physiol. Rev.* **78**: 143.

Forgac, M., 2000, Structure, mechanism and regulation of the clathrin-coated vesicle and yeast vacuolar H(+)-ATPases, *J. Exp. Biol.* **203**: 71.

Fremeau, R. T. Jr., Voglmaier, S., Seal, R. P., and Edwards, R. H., 2004, VGLUTs define subsets of excitatory neurons and suggest novel roles for glutamate, *Trends Neurosci.* **27**: 98.

Fremeau, R. T. Jr., Troyer, M. D., Pahner, I., Nygaard, G. O., Tran, C. H., Reimer, R. J., Bellocchio, E. E., Fortin, D., Storm-Mathisen, J., and Edwards, R. H., 2001, The expression of vesicular glutamate transporters defines two classes of excitatory synapse, *Neuron* **31**: 247.

Fremeau, R. T. Jr., Burman, J., Qureshi, T., Tran, C. H., Proctor, J., Johnson, J., Zhang, H., Sulzer, D., Copenhagen, D. R., Storm-Mathisen, J., Reimer, R. J., Chaudhry, F. A., and Edwards, R. H., 2002, The identification of vesicular glutamate transporter 3 suggest novel modes of signalling by glutamate, *PNAS USA* **99**: 14488.

Freund, T. F., and Gulyas, A. I., 1991, GABAergic interneurons containing calbindin D28K or somatostatin are major targets of GABAergic basal forebrain afferents in the rat neocortex, *J. Comp. Neurol.* **314**: 187.

Fujiyama, F., Furuta, T., and Kaneko, T., 2001, Immunocytochemical localization of candidates for vesicular glutamate transporters in the rat cerebral cortex, *J. Comp. Neurol.* **435**: 379.

Glitsch, M., Llano, I., and Marty, A., 1996, Glutamate as a candidate retrograde messenger at interneuron-Purkinje cell synapses of rat cerebellum, *J. Physiol.* **497**: 531.

Gras, C., Herzog, E., Bellenchi, G. C., Bernard, V., Ravassard, P., Pohl, M., Gasnier, B., Giros, B., and El Mestikawy, S., 2002, A third vesicular glutamate transporter expressed by cholinergic and serotoninergic neurons, *J. Neurosci.* **22**: 5442.

Gupta, A., Wang, Y., and Markram, H., 2000, Organizing principles for a diversity of GABAergic interneurons and synapses in the neocortex, *Science* **287**: 273.

Harkany, T., Härtig, W., Berghuis, P., Dobszay, M. B., Zilberter, Y., Edwards, R. H., Mackie, K., and Ernfors, P., 2003, Complementary distribution of type 1 cannabinoid receptors and vesicular glutamate transporter 3 in basal forebrain suggests input-specific retrograde signalling by cholinergic neurons, *Eur. J. Neurosci.* **18**: 1979.

Harkany, T., Holmgren, C., Härtig, W., Qureshi, T., Chaudhry, F. A., Storm-Mathisen, J., Dobszay, M. B., Berghuis, P., Schulte, G., Fremeau, R. T. Jr., Edwards, R. H., Mackie, K., Ernfors, P., and Zilberter, Y., 2004, Endocannabinoid-independent retrograde signalling at inhibitory synapses in layer 2/3 of neocortex: Involvement of vesicular glutamate transporter 3, *J. Neurosci.* (in press).

Härtig, W., Derouiche, A., Welt, K., Brauer, K., Grosche, J., Mader, M., Reichenbach, A., and Bruckner, G., 1999, Cortical neurons immunoreactive for the potassium channel Kv3.1b subunit are predominantly surrounded by perineuronal nets presumed as a buffering system for cations, *Brain Res.* **842**: 15.

Herkenham, M., Lynn, A. B., Little, M. D., Johnson, M. R., Melvin, L. S., de Costa, B. R., and Rice, K. C., 1990, Cannabinoid receptor localization in brain, *PNAS USA* **87**: 1932.

Holmgren, C., Harkany, T., Svennenfors, B., and Zilberter, Y., 2003, Pyramidal cell communication within local networks in layer 2/3 of rat neocortex, *J. Physiol.* **551**: 139.

Hsieh, C. S., Brown, S., Derleth, C., and Mackie, K., 1999, Rapid internalization and recycling of the CB1 cannabinoid receptor, *J. Neurochem.* **73**: 493.

Jones, E. G, and Peters, A., 1984, Functional properties of cortical cells. in, *Cerebral Cortex*, E. G.Jones, and A. Peters, eds, Plenum Press, New York and London.

Kaiser, K. M. M., Lubke, J., Zilberter, Y., and Sakmann, B., 2004, Postsynaptic calcium influx at single synaptic contacts between pyramidal neurons and bitufted interneurons in layer 2/3 of rat neocortex is enhanced by backpropagating action potentials, *J. Neurosci.* **24**: 1319.

Kaneko, T., Fujiyama, F., and Hioki, H., 2002, Immunohistochemical localization of candidates for vesicular glutamate transporters in the rat brain, *J. Comp. Neurol.* **444**: 39.

Katona, I., Sperlagh, B., Sik, A., Kafalvi, A., Vizi, E. S., Mackie, K., and Freund, T. F., 1999, Presynaptically located CB1 cannabinoid receptors regulate GABA release from axon terminals of specific hippocampal interneurons, *J. Neurosci.* **19**: 4544.

Kawaguchi, Y., 1995, Physiological subgroups of nonpyramidal cells with specific morphological characteristics in layer II/III of rat frontal cortex, *J. Neurosci.* **15**: 2638.

Kawaguchi, Y., and Kubota, S., 1997, GABAergic cell subtypes and their synaptic connections in rat frontal cortex, *Cereb. Cortex* **7**: 476.

Kawaguchi, Y., and Kondo, S., 2002, Parvalbumin, somatostatin and cholecystokinin as chemical markers for specific GABAergic interneuron types in the rat frontal cortex, *J. Neurocytol.* **31**: 277.

Kreitzer, A. C., and Regehr, W. G., 2001a, Retrograde inhibition of presynaptic calcium influx by endogenous cannabinoids at excitatory synapses onto Purkinje cells, *Neuron* **29**: 717.

Kreitzer, A. C., and Regehr, W. G., 2001b, Cerebellar depolarization-induced supression of inhibition is mediated by endogenous cannabinoids, *J. Neurosci.* **21**: RC171.

Li, J. L., Fujiyama, F., Kaneko, T., and Mizuno, N., 2003, Expression of vesicular glutamate transporters, VGluT1 and VGluT2, in axon terminals of nociceptive primary afferent fibers in the superficial layers of the medullary and spinal dorsal horns of the rat, *J. Comp. Neurol.* **457**: 236.

Llano, I., Leresche, N., and Marty, A., 1991, Calcium entry increases the sensitivity of cerebellar Purkinje cells to applied GABA and decreases inhibitory synaptic currents, *Neuron* **6**: 565.

Ludwig, M., and Pittman, Q. J., 2003, Talking back: dendritic neurotransmitter release, *Trends Neurosci.* **26**: 255.

Maejima, T., Hashimoto, K., Yoshida, T., Aiba, A., and Kano, M., 2001, Presynaptic inhibition caused by retrograde signal from metabotropic glutamate to cannabinoid receptors, *Neuron* **31**: 463.

Maletic-Savatic, M., and Malinow, R., 1998, Calcium-evoked dendritic exocytosis in cultured hippocampal neurons. Part I: trans-Golgi network-derived organelles undergo regulated exocytosis, *J. Neurosci.* **18**: 6803.

Markram, H., Wang, Y., and Tsodyks, M., 1998, Differential signalling via the same axon of neocortical pyramidal neurons, *PNAS USA* **95**: 5323.

Marsicano, G., and Lutz, B., 1999, Expression of the cannabinoid receptor CB1 in distinct neuronal subpopulations in the adult mouse forebrain, *Eur. J. Neurosci.* **11**: 4213.

Montana, V., Ni, Y., Sunjara, V., Hua, X., and Parpura, V., 2004, Vesicular glutamate transporter-dependent glutamate release from astrocytes, *J. Neurosci.* **24**: 2633.

Morishita, W., Kirov, S. A., and Alger, B. E., 1998, Evidence for metabotropic glutamate receptors activation in the induction of depolarization-induced suppression of inhibition in hippocampal CA1, *J. Neurosci.* **18**: 4870.

Morozov, Y. M., and Freund, T. F., 2003, Post-natal development of type 1 cannabinoid receptor immunoreactivity in the rat hippocampus, *Eur. J. Neurosci.* **18**: 1213.

Ohno-Shosaku, T., Maejima, T., and Kano, M., 2001, Endogenous cannabinoids mediate retrograde signals from depolarized postsynaptic neurons to presynaptic terminals, *Neuron* **29**: 729.

Pitler, T. A., and Alger, B. E., 1992, Postsynaptic spike firing reduces synaptic $GABA_A$ responses in hippocampal pyramidal cells, *J. Neurosci.* **12**: 4122.

Reyes, A., Lujan, R., Rozov, A., Burnashev, N., Somogyi, P., and Sakmann, B., 1998, Target- cell-specific facilitation and depression in neocortical circuits, *Nat. Neurosci.* **1**: 279.

Roseth, S., Fykse, E. M., and Fonnum, F., 1995, Uptake of L-glutamate into rat brain synaptic vesicles: effect of inhibitors that bind specifically to the glutamate transporter, *J. Neurochem.* **65**: 96.

Rozov, A., Jerecic, J., Sakmann, B., and Burnashev, N., 2001, AMPA receptor channels with long-lasting desensitization in bipolar interneurons contribute to synaptic depression in a novel feedback circuit in layer 2/3 of rat neocortex, *J. Neurosci.* **21**: 8062.

Schäfer, M. K., Varoqui, H., Defamie, N., Weihe, E., and Erickson, J. D., 2002, Molecular cloning and functional identification of mouse vesicular glutamate transporter 3 and its expression in subsets of novel excitatory neurons, *J. Biol. Chem.* **277**: 50734.

Sjöström, P. J., Turrigiano, G. G., and Nelson, S. B., 2003, Neocortical LTD via coincident activation of presynaptic NMDA and cannabinoid receptors, *Neuron* **39**: 641.

Stella, N., Schweitzer, P., and Piomelli, D., 1997, A second endogenous cannabinoid that modulates long-term potentiation, *Nature* **388**: 773.

Takamori, S., Rhee, J. S., Rosenmund, C., and Jahn, R., 2000, Identification of a vesicular glutamate transporter that defines a glutamatergic phenotype in neurons, *Nature* **407**: 189.

Takamori, S., Rhee, J. S., Rosenmund, C., and Jahn, R., 2001, Identification of differentiation-associated brain-specific phosphate transporter as a second vesicular glutamate transporter (VGLUT2), *J. Neurosci.* **21**: RC182.

Takamori, S., Malherbe, P., Broger, C., and Jahn, R., 2002, Molecular cloning and functional characterization of human vesicular glutamate transporter 3, *EMBO J.* **3**: 798.

Trettel, J., and Levine, E. S., 2003, Endocannabinoids mediate rapid retrograde signalling at interneuron - pyramidal neuron synapses of the neocortex, *J. Neurophysiol.* **89**: 2334.

Trettel, J., Fortin, D. A., and Levine, E. S., 2004, Endocannabinoid signalling selectively targets perisomatic inhibitory inputs to pyramidal neurons in juvenile mouse neocortex, *J. Physiol.* **556**: 95.

Tsou, K., Brown, S., Sanudo-Pena, M. C., Mackie, K., and Walker, J. M., 1998, Immunohistochemical distribution of cannabinoid CB1 receptors in the rat central nervous system, *Neuroscience* **83**: 393.

Varma, N., Carlson, G. C., Ledent, C., and Alger, B. E., 2001, Metabotropic glutamate receptors drive the endocannabinoid system in hippocampus, *J. Neurosci.* **21**: RC181.

Wang, Y., Gupta, A., Toledo-Rodriguez, M., Wu, C. Z., and Markram, H., 2002, Anatomical, physiological, molecular and circuit properties of nest basket cells in the developing somatosensory cortex, *Cereb. Cortex* **12**: 395.

Weiser, M., Bueno, E., Sekirnjak, C., Martone, M. E., Baker, H., Hillman, D., Chen, S., Thornhill, W., Ellisman, M., and Rudy, B., 1995, The potassium channel subunit KV3.1b is localized to somatic and axonal membranes of specific populations of CNS neurons, *J. Neurosci.* **15**: 4298.

Wilson, R. I., and Nicoll, R. A., 2001, Endogenous cannabinoids mediate retrograde signalling at hippocampal slices, *Nature* **410**: 588.

Zilberter, Y., 2000, Dendritic release of glutamate suppresses synaptic inhibition of pyramidal neurons in rat neocortex, *J. Physiol.* **528**: 489.

Zilberter, Y., Kaiser, K. M. M., and Sakmann, B., 1999, Dendritic GABA release depresses excitatory transmission between L 2/3 pyramidal and bitufted neurons in rat neocortex, *Neuron* **24**: 979.

9

THE THALAMIC INTERNEURON

S. Murray Sherman[*]

1. INTRODUCTION

In the thalamus, local, GABAergic interneurons play a crucial role in controlling the flow of information relayed to the cortex. Throughout the thalamus, with occasional variations with relay nucleus or species, these interneurons comprise roughly 20-25% of the cells present, the remainder being relay cells (see Arcelli et al., 1997; reviewed in Jones, 1985; Sherman and Guillery, 2001). As is the case with relay cells (see also below), there is probably more than one type of interneuron (Sanchez-Vives et al., 1996; Carden and Bickford, 2002), but this account will focus on the best known and prime exemplar, namely, the interneuron found in the cat's lateral geniculate nucleus, which appears to be widespread throughout the thalamus. Two particularly interesting features of this cell are the nature of its synaptic outputs, which are both axonal and dendritic, and the type of synaptic circuits that are entered into by the dendritic outputs. These two features form the focus of the rest of this narrative.

2. OUTPUTS OF THE INTERNEURON

Figure 1 shows an example each of the typical interneuron found in the cat's lateral geniculate nucleus plus the two relay cell types, X and Y, that the interneuron innervates (Friedlander et al., 1981; Sherman and Friedlander, 1988). The X and Y cells represent the thalamic links in two parallel, largely independent retino-geniculo-cortical streams, an arrangement common to mammals (reviewed in Sherman, 1985). A particularly interesting feature of the interneuron is that it has two different avenues for synaptic output. One is a conventional axon that arborizes within the dendritic arbor, and the other are clusters of terminals that emanate from distal dendrites (Guillery, 1969; Ralston, 1971; Famiglietti and Peters, 1972; Wilson et al., 1984; Hamos et al., 1985). These presynaptic dendritic terminals are shown more clearly in the inset for the interneuron in Figure 1. Electron microscopy reveals that both the axonal and dendritic terminals appear

[*]Department of Neurobiology, State University of New York, Stony Brook, NY 11794-5230

Dendritic Neurotransmitter Release, edited by M. Ludwig
Springer Science+Business Media, Inc., 2005

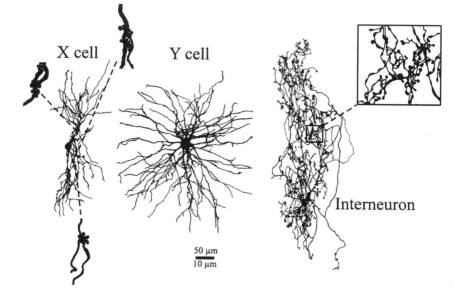

Figure 1. Representative cells of the cat's lateral geniculate nucleus. The cells were labeled intracellularly with horseradish peroxidase (from Friedlander et al., 1981; Sherman and Friedlander, 1988). Shown are relay X and Y cells and an interneuron. The scale bar is 50μm for the main drawings and 10μm for the insets.

to be inhibitory, because they have flattened or pleomorphic vesicles and form symmetric synaptic contacts, but they differ in other respects. One is that the axonal terminal, called *F1* ("F" for flattened vesicle), forms a simple, conventional synaptic contact onto the dendrites of both X and Y cells; whereas the dendritic terminal, called *F2*, is both postsynaptic to various terminals (see below) and presynaptic to X but rarely to Y cells. There are many more F2 outputs per interneuron than F1 outputs. Other differences between F1 and F2 outputs are noted below.

There is also a relationship between the structure of the relay cells and the different interneuron outputs. The Y cell, which receives predominantly F1 (axonal) inputs from the interneuron, has a fairly simple, radiate dendritic arbor with few appendages. The F1 terminals contact proximal dendritic shafts with simple, conventional synapses. The X cell has a bipolar dendritic tree, oriented perpendicular to laminar borders, but more interesting are the clusters of grape-like appendages near primary branch points (see higher power insets in Figure 1). These are not spines, lacking the apparatus found in true spines in cortex and hippocampus, but seem to be simple appendages. What is especially interesting about them is that they mark the postsynaptic target of both F2 (dendritic) terminals of interneurons as well as retinal terminals. F1 terminals onto X cells may contact these same appendages but also frequently contact dendritic shafts in the vicinity. Thus the vast majority of interneuron synapses onto both X and Y cells are onto proximal dendrites.

3. TRIADS AND GLOMERULI

As mentioned, the F2 terminals are both presynaptic and postsynaptic, and they commonly enter into complex synaptic arrangements known as *triads* (see Figure 2). Here, a single retinal terminal contacts both an F2 terminal and the appendage of a relay X cell, and the F2 terminal contacts the same appendage[†]. Thus, three synapses are involved: from the retinal terminal to the relay cell dendritic appendage, from the retinal terminal to the F2 terminal, and from the F2 terminal to the same appendage. Less commonly, an F2 terminal can be postsynaptic to a cholinergic terminal from the brainstem parabrachial region, and the same parabrachial axon (but not the same terminal) contacts the same relay cell (see Figure 2). While not as tightly organized as the common triad, this forms another sort of triadic arrangement, since three synapses are involved. Although not illustrated, occasionally F1 terminals are found presynaptic to F2 terminals (Guillery, 1969; Ralston, 1971; Famiglietti and Peters, 1972; Wilson et al., 1984; Hamos et al., 1985). As shown in Figure 2, F1 terminals from interneuron axons commonly contact dendritic shafts both within and outside of complex synaptic zones known as *glomeruli* (Szentágothai, 1963).

A simplified schema of a glomerulus is shown in Figure 2. Individual synaptic zones in the glomerulus are not juxtaposed to glial processes, but instead the entire synaptic complex is enclosed in a glial sheath. The function of this arrangement is not known, but since glial processes have been implicated in the uptake and regulation of neurotransmitters and other neuroactive substances (Pfrieger et al., 1992; Guatteo et al., 1996), their lack within a glomerulus might affect the extent to which neurotransmitters and other substances remain active.

Glomeruli occur in a range of complexities, from little more than a triad to a mass of tens of synapses. Every triad so far seen in the lateral geniculate nucleus of the cat contains at least one retinal terminal and one triad (and thus one F2 terminal). F1 terminals and parabrachial terminals are also commonly present in glomeruli, but terminals from layer 6 of cortex are virtually never there (Erişir et al., 1997; but see Vidnyanszky and Hamori, 1994). It is worth noting that F1 terminals commonly derive from axons of GABAergic cells, and the other main source of GABAergic innervation to relay cells besides interneurons are cells of the nearby thalamic reticular nucleus. However, while reticular axons focus their contacts onto relay cells rather than interneurons, they contact mostly distal dendrites and are rarely found in glomeruli (Cucchiaro et al., 1991; Wang et al., 2001). It is thus from a process of elimination that we conclude that most F1 terminals in glomeruli and onto proximal dendrites outside of glomeruli emanate from interneurons.

4. CELLULAR PROPERTIES OF THE INTERNEURON

Given that there are two output routes for this thalamic interneuron - axonal and dendritic - the obvious questions arise as to how these are controlled and how they might

---

[†]. Recently, Datskovskaia et al. (2001) have argued that there may be more Y retinal axon innervation of interneurons than previously thought, although Y cell involvement in triads was rarely seen, so this issue of the extent of the limitation to X cells of F2 and triadic innervation needs further resolution.

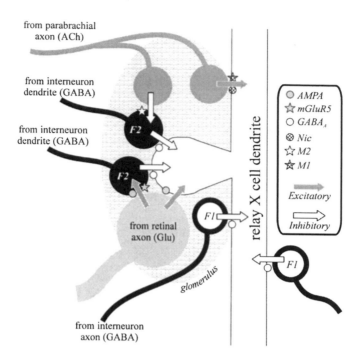

Figure 2. Schematic view of triad and glomerulus in cat lateral geniculate nucleus. Shown are the various synaptic contacts (arrows), whether they are inhibitory or excitatory, and the related postsynaptic receptors. For simplicity, the NMDA receptor on the relay cell postsynaptic to the retinal input has been left off. Abbreviations: *ACh*; acetylcholine; *AMPAR*, (RS)-α-amino-3-hydroxy-5-methyl-4-isoxazolepropionic acid receptor; *F1* and *F2*; two types of synaptic terminal; *GABA*, γ-aminobutyric acid; *GABA$_A$R*, type A receptor for GABA; *Glu*; glutamate; *M1R* and *M2R*, two types of muscarinic receptor; *mGluR5*, type 5 metabotropic glutamate receptor; *Nicotinic*; nicotinic receptor;

relate to one another. The cable properties of the dendritic arbor offer an initial clue. These properties determine, among other things, how current flows through the dendrites and thus how effective synaptic inputs at one site affect membrane voltage at others, including the axon hillock or spike generating region. For practical purposes, we can consider the cell body and axon hillock as isopotential, so part of the problem reduces to a consideration of how synapses at various sites affect membrane voltage at the cell body.

An analysis of cable properties of interneurons compared to relay cells in the cat's lateral geniculate nucleus showed a striking difference (Bloomfield et al., 1987, 1989). Relay cells were found to be fairly compact electrotonically, largely because of their dendritic branching pattern that supports impedance matching across the branches. This minimizes current leakage across the membrane and thus reduces the extent of voltage attenuation as current flows from a distal dendritic location, across several branch points, to the cell body. The conclusion is that even postsynaptic potentials (PSPs; excitatory and inhibitory postsynaptic potentials are, respectively, EPSPs and IPSPs) generated at distal synaptic locations will attenuate by at most ~50% upon arriving at the cell body. In contrast, the branching pattern of interneurons is different: there are more branches, and they seem to create an impedance mismatch, leading to current leakage across branch points. The result is that a PSP generated at a distal dendritic site will be attenuated by more than 90% at the cell body. The F2 terminals themselves are usually appended to the distal sites by long (10μm or more), thin (~0.1μm in diameter) processes (Hamos et al., 1985), which implies that inputs to these F2 terminals are even more isolated from the cell body and also from other F2 terminals on other dendritic branches (see also below).

Before considering further the implications of this cable modeling, there are several important provisos to be considered. Cable modeling requires many assumptions, including the value of certain parameters, such as membrane capacitance, that are basically guesses. Cables themselves act like low pass temporal filters, meaning that slow or sustained voltage changes will attenuate less than faster changes, and faster PSPs will attenuate more than the above modeling suggests. Perhaps most serious is the fact that cable modeling assumes a passive membrane, and in fact, interneurons, like all neurons, possess many ion channels that are dynamically gated (see also below); these will certainly affect cable properties. Thus, for example, the presence of voltage gated Na^+ or Ca^{2+} channels can greatly affect how PSPs are conducted to the cell body. Furthermore, the possibility of back-propagation of action potentials supported by dendritic Na^+ channels could mean that cell spiking will affect F2 terminal transmitter release.

Given these qualifications, cable modeling provides a useful if limited approach to understanding how inputs affect the postsynaptic cell. The picture that emerges for the interneuron is schematically shown in Figure 3, and there are two interesting conclusions drawn. First, because the branched dendritic arbor strongly attenuates PSPs from distal inputs, the main control of the axonal (F1) output is mostly limited to relatively proximal inputs. It is interesting in this context that these interneurons receive many synaptic inputs onto their cell bodies, whereas relay cells rarely have synapses (Hamos et al., 1985). Second, the strong electrotonic isolation of the dendritic (F2) terminals is such that PSPs generated in them will affect their release of GABA but will have little influence on the axonal output; also local clusters of F2 terminals will be functionally isolated from others. Below is offered some indirect evidence for this. In summary, cable modeling suggests that the interneuron multiplexes, with proximal synaptic inputs controlling the axonal output in a conventional manner and inputs onto the F2 terminals controlling them more or less independently, leaving one computational input/output route for the F1 terminals and many other independent ones for the F2 terminals.

However, modeling the dendritic arbor of the interneuron as a passive cable is clearly an oversimplification, since a number of voltage gated channels exist for the cell body and dendrites. Unfortunately, the interneuron has not been thoroughly studied in this regard, and some of the different observations are in conflict and require resolution (for details, see Sherman and Guillery, 2001). For instance, relay cells clearly possess many voltage gated T-type Ca^{2+} channels that, when activated, lead to an inward, depolarizing current, I_T; this results in an all-or-none, low threshold Ca^{2+} spike that propagates through the dendrites and cell body (Deschênes et al., 1984; Jahnsen and Llinás, 1984a,b; Hernández-Cruz and Pape, 1989; McCormick and Feeser, 1990; Scharfman et al., 1990; Bal et al., 1995). However, the situation in interneurons is less clear. Several studies claim that such a low threshold Ca^{2+} spike is rarely, if ever seen in these interneurons (Pape et al., 1994; Pape and McCormick, 1995; Cox et al., 2003; Govindaiah and Cox, 2004), while others conclude that they commonly occur (Zhu et al., 1999a,b). One explanation is that T type channels do exist in interneurons but that the generation of I_T is offset by I_A, which is created by a voltage gated K^+ conductance that hyperpolarizes the cell and has a similar voltage dependency for activation (Pape et al.,

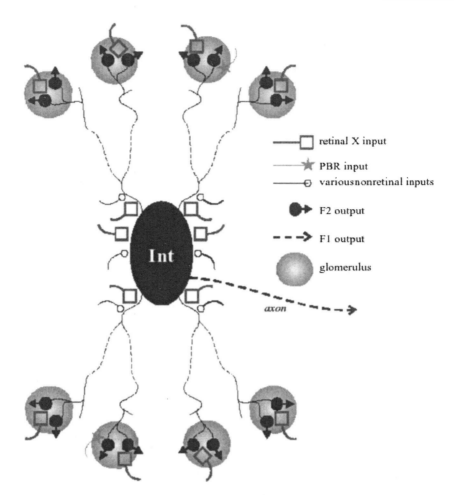

Figure 3. Schematic model for functioning of the interneuron (*Int*). The axon outputs (*F1*) are controlled by inputs onto the cell body and proximal dendrites. The peripheral dendritic outputs (*F2*) are controlled locally by direct inputs, mostly either retinal or from the parabrachial region (*PBR*), and these F2 circuits are within glomeruli. The dashed lines connecting proximal dendrites to the F2 circuits represent 5-10 levels of dendritic branching.

1994). This issue of whether or not a low threshold Ca^{2+} spike commonly exists in interneurons needs to be resolved, because such a spike could affect the dendritic F2 terminal, thereby limiting its isolation. Nonetheless, with this proviso, none of the voltage gated properties of the interneuron described to date, including no evidence for back-propagation of the action potential, require a major revision of the functional model illustrated by Figure 3.

There are several important implications of this model. One is that, if accurate, recordings from the interneuron, which to date have all been from the cell body, reveal properties of synaptic integration involving the axonal output but tell us nothing about integration that involves dendritic output, and this is the cell's major output. Likewise, information about firing rate of the interneuron relates to activity of the axonal but not the dendritic output. Related to this, all of the interneurons recorded *in vivo* from the cat's lateral geniculate nucleus have receptive fields of X cells (Sherman and Friedlander, 1988), but since this analysis depends on judging firing rate of the cell, this tells us about the nature of the retinal inputs to proximal dendrites and the cell body being from retinal X but not Y cells, but reveals nothing about retinal inputs to the F2 terminals involved in triads. However, since these are effectively limited to X and not Y relay cells, these are also from X and not Y retinal axons (but see footnote 1). What is not clear is whether the same individual retinal axon(s) innervate both F2 terminals and proximal dendritic locations on the interneuron.

5. CONTROL OF THE F2 TERMINALS AND FUNCTIONING OF THE TRIAD

Given the importance of the dendritic F2 output of the interneuron, it is of obvious importance to understand how it is controlled and how it functions, particularly with regard to the triad. Part of the answer comes from recent *in vitro* experiments using a combination of agonists and antagonists and from immunocytochemical studies with the electron microscope that together elucidate the postsynaptic receptors found on the F2 terminals (Godwin et al., 1996; Cox and Sherman, 2000; Govindaiah and Cox, 2004). Figure 2 summarizes these data. For the retinal input to the F2 terminal, metabotropic glutamate receptors type 5 (mGluR5s) dominate, and there are also probably ionotropic glutamate receptors (AMPARs and perhaps NMDARs) as well (Godwin et al., 1996; Cox and Sherman, 2000). Activation of either glutamate receptor *increases* release of GABA from the F2 terminal (Cox and Sherman, 2000; Govindaiah and Cox, 2004). For the parabrachial input to the F2 terminal, type 2 muscarinic receptors (M2Rs) exist (Plummer et al., 1999; Carden and Bickford, 1999), and activation of these receptors *decreases* release of GABA from the F2 terminal (Cox and Sherman, 2000). Regarding the inputs of retinal or parabrachial axons onto the relay cell, the retinal input activates ionotropic glutamate receptors (AMPARs and NMDARs) only, and the parabrachial input activates both ionotropic nicotinic and type 1 muscarinic receptors (nicotinic receptors and M1Rs); activation of these various receptors on the relay cell produces EPSPs (reviewed in McCormick, 1992; Sherman and Guillery, 2001). Note that the dominant receptors on the F2 terminal, mGluR5s and M2Rs, are both metabotropic.

The observation that activation of mGluR5s on the F2 terminal increases GABA release is interesting because recordings from the cell body of interneurons has consistently reported no discernable direct affects of application of mGluR agonists (Pape and McCormick, 1995; Cox and Sherman, 2000; Govindaiah and Cox, 2004). If we assume that F2 release is increased from F2 terminals because of their depolarization by mGluR5 activation, then this depolarization is effectively attenuated before reaching the cell body. This provides some indirect evidence for the electrotonic isolation of the F2 terminals from the cell body and axon.

Unfortunately, we have no direct evidence as to how activation of the mGluR5s and M2Rs control the output of the F2 terminal. Nonetheless, we can make several plausible assumptions based on the functioning of these and other metabotropic receptors elsewhere (Nicoll et al., 1990; Mott and Lewis, 1994; Recasens and Vignes, 1995; Pin and Duvoisin, 1995; Conn and Pin, 1997; Brown et al., 1997).

These are:

- They operate by opening or closing a "leak" K^+ channel. Activation of the mGluR5s closes the leak channel and thus depolarizes the terminal, which in turn increases GABA release, whereas activation of the M2R opens the leak channel, thus hyperpolarizing the terminal and decreasing GABA release.
- Compared to ionotropic equivalents such as the AMPA and nicotinic receptors, higher rates of firing in the afferent input are required to activate the metabotropic mGluR5s and M2Rs. This has recently been confirmed for the mGluR5s on the F2 terminal (Govindaiah and Cox, 2004).
- While the ionotropic receptors involved produce PSPs with a latency of only about a msec or so and a duration of 10 msec or somewhat longer, the time course for the mGluR5 and M2R is much longer, with a latency of 10 msec or more and a duration of hundreds of msec or longer. Again, this has recently been confirmed for the mGluR5 on the F2 terminal (Govindaiah and Cox, 2004).

Given these assumptions, we can suggest the following consequences of activation of the various inputs to the F2 terminal. We assume that release of GABA by the F2 terminal has some background rate that the retinal or parabrachial afferents can up- or down-regulate.

5.1. Activation of the Parabrachial Input

These brainstem inputs display levels of activity that seem to correspond to overall levels of arousal and alertness: when the animal is in slow wave sleep, they are mainly silent, when the animal is drowsy, they are moderately active, and when the animal is fully aroused and alert, they are highly active (Steriade and McCarley, 1990; Steriade and Contreras, 1995). It thus follows that, when asleep, the direct excitatory cholinergic input to the relay cell is absent, and this is exacerbated by the absence of inhibition of the F2 terminal's output, which translates to an absence of disinhibition. Together this leads to a depression of the relay. When the animal is drowsy, the moderate level of activity in the parabrachial afferents would activate the nicotinic receptors on the relay cell, producing some excitation, and the activity may be sufficient to further enhance the relay by activating the M1Rs on the relay cell (producing a prolonged depolarization) and the M2Rs on the F2 terminal (producing a prolonged disinhibition). The relay would be more enhanced as the animal becomes more awake and aroused, because the higher activity in the parabrachial afferents will produce more direct nicotinic receptor and M1R excitation of the relay cell and more disinhibition via activation of the M2Rs on the F2 terminal. Thus activity levels among the parabrachial afferents, which correlates to levels of arousal, also correlates to enhancement in the ability of the relay cell to pass on information to cortex from retina, and the circuitry represented by the F2 terminal of the interneuron plays a significant role here.

5.2. Activation of the Retinal Input

The result here is a bit more complicated, because the triadic circuit provides a basis for direct excitation (retinal-to-relay cell) and indirect inhibition (retinal-F2 terminal-relay cell). With the above assumptions for relative activation of ionotropic and metabotropic receptors that have been partially confirmed (Govindaiah and Cox, 2004), the following predictions follow. With low levels of retinal activity, only ionotropic receptors will be active, so EPSPs will be generated in the relay cell. If there is an AMPAR on the F2 terminal, the EPSPs will be abbreviated by disynaptic IPSPs. Unless there is a major difference between sensitivity of the AMPARs on the relay cell versus the F2 terminal, the only role of those on the F2 terminal will be to oppose the EPSP in the relay cell in a fairly linear fashion, but this will not change appreciably with firing level in the retinal afferent. However, as this firing level becomes sufficiently large to activate the mGluR5s on the F2 terminal, an extra dose of GABAergic inhibition will be apparent in the relay cell. There are two interesting features of this latter inhibition based on activating the mGluR5s: first, it appears and grows only after retinal firing exceeds some threshold that appears to be roughly 10Hz, growing with increasing afferent activity to roughly 200Hz (Govindaiah and Cox, 2004); and second, once activated, it will outlast the retinal activity by several seconds, because this is the duration of a mGluR-activated EPSP (Govindaiah and Cox, 2004). Also note that the interesting part of this process does not depend on the possible presence of an AMPAR on the F2 terminal: its presence and level would serve only to add a constant extra amount of disynaptic inhibition in the relay cell that would partly offset the monosynaptic EPSP there. It is the presence of the mGluR5 that adds an interesting extra inhibition that depends on the firing level of the retinal afferent and outlasts it, and several consequences may be contemplated as outlined below.

5.3. Effects on Contrast Gain Control or Adaptation

In general, the firing level of the the retinal axon is monotonically related to contrast in the visual stimulus, and thus the greater the contrast, the higher the firing level. This means that, once contrast exceeds a certain level, the retinal afferent fires sufficiently to activate the mGluR5s on the F2 terminal and thereby increases inhibition in the relay cell. This will reduce the responsiveness, or the gain, of the relay cell to retinal inputs, and, because of the temporal properties of the mGluR5s, this reduced contrast gain will last for several seconds or so even after the retinal afferent firing returns to normal or prior levels. This is a form of contrast gain control or adaption whereby the contrast response of the relay cell adjusts to overall contrast: as contrast increases enough to activate the triad circuit via the mGluR5s, gain or responsiveness in the relay cell reduces, and vice-versa.

There are many examples of such processes operating in the visual system. However, these effects on receptive field properties to date have been found mostly in visual cortex and perhaps in retina, but there has been no evidence of clear effects on the representation of contrast at the level of the lateral geniculate nucleus as suggested here (for recent statements of these issues, see Soloman et al., 2004; Carandini, 2004). This is an issue that seems worth pursuing, especially with regard to an effect that should be seen primarily in X and not Y cells.

5.4. Effects on Voltage Sensitive Properties

The above discussion considers how triadic circuitry might influence the balance of excitation and inhibition in the relay cell, the key being that higher rates of retinal input brings in extra inhibition via recruitment of mGluR5s on the relay cells, and higher rates of parabrachial input does the opposite via recruitment of M2Rs. It also follows that this can have a net, long lasting effect on membrane potential of the relay cell, and this, in turn, can affect the play of the cell's voltage sensitive properties. Many voltage gated conductances exist that can be so affected, but a detailed accounting of these is beyond the scope of the present account (for reviews, see McCormick and Huguenard, 1992; Sherman and Guillery, 2001). As an example, consider the role of the voltage-gated T-type Ca^{2+} channels, which when activated, lead to an inward current, I_T, and an all-or-none, low threshold Ca^{2+} spike propagating through the dendritic tree and producing a burst of action potentials. When the cell is relatively depolarized, I_T is inactivated, no low threshold spike is produced, and the cell responds to suprathreshold inputs with a steady stream of unitary action potentials: this is *tonic* firing. When the cell is relatively hyperpolarized, inactivation of I_T is removed, and now an excitatory input will produce a low threshold spike and *burst* firing. The firing mode of the relay cell has important implications for functioning of the relay (Sherman, 2001). The point here is that triadic function may affect I_T and other such voltage sensitive properties in interesting ways. This clearly needs more study.

6. CONCLUSIONS

The thalamic interneuron exemplified by that found in the cat's lateral geniculate nucleus provides a potent GABAergic and thus inhibitory input to relay cells. It thus plays a key role in controlling the flow of information to cortex. The synaptic inputs from these interneurons to relay cells are particularly interesting for two reasons. First, the interneuron appears to employ two independent input/output routes: a conventional axonal one that integrates inputs onto the cell body and proximal dendrites; and an unconventional one involving dendritic outputs that are both presynaptic and postsynaptic. Cable modeling suggests that the dendritic output route is organized into numerous, functionally independent streams that are also independent of the cell body and thus action potential generation. Second, the dendritic outputs in addition to being presynaptic to relay cell dendrites, are postsynaptic chiefly to either retinal or parabrachial inputs. Details of these output synapses, which involve complex circuits known as triads found widely throughout thalamus, lead to rather speculative but testable ideas regarding how interneurons help modulate relay cell activity. It is hoped that following through some of these ideas will provide genuine insights into the functioning of this circuitry that appears key to thalamic relays.

7. REFERENCES

Arcelli, P., Frassoni, C., Regondi, M. C., De Biasi, S., and Spreafico, R., 1997, GABAergic neurons in mammalian thalamus: A marker of thalamic complexity? *Brain Res. Bull.* **42**: 27.
Bal, T., Von Krosigk, M., and McCormick, D. A., 1995, Synaptic and membrane mechanisms underlying synchronized oscillations in the ferret lateral geniculate nucleus *in vitro*, *J. Physiol.* **483**: 641.
Bloomfield, S. A., Hamos, J. E., and Sherman, S. M., 1987, Passive cable properties and morphological correlates of neurones in the lateral geniculate nucleus of the cat, *J. Physiol.* **383**: 653.
Bloomfield, S. A., and Sherman, S. M., 1989, Dendritic current flow in relay cells and interneurons of the cat's lateral geniculate nucleus, *PNAS USA* **86**: 3911.
Brown, D. A., Abogadie, F. C., Allen, T. G., Buckley, N. J., Caulfield, M. P., Delmas, P., Haley, J. E., Lamas, J. A., and Selyanko, A. A., 1997, Muscarinic mechanisms in nerve cells, *Life Sci.* **60**: 1137.

Carandini, M., 2004, Receptive fields and suppressive fields in the early visual system, in: *The Cognitive Neurosciences*, M. S. Gazzaniga, ed., MIT Press, Cambridge *(in press)*.

Carden, W. B., and Bickford, M. E., 1999, Location of muscarinic type 2 receptors within the synaptic circuitry of the cat visual thalamus, *J Comp. Neurol.* **410**: 431.

Carden, W. B., and Bickford, M. E., 2002, Synaptic inputs of class III and class V interneurons in the cat pulvinar nucleus: Differential integration of RS and RL inputs, *Visual Neurosci.* **19**: 51.

Conn, P. J., and Pin, J. P., 1997, Pharmacology and functions of metabotropic glutamate receptors, *Ann. Rev. Pharmacol. Toxicol.* **37**: 205.

Cox, C. L., Reichova, I., and Sherman, S. M., 2003, Functional synaptic contacts by intranuclear axon collaterals of thalamic relay neurons, *J. Neurosci.* **23**: 7642.

Cox, C. L., and Sherman, S. M., 2000, Control of dendritic outputs of inhibitory interneurons in the lateral geniculate nucleus, *Neuron* **27**: 597.

Cucchiaro, J. B., Uhlrich, D. J., and Sherman, S. M., 1991, Electron-microscopic analysis of synaptic input from the perigeniculate nucleus to the A-laminae of the lateral geniculate nucleus in cats, *J. Comp. Neurol.* **310**: 316.

Datskovskaia, A., Carden, W. B., and Bickford, M. E., 2001, Y retinal terminals contact interneurons in the cat dorsal lateral geniculate nucleus, *J. Comp. Neurol.* **430**: 85.

Deschênes, M., Paradis, M., Roy, J. P., and Steriade, M., 1984, Electrophysiology of neurons of lateral thalamic nuclei in cat: resting properties and burst discharges, *J. Neurophysiol.* **51**: 1196.

EriÕir, A., Van Horn, S. C., Bickford, M. E., and Sherman, S. M., 1997, Immunocytochemistry and distribution of parabrachial terminals in the lateral geniculate nucleus of the cat: A comparison with corticogeniculate terminals, *J. Comp. Neurol.* **377**: 535.

Famiglietti, E. V. J., and Peters, A., 1972, The synaptic glomerulus and the intrinsic neuron in the dorsal lateral geniculate nucleus of the cat, *J. Comp. Neurol.* **144**: 285.

Friedlander, M. J., Lin, C.-S., Stanford, L. R., and Sherman, S. M., 1981, Morphology of functionally identified neurons in lateral geniculate nucleus of the cat, *J. Neurophysiol.* **46**: 80.

Godwin, D. W., Van Horn, S. C., EriÕir, A., Sesma, M., Romano, C., and Sherman, S. M., 1996, Ultrastructural localization suggests that retinal and cortical inputs access different metabotropic glutamate receptors in the lateral geniculate nucleus, *J. Neurosci.* **16**: 8181.

Govindaiah, and Cox, C. L., 2004, Synaptic activation of metabotropic glutamate receptors regulates dendritic outputs of thalamic interneurons, *Neuron* **41**: 611.

Guatteo, E., Franceschetti, S., Bacci, A., Avanzini, G., and Wanke, E., 1996, A TTX-sensitive conductance underlying burst firing in isolated pyramidal neurons from rat neocortex, *Brain Res.* **741**: 1.

Guillery, R. W., 1969, The organization of synaptic interconnections in the laminae of the dorsal lateral geniculate nucleus of the cat, *Z. Zellforsch.* **96**: 1.

Hamos, J. E., Van Horn, S. C., Raczkowski, D., Uhlrich, D. J., and Sherman, S. M., 1985, Synaptic connectivity of a local circuit neurone in lateral geniculate nucleus of the cat, *Nature* **317**: 618.

Hernández-Cruz, A., and Pape, H.-C., 1989, Identification of two calcium currents in acutely dissociated neurons from the rat lateral geniculate nucleus, *J. Neurophysiol.* **61**: 1270.

Jahnsen, H., and Llinás, R., 1984a, Electrophysiological properties of guinea-pig thalamic neurones: an *in vitro* study, *J. Physiol.* **349**: 205.

Jahnsen, H., and Llinás, R., 1984b, Ionic basis for the electroresponsiveness and oscillatory properties of guinea-pig thalamic neurones *in vitro*, *J. Physiol.* **349**: 227.

Jones, E. G., 1985, *The Thalamus*. Plenum Press. New York.

McCormick, D. A., 1992, Neurotransmitter actions in the thalamus and cerebral cortex and their role in neuromodulation of thalamocortical activity, *Prog. Neurobiol.* **39**: 337.

McCormick, D. A., and Feeser, H. R., 1990, Functional implications of burst firing and single spike activity in lateral geniculate relay neurons, *Neuroscience* **39**: 103.

McCormick, D. A., and Huguenard, J. R., 1992, A model of the electrophysiological properties of thalamocortical relay neurons, *J. Neurophysiol.* **68**: 1384.

Mott, D. D., and Lewis, D. V., 1994, The pharmacology and function of central GABAB receptors, *Int. Rev. Neurobiol.* **36**: 97.

Nicoll, R. A., Malenka, R. C., and Kauer, J. A., 1990, Functional comparison of neurotransmitter receptor subtypes in mammalian central nervous system, *Physiol. Rev.* **70**: 513.

Pape, H.-C., Budde, T., Mager, R., and Kisvárday, Z. F., 1994, Prevention of Ca^{2+}-mediated action potentials in GABAergic local circuit neurones of rat thalamus by a transient K^+ current, *J. Physiol.* **478**: 403.

Pape, H.-C., and McCormick, D. A., 1995, Electrophysiological and pharmacological properties of interneurons in the cat dorsal lateral geniculate nucleus, *Neuroscience* **68**: 1105.

Pfrieger, F. W., Veselovsky, N. S., Gottmann, K., and Lux, H. D., 1992, Pharmacological characterization of calcium currents and synaptic transmission between thalamic neurons in vitro, *J. Neurosci.* **12**: 4347.

Pin, J. P., and Duvoisin, R., 1995, The metabotropic glutamate receptors: structure and functions, *Neuropharmacology* **34**: 1.

Plummer, K. L., Manning, K. A., Levey, A. I., Rees, H. D., and Uhlrich, D. J., 1999, Muscarinic receptor subtypes in the lateral geniculate nucleus: A light and electron microscopic analysis, *J. Comp. Neurol.* **404**: 408.

Ralston, H. J., 1971, Evidence for presynaptic dendrites and a proposal for their mechanism of action, *Nature* **230**: 585.

Recasens, M., and Vignes, M., 1995, Excitatory amino acid metabotropic receptor subtypes and calcium regulation, *Ann. NY Acad. Sci.* **757**: 418.

Sanchez-Vives, M. V., Bal, T., Kim, U., Von Krosigk, M., and McCormick, D. A., 1996, Are the interlaminar zones of the ferret dorsal lateral geniculate nucleus actually part of the perigeniculate nucleus?, *J. Neurosci.* **16**: 5923.

Scharfman, H. E., Lu, S.-M., Guido, W., Adams, P. R., and Sherman, S. M., 1990, *N*-methyl-D-aspartate (NMDA) receptors contribute to excitatory postsynaptic potentials of cat lateral geniculate neurons recorded in thalamic slices, *PNAS USA* **87**: 4548.

Sherman, S. M., 1985, Functional organization of the W-,X-, and Y-cell pathways in the cat: a review and hypothesis, in: *Progress in Psychobiology and Physiological Psychology*, Vol. 11, J. M. Sprague and A. N. Epstein, eds., Academic Press, Orlando, pp 233-314.

Sherman, S. M., 2001, Tonic and burst firing: dual modes of thalamocortical relay, *Trends Neurosci.* **24**: 122.

Sherman, S. M., and Friedlander, M. J., 1988, Identification of X versus Y properties for interneurons in the A-laminae of the cat's lateral geniculate nucleus, *Exp. Brain Res.* **73**: 384.

Sherman, S. M., and Guillery, R. W., 2001, *Exploring the Thalamus*, Academic Press, San Diego.

Soloman, S. G., Pierce, J. W., Dhruv, N. T., and Lennie, P., 2004, Profound contrast adaptation early in the visual pathway. *Neuron . In press.*

Steriade, M., and Contreras, D., 1995, Relations between cortical and thalamic cellular events during transition from sleep patterns to paroxysmal activity, *J. Neurosci.* **15**: 623.

Steriade, M., and McCarley, R. W., 1990, *Brainstem Control of Wakefulness and Sleep*, Plenum Press, New York.

Szentágothai, J., 1963, The structure of the synapse in the lateral geniculate nucleus, *Acta Anat.* **55**: 166.

Vidnyanszky, Z., and Hamori, J., 1994, Quantitative electron microscopic analysis of synaptic input from cortical areas 17 and 18 to the dorsal lateral geniculate nucleus in cats, *J. Comp. Neurol.* **349**: 259.

Wang, S., Bickford, M. E., Van Horn, S. C., EriÖir, A., Godwin, D. W., and Sherman, S. M., 2001, Synaptic targets of thalamic reticular nucleus terminals in the visual thalamus of the cat, *J. Comp. Neurol.* **440**: 321.

Wilson, J. R., and Friedlander, M. J., and Sherman, S. M., 1984, Fine structural morphology of identified X- and Y-cells in the cat's lateral geniculate nucleus, *Proc. Roy. Soc. Lond. B* **221**: 411.

Zhu, J. J., Lytton, W. W., Xue, J. T., and Uhlrich, D. J., 1999a, An intrinsic oscillation in interneurons of the rat lateral geniculate nucleus, *J. Neurophysiol.* **81**: 702.

Zhu, J. J., Uhlrich, D. J., and Lytton, W. W., 1999b, Burst firing in identified rat geniculate interneurons, *Neuroscience* **91**: 1445.

10

RELEASE OF NORADRENALINE IN
THE LOCUS COERULEUS

Olga L. Pudovkina and Ben H.C. Westerink[1]

1. INTRODUCTION

The locus coeruleus (LC) is a small cluster of 1600 noradrenergic cells, located near the fourth ventricle in the pontine brainstem. The LC plays an important role in vigilance, response to novelty and stress, the sleep-wake cycle, attention, memory, antinociception, and autonomic control (for review see Aston-Jones et al., 1995; Singewald and Philippu, 1998; Berridge and Waterhouse, 2003).

The LC consists of several types of noradrenergic neurons among which medium-sized fusiform cells, multipolar neurons, and ovoid-shaped cells can be distinguished (Shimizu and Imamoto, 1970; Swanson, 1976; Cintra et al., 1982). Morphological heterogeneity of these cells is further manifested by the fact that they have differently oriented dendrites and distinct sets of efferent fibres that provide widespread innervation throughout the brain (Aston-Jones et al., 1995). Dendritic processes are medium size in diameter (1.5-2.5 μm) and are varicose. The remarkable feature of LC dendrites is that they extend beyond the nucleus proper into the rostromedial and the caudal juxtaependymal regions to 400-600 μm. Axons of LC neurons are thin (about 3 μm) and often indistinguishable from dendritic processes. On their way to other brain areas they give numerous collaterals but only a few of them within the LC (Shipley et al., 1996; Swanson, 1976).

Concentration of noradrenaline in the LC is about 15 ng/mg protein (Lambas-Senas et al., 1986). Besides noradrenaline the LC cell bodies contain several neuropeptides, such as galanin, neuropeptide Y, neurotensin, substance P, and the vasoactive intestinal protein. Furthermore, non-noradrenergic cells containing corticotropin releasing factor, acetylcholine, serotonin, and GABA as neurotransmitter were also found in the pericerulear region of the LC.

Major afferents that form contacts mainly with the core and to somewhat lesser extent with peripheral dendrites of the LC, arise from two nuclei in the rostral medulla: the nucleus paragigantocellularis (PGi) and the nucleus prepositus hypoglossus (PrH).

[1] Ben H.C. Westerink, Department of Biomonitoring & Sensoring, University Centre for Pharmacy, 9713AV, Groningen, The Netherlands.
Dendritic Neurotransmitter Release, edited by M. Ludwig
Springer Science+Business Media, Inc., 2005

Projections from PGi possess glutamate, corticotrophin-releasing factor, and adrenaline, whereas projections from PrH contain GABA (Aston-Jones et al., 1991). Both PGi and PrH also send enkephalin-containing projections to the LC. Afferents coming from the prefrontal and infralimbic cortices, central nucleus of the amygdala, dorsal raphe, lateral hypothalamus, bed nucleus of the stria terminalis, Kölliker-Fuse nucleus, lateral preoptic area, the Barrington's nucleus, and nucleus of solitary tract make synaptic contacts preferentially with extranuclear LC dendrites (Aston-Jones et al., 1991; Luppi et al., 1995).

2. ELECTROPHYSIOLOGICAL PROPERTIES OF LC NEURONS

LC neurons display primarily two modes of discharge activity: tonic and phasic. Tonic activity is characterized by a sustained and highly regular discharge pattern (up to 5.0 Hz in rat), the firing rate of which is state dependent (Aston-Jones and Bloom, 1981). For example, during sleep the LC activity is profoundly suppressed by GABAergic inhibitory tone which is always present in the LC but reaches maximum values during rapid eye movement (REM) sleep. Cognitive activity and productive learning especially require appropriate noradrenergic neuronal firings to maintain the necessary level of attention. As it was shown by Aston-Jones and colleagues (1999), information processing during behavioral performance is optimal with moderate levels of tonic LC activity and diminished with low or high tonic LC discharge.

The results of *in vitro* studies have indicated that LC neurons can be considered as pacemakers, as they are able to fire spontaneously (0.18-3.6 Hz) in the absence of synaptic input (Alreja and Aghajanian, 1991; Williams et al., 1991). Furthermore, the spontaneous activity of LC neurons in the newborn rat is synchronized by electronic coupling between dendrites (Nakamura et al., 1987; Ishimatzu and Williams, 1996), though after the third postnatal week coupling is dramatically down-regulated.

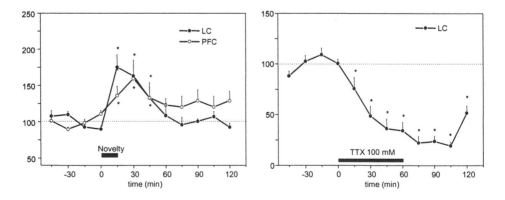

Figure 1: Left panel. Effect of novelty for 15 min (black bar) on extracellular noradrenaline in the LC (closed circles) and ipsilateral PFC (open circles). Data are given as % of basal levels ± S.E.M. (n= 5). * p < 0.05 compared to basal levels.
Right panel. Effect of tetrodotoxin (TTX) infusion (1 μM, black bar) in the LC during 60 min on extracellular noradrenaline in the LC. Data are given as % of basal levels ± S.E.M. (n=7). * p < 0.05 compared to basal levels.

A phasic burst discharge mode occurs in response to excitatory stimuli and is characterized by a brief train of 2-4 action potentials of up to 20 Hz that is followed by a prolonged (200-500 ms) period of inhibition (Aston-Jones and Bloom, 1981). The propagation o f a ction p otentials a long t he a xons o f LC neurons i nduces t he r elease o f noradrenaline from the terminals and collaterals. A study that combined extracellular recordings and microdialysis has revealed that noradrenaline released in the major projection area of the LC, the prefrontal cortex (PFC), is linearly related to LC electrical activity (Berridge and Abercrombie, 1999).

3. RELEASE OF NORADRENALINE FROM THE LC

Noradrenaline is not only released in terminal fields of the LC, but also from dendrites/axon collaterals within the somatodendritic area to provide recurrent collateral inhibition via autoadrenoceptors (Aghajanian et al., 1977; Swanson, 1976). *In vitro* experiments showed that α_2-adrenoceptors mediate the suppression of the firing rate of the noradrenergic cells after direct application of noradrenaline to the LC (Svensson et al, 1975). Although the fact that noradrenaline can be released from dendrites was not directly proven, small granular vesicles were found in LC dendrites, and dendro-dendritic contacts have been observed (Shimizu et al., 1979; Groves and Wilson, 1980). In addition, s omatodendritic noradrenaline uptake s ites i n t he L C h ave b een d escribed a s well (Ordway et al., 1997).

Using push-pull or microdialysis methods several investigators have studied the release of noradrenaline within the LC (Singewald and Philippu, 1993, 1998; Singewald et al., 1994; 1999; Thomas et al., 1994; Mateo et al., 1998; Kawahara et al., 1999; Pudovkina et al., 2001). These studies have shown that – similar to terminal areas – noradrenaline in LC responds strongly to arousing stimuli such as novelty and stress but also cardiovascular stimuli such as hypotension. Fig. 1 shows a dual-probe microdialysis experiment in which one probe was implanted in the LC and a second probe in the PFC. It demonstrates that the release of noradrenaline in both areas increased during exposure to a novel environment.

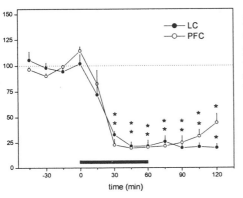

Figure 2. Effect of the α_2-adrenoceptor agonist clonidine infusion (100 μM, black bar) in the LC during 6 0 min on extracellular noradrenaline in t he LC (closed circles) and PFC (open circles). Data are given as % of basal levels ± S.E.M. (n=7). * p < 0.05 compared to basal levels.

The origin of the release of noradrenaline in the LC is still unknown. The finding that omission of calcium from the perfusion fluid decreased noradrenaline in the LC (Pudovkina et al., 2001), suggests that somatodendritic noradrenaline is released by a classical release mechanism involving exocytosis. Singewald et al. (1994) reported that - in a push-pull study on anesthetised cats - a major part of noradrenaline was tetrodotoxin-insensitive. However in conscious rats infusion of tetrodotoxin into the LC strongly decreased the levels of noradrenaline in the LC (Fig. 1b) as well as in the ipsilateral LC (Van Gaalen et al., 1997).

4. ROLE OF AUTOADRENOCEPTORS IN THE LC

4.1. Alpha$_2$-Adrenoceptors

Dual-probe microdialysis experiments have demonstrated that local infusion of the α_2-adrenoceptor agonist clonidine into the LC decreases extracellular levels of noradrenaline in the LC as well PFC to about 25% of controls (Fig. 2). The observed decreases reflect the inhibition of impulse flow of noradrenergic neurons (Svensson et al., 1975). Other studies have revealed that the release of noradrenaline can also be regulated presynaptically by α_2-adrenoceptors on noradrenergic axon terminals. Microdialysis experiments have shown that infusion of clonidine into projection areas such as the PFC decreases the levels of noradrenaline in this region (Van Veldhuizen et al., 1993). A study using fast cyclic voltammetry demonstrated that the release of noradrenaline in the LC and terminal regions is mainly regulated by α_{2A}- but not by $\alpha_{2B/C}$-adrenoceptors (Callado and Stamford, 1999).

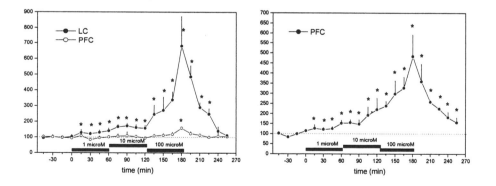

Figure 3: Left panel. Effect of α_{2A}-adrenoceptor antagonist BRL 44408 infusion (10, 100, 500 μM, black bars) in the LC during 60 min each dose, on extracellular noradrenaline in the LC (closed circles) and ipsilateral PFC (open circles). Data are given as % of basal levels ± S.E.M. (n= 5). * p < 0.05 compared to basal levels.
Right panel. Effect of BRL 44408 infusion (10, 100, 500 μM, black bars) in the PFC during 60 min each dose, on extracellular noradrenaline in the PFC (closed circles). Data are given as % of basal levels ± S.E.M. (n= 5). * p < 0.05 compared to basal levels.

In this regard it is of interest to study the effects of specific α_2-adrenoceptor compounds in the LC. Dual-probe microdialysis experiments revealed that local infusion of the α_{2A}-adrenoceptor antagonist BRL 44408 increased extracellular noradrenaline in the LC as well as in the PFC (Fig. 3). However the increase in noradrenaline release in the PFC reached only 150% of controls (Fig. 3). This result- although limited in size – favours the idea of tonic inhibition of LC electrical activity by somatodendritic noradrenaline release. Similar conclusions were drawn by Mateo and Meana (1999). Much stronger effects on noradrenaline release were seen when BRL4 44408 – in a single probe experiment - was directly infused into the PFC (Fig. 3).

The larger response of noradrenaline in the LC and the PFC after local application could be due to summation of two effects: blockade of inhibitory α_2 release-controlling adrenoceptors on dendrites/axon collateral and stimulation of excitatory α_1 release-controlling adrenoceptors by released noradrenaline (discussed below). The latter mechanisms act apparently independent of impulse flow of LC neurons.

4.2. Alpha$_1$-Adrenoceptors

While stimulation of α_2-adrenoceptors always mediate hyperpolarization of LC neurons, the function of α_1-adrenoceptors depends on the developmental stage of the brain. Low doses of noradrenaline applied iontophoretically to the LC produce excitation of noradrenergic neurons in the rat during the first postnatal weeks via α_1-adrenoceptors, whereas high doses of noradrenaline results in suppression of LC activity via α_2-adrenoceptors (Nakamura et al., 1988). Apparently, during early development noradrenaline has higher affinity for α_1-adrenoceptors as compared to α_2-adrenoceptors. In mature rats, the affinity of noradrenaline for α_2-adrenoceptors (56 nM) predominates

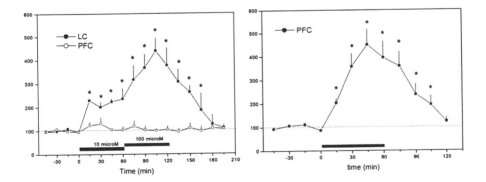

Figure 4. Left panel. Effect of the α_1-adrenoceptor agonist cirazoline infusion (10 and 100 µM, black bars) in the LC during 60 min each dose on extracellular noradrenaline in the LC (closed circles) and ipsilateral PFC (open circles). Data are given as % of basal levels ± S.E.M. (n= 5). * p < 0.05 compared to basal levels.
Right panel. Effect of cirazoline infusion (100 µM, black bars) in the PFC during 60 min on extracellular noradrenaline in the PFC (closed circles). Data are given as % of basal levels ± S.E.M. (n= 5). * p < 0.05 compared to basal levels.

over that for α_1-adrenoceptors (330 nM) and hyperpolarizing effects of α_1-adrenoceptors become hardly detectable (Nakamura et al., 1988). Nevertheless, there is some evidence that in the mature rat noradrenaline released in the LC from dendrites/collaterals acts upon α_1-adrenoceptors to regulate the activity of LC neurons. An electrophysiological *in vitro* study from Ivanov and Aston-Jones (1995) demonstrated that the excitatory effect of an α_2-adrenoceptor antagonists on LC neurons could be abolished by infusion of an α_1-adrenoceptor antagonist (prazosin). Moreover, prazosin administered in the LC alone was able to decrease the spontaneous firing rate of the LC (Ivanov and Aston-Jones, 1995). Further support for a significant role of α_1-adrenoceptors in the LC was provided by an *in vitro* study from Osborne and colleagues (2002), who found that activation of α_1-adrenoceptors suppresses inhibitory potassium currents induced by stimulation of α_2-adrenoceptors.

A dual-probe microdialysis experiment from our laboratory carried out in conscious rats brought new evidence about the excitatory role of α_1-adrenoceptors in the LC. Infusion of the α_1-adrenoceptor agonist cirazoline into the LC induced a large increase of noradrenaline release that was comparable with the increase observed after local application of cirazoline in the PFC (Fig. 4a, b). In turn, administration of prazosin in the LC induced an opposite effect: a decrease of noradrenaline (results not shown).

Application of prazosin directly to the PFC also significantly decreased the release of noradrenaline in this structure (results not shown). These observations demonstrate that α_1-adrenoceptors located on dendrites/collaterals in the LC and presynaptically on noradrenergic nerve terminals in the PFC are able to regulate the release of noradrenaline.

The fact that during infusion of cirazoline in the LC no changes in noradrenaline concentrations were observed in the PFC, questions the role of α_1-adrenoceptors in the regulation of impulse flow activity. However, the large increase of noradrenaline in the LC during cirazoline infusion might have decreased noradrenaline levels in the PFC, via

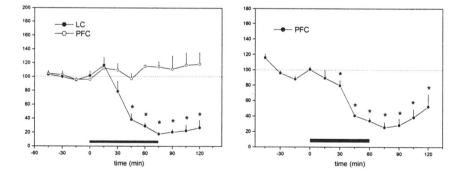

Figure 5: Left panel. Effect of infusion of the 5-HT$_{1A}$ agonist flesinoxan (1 µM, black bar) in the LC during 75 min on extracellular noradrenaline in the LC (closed circles) and ipsilateral PFC (open circles). Data are given as % of basal levels ± S.E.M. (n= 7). * p < 0.05: compared to basal levels.
Right panel. Effect of flesinoxan infusion (1 µM, black bar) in the PFC during 60 min on extracellular noradrenaline in the PFC (open circles). Data are given as % of basal levels ± S.E.M. (n= 7). * p < 0.05 compared to basal levels.

activation of α_2-autoadrenoceptors. Although not fully evaluated, this evidence supports the hypothesis that the role of α_1-adrenoceptors in the LC is to prevent the complete suppression of LC cell firing via α_2-adrenoceptors.

5. AFFERENT REGULATION OF NORADRENALINE RELEASE

The excitatory glutamatergic and inhibitory GABAergic afferents are robust projections that contribute to the regulation of the firing activity of the LC. Stimulation of glutamate receptors in the LC was reported to enhance the release of noradrenaline in the LC (Singewald et al., 1999) and the ipsilateral PFC (Kawahara et al., 2000). Local application of the GABA$_A$ receptor antagonist bicuculline in the LC increased the release of noradrenaline in the LC (Singewald et al., 1999) as well as in the PFC (Kawahara et al., 2000). Stimulation of muscarinic receptors in the LC led to an increase in noradrenaline release in both the LC and various projection areas (Kawahara et al., 2001; Pudovkina et al., 2002). Taken together these findings suggest that the afferent regulation of LC activity modifies the release of noradrenaline in the LC as well as in the terminal areas in a similar fashion.

An example of a different type of interaction with LC neurons was observed for serotonergic afferents. It is well known that the LC receives a serotonergic input from the dorsal raphe nucleus (Palkovits et al., 1977). In a dual-probe microdialysis experiment the specific 5-HT$_{1A}$ receptor agonist flesinoxan was infused into the LC. This infusion caused a profound decrease of noradrenaline in the LC, but it was without effect in the ipsilateral PFC (Fig. 5a). The decrease in noradrenaline release induced by the 5-HT$_{1A}$ agonist was apparently not related to a change in the impulse flow of the LC cells. Interestingly flesinoxan also decreased extracellular noradrenaline when infused locally in terminal areas such as the PFC (Fig. 5b).

6. THE FUNCTION OF NORADRENALINE IN THE LC

6.1. Dendritic Release: Impulse Flow Driven or Not?

The first reports describing dendritic neurotransmitter release were concerned with dopamine release in the substantia nigra (Geffen et al., 1976; Korf et al., 1976). One of the properties of dendritic release is its impulse flow driven nature, as electrical stimulation of the medial forebrain bundle induced dopamine metabolism in the substantia nigra as well as the striatum (Korf et al., 1977). Microdialysis experiments in which carbachol or clonidine were infused into the LC indeed suggests that changes in electrical impulse flow of LC neurons parallel changes in local release of noradrenaline. However other dual-probe experiments described in this Chapter, where cirazoline, flesinoxan or BRL44408 were infused into the LC, indicated that certain receptor interactions are able to modify strongly the release of noradrenaline in the LC without affecting the electrical activity of these neurons.

If we define dendritic release as impulse driven, it is evident that in the experiments described above, the drug-induced release of noradrenaline in the LC did not fulfil this criterion. On the other hand, impulse flow independent release modulation might be functional as it has been recently shown that the release of vasopressin and oxytocin from dendrites in the hypothalamic supraoptic nucleus and paraventricular nuclei can be regulated independently from impulse-flow activity (Ludwig and Pittman, 2003).

6.2. Other Possible Functions of Noradrenaline Released in the LC

Much evidence suggests that the major role of noradrenaline release in the LC is to regulate the electrical activity of its own neurons. We have shown that excitatory α_1- and inhibitory α_2-autoceptors are both implicated in this effect. However other mechanisms should be considered as well. Alpha$_2$-adrenoceptors were also described on presynaptic non-catecholaminergic processes within the LC, suggesting that dendritic release of noradrenaline could be a mechanism by which noradrenergic cells interact with afferent terminals (Lee et al., 1998). In addition, extracellular noradrenaline in the LC may also stimulate glial processes via α_2-adrenoceptors (Lee et al., 1998). In this regard it is of interest to note that the LC might be involved in neuronal plasticity by mechanisms such as electronic coupling with glial cells (Alvarez-Maubecin et al., 2000).

This review is an attempt to trace the major features and possible role of the somatodendritic release of noradrenaline in the LC. Although the results of the presented studies clarify important aspects regarding the mechanisms of noradrenergic efflux and its relation to the electrical activity of the LC, several issues remain to be investigated. Moreover a clear-cut distinction between collateral and somatodendritic noradrenaline release has not yet been made. However the contribution of collaterals to noradrenaline release in the LC might be less important as Lee and co-workers (1998) have reported that the noradrenergic axon collaterals within the LC lack α_2-adrenoceptors. Whether this finding is due to imperfection of the technique, as the authors assumed, needs further clarification.

7. REFERENCES

Aghajanian, K., Cedarbaum, J. M., and Wang, R. Y., 1977, Evidence for norepinefrine-mediated collateral inhibition of locus coeruleus neurons, Brain Res. 136: 570.
Alreja, M., and Aghajanian, G. K., 1991, Pacemaker activity of locus coeruleus neurons: whole-cell recordings in brain slices show dependence on cAMP and protein kinase A, Brain Res. 556: 339.
Alvarez-Maubecin, V., Garcia-Hernandez, F., Williams, J. T., and Van Bockstaele, E. J., 2000, Functional coupling between neurons and glia, J. Neurosci. 20: 4091.
Aston-Jones, G., and Bloom, F. E., 1981, Activity of norepinephrine-containing locus coeruleus neurons in behaving rats anticipates fluctuations in the sleep-waking cycle, J. Neurosci. 1: 876.
Aston-Jones, G., Rajkowski, J., and Cohen, J., 1999, Role of locus coeruleus in attention and behavioral flexibility, Biol. Psychiatry. 46: 1309.
Aston-Jones, G., Shipley, M. T., Chouvet, G., Ennis, M., Van Bockstaele, E., Pieribone, V., Shiekhattar, R., Akaoka, H., Drolet, D., and Astier, B., 1991, Afferent regulation of locus coeruleus neurons: anatomy, physiology and pharmacology, Prog. Brain Res. 88: 47.
Aston-Jones, G., Shipley, M. T., and Grzanna, R., 1995, The locus coeruleus, A5 and A7 noradrenergic cell groups, in: The Rat Nervous System, G. Paxinos, ed., Academic Press, San Diego, CA, pp.183-213.
Berridge, C. W., and Abercrombie, E. D., 1999, Relationship between locus coeruleus discharge rates and rates of norepinephrine release within neocortex as assessed by in vivo microdialysis, Neuroscience 93: 1263.
Berridge, C. W., and Waterhouse, B. D., 2003, The locus coeruleus-noradrenergic system: modulation of behavioral state and state-dependent cognitive processes, Brain Res. Rev. 42: 33.
Callado, L. F., and Stamford, J. A., 1999, Alpha2A- but not alpha2B/C-adrenoceptors modulate noradrenaline release in rat locus coeruleus: voltammetric data, Eur. J. Pharmacol. 366: 35.

Cintra, L., Diaz-Cintra, S., Kemper, T., and Morgane. P. J., 1982, Nucleus locus coeruleus: a morphometric Golgi study in rats of three age groups, *Brain Res.* **247**: 17.

Geffen, L. B., Jessell, T. M., Cuello, A. C., and Iversen, L. L., 1976, Release of dopamine from dendrites in rat substantia nigra, *Nature* **260**: 258.

Groves, P. M., and Wilson, C. J., 1980, Monoaminergic presynaptic axons and dendrites in rat locus coeruleus seen in reconstructions of serial sections, *J. Comp. Neurol.* **193**: 853.

Ishimatsu, M., and Williams, J. T., 1996, Synchronous activity in locus coeruleus results from dendritic interactions in pericoerulear regions, *J. Neurosci.* **16**: 5196.

Ivanov, A., and Aston-Jones, G., 1995, Extranuclear dendrites of locus coeruleus neurons: activation by glutamate and modulation of activity by alpha adrenoceptors, *J. Neurophysiol.* **74**: 2427.

Kawahara, Y., Kawahara, H., and Westerink, B. H., 1999, Comparison of effects of hypotension and handling stress on the release of noradrenaline and dopamine in the locus coeruleus and medial prefrontal cortex of the rat, *Naunyn Schmiedebergs Arch. Pharmacol.* **360**: 42.

Kawahara, H., Kawahara, Y., and Westerink, B. H., 2000, The role of afferents to the locus coeruleus in the handling stress-induced increase in the release of noradrenaline in the medial prefrontal cortex: a dual-probe microdialysis study in the rat brain, *Eur. J. Pharmacol.* **387**: 279.

Kawahara, H., Kawahara, Y., and Westerink, B. H., 2001, The noradrenaline-dopamine interaction in the rat medial prefrontal cortex studied by multi-probe microdialysis, *Eur. J. Pharmacol* **418**: 177.

Korf, J., Zieleman, M., and Westerink, B. H., 1976, Dopamine release in substantia nigra? *Nature* **260**: 257.

Korf, J., Zieleman, M., and Westerink, B. H., 1977, Metabolism of dopamine in the substantia nigra after antidromic activation, *Brain Res.* **120**: 184.

Lambas-Senas, L., Chamba, G., Fety, R., and Renaud, B., 1986, Comparative responses of the central adrenaline- and noradrenaline-containing neurons after reserpine injections, *Biochem. Pharmacol.* **35**: 2207.

Lee, A., Rosin, D. L., and Van Bockstaele, E. J., 1998, Alpha2A-adrenergic receptors in the rat nucleus locus coeruleus: subcellular localization in catecholaminergic dendrites, astrocytes, and presynaptic axon terminals, *Brain Res.* **795**: 157.

Ludwig, M., and Pittman, Q. J., 2003, Talking back: dendritic neurotransmitter release, *Trends Neurosci.* **26**: 255.

Luppi, P. H., Aston-Jones, G., Akaoka, H., Chouvet, G., and Jouvet, M., 1995, Afferent projections to the rat locus coeruleus demonstrated by retrograde and anterograde tracing with cholera-toxin B subunit and Phaseolus vulgaris leucoagglutinin, *Neuroscience* **65**: 119.

Mateo, Y., and Meana, J. J., 1999, Determination of the somatodendritic alpha2-adrenoceptor subtype located in rat locus coeruleus that modulates cortical noradrenaline release *in vivo*, *Eur. J. Pharmacol.* **379**: 53.

Mateo, Y., Pineda, J., and Meana, J. J., 1998, Somatodendritic alpha2-adrenoceptors in the locus coeruleus are involved in the *in vivo* modulation of cortical noradrenaline release by the antidepressant desipramine, *J. Neurochem.* **71**: 790.

Nakamura, S., Kimura, F., and Sakaguchi, T., 1987, Postnatal development of electrical activity in the locus ceruleus, *J. Neurophysiol.* **58**: 510.

Nakamura, S., Sakaguchi, T., Kimura, F., and Aoki, F., 1988, The role of alpha 1-adrenoceptor-mediated collateral excitation in the regulation of the electrical activity of locus coeruleus neurons, *Neuroscience* **27**: 921.

Ordway, G. A., Stockmeier, C. A., Cason, G. W., and Klimek, V., 1997, Pharmacology and distribution of norepinephrine transporters in the human locus coeruleus and raphe nuclei, *J. Neurosci.* **17**: 1710.

Osborne, P. B., Vidovic, M., Chieng, B., Hill, C. E., and Christie M. J., 2002, Expression of mRNA and functional alpha (1)-adrenoceptors that suppress the GIRK conductance in adult rat locus coeruleus neurons, *Br. J. Pharmacol.* **135**: 226.

Palkovits, M., Saavedra, J. M., Jacoboqitz, D. M., Kizer, J. S., Zaborszky, L., and Brownstein, M. J., 1977, Serotonergic innervation of the forebrain: effect of lesions on serotonin and tryptophan hydroxylase levels, *Brain Res.* **130**: 121.

Pudovkina, O. L., Cremers, T. I. F. H., and Westerink, B. H., 2002, The interaction between the locus coeruleus and dorsal raphe nucleus studied with dual probe microdialysis, *Eur. J. Pharmacol.* **445**: 37.

Pudovkina, O. L., Kawahara, Y., de Vries, J., and Westerink, B. H., 2001, The release of noradrenaline in the locus coeruleus and prefrontal cortex studied with dual-probe microdialysis, *Brain Res.* **906**: 38.

Shimizu, N., and Imamoto, K., 1970, Fine structure of the locus coeruleus in the rat, *Arch Histol. Jpn.* **31**: 229.

Shimizu, N., Katoh, Y., Hida, T., and Satoh, K., 1979, The fine structural organization of the locus coeruleus in the rat with reference to noradrenaline contents, *Exp. Brain Res.* **37**: 139.

Shipley, M. T., Fu, L., Ennis, M., Liu, W.L., and Aston-Jones, G., 1996, Dendrites of locus coeruleus neurons extend preferentially into two pericoerulear zones, *J. Comp. Neurol.* **365**: 56.

Singewald, N., Kaehler, S.T., Philippu, A., 1999, Noradrenaline release in the locus coeruleus of conscious rats is triggered by drugs, stress and blood pressure changes, *Neuroreport* **10**:1583

Singewald, N., and Philippu, A., 1993, Catecholamine release in the locus coeruleus is modified by experimentally induced changes in haemodynamics, *Naunyn Schmiedebergs Arch. Pharmacol.* **347**: 21.

Singewald, N., and Philippu, A., 1998, Release of neurotransmitters in the locus coeruleus, *Prog. Neurobiol.* **56**: 237.

Singewald, N., Schneider, C., Pfitscher, A., and Philippu, A., 1994, *In vivo* release of catecholamines in the locus coeruleus, *Naunyn Schmiedebergs Arch. Pharmacol.* **350**: 339.

Svensson, T. H., Bunney, B. S., and Aghajanian, G. K., 1975, Inhibition of both noradrenergic and serotonergic neurons in brain by the alpha-adrenergic agonist clonidine, *Brain Res.* **92**: 291.

Swanson, L. W., 1976, The locus coeruleus: a cytoarchitectonic, Golgi and immunohistochemical study in the albino rat, *Brain Res.* **110**: 39.

Thomas, D. N., Post, R. M., and Pert, A., 1994, Focal and systemic cocaine differentially affect extracellular norepinephrine in the locus coeruleus, frontal cortex and hippocampus of the anaesthetized rat, *Brain Res.* **645**: 135.

Van Gaalen, M., Kawahara, H., Kawahara, Y., and Westerink, B. H., 1997, The locus coeruleus noradrenergic system in the rat brain studied by dual-probe microdialysis, *Brain Res.* **763**: 56.

Van Veldhuizen, M. J., Feenstra, M. G., Heinsbroek, R. P., and Boer, G. J., 1993, *In vivo* microdialysis of noradrenaline overflow: effects of alpha-adrenoceptor agonists and antagonists measured by cumulative concentration-response curves, *Br. J. Pharmacol.* **109**: 655.

Williams, J. T., Bobker, D. H., and Harris, G. C., 1991, Synaptic potentials in locus coeruleus neurons in brain slices, *Prog. Brain Res.* **88**: 167.

11

REGULATION OF SOMATODENDRITIC SEROTONIN RELEASE IN THE MIDBRAIN RAPHE NUCLEI OF THE RAT

Laszlo G. Harsing, Jr.[1]

1. INTRODUCTION

Since the early discovery of Dahlstrom and Fuxe (1964), several lines of evidence indicate that the cell bodies of 5-HT neurons are concentrated in the midbrain. Of the 5-HT cell groups, the dorsal and median raphe nuclei have been most studied; at least half of all 5-HT neurons in the CNS are located in the dorsal raphe and almost 70% of the cells in the raphe contain 5-HT (Wiklund and Bjorklund, 1980; Wiklund et al., 1981). 5-HT neurons are medium sized cells with spiny dendritic arborization, and 5-HT-containing dendrites synapse upon other 5-HT-containing dendrites. Serotonergic interconnections between various raphe nuclei have also been demonstrated (Chazal and Ralston, 1987). Besides dendrodendritic interactions, the activity of 5-HT neurons may also be regulated locally by recurrent axon collaterals (Harandi et al., 1987), and synaptic afferents found on 5-HT dendrites and cell bodies are often serotonergic. Few serotonergic nerve endings have synaptic specializations, suggesting that most 5-HT is released in a non-synaptic manner (Harandi et al., 1987).

5-HT neurons in the raphe fire spontaneously at 1-5 spikes/s (Vandermaelen and Aghajanian, 1983), and this activity is determined by at least three different processes: spontaneous, slow firing activity; 5-HT mediated autoinhibition; and afferent inputs mediated by non-5-HT receptors (Pineyro and Blier, 1999). Autoinhibition includes regulation of transmitter release; and synthesis and neural firing which are determined by various metabotropic ($5-HT_{1A/1B/1D}$, $5-HT_{2A/2C}$, $5-HT_7$) and ionotropic ($5-HT_3$) 5-HT receptors (Zifa and Fillion, 1992; Barnes and Sharp, 1999). Different $5-HT_1$ receptor subtypes are present on 5-HT neurons in the raphe and they inhibit 5-HT release (Davidson and Stamford, 1995; Sprouse and Aghajanian, 1987; Starkey and Skingle, 1994; Pineyro and Blier, 1996) although their regulatory functions are different. In the raphe nuclei, few data are available about the 5-HT receptors that mediate actions on

---

[1] Laszlo G. Harsing, Jr., EGIS Pharmaceuticals Ltd., Budapest, Hungary.

Dendritic Neurotransmitter Release, edited by M. Ludwig
Springer Science+Business Media, Inc., 2005

neurons other than 5-HT neurons. Postsynaptic 5-HT receptors are the target of 5-HT released from dendritic varicosities into the synaptic gap or into the extracellular space, as both synaptic and non-synaptic 5-HT neurotransmission may occur (Chazal and Ralston, 1987; Harandi et al., 1987).

5-HT release in the raphe originates from the somata/dendrites and from recurrent axon collaterals, and the regulation of release from these sites may be different. Dendritic release originates from 5-HT neurons with vesicles in the dendrites (Hery et al., 1982) and may provide the neurochemical basis for synchronous firing among 5-HT cells. Activation of ascending 5-HT pathways inhibits 5-HT cells of the dorsal raphe, an effect mediated by 5-HT (Hajos et al., 1998); iontophoretically applied 5-HT also inhibits raphe neurons. The cycles of hyper- and depolarization of 5-HT cells are altered by excitatory and inhibitory neuronal projections and interneurons besides 5-HT released from 5-HT neurons (Pineyro and Blier, 1996). The neural circuitry of the raphe converts incoming signals into output messages. The 5-HT projection neurons, GABA interneurons and glutamatergic axon terminals form complex excitatory-inhibitory connections by which incoming excitatory signals are converted into inhibitory output and sent back to the areas where stimulation was generated. The functional state of this neural network is determined b y neurotransmitters r eleased b oth synaptically a nd non-synaptically (Vizi, 2000).

Ascending 5-HT projections influence several functions in different forebrain structures (Parent et al., 1981; Steinbusch, 1981). The raphe-hippocampal pathway is one of numerous ascending 5-HT projections (El Mansari and Blier, 1996), and the caudate nucleus is another important projection field for 5-HT neurons of the raphe nuclei (Van der Kooy and Hattori, 1980; Vizi et al., 1981). 5-HT neurons of the raphe also form important projections to the median cerebral cortex (Kidd et al., 1991; Hajos et al., 1998), mediating inhibition in the cortical neural network (Steinbusch, 1981; O'Hearn and Molliver, 1984). In many cases, forebrain structures that receive 5-HT innervation from the raphe send back reciprocal projections; for example, there is a glutamate pathway from the median prefrontal cortex to the raphe nuclei (Behzadi et al., 1990). GABA interneurons i n t he r aphe nuclei a re t he p rimary t arget for t he c ortico-raphe glutamate neurons, and their stimulation by glutamate leads to inhibition of 5-HT cells (Hajos et al., 1998). 5-HT neurons in the raphe are regulated by a number of efferent projections. The habenulo-raphe projection represents a major GABA inhibition of raphe neurons (Stern et al., 1981; Wang et al., 1992), this pathway also contains excitatory transmitters and the presence of substance P-containing and glutamatergic fibers has been reported. Another excitatory input to raphe 5-HT neurons is noradrenergic, acting through postsynaptic α_2-adrenoceptors (Baraban and Aghajanian, 1981).

This review discusses the functional relevance of somatodendritic 5-HT release and it contrasts the release properties between axon terminal and mindbrain raphe nuclei, and considers possible extrasynaptic targets.

2. SOMATODENDRITIC 5-HT RELEASE

An *in vitro* superfusion technique has been developed to characterize neurotransmitter release from brain slices. Wistar rats were decapitated, and brain slices containing the raphe nuclei were prepared as described previously (Kerwin and Pycock,

1979). Slices were incubated with [^{3}H]5-HT (1.25 µCi/ml) in Krebs-bicarbonate buffer for 30 min at 36 °C, and then transferred into low volume superfusion chambers (Juranyi and Harsing, 2004) and superfused with aerated (95% O_2/5% CO_2) preheated (37 °C) Krebs-bicarbonate buffer at 1 ml/min.. The superfusate was discarded for the first 60 min, then fractions were collected at 3-min intervals. Biphasic electrical field stimuli were delivered by an electrostimulator to stimulate [^{3}H]5-HT efflux. The efflux, expressed as Bq/g/fraction or as a fractional rate, was calculated for the first (S1) and second (S2) stimulations, and the ratio S2/S1 was calculated to determine the effects of drugs on release. The time-course of [^{3}H]5-HT efflux from raphe slices is shown in Figure 1. The [^{3}H]5-HT content in raphe and hippocampal tissues determined at the beginning and the end of superfusion as well as the total efflux during a 66-min superfusion are shown in Table 1. We also calculated efflux from the central and peripheral parts of 5-HT neurons at rest and during electrical stimulation. Basal and evoked release did not differ significantly in raphe nuclei and in hippocampal slices (Table 1) although net release seems to be higher at the projection field.

Table 1. Efflux of [^{3}H]5-HT from hippocampal and raphe slices of the rat

	Hippocampus	Raphe nuclei
Total content (kBq/g)	163±10	218±14
Total efflux (kBq/g)	72±6	89±4
Remaining tissue content (kBq/g)	91±5	129±10
Basal efflux		
(% of content/3 min)	2±0.05	2±0.1
(kBq/g/3 min)	3.8±0.6	4.1±0.17
Stimulated efflux		
(% of content/3 min)	5.8±0.2	4.3±0.1
(kBq/g/3 min)	8.5±0.7	8.5±0.8
Net efflux(% of content/stimulus)	7.6±0.4	3.7±0.4
S2/S1	0.98±0.06	0.9±0.03

Slices loaded with [^{3}H]5-HT were stimulated electrically (20 V, 2 Hz, 2-ms for 2 min. Fractional efflux is a percentage of the tissue radioactivity at the time the efflux is measured. Net efflux is the area under the curve from fractions 4 to 8. Mean±S.E.M., n=3-4.

2.1. Ion Channels in the Regulation of Somatodendritic 5-HT Release

The voltage-sensitive K$^+$ channel blocker 4-aminopyridine (4-AP, Rudy, 1988) increased electrical stimulation-induced [^{3}H]5-HT efflux from raphe slices (Bagdy and Harsing, 1995). The effect of 4-AP on transmitter release has been extensively studied at a variety of synapses (Illes, 1986); our experiments indicate that 4-AP-sensitive channels are involved not only in axon terminal transmitter release (Hu and Fredholm, 1989) but also in that which originates from the somatodendritic areas of 5-HT cells. On the other hand neither glibenclamide nor tolbutamide, drugs blocking ATP-dependent K^+ channels, affected [^{3}H]5-HT efflux in raphe slices. Thus ATP-sensitive K^+ channels are apparently not involved in somatodendritic 5-HT release (Bagdy and Harsing, 1995).

The voltage-dependent Na^+ channel blocker tetrodotoxin (TTX) inhibits electrically evoked [^{3}H]5-HT efflux presumably by blocking action potential propagation membrane (Starke, 1987). TTX does not block [^{3}H]5-HT efflux evoked by elevated [KCl], a depolarizing stimulus that acts directly on axon terminals or dendritic boutons (Harsing and Zigmond, 1997). Therefore, elevated [KCl] instead of electrical stimulation can be used with TTX to localize receptors in raphe neural circuitry (Bagdy et al., 2000; Harsing et al., 2004).

To demonstrate the role of external Ca^{2+}, raphe slices were superfused with Ca^{2+}-free buffer. Omission of Ca^{2+} blocked electrically-evoked [^{3}H]5-HT efflux but basal efflux was not affected (Fig. 1). This indicates that somatodendritic 5-HT release could be of vesicular origin. Since both N- and L-type Ca^{2+} channels are present in neuronal cell bodies and dendrites (de Erausquin et al., 1992) we applied selective blockers. Addition of Cd^{2+} and ω-conotoxin GVIA inhibited electrically-evoked [^{3}H]5-HT efflux (Bagdy and Harsing, 1995). In contrast, the dihydropyridine diltiazem failed to inhibit evoked release. The experiments suggest that N- rather than L-type Ca^{2+} channels are involved in somatodendritic 5-HT release. The regulatory role of N-type Ca^{2+} channels in axon terminal release of dopamine and norepinephrine has been shown in several studies (Zimanyi et al., 1985; Herdon and Nahorski, 1989; Harsing et al., 1992), so our findings may suggest common mechanism for regulation of Ca^{2+} entry during transmitter efflux

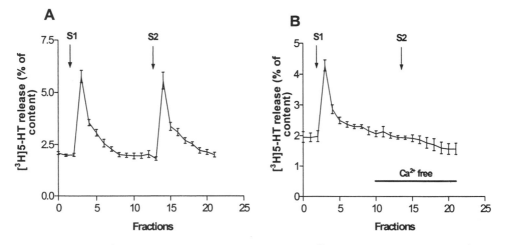

Figure 1. Efflux of [^{3}H]5-HT from raphe slices (A). Effect of Ca^{2+} removal (B). Slices were stimulated electrically (20 V, 2 Hz, 2-ms for 2 min) in fractions 4(S1) and 15(S2) and the ratio S2/S1 was calculated. Ca^{2+}-free superfusion buffer containing 1 mM EGTA was added from fraction 10 and maintained through the experiment. Mean±S.E.M., n=4.

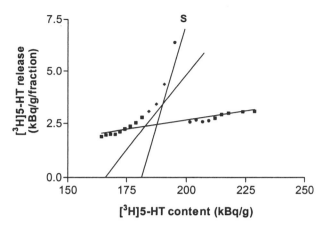

Figure 2. Slices containing the raphe nuclei loaded with [³H]5-HT were stimulated (S) electrically (20 V, 2 Hz, 2-ms for 2 min) in fraction 10. The efflux was plotted as a function of tissue content. Desaturation curves with different slopes yield to distinguish various release compartments.

from dendrites and nerve axon terminals. Other types of voltage-sensitive Ca^{2+} channels (P- and Q-type) may also be important in dendritic and nerve ending transmitter release (Gauer et al., 1994; Dunlap et al., 1995).

2.2. Compartmentalization of [³H]5-HT in Raphe Slices

[³H]5-HT storage in and efflux from raphe nuclei were characterized by compartmental analysis of the radioactivity. Accumulated [³H]5-HT distributes mainly as an efflux compartment from where release occurs and a bound fraction (slowly exchanging compartment) that does not contribute to efflux (de Langen and Mulder, 1979; 1980). The bound fraction was estimated by plotting the rate of efflux as a function of radioactivity remaining in the tissue (Fig. 2). The desaturation curve obtained in the de Langer-Mulder plot is linear and the intercept of the regression line with the abscissa estimates the size of the bound fraction. Substraction of the bound fraction from the tissue content of radioactivity determined at the beginning of superfusion reveals the efflux compartment (Table 2).

About 76% of the [³H]5-HT taken up represents the efflux compartment and the remaining 24% contributes to the bound fraction. About 29% of radioactivity in the efflux compartment is released during a 66-min superfusion. Compartmental analysis also indicates that the efflux compartment consists of various pools that vary in their rate constant, turnover rate, and turnover time. The efflux compartment comprises at least three pools from which fast and slow efflux of [³H]5-HT occur in response to electrical stimulation and a third pool which represents the source of basal efflux of [³H]5-HT (Fig. 2). Table 2 shows the calculated pool sizes. The different pools may represent vesicular and cytoplasmic stores from which release takes place with various velocities, but may also reflect pools in the somatodendritic and axon terminal parts of neurons.

Table 2. Compartmentalization of [^{3}H]5-HT in raphe slices

Tissue compartment	Pool size	
	(kBq/g)	(Percent)
Tissue content	173 ± 37	100
Non-releasable pool	40 ± 10	24 ± 3
Efflux compartment	132 ± 29	76 ± 3
Total efflux during a 66-min superfusion	50 ± 11	29 ± 2
Fast efflux compartment	21± 5	16 ±1.5
Slow efflux compartment	37.5 ± 9.5	28 ± 3
Resting efflux compartment	73 ± 18	55 ± 4.4

Raphe slices were loaded with [^{3}H]5-HT, superfused, and stimulated electrically (40 V, 2 Hz, 2-ms for 2 min) and the efflux and tissue content of [^{3}H]5-HT was determined. Mean±S.E.M., n=3.

3. 5-HT RECEPTOR-REGULATION OF 5-HT RELEASE

3.1. Feedback Inhibition of 5-HT Release: 5-HT$_1$ Receptor Subtypes

Stimulation of ascending 5-HT pathways inhibits dorsal raphe 5-HT cells (Wang and Aghajanian, 1977) through 5-HT released from 5-HT neurons acting at release-inhibitory autoreceptors. 5-HT may be released from vesicles of the dendrites (Hery et al., 1982; Saavedra et al., 1986) and from the recurrent axon collaterals of 5-HT neurons but the regulation of the release from these two sites may be different (Davidson and Stamford, 1995). Further evidence for the presence of 5-HT receptors on 5-HT neurons is that iontophoretically applied 5-HT inhibits raphe neurons.

The 5-HT receptors involved in regulation of 5-HT neurons, belong to the 5-HT$_1$ receptor family (Barnes and Sharp, 1999). Different 5-HT$_1$ receptors including 5-HT$_{1A}$, 5-HT$_{1B}$ and 5-HT$_{1D}$ are present on 5-HT neurons in the raphe and they inhibit 5-HT release (Starkey and Skingle, 1994; Sprouse and Aghajanian, 1987; Davidson and Stamford, 1995; Pineyro and Blier, 1996) although their regulatory function in this process is different. Activation of 5-HT$_{1A}$ receptors on 5-HT neurons causes hyperpolarization (Arborelius et al., 1994; Hajos et al., 1995) and also inhibits 5-HT synthesis (Hjorth and Magnusson, 1988). Changes in 5-HT neuron firing activity is the consequence of 5-HT$_{1A}$ receptor activation and firing-dependent control of 5-HT release occurs in the cell body level mediated by 5-HT$_{1A}$ receptors. 5-HT$_{1A}$ autoreceptors inhibit 5-HT release in the raphe but 5-HT$_{1A}$ receptors in forebrain projection regions do not influence 5-HT release. Whereas release-inhibitory 5-HT$_{1A}$ receptors are located on the somatodendritic part of 5-HT neurons (Palacios et al, 1987; O'Connor and Kruk, 1992; Davidson and Stamford, 1995; Bagdy and Harsing, 1995), 5-HT$_{1B}$ receptors are situated on axon terminals (Engel et al., 1986; Davidson and Stamford, 1995; El Mansari and Blier, 1996). The functions of terminal and somatodendritic 5-HT autoreceptors exhibit a number of differences: the former primarily govern impulse-mediated 5-HT release of vesicular origin and the latter regulate dendritic release and action potential generation (Barnes and Sharp, 1999;

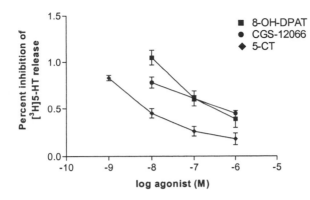

Figure 3. Effects of 8-OH-DPAT, CGS-12066 and 5-CT on electrically-evoked [³H]5-HT efflux from raphe slices. Slices were stimulated electrically (20 V, 2 Hz, 2-ms for 2 min in fractions 4(S1) and 15(S2) and the ratio S2/S1 was calculated. Drugs were added from fraction 10. Mean±S.E.M., n=5.

Pineyro and Blier, 1999). 5-HT release in the midbrain that is regulated by 5-HT$_{1B}$ receptors involves release from recurrent axon collaterals and 5-HT terminals. 5-HT$_{1B}$ receptors exert only modest effects on 5-HT release because the amount of neurotransmitter is small compared with the massive release from the somatodendritic areas. 5-HT$_{1D}$ receptors also control 5-HT release (Starkey and Skingle, 1994; Haj-Dahmane et al., 1991; El Mansari and Blier, 1996); 5-HT$_{1D}$ receptors in the raphe are proposed to be autoreceptors and control the larger somatodendritic 5-HT release. 5-HT$_{1D}$ receptors can alter 5-HT release in cell body level and on axon collaterals without modifying firing activity of 5-HT neurons. The axon terminals of these cells also possess 5-HT$_{1B}$ or 5-HT$_{1D}$ receptor subtypes which serve as inhibitory release-mediating autoreceptors (Haj-Dahmane et al., 1991; Martin et al., 1992; Bagdy et al., 1994; El Mansari and Blier, 1996).

Using slice preparations, it is possible to demonstrate the inhibitory effects of 5-HT$_{1A}$, 5-HT$_{1B}$, and/or 5-HT$_{1D}$ receptors on 5-HT release (Bagdy and Harsing, 1995). The 5-HT$_{1A}$ agonist 8-OH-DPAT (Middlemiss and Fozard, 1983) inhibits electrically-evoked [³H]5-HT efflux from raphe slices and this inhibition is concentration dependent (Fig. 3). In contrast, 8-OH-DPAT failed to alter electrically-evoked [³H]5-HT efflux from hippocampal slices. Our results are further indication that somatodendritic autoreceptors in the raphe nuclei belong to 5-HT$_{1A}$ subtypes (Verge et al., 1985), whereas 5-HT$_{1A}$ receptors in the hippocampus do not serve as presynaptic homologous autoreceptors.

The 5-HT$_{1B}$ agonist CGS-12066 (Neale et al., 1987) also inhibited electrically-evoked [³H]5-HT efflux in raphe slices (Fig. 3). CGS-12066 also inhibits [³H]5-HT efflux in hippocampal slices, supporting the view that the release-mediating 5-HT receptors at nerve endings is a 5-HT$_{1B}$ subtype (Middlemiss, 1984). CGS-120066 has low selectivity in stimulating 5-HT$_{1A}$ and 5-HT$_{1B}$ receptors (Schoeffter and Hoyer, 1989) and this may explain why CGS-120066 inhibited [³H]5-HT efflux not only in the hippocampus but also in the raphe. Our data further confirm that 5-HT$_{1A}$ and 5-HT$_{1B}$ receptors control somatodendritic and axon terminal 5-HT release in the raphe and

Table 3. Effects of 5-HT antagonists on [^{3}H]5-HT efflux from raphe slices of the rat

Compounds	Receptor	Concentration (μM)	[^{3}H]5-HT efflux (S2/S1)
1. Control	-	-	0.8 ± 0.02
2. (+)WAY-100135	5-HT$_{1A}$	1	1.2 ± 0.15*
3. WAY-100635	5-HT$_{1A}$	10	0.9 ± 0.06
4. SB-216641	5-HT$_{1B/1D}$	1	1.1 ± 0.09*
5. Ondansetron	5-HT$_3$	1	0.95 ± 0.07
6. SB-271046	5-HT$_6$	10	0.98 ± 0.03
7. SB-258719	5-HT$_7$	10	0.86 ± 0.1

Raphe slices loaded with [^{3}H]5-HT were stimulated electrically (20 V, 2 Hz, 2-ms for 2 min) in fractions 4(S1) and 15(S2) and the efflux of [^{3}H]5-HT was expressed by the S2/S1 ratio. Drugs were added from fraction 9. ANOVA and Dunnett's test, *P<0.05, mean±S.E.M., n=4-8.

hippocampus, respectively. It is not clear where these receptors are situated, but the location of 5-HT$_{1A}$ receptors on the somatodendritic part and the location of 5-HT$_{1B/1D}$ receptors on recurrent axon collaterals have been suggested (Davidson and Stamford, 1995) and thus these might be the sites for the action of 8-OH-DPAT and CGS-12066.

5-Carboxamidotryptamine (5-CT), which is an agonist at 5-HT$_1$ receptors (Voigt et al., 1991; Boddeke et al., 1992; Hamblin et al., 1992) and 5-HT$_7$ receptors (Boess and Martin, 1994; Hoyer et al., 1994), also inhibited electrically-evoked [^{3}H]5-HT efflux (Fig. 3) from raphe slices. The IC$_{50}$ value for 5-CT concentration that induced a 50% decrease in efflux, was 3.3 ± 0.4 nM (Harsing et al., 2004).

Of the 5-HT$_{1A}$ antagonists (+)WAY-100135 (Cliffe et al., 1993) but not WAY-100635 (Fletcher et al., 1996) increased [^{3}H]5-HT efflux (Table 3). SB-216641, a 5-HT$_{1B}$ antagonist with approximately 25-fold selectivity over 5-HT$_{1D}$ receptors (Price et al., 1997; Schlicker et al., 1997), increased electrically-evoked [^{3}H]5-HT efflux in raphe slices. The non-selective 5-HT$_{1D}$ antagonist methiothepine and the selective 5-HT$_{1D}$ antagonist GR127935 increased [^{3}H]5-HT efflux in guinea-pig cerebral cortex slices (Roberts et al., 1996).

(+)WAY-100135 dose-dependently antagonized the inhibitory effect of 8-OH-DPAT on [^{3}H]5-HT release. WAY-100635 only slightly influenced the inhibitory effect of 5-CT and the apparent pA$_2$ value for WAY-100635 against 5-CT was 4.99. SB-216641 produced a marked concentration-dependent rightward shift of the concentration-inhibition curve of 5-CT with no significant alteration in the maximal 5-CT response. The calculated apparent pA$_2$ value for SB-216641 on 5-CT-induced inhibition of [^{3}H]5-HT overflow was 7.12. This is in accord with the general view that 5-HT$_{1B}$ and 5-HT$_{1D}$ receptors are inhibitory autoreceptors on raphe neurons (Sprouse et al., 1997) although different regulatory role of the two receptors was also suggested (Starkey and Skingle, 1994). The combination of WAY-100635 with SB-216641 resulted in an additional inhibition of [^{3}H]5-HT efflux further pointing out the role of 5-HT$_1$ subtypes in 5-CT-induced 5-HT release inhibition (Fig. 4).

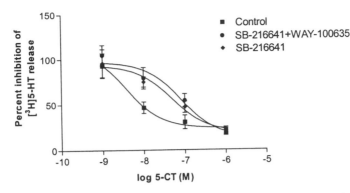

Figure 4. Effects of SB-216641 (1 μM), SB-216641 (1 μM) and WAY-100635 (10 μM) on 5-CT-induced inhibition of [³H]5-HT efflux from raphe slices. The slices were stimulated electrically (20 V, 2 Hz, 2-ms for 2 min) in 4(S1) and 15(S2) and the ratio S2/S1 was calculated. 5-CT was added 9 min and the 5-HT antagonists were added 18 min before S2 and maintained through the experiment. The IC$_{50}$ value of 5-CT to inhibit [³H]5-HT efflux was 3.3 ± 0.4 nM, it was 47.8 ± 4.7 nM in the presence of SB-2166410 and it was 80 ± 5 nM in the presence of SB-216641 plus WAY-100635. Mean±S.E.M., n=3-4.

3.2. Feedback Stimulation of 5-HT Release: the Role of 5-HT$_3$ Receptors

Although different receptor subtypes mediate autoinhibition of 5-HT release in different of brain areas, feedback stimulation of release by 5-HT$_3$ receptors has also been reported (Martin et al., 1992). Moreover, inhibitory 5-HT$_{1D}$ autoreceptors and facilitatory 5-HT$_3$ autoreceptors may interact in the same axon terminal (Martin et al., 1992). 5-HT$_3$ receptors may participate in the feedback stimulation of 5-HT release as autoreceptors (Bagdy et al., 1998) whereas 5-HT$_3$ receptors in the hypothalamus influence 5-HT release as heteroreceptors (Blier et al., 1993). 5-HT$_3$ stimulation also increases electrically evoked 5-HT release from guinea-pig cerebral cortex or hippocampus, brain areas that contain axon terminals of 5-HT neurons (Blier and Bouchard, 1993; Galzin et al., 1990).

The 5-HT$_3$ agonist 2-methyl-5-HT increased resting [³H]5-HT efflux from raphe slices, the calculated concentration of 2-methyl-5-HT that increased release by 50% was 5.3 μmol/l (Fig. 5). The facilitatory effect of 2-methyl-5-HT on basal efflux was shown not only in the raphe but also in the hippocampus, brain areas that contain central and peripheral parts of 5-HT neurons, respectively (Bagdy et al., 1998). 2-Methyl-5-HT (1.15 μmol/l) increased basal efflux by 50% in hippocampal slices. 2-Methyl-5-HT was able to enhance not only basal ³H]5-HT efflux from slices (Blier and Bouchard, 1993) but also the basal efflux of [³H]dopamine in striatal slices (Benuck and Reith, 1992; Sershen et al., 1995) and the latter effect was explained by reversed operation of dopamine carrier.

2-Methyl-5-HT also increased electrically evoked [³H]5-HT efflux from raphe slices (Fig. 5). The calculated concentration of 2-methyl-5-HT that induced 50% increase in efflux were 0.56 and 0.25 μmol/l in raphe slices and hippocampal slices, respectively. Galzin *et al.* (1990) found that 2-methyl-5-HT increased depolarization-induced [³H]5-HT efflux in guinea-pig cerebral cortex at concentrations that were ineffective on basal efflux. The *in vitro* stimulatory effect of 2-methyl-5-HT on electrically evoked [³H]5-HT efflux observed in our experiments may be consonant with previous *in vivo* findings: Martin and coworkers (1992) reported that 2-methyl-5-HT increases 5-HT release from hippocampus and the 5-HT₃ antagonist MDL 72222 antagonized this effect.

The competitive 5-HT₃ antagonist ondansetron when added alone did not influence electrically evoked [³H]5-HT efflux (Table 3). The ineffectiveness of 5-HT₃ antagonists may be explained by an extrasynaptic location of 5-HT₃ receptors; endogenous 5-HT may reach only low concentrations at the vicinity of these receptors. When the stimulation frequency was increased from 2 Hz to 10 Hz, ondansetron reduced electrically evoked [³H]5-HT efflux. The S2/S1 ratio decreased from 1.15 ± 0.07 at 2 Hz to 0.63 ± 0.07 at 10 Hz stimulation frequencies in ondansetron-treated raphe slices.

The selective 5-HT₃ antagonist ondansetron did not antagonize the stimulatory effect of 2-methyl-5-HT on basal [³H]5-HT efflux in raphe slices (Fig. 5), suggesting that this effect is not a receptor-mediated event. On the contrary, ondansetron reversed the stimulatory effect of 2-methyl-5-HT on electrically evoked [³H]5-HT efflux (Fig. 5) and the calculated pA₂ value was 6.51. The fact that the stimulatory effect of 2-methyl-5-HT on depolarization-induced but not on basal 5-HT release was antagonized by ondansetron, suggests the involvement of 5-HT₃ receptors in the facilitatory effect of 2-methyl-5-HT on depolarization-induced 5-HT release. Ondansetron also antagonized the stimulatory effect of 2-methyl-5-HT on depolarization-induced [³H]5-HT efflux in hippocampal slices (Bagdy et al., 1998).

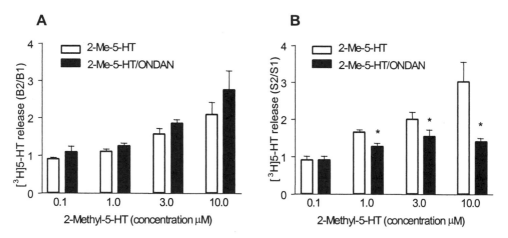

Figure 5. The stimulatory effect of 2-methyl-5-HT (0.1 to 10 μmol/L) on (A) resting and (B) electrically stimulated [³H]5-HT efflux from raphe slices. Fractional outflow of [³H]5-HT was measured in fractions 3(B1) and 14(B2) and was expressed as B2/B1. The slices were stimulated electrically (40 V, 2 Hz, 2-ms for 2 min) in fractions 4(S1) and 15(S2) and S2/S1 was calculated. 2-Methyl-5-HT was added from fraction 12 and the 5-HT₃ antagonist ondansetron (Ondan, 1 μmol/l) was added from fraction 8. ANOVA and Dunnett's test, *P<0.05, mean±S.D., n=3-4.

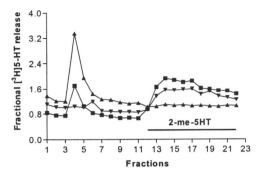

Figure 6. 2-Methyl-5-HT-induced [³H]5-HTefflux in normal conditions (■), in Ca^{2+}-free buffer containing EGTA (1 mmol/l) (●) and in the presence of fluoxetine (1 μmol/l). (♦). Raphe slices loaded with [³H]5-HT, were electrically stimulated (40 V, 2 Hz, 2-ms, 2 min) in fraction 4. 2-Methyl-5-HT (10 μmol/l) was added from fraction 12. Data are means of 4 experiments in each group, the average S.D. was 14% of the means.

Although 2-methyl-5-HT increased both basal and electrically evoked [³H]5-HT efflux, the mechanisms involved may be different. The increased basal efflux is probably due to exchange diffusion whereas the increase in stimulation-induced release is likely to be the result of mobilization of 5-HT stored in vesicles. This conclusion was supported by the observation that the activity of 2-methyl-5-HT in enhancing basal release did not require external Ca^{2+}. (Fig. 6). Moreover, the 5-HT uptake inhibitor fluoxetine blocked the stimulatory effect of 2-methyl-5-HT on basal efflux, indicating the involvement of 5-HT transporters in the effect of 2-methyl-5-HT (Fig. 6). 2-Methyl-5-HT may be a substrate for the 5-HT transporter promoting exchange diffusion by reversal of the carrier that transports 5-HT from the extracellular space into the cytoplasmic storage sites. It is less possible that 2-methyl-5-HT acts as blocker of 5-HT transporter and increases outflow of radioactivity by preventing uptake of released 5-HT. 2-Methyl-5-HT-altered 5-HT efflux was also blocked by the 5-HT uptake blocker paroxetine in guinea-pig hypothalamic slices (Blier and Bouchard, 1993).

4. SIGNAL TRANSDUCTION MECHANISMS IN THE REGULATION OF 5-HT RELEASE IN RAPHE NUCLEI

Although 5-HT$_1$ receptor subtypes on central 5-HT neurons belong to the metabotropic receptor family and are coupled to adenylate cyclase and phosphoinositide pathways, connection between these signal transduction mechanisms and 5-HT release is a question of debate (Harsing et al., 2000). We employed the 5-HT$_{1A}$ agonist 8-OH-DPAT and the 5-HT$_{1B}$ agonist CGS-12066 to inhibit electrically evoked [³H]5-HT efflux in conditions when various modulatory drugs manipulated signal transduction pathways.

4.1. The Role of K$^+$ Channels in Autoregulation of Somatodendritic 5-HT Release

The inhibitory effect of 8-OH-DPAT on [^{3}H]5-HT efflux in raphe slices was abolished in the presence of 4-AP whereas the effect of CGS-12066 on was not influenced (Table 4). Glibenclamide, on the other hand, failed to affect the inhibitory effect of either 8-OH-DPAT or CGS-12066. This suggests that stimulation by 8-OH-DPAT of somatodendritic 5-HT$_{1A}$ autoreceptors opens voltage-sensitive K$^+$ channels and the consequent hyperpolarization may then cause decreased [^{3}H]5-HT release (Bagdy and Harsing, 1995). It is possible; however, that stimulation of 5-HT$_{1A}$ and 5-HT$_{1B}$ receptors leads to opening of different K$^+$ channels, as opposed to 5-HT$_{1A}$ receptors, the voltage-sensitive K$^+$ channels did not interfer with 5-HT$_{1B}$ receptor-mediated 5-HT release inhibition. In contrast to 4-AP, the ATP-sensitive K$^+$ channel inhibitor glibenclamide did not modify either 8-OH-DPAT- or CGS-12066-induced inhibition of [^{3}H]5-HT release. Thus, ATP-sensitive K$^+$ channels are probably not involved in the 5-HT$_{1A}$ and 5-HT$_{1B}$ receptor-mediated autoinhibition of somatodendritic 5-HT release.

4.2. The Role of Ca^{2+} Channels in Autoregulation of Somatodendritic 5-HT Release

Since autoreceptor mediated inhibition of neurotransmitter release originates from the vesicular stores but not from the cytoplasm (Vizi et al., 1985) we have speculated that Ca^{2+} availability may be a key factor for autoregulation of somatodendritic 5-HT release. The somatodendritic N-type Ca^{2+} channels substantially regulate neurotransmitter release: closing of voltage-sensitive Ca^{2+} channels will lead to a decrease in [Ca^{2+}]$_i$ on which exocytotic release is highly dependent. Membrane hyperpolarization after 5-HT$_{1A}$ receptor stimulation and K$^+$ channel opening leads to closing of voltage-sensitive Ca^{2+} channel. 5-HT$_{1A}$ receptor stimulation results in reduction of high threshold Ca^{2+} current in current clamp studies of raphe neurons (Penington and Kelly, 1990). Thus, both opening of K$^+$ channels and closing of Ca^{2+} channels may be involved in 5-HT$_{1A}$ autoreceptor-mediated feedback inhibition of [^{3}H]5-HT release in raphe slices.

5-HT$_3$ receptors may stimulate 5-HT release by receptor-coupled ion channels as well as voltage-sensitive Ca^{2+} channels. When 5-HT$_3$ receptor is activated, the receptor-linked ion channel permeable to Na$^+$, Ca^{2+} and K$^+$ opens and Na$^+$ influx will lead to local depolarization (Derkach et al., 1989; Yakel and Jackson, 1988). The consequent increase in intracellular [Na$^+$] will reverse the operation of 5-HT transporter and 5-HT outflow from the cytoplasmic pool to the synaptic cleft will occur. In addition, depolarization opens voltage-sensitive Ca^{2+} channels and the Ca^{2+} influx will lead to exocytotis of 5-HT from the vesicular pool. The 5-HT$_3$ agonist 2-methyl-5-HT increases both basal and electrically evoked [^{3}H]5-HT efflux from raphe slices, releases that originate from cytoplasmic and vesicular transmitter pools respectively (Bagdy et al., 1998).

4.3. Signal Transduction Processes in the Autoregulation of 5-HT Release

8-OH-DPAT and CGS-12066 decreased electrically evoked [^{3}H]5-HT efflux from rat raphe slices and this was blocked by pretreatment with the G protein inhibitor N-ethylmaleimide (Allgaier et al., 1986) (Table 4). These data indicate that the modulation of 5-HT release by both 5-HT$_{1A}$ and 5-HT$_{1B}$ receptors involves G$_{i/o}$- protein-linked transduction mechanisms when they inhibit 5-HT release (Innis and Aghajanian, 1987; Blier et al., 1993a). N-ethylmaleimide did not affect [^{3}H]5-HT release by itself (Table 4).

To investigate the role of cAMP, raphe slices were incubated with forskolin, which stimulates adenylate cyclase (Seamon and Daly, 1981), or with 8-Br-cAMP, a membrane-permeable analogue of cAMP (Innis et al., 1988). Neither compound influenced the inhibitory effects of 8-OH-DPAT or CGS-12066 on [^{3}H]5-HT efflux (Table 4). We

Table 4. Effects of drugs altering cellular signal transduction on 5-HT$_{1A}$ and 5-HT$_{1B}$ receptor-mediated [^{3}H]5-HT release inhibition in raphe nuclei slices of the rat

Compounds	Concentration (µM)	[^{3}H]5-HT release (S2/S1)
None	-	0.92±0.02
8-OH-DPAT	-	0.42±0.04
CGS-12066	-	0.50±0.06
4-AP	50	1.91±0.33
4-AP + 8-OH-DPAT	50	2.19±0.26
4-AP + CGS-12066	50	0.87±0.14*
Glibenclamide	10	1.30±0.04
Glibenclamide + 8-OH-DPAT	10	0.75±0.08*
Glibenclamide + CGS-12066	10	0.43±0.09*
NEM	30	0.89±0.12
NEM + 8-OH-DPAT	30	0.80±0.04
NEM + CGS-12066	30	0.90±0.05
Forskolin	3	1.53±0.12
Forskolin + 8-OH-DPAT	3	0.54±0.12*
Forskolin + CGS-12066	3	0.51±0.10*
8-Br-cAMP	100	1.14±0.12
8-Br-cAMP + 8-OH-DPAT	100	0.56±0.12*
8-Br-cAMP + CGS-12066	100	0.51±0.10*
PMA	1	1.05±0.05
PMA + 8-OH-DPAT	1	0.87±0.11
PMA + CGS-12066	1	0.89±0.09
R-59022	0.5	0.72±0.08
R-59022 + 8-OH-DPAT	0.5	0.78±0.19
R-59022 + CGS-12066	0.5	0.96±0.14

Raphe nuclei slices were loaded with [^{3}H]5-HT, superfused, and stimulated electrically (20 V, 2 Hz, 2-msec for 2 min) in fractions 4(S1) and 15(S2) and the release of [^{3}H]5-HT was expressed by the S2/S1 ratio. 8-OH-DPAT (1 µM) or CGS-12066 (1 µM) was added 9 min, drugs acting on the signal transduction pathways were added 15 min before S2 in concentrations shown in the table. NEM, N-ethylmaleimide; PMA, phorbol 12-myristate 13-acetate, 4-AP, 4-aminopyridine. ANOVA and Dunnett's test, *P<0.05, mean±S.E.M., n=3-8.

concluded that although both 5-HT_{1A} and 5-HT_{1B} receptors are negatively coupled to adenylate cyclase (Schoeffter and Hoyer, 1989), this inhibition may not be directly linked to alteration of 5-HT release (Harsing et al., 2000). It is not clear, however, whether 5-HT_7 receptors that are positively coupled to adenylate cyclase (Bard et al., 1993; Lovenberg et al., 1993), inhibit 5-HT release through cAMP production or whether this inhibition is also mediated by other second messenger systems.

To investigate the involvement of the phosphoinositol system, raphe slices were incubated with phorbol 12-myristate 13-acetate (PMA), an activator of protein kinase C (PKC) (Musgrave and Majewski, 1989). This blocked the inhibitory effect of 8-OH-DPAT and CGS-12066 on $[^3\text{H}]$5-HT efflux (Table 4). Moreover, R-59022, a drug known to inhibit diacylglycerol (DAG) kinase and reduce DAG degradation (Burt et al., 1996), suspended the 5-HT release-inhibitory effect of 8-OH-DPAT and CGS-12066 (Table 4). These data indicate that the Gq protein-mediated phosphoinositide pathway is involved in the 5-HT release inhibition elicited by 5-HT_{1A} and 5-HT_{1B} receptor stimulation. Moreover stimulation of $5\text{-HT}_{1A/1B}$ receptors may activate the Gq protein-mediated phosphoinositide pathway and the increased PKC activity leads to phosphorylation of Ca^{2+} channels and the increase in Ca^{2+} influx terminates 5-HT release autoinhibition.

4.4. Cellular Signal Transduction Processes Regulate Development and Suspension of 5-HT Release Autoinhibition

Data summarized above indicate that, while both 5-HT_{1A} and 5-HT_{1B} receptors are negatively coupled to adenylate cyclase, it is the Gq protein-mediated phosphoinositide pathway rather than the cAMP-mediated second messenger system which may be involved in the 5-HT release-inhibitory effects of 5-HT_{1A} and 5-HT_{1B} receptors in the raphe nuclei. Thereby, a sequence of events can be hypothesized to be involved in development and suspension of 5-HT release autoinhibition. Extracellular 5-HT is detected by cell surface $5\text{-HT}_{1A/1B}$ receptors triggering activation of G proteins on the cytoplasmic surface of the plasma membrane. Activation of 5-HT_1 receptors promotes release of GDP from G proteins allowing entry of GTP onto the nucleotide binding site. When GTP-bound G protein molecule is generated, the α subunit dissociates from the β/γ subunits. The free β/γ complex will then lead to opening voltage-dependent K^+ channels and the membrane is hyperpolarized. Membrane hyperpolarization will decrease 5-HT release and the autoinhibition of transmitter release occurs.

In the next step, free α subunit associated with the bound GTP stimulates membrane enzyme phospholipase C-β (PLC) which specifically hydrolyses phosphatidylinositol-4,5-bisphosphate (PIP_2). PIP_2 splits into inositol-1,4,5-triphosphate (IP_3) and DAG. IP_3 triggers the release of Ca^{2+} from internal stores whereas the other messenger, DAG will activate Ca^{2+}-sensitive PKC. The activated PKC then phosphorylates voltage-sensitive Ca^{2+} channels, they open up and a Ca^{2+} influx occurs through the ion permeable channels. Entry of Ca^{2+} will decrease the membrane potential, which leads to increase of transmitter release and the negative feedback inhibition of 5-HT release is suspended. This will lead to termination of negative feedback inhibition of 5-HT release (Fig. 7).

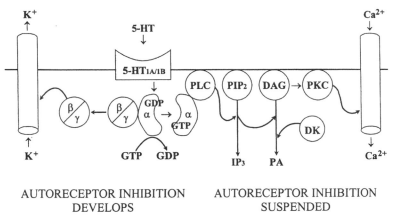

AUTORECEPTOR INHIBITION   AUTORECEPTOR INHIBITION
        DEVELOPS   SUSPENDED

Figure 7. The cellular mechanisms proposed to be involved in 5-HT$_{1A/1B}$ receptor-mediated inhibition of 5-HT release in the raphe nuclei. Negative feedback inhibition of 5-HT release may be developed by opening voltage-sensitive K$^+$ channels that is followed by opening of voltage-sensitive Ca^{2+} channels in the somatodendritic area of 5-HT neurons. We have presented evidence that the 5-HT release-inhibitory effect of 5-HT$_{1A}$ and 5-HT$_{1B}$ receptors is mediated by the phosphoinositide pathway, whereas inhibition of adenylate cyclase may not be directly linked to modulation of transmitter release. Abbreviations: G protein guanine nucleotide-binding protein; GTP, guanosine triphosphate; GDP, guanosine diphosphate; PLC, phospholipase C-β; PIP$_2$, phosphatidylinositol-4,5-bisphosphate; IP$_3$, inositol-1,4,5-triphosphate; DAG, diacylglycerol; PKC, protein kinase C.

5. CONCLUSION

Raphe nuclei possess at least two autoreceptor mechanisms that control 5-HT release: one directly mediated by 5-HT$_{1B}$ and/or 5-HT$_{1D}$ receptors and one indirectly mediated by 5-HT$_{1A}$ autoreceptors that also modulate neural firing rate. The release inhibitory 5-HT$_{1B}$ receptors are situated on recurrent axon collaterals and axon terminals within the raphe nuclei, these receptors have little effect on the firing of these neurons. Inhibition of 5-HT$_{1D}$ receptor-mediated 5-HT release in the raphe nuclei is also independent from regulation of neuronal firing. 5-HT$_{1D}$ receptors are autoreceptors in the raphe nuclei and may control larger somatodendritic 5-HT release. Moreover, changes in 5-HT neuronal firing activity are the consequence of 5-HT$_{1A}$ receptor activation and firing-dependent control of 5-HT release occurs in the cell body level mediated by 5-HT$_{1A}$ receptors. The release-inhibitory 5-HT$_{1A}$ receptors are located on the somatodendritic part of 5-HT neurons and control the substantial, possibly non-synaptic release of 5-HT. It has been shown that changes in 5-HT neuronal firing activity are the consequence of 5-HT$_{1A}$ receptor activation and firing-dependent control of 5-HT release occurs in the cell body level mediated by 5-HT$_{1A}$ receptors. 5-HT$_{1B/1D}$ receptors may be under tonic influence of 5-HT released whereas 5-HT$_{1A}$ receptors may influence raphe 5-HT release with a more phasic mode.

Involvement of other types of 5-HT receptors in 5-HT release and neuronal firing rate has also been postulated. Stimulation of 5-HT$_3$ receptors leads to increase of 5-HT release whereas inhibition of 5-HT release by 5-HT$_7$ heteroreceptors was suggested as well (Harsing et al., 2004). There may be interaction between various 5-HT receptors in regulation of 5-HT release in raphe nuclei. We have made the following speculation for

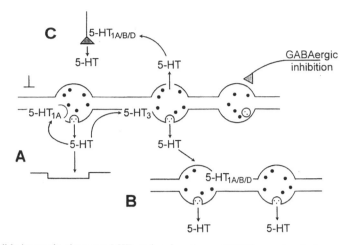

Figure 8. Possible interaction between 5-HT$_3$ and various types of 5-HT$_1$ receptors in the regulation of 5-HT release in the raphe nuclei. Vesicle-containing dendrites of 5-HT neurons are invaded by depolarization with the results that they release 5-HT. 5-HT release in dendritic synapse (A) is inhibited by presynaptic 5-HT$_{1A}$ autoreceptors. If released 5-HT diffuses out into the biophase, it may reach extrasynaptic 5-HT$_3$ receptors. Non-synaptic 5-HT release from dendrites is facilitated by 5-HT$_3$ receptors and 5-HT concentrations in the extrasynaptic space increase further. As a consequence, dendritic membranes around 5-HT$_3$ receptors will be depolarized and inhibitory GABA synaptic inputs from distal dendrites will not propagate proximally to the dendritic shaft and cell body of 5-HTneurons. High concentrations of 5-HT in the biophase will activate inhibitory 5-HT$_1$ receptors of A, B and D subtypes sited on parallel dendrites (B) and on serotonergic recurrent axon collaterals (C). The release of 5-HT will then be inhibited in many dendritic branches and axon terminals. In the raphe nuclei, 5-HT$_7$ receptors modulate 5-HT release as heteroreceptors located on glutamatergic axon terminals (Harsing et al., 2004).

the functional interaction of 5-HT. 5-HT is released as action potential propagation invades dendrites or recurrent axon collaterals of 5-HT neurons. As a consequence of transmitter release, presynaptic 5-HT$_{1A}$ autoreceptors in the dendritic synapses are activated, inhibiting further release of 5-HT. If the autoreceptor-mediated inhibition is attenuated (autoreceptor is desensitized by 5-HT or 5-HT transporter is down-regulated) or initiated with a delay, the concentration of 5-HT in the gap of dendritic synapse may reach high enough concentrations to diffuse into the extrasynaptic space. 5-HT may then activate extrasynaptic 5-HT$_3$ receptors and as a consequence, changes in electrical properties of dendritic membranes will occur. 5-HT$_3$ receptors may be located more proximally on dendritic tree and therefore, depolarization will not allow propagation of inhibitory inputs from distal dendrites to the dendritic shaft and cell body. GABA may be a possible candidate for this kind of inhibition, as GABA neurons in the raphe nuclei synapse with 5-HT neurons (Fig. 8).

6. ACKNOWLEDGMENTS

This research was supported in part by the Research Council for Health Sciences, Hungarian Ministry of Health and Welfare (ETT-482/2003) and the Hungarian Science Research Fund (OTKA) grant No. T-43511. The authors acknowledge the collaboration of Drs. Erzsebet Bagdy and Zsolt Juranyi, the excellent technical assistance of Mrs. Erika Hajdune-Gosi and Zsuzsa Major and the editorial work of Ms. Judit Puskas.

7. REFERENCES

Allgaier, C., Feuerstein, T. J., and Hertting, G., 1986, N-ethyl-maleimide (NEM) diminishes alpha2-adrenoceptor mediated effects on noradrenaline release, *Naunyn-Schmiedeberg's Arch. Pharmacol.* **331**:235.

Arborelius, L., Backlund Hook, B., Hacksell, U., and Svensson, T. H., 1994, The 5-HT$_{1A}$ receptor antagonist (S)-UH-301 blocks the (R)-8-OH-DPAT-induced inhibition of serotonergic dorsal raphe cell firing in the rat, *J. Neural. Transm.* **96**:179.

Bagdy, E., Horvath, E., Sziraki, I., Kiraly, I., and Harsing, L. G., 1994, Further evidence on 5-HT$_{1A}$ agonist action of 8-OH-DPAT, *in vitro* release studies *17th Annual Meeting of the European Neuroscience Association* **Abstr. 51**:19.

Bagdy, E., and Harsing, L. G., Jr., 1995, The role of various calcium and potassium channels in the regulation of somatodendritic serotonin release, *Neurochem. Res.* **20**:1409.

Bagdy, E., Solyom,S., and Harsing,L. G., Jr., 1998, Feedback stimulation of somatodendritic serotonin release: a 5-HT$_3$ receptor-mediated effect in the raphe nuclei of the rat, *Brain Res. Bulletin* **45**:203.

Bagdy, E., Kiraly, I., and Harsing, L. G., Jr., 2000, Reciprocal innervation between serotonergic and GABAergic neurons in the raphe nuclei of the rat, *Neurochem. Res..* **25**:1465.

Baraban, J. M., and Aghajanian, G. K., 1981, Noradrenergic innervation of serotonergic neurons in the dorsal raphe: Demonstration by electron microscopic autoradiography, *Brain Res.* **204**:1.

Bard, J. A., Zgombick, J., Adham, N., Vaysee, P., Branchek, T. A., and Weinshank, R. L., 1993, Cloning of a novel human serotonin receptor (5-HT$_7$) positively linked to adenylate cyclase, *J. Biol. Chem.* **268**:23422.

Barnes, N. M., and Sharp, T., 1999, A review of central 5-HT receptors and their function, *Neuropharmacology* **38**:1083.

Behzadi, G., Kalen, P., Parvapassu, F., and Wiklund, L., 1990, Afferents to the median raphe nucleus of the rat: retrograde choleratoxin and wheat germ-conjugated horseradish peroxidase tracing and selective D-[^{3}H]aspartate labelling of possible excitatory amino acid inputs, *Neuroscience* **37**:77.

Benuck, M., and Reith, M. E. A., 1992, Dopamine releasing effect of phenylbiguanide in rat striatal slices, *Naunyn-Schmiedeberg's Arch. Pharmacol.* **345**:666.

Blier, P., and Bouchard, C., 1993, Functional characterization of a 5-HT$_3$ receptor which modulates the release of 5-HT in the guinea-pig brain, *Br. J. Pharmacol.* **108**:13.

Blier, P., Monroe, P. J., Bouchard, C., Smith, D. L., and Smith, D. J., 1993, 5-HT$_3$ receptors which modulate [^{3}H]5-HT release in the guinea-pig hypothalamus are not autoreceptors, *Synapse* **15**:143.

Blier, P., and Bouchard, C., and de Montigny, C., 1993a, Differential properties of pre- and postsynaptic 5-hydroxytryptamine1A receptors in the dorsal raphe and hippocampus. II. Effect of pertussis and cholera toxins, *J. Pharm. Exp. Ther.* **265**:16.

Boddeke, H. W. G. M., Fargin, A., Raymond, J. R., Schoeffter, P., and Hoyer, D., 1992, Agonist/antagonist interactions with cloned human 5-HT$_{1A}$ receptors: Variations in intrinsic activity studies in transfected HeLa cells, *Naunyn-Schmiedeberg's Arch. Pharmacol.* **345**:257.

Boess, F. G., and Martin, I. L., 1994, Molecular biology of 5-HT receptors, *Neuropharmacology* **33**:275.

Burt, R. P., Chapple, C. R., and Marshall, I., 1996, The role of diacylglycerol and activation of protein kinase C in α1A-adrenoceptor-mediated contraction to noradrenaline of rat isolated epidymal vas deferens, *Br. J. Pharmacol.* **117**:224.

Chazal, G., and Ralston, H. J., 1987, 5-HT-containing structures in the nucleus raphe dorsalis of the cat: an ultrastructural analysis of dendrites, presynaptic dendrites, and axon terminals, *J. Comp. Neurol.* **259**:259.

Cliffe, I. A., Brightwell, C. I., Fletcher, A., Forster, E. A., Mansell, H. L., Reilly, Y., Routledge, C., and White, A. C., 1993, (S)-N-tert-Butyl-3-(4-(2-methoxyphenyl)-piperazin-1-yl)-2- phenylpropanamide [(S)-WAY-100135]: a selective antagonist at presynaptic and postsynaptic 5-HT$_{1A}$ receptors, *J. Med. Chem.* **36**:1509.

Dahlstrom, A., and Fuxe, K., 1964, Evidence for the existence of monoamine containing neurons in the central nervous system 1. Demonstration of monoamines in the cell bodies of brain stem neurons, *Acta Physiol. Scand. Suppl.* **232**:1.

Davidson, C., and Stamford, J. A., 1995, Evidence that 5-hydroxytryptamine release in rat dorsal raphe nucleus is controlled by 5-HT$_{1A}$ 5-HT$_{1B}$ and 5-HT$_{1D}$ autoreceptors, *Br. J. Pharmacol.* **114**:1107.

de Erausquin, G., Brooker, G., and Hanbauer, I., 1992, K$^+$-evoked dopamine release depends on a cytosolic Ca^{2+} pool regulated by N-type Ca^{2+} channels, *Neurosci. Lett.* **145**:1211.

de Langen, C. D. J., and Mulder, A. H., 1979, Compartmental analysis of the accumulation of ^{3}H-dopamine in synaptosomes from rat corpus striatum, *Naunyn-Schmiedeberg's Arch. Pharmacol.* **308**:31.

de Langen, C. D. J., and Mulder, A. H., 1980, Effects of psychotropic drugs on the distribution of ^{3}H-dopamine into compartments of rat striatal synaptosomes and on subsequent depolarization-induced ^{3}H-dopamine release. *Naunyn-Schmiedeberg's Arch. Pharmacol.* **311**:169.

Derkach, V., Surprenant, A., and North, R. A., 1989, 5-HT$_3$ receptors are membrane ion channels, *Nature* **339**:706.

Dunlap, K., Luebke, J. I., and Turner, T. J., 1995, Exocytotic Ca^{2+} channels in mammalian central neurons, *Trends Neurosci.* **18**:89.

El Mansari, M., and Blier, P., 1996, Functional characterization of 5-HT$_{1D}$ autoreceptors on the modulation of 5-HT release in the guinea-pig mesencephalic raphe, hippocampus and frontal cortex, *Br. J. Pharmacol.* **118**:681.

Engel, G., Gothert, M., Hoyer, D., Schlicker, E., and Hillenbrand, K., 1986, Identity of inhibitory presynaptic 5-hydroxytryptamine (5-HT) autoreceptors in the rat cortex with 5-HT$_{1B}$ binding sites, *Naunyn-Schmiedeberg's Arch. Pharmacol.* **332**:1.

Fletcher, A., Forster, E. A., Bill, D. J., Brown, G., Cliffe, I. A., Hartley, J. E., Jones, D. E., McLenachan, A., Stanhope, K. J., Critchley, D. J. P., Childs, K. J., Middlefell, V. C., Lanfumey, L., Corradetti, R., Laporte, A., Gozlan, H., Hamon, M., and Dourish, C. T., 1996, Electrophysiological, biochemical, neurohormonal and behavioural studies with WAY-100635, a potent, selective and silent 5-HT$_{1A}$ receptor antagonist, *Behav. Brain Res.* **73**:337.

Galzin, A. M., Poncet, V., and Langer, S. Z., 1990, 5-HT$_3$ receptor agonists enhance the electrically-evoked release of [^{3}H]-5-HT in guinea-pig frontal cortex slices, *Br. J. Pharmacol.* **101**:307P.

Gauer, S., Newcomb, R., Rivnay, B., Bell, J. R., Yamashiro, D., Ramachandran, J., and Miljanich, G. P., 1994, Calcium channel antagonist peptides define several components of transmitter release in the hippocampus, *Neuropharmacology* **33**:1211.

Haj-Dahmane, S., Hamon, M., and Lanfumey, L., 1991, K$^+$ channel and 5-hydroxytryptamine 1A autoreceptor interactions in the rat dorsal raphe nucleus: an *in vitro* electrophysiological study, *Neuroscience* **41**:495.

Hajos, M., Gartside, S. E., Villa, A. E. P., and Sharp, T., 1995, Evidence for a repetitive (burst) firing pattern in a sub-population in the dorsal and median raphe nuclei of the rat, *Neuroscience* **69**:189.

Hajos, M., Richards, C. D., Szekely, A. D., and Sharp, T., 1998, An electrophysiological and neuroanatomical study of the medial prefrontal cortical projection to the midbrain raphe nuclei in the rat, *Neuroscience* **87**:95.

Hamblin, M. W., McGuffin, R. W., Metcalf, M. A., Dorsa, D. M., and Merchant K. M., 1992, Distinct 5-HT$_{1B}$ and 5-HT$_{1D}$ serotonin receptors in rat: structural and pharmacological characterization of the two cloned receptors, *Mol. Cell Neurosci.* **3**:578.

Harandi, M., Aguera, M., Gamrani, H., Didier, M., Maitre, M., Calas, A., and Belin, M. F., 1987, γ-Aminobutyric acid and 5-hydroxytryptamine interrelationship in the rat nucleus raphe dorsalis: combination of radiographic and immunocytochemical techniques at light and electron microscopy levels, *Neuroscience* **21**:237.

Harsing, L. G., Jr., Sershen, H., Vizi, E. S., and Lajtha, A., 1992, N-type calcium channels are involved in the dopamine releasing effect of nicotine, *Neurochem. Res.* **17**:729.

Harsing, L. G., Jr., and Zigmond, M. J., 1997, Influence of dopamine on GABA release in striatum: Evidence for D1-D2 interactions and non-synaptic influences, *Neuroscience* **77**:419.

Harsing, L. G., Jr., Kiraly, I., and Bagdy, E., 2000, Signal transduction pathways coupled to serotonin release-inhibitory 5-HT$_{1A}$ and 5-HT$_{1B}$ receptors in rat raphe nuclei slices, *J. Physiol. Proceedings* **526**:67P.

Harsing, L. G., Prauda, I., Barkoczy, J., Matyus, P., and Juranyi, Zs., 2004, 5-HT$_7$ heteroreceptor-mediated inhibition of [^{3}H]serotonin release in raphe nuclei slices of the rat: evidence for a serotonergic-glutamatergic interaction, *Neurochem. Res.* in press.

Herdon, H., and Nahorski, S. R., 1989, Investigation of the roles of dihydropyridine and -conotoxin-sensitive calcium channels in mediating depolarisation-evoked endogenous dopamine release from striatal slices, *Naunyn-Schmiedeberg's Arch. Pharmacol.* **340**:36.

Hery, F., Faudon, M., and Ternaux, J. P., 1982, *In vivo* release of 5-HT in two raphe nuclei (raphe dorsalis and magnus) of cat, *Brain Res. Bull.* **8**:123.

Hjorth, S., and Magnusson, T., 1988, The 5-HT$_{1A}$ receptor agonist, 8-OH-DPAT, preferentially activates cell body 5-HT autoreceptors in rat brain *in vivo*, *Naunyn-Schmiedeberg's Arch. Pharmacol.* **338**:463.

Hoyer, D., Clarke, D. E., Fozard, J. E., Hartig, P. R., Martin, G. R., Mylecharane, E. J., Saxena, P. R., and Humbrey, P. P. A., 1994, International Union of Pharmacology classification of receptors for 5-hydroxytryptamine (serotonin), *Pharmacol. Rev.* **46**:157.

Hu, P.-S., and Fredholm, B. B., 1989, Alpha2-adrenoceptor agonist-mediated inhibition of [³H]noradrenaline release from rat hippocampus is reduced by 4-aminopyridine, but that caused by an adenosine analogue or omega-conotoxin is not, *Acta Physiol. Scand.* **136**:347.

Illes, P., 1986, Mechanisms of receptor-mediated modulation of transmitter release in noradrenergic, cholinergic and sensory neurons, *Neuroscience* **14**:909.

Innis, R. B., and Aghajanian, G. K., 1987, Pertussis toxin blocks 5-HT$_{1A}$ and GABA$_B$ receptor mediated inhibition of serotonergic neurons, *Eur. J. Pharmacol.* **143**:195.

Innis, R. B., Nestler, E. J., and Aghajanian, G. K., 1988, Evidence for G protein mediation of serotonin and GABA$_B$-induced hyperpolarization in rat dorsal raphe neurons, *Brain Res.* **459**:27.

Juranyi, Zs., and Harsing, L. G., Jr., 2004, Brain slice chambers designed for *in vitro* experiments with nervous tissue. in: *Monoamine Oxidase Inhibitors,* T. L. Torok, ed., Medicina Publishing House Co, Budapest, in press.

Kerwin, R. W., and Pycock, C. J., 1979, The effect of some putative neurotransmitters on the release of 5-hydroxytryptamine and gamma-aminobutyric acid from slices of the rat midbrain raphe area, *Neuroscience* **4**:1359.

Kidd, E. J., Garratt J. C., and Marsden, C. A., 1991, Effects of repeated treatment with 1-(2,3-dimethoxy-4-iodophenyl)-2-aminopropane (DOI) on the autoregulatory control of dorsal raphe 5-HT neuronal firing and cortical 5-HT release, *Eur. J. Pharmacol.* **200**:131.

Lovenberg, T. W., Baron, B. M., de Lecea, L., Miller, J. D., Prosser, R. A., Rea, M. A., Foye, P. E., Racke, M., Slone, A. L., Siegel, B. W., Danielson, P. A., Sutcliffe, J. G., and Erlander, M. G., 1993, A novel adenylyl cyclase-activating serotonin receptor (5-HT$_7$ implicated in the regulation of mammalian circadian rhythms, *Neuron* **11**:449.

Martin, K. F., Hannon, S., Phillips, I., and Heal, D. J., 1992, Opposing roles for 5-HT$_{1B}$ and 5-HT$_3$ receptors in the control of 5-HT release in rat hippocampus *in vivo, Br. J. Pharmacol.* **106**:139.

Middlemiss, D. N., and Fozard, J. R., 1983, 8-Hydroxy-2-(di-n-propylamino) tetralin discriminates between subtypes of the 5-HT$_1$ recognition site, *Eur. J. Pharmacol.* **90**:151.

Middlemiss, D. N., 1984, Stereoselective blockade at [³H]5-HT binding sites and at the 5-HT autoreceptor by (-)propranolol, *Eur. J. Pharmacol.* **101**:289.

Musgrave, I. F., and Majewski, H., 1989, Effect of phorbol ester and polymyxin B on modulation of noradrenaline release in mouse atria *Naunyn-Schmiedeberg's Arch. Pharmacol.* **339**:48.

Neale, R. F., Fallon, S. L., Royar, W. C., Wasley, J. W. F., Martin, L. L., Stone, G. A., Glaester, B. S., Sinton, C. M., and Williams, M., 1987, Biochemical and pharmacological characterization of CGS-12066, a selective 5-HT$_{1B}$ agonist, *Eur. J. Pharmacol.* **136**:1.

O'Connor, J. J., and Kruk, Z. L., 1992, Pharmacological characteristics of 5-hydroxytryptamine autoreceptors in rat brain slices incorporating the dorsal raphe or the suprachiasmatic nucleus, *Br. J. Pharmacol.* **106**:524.

O'Hearn, E., and Molliver, M. E., 1984, Organization of raphe-cortical projections in the rat: a quantitative retrograde study, *Brain Res. Bull.* **13**:709.

Palacios, J. M., Pazos, A., and Hoyer, D., 1987, Characterization and mapping of 5-HT$_{1A}$ sites in the brain of animals and man. in: *Brain 5-HT$_{1A}$ Receptors,* C. T. Dourish, S. Ahlenins, P. H. Hutson, eds., Ellis Horwood, Chichester, p. 67.

Parent, A., Descarries, L., and Beaudet, A., 1981, Organization of ascending serotonin systems in the adult rat. A radioautographic study after intraventricular administration of [³H]5-hydroxytryptamine, *Neuroscience* **6**:115.

Penington, N. J., and Kelly, J. S., 1990, Serotonin receptor activation reduces calcium receptor activation reduces calcium current in an acutely dissociated adult central neuron, *Neuron* **4**:751.

Pineyro, G., and Blier, P., 1996, Regulation of 5-hydroxytryptamine release from rat midbrain raphe nuclei by 5-hydroxytryptamine1D receptors: Effect of tetrodotoxin, G protein inactivation and long-term antidepressant administration, *J. Pharm. Exp. Ther.* **276**:697.

Pineyro, G., and Blier, P., 1999, Autoregulation of serotonin neurons: role in antidepressant drug action, *Pharmacol. Rev.* **51**:533.

Price, G. W., Burton, M. J., Collin, L. J., Duckworth, M., Gaster, L., Gothert, M., Jones, B. J., Roberts, C., Watson, J. M., and Middlemniss, D. N., 1997, SB-216641 and BRL-15572-compounds to pharmacologically discriminate h5-HT$_{1B}$ and h5-HT$_{1D}$ receptors, *Naunyn-Schmiedeberg's Arch. Pharmacol.* **356**:312.

Roberts, C., Watson, J., Burton, M., Price, G. W., and Jones, B. J., 1996, Functional characterization of the 5-HT terminal autoreceptor in the guinea-pig brain cortex, *Br. J. Pharmacol.* **117**:384.

Rudy, B., 1988, Diversity and ubiquity of K channels, *Neuroscience* **25**:729.

Saavedra, J. P., Brusco, A., Pressini, S., and Olivia, D., 1986, A new case for the presynaptic role of dendrites: an immunocytochemical study of the N. Raphe Dorsalis. *Neurochem. Res.* **11**:997.

Schlicker, E., Fink, K., Moldering, G. J., Prince, G. W., Duckworth, M., Gaster, L., Middlemniss, D. N., Zentner, J., Likungu, J., and Gothert, M., 1997, Effects of selective h5-HT$_{1B}$ (SB-216641) and h5-HT$_{1D}$ (BRL-15572) receptor ligands on guinea-pig and human auto- and heteroreceptors, *Naunyn-Schmiedeberg's Arch. Pharmacol.* **356**:321.

Schoeffter, P., and Hoyer, D., 1989, 5-Hydroxytryptamine 5-HT$_{1B}$ and 5-HT$_{1D}$ receptors mediating inhibition of adenylate cyclase activity. Pharmacological comparison with special reference to the effects of yohimbine, rauwolscine and some beta-adrenoceptor antagonists, *Naunyn-Schmiedbergs's Arch. Pharmacol.* **340**:285.

Seamon, K. B., and Daly, J.W., 1981, Forskolin: an unique diterpene activator of cyclic AMP-generating systems. *J. Glch. Nucleotide Res.* **7**:201.

Sershen, H., Hashim, A., and Lajtha, A., 1995, The effect of ibogaine on kappa-opioid and 5-HT$_3$-induced changes in stimulation-evoked dopamine release *in vitro* from striatum of C57/BL/6By mice, *Brain Res. Bull.* **36**:587.

Sprouse, J. S., and Aghajanian, G. K., 1987, Electrophysiological responses of serotonergic dorsal raphe neurons to 5-HT$_{1A}$ and 5-HT$_{1B}$ agonists, *Synapse* **1**:3.

Sprouse, J., Reynolds, L., and Rollema, H., 1997, Do 5-HT$_{1B/1D}$ autoreceptors modulate dorsal raphe cell firing? *In vivo* electrophysiological studies in guinea-pigs with GR127935, *Neuropharmacology* **36**:559.

Starke, K., 1987, Presynaptic alpha-adrenoceptors, *Rev. Physiol. Biochem. Pharmac.* **107**:73.

Starkey, S. J., and Skingle, M., 1994, 5-HT$_{1D}$ as well as 5-HT$_{1A}$ autoreceptors modulate 5-HT release in the guinea-pig dorsal raphe nucleus, *Neuropharmacology* **33**:395.

Steinbusch, H. W. M., 1981, Distribution of serotonin-immunoreactivity in the central nervous system of the rat-Cell bodies and terminals, *Neuroscience* **6**:557.

Stern, W. C., Johnson, A., Bronzino, J. D., and Morgane, J. P. 1981, Neuropharmacology of the afferent projection from the lateral habenula and substantia nigra to the anterior raphe in the rat, *Neuropharmacology* **20**:979.

Van der Kooy, D., and Hattori, T., 1980, Dorsal raphe cells with collateral projections to the caudate-putamen and substantia nigra: a fluorescent retrograde double labeling study in the rat, *Brain Research* **186**:1.

Vandermaelen, C. P., and Aghajanian, G. K., 1983, Electrophysiological and pharmacological characterization of serotonin dorsal raphe neurons recorded extracellularly and intracellularly in rat brain slices, *Brain Res.* **289**:109.

Verge, D., Daval, G., Patey, A., Gozlan, H., El Mestikawy, S., and Hamon, M., 1985, Presynaptic 5-HT autoreceptors on serotonergic cell bodies and/or dendrites but not terminals are of the 5-HT$_{1A}$ subtype, *Eur. J. Pharmacol.* **113**:463.

Vizi, E. S., Harsing, L. G., Jr., and Zsilla, G., 1981, Evidence of the modulatory role of serotonin in acetylcholine release from striatal interneurons, *Brain Res.* **212**:89.

Vizi, E. S., Somogyi, G. T., Harsing, L. G., and Zimanyi I., 1985, External Ca-independent release of norepinephrine by sympathomimietics and its role in negative feedback modulation, *PNAS* **82**:8775.

Vizi, E. S., 2000, Role of high-affinity receptors and membrane transporters in nonsynaptic communication and drug action in the CNS, *Pharmacol. Reviews* **52**:63.

Voigt, M. M., Laurie, D. J., Seeburgh, P. H., and Bach, A., 1991, Molecular cloning and characterization of a rat brain cDNA encoding a 5-hydroxytryptamine1B receptor, *EMBO J.* **10**:4017.

Wang, R. Y., and Aghajanian, G. K., 1977, Physiological evidence for habenula as major link between forebrain and midbrain raphe, *Science* **197**:89.

Wang, Q. P., Ochiaia, P. G., and Nakai, Y., 1992, GABAergic innervation of serotonergic neurons in the dorsal raphe nucleus of the rat studied by electron microscopy double immunostaining, *Brain Res. Bull.* **29**:943.

Wiklund, L., and Bjorklund, A., 1980, Mechanism of regrowth in the bulbospinal 5-HT system following 5,6-dihydroxytryptamine induced axotomy. II. Fluorescence histochemical observations, *Brain Res.* **191**:129.

Wiklund, L., Leger, L., and Persson, M., 1981, Monoamine cell distribution in the cat brain stem. A fluorescence histochemical study with quantification of indolaminergic and locus coeruleus cell groups, *J. Comp. Neurol.* **203**:613.

Yakel, J. L., and Jackson, M. B., 1988, 5-HT$_3$ receptors mediate rapid responses in cultured hippocampal and clonal cell line, *Neuron* **1**:615.

Zifa, E., and Fillion, G., 1992, 5-Hydroxytryptamine receptors, *Pharmacol. Rev.* **44**:401.

Zimanyi, I., Somogyi, G. T., Harsing, L. G., Jr., and Vizi, E. S., 1985, Release of ^{3}H-noradrenaline by 4-aminopyridine and alpha-2 adrenoceptor agonists, in: *Pharmacology of Adrenoceptors,* E. Szabadi, C. M. Bradshaw, S. R. Nahorski, eds., MacMillan Press Ltd., London, p. 334.

12

EXTRASYNAPTIC RELEASE OF DOPAMINE AND VOLUME TRANSMISSION IN THE RETINA

Michelino Puopolo*, Spencer E. Hochstetler[‡], Stefano Gustincich*, R. Mark Wightman[‡] and Elio Raviola*[1]

1. INTRODUCTION

In the retina, dopamine is a catecholamine modulator responsible for many of the events that lead to neural adaptation to light (see Witkovsky and Dearry, 1991; Djamgoz and Wagner, 1992). By acting on numerous cell types in the retina, dopamine sets the gain of the retinal networks for vision in the light. Measurements of the amount of dopamine released by the retina showed that there is a basal efflux of dopamine in the dark; the efflux is increased by steady illumination and maximal with flickering light (see Witkovsky and Dearry, 1991; Djamgoz and Wagner, 1992). Dopamine acts at multiple levels in the retina: in teleosts, acting at D_2 receptors it induces contraction of cones and movement of melanin granules in pigment epithelial cells (Dearry and Burnside, 1986, 1989). In photoreceptors, acting at D_4 receptors, it inhibits the Na/K ATPase of rods (Shulman and Fox, 1996). Acting at D_2 receptors, it decreases voltage-gated Ca^{2+} current in cones and increases it in rods (Stella and Thoreson, 2000), where, in addition, it inhibits the hyperpolarization-activated current I_h (Akopian and Witkovsky, 1996). Dopamine has complex effects on horizontal and bipolar cells, all mediated by D_1 receptors. It decreases the conductance of the gap junctions between horizontal cells (Piccolino et al., 1984; Lasater and Dowling, 1985; DeVries and Schwartz, 1989), thus reducing the size of their receptive field; in both horizontal and bipolar cells, it potentiates the activity of ionotropic glutamate receptors (Knapp and Dowling, 1987; Maguire and Werblin, 1994), inhibits $GABA_C$-mediated responses (Dong and Werblin, 1994) and modulates voltage-gated calcium (Pfeiffer-Linn and Lasater, 1998) and potassium (Fan and Yazulla, 1999, 2001) channels. Thus, the overall effect of dopamine in the outer retina is an improvement of contrast detection (Hare and Owen, 1995; see Witkovsky, 2004). In the inner plexiform layer (IPL), dopamine, acting at D_1 receptors,

---

[1] *Department of Neurobiology, Harvard Medical School, Boston, MA 02115, ‡Department of Chemistry, University of North Carolina, Chapel Hill, NC 27599

Dendritic Neurotransmitter Release, edited by M. Ludwig
Springer Science+Business Media, Inc., 2005

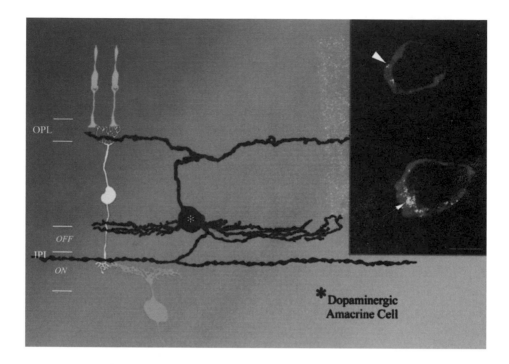

Figure 1. The perikaryon of the DA cell gives rise to a dense dentritic plexus in the scleral stratum of the inner plexiform layer (IPL). From this plexus arise long processes that descend into the middle stratum of the IPL. In addition, a process ascends toward the outer plexiform layer (OPL), where it forms a much looser plexus. Inset: two confocal optical sections of a DA cell double-immunostained for VMAT2 (green) and TH (red). A small number of organelles of varying diameter are scattered throughout the cytoplasm and, occasionally, they are very close to the cell surface (arrowhead). Scale bar: 1 μm.

(A color version of this figure appears in the signature between pp. 256 and 257.)

uncouples AII amacrine cells (Hampson et al., 1992; Mills and Massey, 1995), potentiates the GABA-induced chloride current of amacrine cells (Feigenspan and Bormann, 1994) and relieves $GABA_C$ inhibition of transmitter release by the synaptic endings of bipolar cells (Wellis and Werblin, 1995). Ultimately, it decreases the sensitivity of ganglion cells and modifies both their spontaneous discharge and the center-surround balance of their receptive fields (Jensen and Daw, 1984, 1986; Jensen 1989, 1991).

Retinal dopamine is synthesized by a type of neuron characterized by a large, spherical perikaryon, situated in the vitreal tier of cell bodies of the inner nuclear layer (Fig. 1). Their dendrites form a dense plexus in the scleral stratum (S1) of the inner plexiform layer (IPL); in addition, they give rise to thin, varicose processes, that in the cat and mouse descend into the middle stratum (S3) of the IPL, where they travel long distances before returning to S1 (Kolb et al., 1990). In some species, such as the rabbit, the dopaminergic neurons are typical amacrine cells, i.e. their processes do not extend beyond the IPL (Tauchi et al., 1990). In other species, they send additional processes to the outer plexiform layer (OPL), where they form a loose plexus intermeshed with the dendrites of horizontal and bipolar cells. Because of this plexus, whose richness varies greatly in different animal species, the dopaminergic neurons were called interplexiform cells (see Nguyen-Legros, 1988). They are also named type 1 catecholaminergic amacrines, because uptake of catecholamines or antibodies to tyrosine hydroxylase label a second type of retinal neuron, the type 2 catecholaminergic amacrines, that exhibit a smaller perikaryon and possess a dendritic plexus situated in the S3 stratum of the IPL. Very little is known about this second type of catecholaminergic neuron and the nature of its transmitter(s) is unclear. They will not be considered further in this chapter.

Thus, dopamine receptors are present throughout the retina, often at a considerable distance from the processes of DA cells (Bjielke et al.,1996; Veruki and Wässle, 1996; Derouiche and Asan, 1999). It was this mismatch between the distribution of the receptors and the localization of the branches of DA cells that prompted the idea that dopamine acts on its targets by volume transmission (Witkovsky et al., 1993).

It is generally accepted that dopamine is released upon photopic illumination (see Witkovsky and Dearry, 1991), an indication that DA cells receive input from ON-cone bipolars. Surprisingly, DA cells appeared to be postsynaptic to bipolar endings situated in stratum S1 of the IPL (Hokoç and Mariani, 1987; Kolb et al., 1990; Gustincich et al., 1997), which is occupied by axonal arborizations of OFF-cone bipolars. We have recently observed that DA cells also receive a bipolar input on the processes that course in stratum S3 of the IPL (Contini and Raviola, unpublished observations). Since this stratum is occupied by the axonal arborization of ON-bipolar cells, our finding confirms the expectation that DA cells are excited by illumination of the retina. The significance of the bipolar synapses in S1 remains to be elucidated. On the other hand, OFF-bipolars have an important role in the excitation of the GABAergic amacrine cell(s) that inhibits the release of dopamine in the dark. Relief of this inhibition contributes to dopamine release upon illumination of the retina.

After the discovery of a genetically programmed circadian oscillator in the retina that regulates the synthesis of melatonin (Tosini and Menaker, 1996), DA cells represent an obvious target for the circadian modulator and, indeed, melatonin inhibits release of dopamine in the rabbit (Dubocovich, 1983, 1985; Dubocovich and Hensler, 1986). In *Xenopus*, this effect is blocked by $GABA_A$ antagonists (Boatright et al., 1994), which led to the suggestion that melatonin receptors reside on GABAergic amacrines. DA cells do receive a substantial input from GABAergic amacrines over the vitreal aspect of the perikaryon and their entire dendritic tree in S1 (Kolb et al., 1991; Gustincich et al., 1999), and this input is probably responsible for inhibition of dopamine release in the dark. Melatonin receptors, however, are also present on DA cells (Fujieda et al., 2000). The major output of DA cells is on amacrine cell types which are part of the rod pathway: DA cells establish synapses on AII amacrines (Pourcho, 1982; Voigt and Wässle, 1987; Kolb, et al. 1990, 1991; Strettoi et al., 1992), a neuronal type inserted in series along the pathway that carries rod signals to ganglion cells, and S1/S2 amacrine cells (Kolb et al., 1990), that are responsible for an inhibitory feedback onto rod bipolars. The neurotransmitter released at these synapses was not known, but, in addition to dopamine, GABA was a candidate, because both this molecule and its synthetic enzyme glutamic acid decarboxylase are present in the perikarya of DA cells (Kosaka et al., 1987; Wässle and Chun, 1988; Wulle and Wagner 1990). By using triple-label immunocytochemistry and confocal microscopy, we identified a cluster of $GABA_A$ receptors at the postsynaptic active zone of the DA-to-AII amacrine cell synapses (Contini and Raviola, 2003). Since both the GABA vesicular transporter and the vesicular monoamine transporter-2

(VMAT2) are present in the endings of DA cells, we suggested that the synapses between retinal dopaminergic neurons and AII amacrine cells are GABAergic and that both GABA and dopamine are released by the presynaptic endings.

Finally, a study of global gene expression in single DA cells showed that these neurons secrete insulin, the neuropeptide CART, the cytokine interferon α, and the chemokine monocyte chemoattractant protein-1. They also contain the most common circadian clock-related proteins. Interestingly, cryptochrome 2 is only localized in DA cells and a subset of ganglion cells, supporting the idea that DA cells have a role in the retinal internal clock (Witkovsky et al., 2003; Gustincich et al., 2004).

Thus, at the present state of our knowledge, DA cells appear to carry out multiple functions in the retina, each characterized by a different time course. 1) Through their fast, possibly GABAergic synapses on AII amacrine cells and S1/S2 amacrine cells, DA cells inhibit the transfer of rod signals to ganglion cells on a time scale in the order of a millisecond. 2) They release dopamine, that acts at a distance by volume transmission on a large number of retinal neurons, presiding over the process of transition from the dark-adapted to the light-adapted state. Dopamine also influences gene expression by inducing synthesis of c-fos in amacrine cells (Koistinaho and Sagar, 1995). These functions take place over a time scale of seconds to minutes. 3) DA cells contain the most common circadian clock-related proteins suggesting a role in circadian regulation of retinal function over a time scale of hours. 4) In addition to dopamine, they synthesize four secreted neuroactive molecules whose precise function in the economy of the retina is unknown.

2. EXTRASYNAPTIC RELEASE OF DOPAMINE

The retina represents one of the districts of the nervous system in which convincing evidence documents the physiological importance of volume transmission, because cells that carry dopamine receptors reside at a considerable distance from the processes of the dopamine-synthesizing neurons and the effects of dopamine on remote retinal targets are well-understood. Clearly, dopamine, once released by the DA cells, diffuses along the narrow intercellular clefts of the retina, only partially removed by uptake and metabolism, and reaches cells, such as those of the pigment epithelium, that are located at a distance of several tens of micrometers from the release sites. One source of the dopamine could be the synapses established by DA cells on the cell body of AII amacrines in the most superficial stratum of the IPL. In addition, dopamine could be released over the entire surface of DA cells, as is the case for the somatodendritic arbor of the nigral neurons (see Cheramy et al.,1981; Garris et al., 1994; Rice et al., 1997; Jaffe et al., 1998; Bunin and Wightman, 1999). DA cell perikarya represent an ideal object to investigate extrasynaptic release, because a large number of electron microscopic studies have shown that, as a rule, presynaptic active zones are absent from the cell bodies of amacrine cells. We therefore applied amperometry to solitary DA cell bodies, isolated by enzymatic digestion and mechanical trituration of the retina. DA cells could be identified in the living state because they were obtained from the retina of transgenic mice in which neurons containing tyrosine hydroxylase (TH) express human placental alkaline phosphatase (Gustincich et al., 1997). Because of the presence of the enzyme on the outer aspect of the plasma membrane, DA cells were stained after dissociation with a monoclonal antibody to human placental alkaline phosphatase, directly conjugated to the fluorochrome Cy3. Dopamine release was measured using beveled carbon fiber electrodes with a diameter of 5µm placed in contact with the surface of DA cells. Oxidation currents were recorded by applying a potential to the carbon fiber electrode sufficient to oxidize dopamine (+650 mV with respect to a Ag/AgCl reference electrode). Although 5-hydroxytryptamine can also be detected in these conditions, it is accumulated by different types of amacrine cells (Sandell and Masland, 1986). In addition, amacrine cells that express TH do not take up serotonin (Tauchi et al., 1990). Often, the DA cell was simultaneously patch clamped in the cell-attached or whole-cell configuration (Fig. 2).

When removed from synaptic influences, DA cells spontaneously generate action potentials (Fig.3) in a rhythmic fashion (Feigenspan et al., 1998), a behavior similar to that of their counterparts in the midbrain (Grace and Bunney, 1983a and b). This pacemaker activity, driven by both a steady state "persistent' sodium current and a subthreshold calcium current, has a frequency of about 6 Hz at room temperature and 13 Hz at 35°C. Simultaneous carbon fiber recordings showed that DA cells generate amperometric spikes with current sizes varying from 0.1 to several pA and variable, but low frequency (about 0.2 Hz at room temperature and 13 Hz at 35°C). They were rapid (half-widths ~0.6 ms) and most of them resembled the events of exocytosis observed in other cells (Bruns and Jahn, 1995), with a rapid rising phase followed by an exponential-like decay (Fig.4). Occasionally, a "foot" signal was observed in the rising phase,

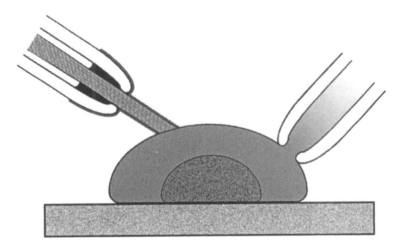

Figure 2. Solitary DA cells, obtained by enzymatic digestion and mechanical trituration of the mouse retina were patch clamped either in the cell-attached or whole cell configuration. Simultaneously oxidation currents were recorded by placing a carbon fiber electrode in contact with the surface.

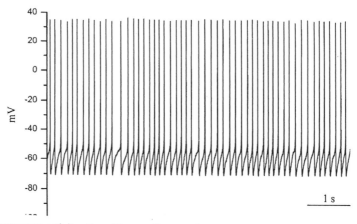

Figure 3. Spontaneous activity of a solitary perikaryon of a DA cell measured in whole-cell current clamp with no current applied.

suggesting an initial trickle of transmitter through a small fusion pore (Chow et al., 1992; Albillos et al., 1997).

A correlation between amperometric spikes and action potentials was not obvious: this is to be expected if the events are rare and a considerable delay separates the electrical and the secretory episodes. Thus, we tested in two ways the dependence of dopamine release on the pacemaker activity of DA cells: we exposed the cells to the sodium channel blocker TTX (Fig.5), to abolish spontaneous firing, and silenced the cell by injecting negative current (Fig.6). In both instances, amperometric spikes were reversibly abolished. Thus, spontaneous dopamine release is triggered by the action potentials that are generated by DA cells in a rhythmic fashion. Dopamine is released by the retina in a tonic fashion (see Witkovsky and Dearry, 1991) and indirect evidence suggests that DA cells also fire action potentials *in vivo* (Piccolino et al., 1987). Furthermore, tracer coupling experiments have demonstrated a relationship between adaptational state of the retina and the coupling of AII amacrine and horizontal cells, which requires a steady presence of dopamine in the intercellular space (Bloomfield et al., 1997; Xin and Bloomfield, 1999). All these observations are consistent with the idea that light modulates a tonic release of dopamine by influencing the pacemaker activity of DA cells. This also suggests that extrasynaptic release occurs *in vivo*, as a result of the physiological activity of the cell.

3. MECHANISM OF RELEASE

Because of the rarity of the spontaneous events and the difficulty in eliciting amperometric spikes by current injection in such a small cell, we resorted to application

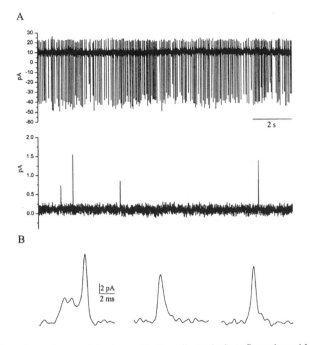

Figure 4. A: a solitary DA cell was patch clamped in the cell-attached configuration with the pipette potential held at 0 mV: the cell generated spontaneous action currents. Simultaneously, a carbon fiber electrode recorded four amperometric spikes, due to events of dopamine release. B: the waveform of the release events exhibits a rapidly rising phase followed by a quasi-exponential decay. In the event on the left, a foot signal is visible at the onset of the rising phase.

of elevated K^+ (63 mM) or 100 µM kainate (Fig.7A) by a pressure-ejection micropipette to study in detail the characteristics of dopamine release. In fact, we had shown in a previous study that DA cells respond to glutamate and to kainate, an agonist at the ionotropic AMPA-kainate receptors, by increasing the frequency of their discharge of action potentials (Gustincich et al., 1997). Both treatments caused a sustained burst of amperometric spikes with considerable latency (0.5 to 27 s). As expected, when high K^+ stimulation was repeated at a higher temperature (35°C), the frequency of the discharge increased significantly and its latency was halved, suggesting that the mechanism that delivers dopamine to the cell surface is temperature-sensitive.

Release from DA cells was dependent on the influx of extracellular Ca^{2+}. Stimulation with high K^+ did not induce dopamine release when Ni^{2+}, a Ca^{2+}-channel blocker, was included in the Ca^{2+}-containing secretagogue solution. In a similar fashion, the response to kainate was suppressed when extracellular Ca^{2+} was replaced by Co^{2+} (Fig.7A). Finally, administration of nimodipine at a 1 µM concentration suppressed release in a reversible fashion (Fig.7B), suggesting the participation of L-type Ca^{2+}

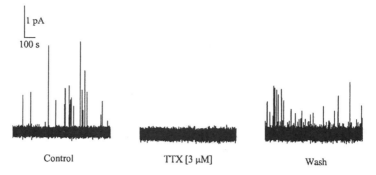

Figure 5. Application of TTX to the superfusion medium reversibly abolished the spontaneous events of dopamine release.

channels that inactivate only slowly (Bean and Mintz, 1994): this explains why Ca^{2+}-entry can be sustained for extended periods of time. Other types of Ca^{2+} channels are also present in DA cells and they exhibit a different distribution in the perikaryon and processes (Tamura et al., 1995; Xu et al., 2003). We confirmed by ratiometric fura-2 measurements that the increase in $[Ca^{2+}]_i$ was primarily due to influx of extracellular Ca^{2+} rather than to Ca^{2+}-induced Ca^{2+} release from intracellular stores. In the presence of thapsigargin, a highly selective, irreversible inhibitor of the sarco(endo)plasmic reticulum Ca^{2+}-ATPase (SERCA) pump (Garaschuk et al., 1997), the internal Ca^{2+} stores were depleted by caffeine. Then, the cells were challenged with high K^+: an elevation of $[Ca^{2+}]_i$ of the same magnitude as before treatment with caffeine was observed, and this increase was abolished by treatment with Ni^{2+} or Cd^{2+}, thus confirming the extracellular origin of the elevated $[Ca^{2+}]_i$. Thus, dopamine release by DA cells exhibited all of the properties expected for exocytosis. Both the Ca^{2+} dependency and a rate of flux four orders of magnitude larger distinguished the events of dopamine exocytosis from a transporter-mediated reverse uptake: [$0.4-1 \times 10^8$ as compared to 3.3×10^4 molecules s^{-1} (Galli et al., 1998)].

Striking features of extrasynaptic dopamine exocytosis are the rarity of the spontaneous quantal events, compared with the frequency of the cell's pacemaker activity, and the long latency of the discharge following stimulation. The broad range of latencies may reflect the slow recruitment of secretory organelles that are few in number and distant from the cell surface. In fact, electron microscopy showed that 50 nm clear vesicles are scattered at random in the cytoplasm at considerable distances from one another and only occasionally situated near the plasma membrane. In addition, as is the case for chromaffin cells (Chow et al., 1996), Ca^{2+} must diffuse far from its entry point before reaching the site of exocytosis, because of the absence of presynaptic active zones where Ca^{2+} channels are associated with the fusion machinery. We confirmed this hypothesis by speeding up the time course of $[Ca^{2+}]_I$ decay by introducing a slow Ca^{2+} buffer into the cells. Finally, an explanation has to be sought for a puzzling anomaly: one

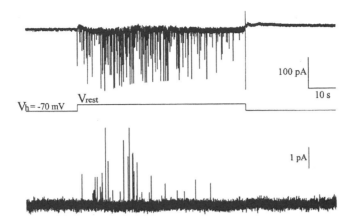

Figure 6. A DA cell was patch clamped in the cell-attached configuration and hyperpolarized to suppress its pacemaker activity. At the offset of the current injection, the cell generated a train of action currents and this caused a burst of events of dopamine release.

would expect that in the case of longer delays between stimulation and release, the number of events would be smaller, reflecting the increasing distance between secretory organelles and cell surface, but this is often not the case. An explanation for this discrepancy could be that each organelle docked to the cell membrane undergoes multiple cycles of exocytosis and refilling during each episode of stimulation, because of the low frequency of the discharge. This suggestion is supported by the observation that the quantal size is larger in spontaneous events of dopamine exocytosis than during the cell response to high K^+.

4. QUANTUM SIZE

By integrating the spontaneously occurring amperometric spikes, we measured the charge generated by the oxidation of the dopamine released at each event of exocytosis: assuming that 2 electrons are transferred for every molecule of dopamine oxidized, we determined that each amperometric spike was caused, on average, by 4×10^4 molecules, a value of the same order of magnitude as that reported in cultures of superior cervical ganglion (Zhou and Misler, 1995; Koh and Hille, 1997) and midbrain neurons (Jaffe et al., 1998). Individual events, however, exhibited a broad spectrum of sizes, ranging from 10^3 to 10^5 molecules, and their distribution was unimodal, but skewed toward smaller events. The charge is a measure of the dopamine contents, and therefore the volume, of the secretory organelles, and its cube root ($Q^{1/3}$) is a measure of the radius of the organelles. The distribution of $Q^{1/3}$ was Gaussian, an indication that a single population of organelles of uniform radius undergoes exocytosis. Indeed, staining with the antibody to VMAT2 labeled a small number of organelles scattered throughout the cytoplasm of the DA cell perikarya (Fig.1, inset). Examination with the electron microscope identified

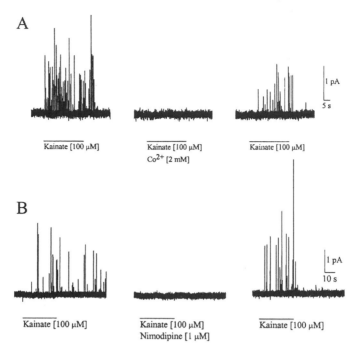

Figure 7. A: Application of kainate, an agonist at glutamate receptors of the AMPA-kainate type, induced a burst of events of dopamine release. Kainate was ineffective when Co^{2+} was substituted for Ca^{2+} in the superfusion medium. The kainate-induced discharge was restored upon washout of the Co^{2+}. This experiment indicates that the release of dopamine is Ca^{++}-dependent. B: Addition of 1 μM nimodipine to the superfusion medium reversibly abolished dopamine release. At this concentration, nimodipine blocks L-type Ca^{++} channels.

three types of secretory organelles: 50 nm clear or agranular vesicles, 80-125 nm dense core or granular vesicles and 0.3 μm granules. Clear vesicles seem a likely candidate for the secretory organelles, because of the absence of a bimodal distribution of the charge and the fact that the release events are briefer ($\leq$ 1 ms) than those reported for dense core vesicles (3.6 ms; Bruns and Jahn, 1995). Another argument in favor of this hypothesis, is the size of the dopamine quanta: 4×10^4 molecules as compared with 1.5×10^5 for 100 nm dense core vesicles in carotid glomus cells (Ureña et al., 1994), and 3×10^6 in 250-300 nm granules of chromaffin cells (Finnegan et al., 1996). If we postulate that the secretory organelles contain a uniform concentration of dopamine, the charge values measured are consistent with a 0.5 M dopamine concentration in the interior of the vesicles. This is nearly identical to that measured in chromaffin cells that have much larger secretory granules [r= 150 nm (Wightman et al., 1991)].

5. CONTROL OF EXTRASYNAPTIC RELEASE

In contrast to the wealth of information available on the pharmacological effects of dopamine, little is known about the mechanisms that control its release in the dark and light. Several experiments were carried out to measure the amount of dopamine synthesized or released by the intact retina in different physiological conditions or after addition of various pharmacological agents (see Djamgoz and Wagner, 1992). These studies have shown that the release of dopamine can be manipulated with agonists and antagonists at the $GABA_A$ receptors. GABA and its agonist muscimol can in fact block the release of dopamine evoked by light (Morgan and Kamp, 1980, 1983; Kirsch and Wagner, 1989). Furthermore, antagonists like bicuculline and picrotoxinin induce dopamine release in dark-adapted retinas (Kamp and Morgan, 1981; Morgan and Kamp, 1983; O'Connor et al., 1987; Piccolino et al., 1987; Ishita et al., 1988; Kirsch and Wagner, 1989; Critz and Marc, 1992; Kolbinger and Weiler, 1993). These results suggest that GABAergic inhibition plays an important role in the regulation of dopamine release in both light and darkness. However, the identification of the target cell was left unsolved, because DA cells receive multiple inputs and the pathway responsible for the control of dopamine release consists of multiple neurons. We demonstrated that the spontaneous firing of solitary perikarya of DA cells is suppressed by administration of the inhibitory transmitter GABA (Gustincich et al., 1997) and characterized the pharmacological properties of $GABA_A$ receptors of solitary DA cells (Feigenspan et al., 2000), seeking evidence for the functional expression of the repertory of subunits that we had identified by single-cell RT-PCR. We had found that DA cells express seven different $GABA_A$ receptor subunits ($\alpha 1$, $\alpha 3$, $\alpha 4$, $\beta 1$, $\beta 3$, $\gamma 1$, $\gamma 2_S$ and $\gamma 2_L$) (Gustincich et al., 1999) and provided evidence that in DA cells at least the $\alpha 1$, $\beta 3$ and $\gamma 2$ subunits are assembled into functional $GABA_A$ receptors (Feigenspan, et al., 2000). Then, we tested for an effect of GABA on the spontaneous release of dopamine. We observed that addition of 10 μM GABA to the superfusion medium reversibly abolished dopamine exocytosis. Since immunocytochemistry showed that a large repertory of subunits of the $GABA_A$ receptor is present throughout the surface of DA cells and one of the subunits, $\alpha 4$, is exclusively extrasynaptic (Gustincich et al., 1999), we concluded that GABA regulates extrasynaptic dopamine exocytosis by acting on extrasynaptic $GABA_A$ receptors.

As described above, extrasynaptic dopamine release is stimulated by kainate, an agonist at glutamate receptors of AMPA-kainate type. Glutamate is the neurotransmitter released by the bipolar cell synapses in the IPL, but electron microscopy shows that glutamatergic synapses are absent on the cell body of DA cells. Thus, the kainate effect on dopamine release by solitary DA cell perikarya is by necessity mediated by extrasynaptic receptors.

Finally, studies on the intact retina maintained *in vitro* have pointed out that dopamine, acting at D_2 receptors, blocks a stimulation-evoked increase in dopamine (see Witkovsky and Dearry, 1991). In solitary perikarya of DA cells, we showed that the agonist at the D_2 receptors quinpirole, which is not oxidized at the electrical potential applied to the carbon electrode, reversibly abolished the burst of events of dopamine exocytosis evoked by application of kainate. Since immunocytochemical studies have demonstrated the presence of D_2 receptors over the entire surface of DA cells (Veruki, 1997; Derouiche and Asan, 1999), extrasynaptic release of dopamine is also controlled by extrasynaptic D_2 receptors.

Our observation that extrasynaptic release is controlled by extrasynaptic receptors suggests that the response of the unspecialized cell surface to the neuroactive molecules present in the extracellular fluid represents an essential component of neuronal signaling in DA cells.

6. CONCLUSIONS

Retinal DA cells share with the neurons of the substantia nigra the property of releasing dopamine in absence of presynaptic active zones. As in the midbrain, the released dopamine acts on receptors at locations distant from the release sites, a modality of communication called volume or paracrine transmission. This extrasynaptic release is controlled by the electrical activity of the cell and by extrasynaptic receptors for the neurotransmitters GABA, glutamate and dopamine. Thus, in addition to the synaptic inputs onto DA cells, dopaminergic volume transmission is regulated by the integration of the passive and active properties of the DA cells surface as well as the presence of neuroactive molecules in the extracellular space.

7. ACKNOWLEDGEMENTS

Supported by National Institutes of Health grants EY01344 (E.R.) and NS38879 (R.M.W.).

8. REFERENCES

Akopian, A., and Witkovsky, P., 1996, D2 dopamine receptor-mediated inhibition of a hyperpolarization-activated current in rod photoreceptors, *J. Neurophysiol.* **76:** 1828.

Albillos, A., Dernick, G., Horstmann, H., Almers, W., Alvarez de Toledo, G., and Lindau, M., 1997, The exocytotic event in chromaffin cells revealed by patch amperometry, *Nature* **389:** 509.

Bean, B. P., and Mintz, I. M., 1994, Pharmacology of different types of calcium channels in rat neurons, in: *Handbook of Membrane Channels. Molecular and Cellular Physiology.* C. Peracchia, ed., Academic Press, San Diego, CA, pp.199-210.

Bjelke, B., Goldstein, M., Tinner, B., Andersson, C., Sesack, S. R., Steinbusch, H. W. M., Lew, J. Y., He, X., Watson, S., Tengroth, B., and Fuxe, K., 1996, Dopaminergic transmission in the rat retina: evidence for volume transmission, *J. Chem. Neuroanat.* 12: 37.

Bloomfield, S. A., Xin, D., and Osborne, T., 1997, Light-induced modulation of coupling between AII amacrine cells in the rabbit retina, *Vis. Neurosci.* **14:** 565.

Boatright, J. H., Rubim, N. M., and Iuvone, P. M., 1994, Regulation of endogenous dopamine release in amphibian retina by melatonin: the role of GABA, *Vis. Neurosci* **11:** 1013.

Bruns, D., and Jahn, R., 1995, Real-time measurement of transmitter release from single synaptic vesicles, *Nature* **377:** 62.

Bunin, M. A., and Wightman, R. M., 1999, Paracrine neurotransmission in the CNS: Involvement of 5-HT, *Trends Neurosci.* **22:** 377.

Cheramy, A., Leviel, V., and Glowinski, J., 1981, Dendritic release of dopamine in the substantia nigra, *Nature* **289:** 537.

Chow, R. H., von Rüden, L., and Neher, E., 1992, Delay in vesicle fusion revealed by electrochemical monitoring of single secretory events in adrenal chromaffin cells, *Nature* **356:** 60.

Chow, R. H., Klingauf, J., Heinemann, C., Zucker, R. S., and Neher, E., 1996, Mechanisms determining the time course of secretion in neuroendocrine cells, *Neuron* **16:** 369.

Contini, M., and Raviola, E., 2003, GABAergic synapses made by a retinal dopaminergic neuron, *PNAS USA* **100:** 1358.

Critz, S. D., and Marc, R. E., 1992, Glutamate antagonists that block hyperpolarizing bipolar cells increase the release of dopamine from turtle retina, *Vis. Neurosci.* **9:** 271.

Dearry, A., and Burnside, B., 1986, Dopaminergic regulation of cone retinomotor movement in isolated teleost retinas: I. Induction of cone contraction is mediated by D2 receptors, *J. Neurochem.* **46:** 1006.

Dearry, A., and Burnside, B., 1989, Light-induced dopamine release from teleost retinas acts as a light-adaptive signal to the retinal pigment epithelium, *J. Neurochem.* **53:** 870.

DeVries, S. H., and Schwartz, E. A., 1989, Modulation of an electrical synapse between solitary pairs of catfish horizontal cells by dopamine and second messengers, *J. Physiol.* **414:** 351.

Derouiche, A., and Asan, E., 1999, The dopamine D_2 receptor subfamily in rat retina: ultrastructural immunogold and *in situ* hybridization studies, *Eur. J. Neurosci.* **11**: 1391.

Djamgoz, M. B. A., and Wagner, H.-J., 1992, Localization and function of dopamine in the adult vertebrate retina, *Neurochem. Int.* **20**: 139.

Dong, C.-J., and Werblin, F. S., 1994, Dopamine modulation of $GABA_c$ receptor function in an isolated retinal neuron, *J. Neurophysiol.* **71**: 1258.

Dubocovich, M. L., 1983, Melatonin is a potent modulator of dopamine release in the retina, *Nature* **306**: 782.

Dubocovich, M. L., 1985, Characterization of a retinal melatonin receptor, *J. Pharmacol. Exp. Ther.* **234**: 395.

Dubocovich, M. L., and Hensler, J. G., 1986, Modulation of [3H]-dopamine released by different frequencies of stimulation from rabbit retina, *Brit. J. Pharmacol.* **88**: 51.

Fan, S.-F., and Yazulla, S., 1999, Modulation of voltage-dependent K^+ currents ($I_{K(V)}$) in retinal bipolar cells by ascorbate is mediated by dopamine D1 receptors, *Vis. Neurosci.* **16**: 923.

Fan, S.-F., and Yazulla, S., 2001, Dopamine depletion with 6-OHDA enhances dopamine D1 receptor modulation of potassium currents in retinal bipolar cells, *Vis. Neurosci.* **18**: 327.

Feigenspan, A., and Bormann, J., 1994, Facilitation of GABAergic signaling in the retina by receptors stimulating adenylate cyclase, *PNAS USA* **91**: 10893.

Feigenspan, A., Gustincich, S., Bean, B. P., and Raviola, E., 1998, Spontaneous activity of solitary dopaminergic cells of the retina, *J. Neurosci.* **18**: 6776.

Feigenspan, A., Gustincich, S., and Raviola, E., 2000, Pharmacology of $GABA_A$ receptors of retinal dopaminergic neurons, *J. Neurophysiol.* **84**: 1697.

Finnegan, J. M., Pihel, K., Cahill, P. S., Huang, L., Zerby, S. E., Ewing, A. G., Kennedy, R. T., and Wightman, R. M., 1996, Vesicular quantal size measured by amperometry at chromaffin, mast, pheochromocytoma, and pancreatic ß-cells, *J. Neurochem.* **66**: 1914.

Fujieda, H., Scher, J., Hamadanizadeh, S. A., Wankiewicz, E., Pang, S. F., and Brown, G. M., 2000, Dopaminergic and GABAergic amacrine cells are direct targets of melatonin: immunocytochemical study of mt_1 melatonin receptor in guinea pig retina, *Vis. Neurosci.* **17**: 63.

Galli, A., Blakely, R. D., and DeFelice, L. J., 1998, Patch-clamp and amperometric recordings from norepinephrine transporters: Channel activity and voltage-dependent uptake, *PNAS USA* **95**: 13260.

Garaschuk, O., Yaari, Y., and Konnerth, A., 1997, Release and sequestration of calcium by ryanodine-sensitive stores in rat hippocampal neurons, *J. Physiol.* **502**: 13.

Garris, P. A., Ciolkowski, E. L., Pastore, P., and Wightman, R. M., 1994, Efflux of dopamine from the synaptic cleft in the nucleus accumbens of the rat brain, *J. Neurosci.* **14**: 6084.

Grace, A. A., and Bunney, B. S., 1983a, Intracellular and extracellular electrophysiology of nigral dopaminergic neurons-1. Identification and characterization, *Neuroscience* **10**: 301.

Grace, A. A., and Bunney, B. S., 1983b, Intracellular and extracellular electrophysiology of nigral dopaminergic neurons-2. Action potential generating mechanisms and morphological correlates, *Neuroscience* **10**: 317.

Gustincich, S., Feigenspan, A., Wu, D.-K., Koopman, L. J., and Raviola, E., 1997, Control of dopamine release in the retina: a transgenic approach to neural networks, *Neuron* **18**: 723.

Gustincich, S., Feigenspan, A., Sieghart, W., and Raviola, E., 1999, Composition of the $GABA_A$ receptors of retinal dopaminergic neurons, *J. Neurosci.* **19**: 7812.

Gustincich, S., Contini, M., Gariboldi, M., Puopolo, M., Kadota, K., Bono, H., LeMieux, J., Walsh, P., Carninci, P., Hayashizaki, Y., Okazaki, Y., and Raviola, E., 2004, Gene discovery in genetically labeled single dopaminergic neurons of the retina, *PNAS USA* **101**: 5069.

Hampson, E. C. G. M., Vaney, D. I., and Weiler, R., 1992, Dopaminergic modulation of gap junction permeability between amacrine cells in mammalian retina, *J. Neurosci.* **12**: 4911.

Hare, W. A., and Owen, W. G., 1995, Similar effects of carbachol and dopamine on neurons in the distal retina of the tiger salamander, *Vis. Neurosci.* **12**: 443.

Hokoç J. N., and Mariani A. P., 1987, Tyrosine hydroxylase immunoreactivity in the rhesus monkey retina reveals synapses from bipolar cells to dopaminergic amacrine cells, *J. Neurosci.* **7**: 2785.

Ishita, S., Negishi, K., Teranishi, T., Shimada, Y., and Kato, S., 1988, GABAergic inhibition on dopamine cells of the fish retina: a [3H]dopamine release study with isolated fractions, *J. Neurochem.* **50**: 1.

Jaffe, E. H., Marty, A., Schulte, A., and Chow, R. H., 1998, Extrasynaptic vesicular transmitter release from the somata of substantia nigra neurons in rat midbrain slices, *J. Neurosci.* **18**: 3548.

Jensen, R. J., and Daw, N. W., 1984, Effects of dopamine antagonists on receptive fields of brisk cells and directionally selective cells in the rabbit retina, *J. Neurosci.* **4**: 2972.

Jensen, R. J., and Daw, N. W., 1986, Effects of dopamine and its agonists and antagonists on the receptive field properties of ganglion cells in the rabbit retina, *Neuroscience* **17**: 837.

Jensen, R. J., 1989, Mechanism and site of action of a dopamine D1 antagonist in the rabbit retina, *Vis. Neurosci.* **3**: 573.

Jensen, R. J., 1991, Involvement of glycinergic neurons in the diminished surround activity of ganglion cells in the dark-adapted rabbit retina, *Vis. Neurosci.* **6:** 43.

Kamp, C. W., and Morgan, W. W., 1981, GABA antagonists enhance dopamine turnover in the rat retina *in vivo, Eur. J. Pharmacol.* **69:** 273.

Kirsch, M., and Wagner, H. J., 1989, Release pattern of endogenous dopamine in teleost retinae during light adaptation and pharmacological stimulation, *Vision. Res.* **29:** 147.

Knapp, A. G., and Dowling, J. E., 1987, Dopamine enhances excitatory amino acid-gated conductances in cultured retinal horizontal cells, *Nature* 325: 437.

Koh, D.-S., and Hille, B., 1997, Modulation by neurotransmitters of catecholamine secretion from sympathetic ganglion neurons detected by amperometry, *PNAS USA* **94:** 1506.

Koistinaho, J., and Sagar, S. M., 1995, Light-induced c-*fos* expression in amacrine cells in the rabbit retina, *Mol. Brain Res.* **29:** 53.

Kolb, H., Cuenca, N., Wang, H.-H., and Dekorver, L., 1990, The synaptic organization of the dopaminergic amacrine cell in the cat retina, *J. Neurocytol.* **19:** 343.

Kolb, H., Cuenca, N., and Dekorver, L., 1991, Postembedding immunocytochemistry for GABA and glycine reveals the synaptic relationships of the dopaminergic amacrine cell of the cat retina, *J. Comp. Neurol.* **310:** 267.

Kolbinger, W., and Weiler, R., 1993, Modulation of endogenous dopamine release in the turtle retina: effects of light, calcium, and neurotransmitters, *Vis. Neurosci.* **10:** 1035.

Kosaka, T., Kosaka, K., Hataguchi, Y., Nagatsu, I., Wu, J.-Y., Ottersen, O. P., Storm-Mathisen, J., and Hama, K., 1987, Catecholaminergic neurons containing GABA-like and/or glutamic acid decarboxylase-like immunoreactivities in various brain regions of the rat, *Exp. Brain Res.* **66:** 191.

Lasater, E. M., and Dowling, J. E., 1985, Dopamine decreases conductance of the electrical junctions between cultured retinal horizontal cells, *PNAS USA* **82:** 3025.

Maguire, G., and Werblin, F., 1994, Dopamine enhances a glutamate-gated ionic current in OFF bipolar cells of the tiger salamander retina, *J. Neurosci.* **14:** 6094.

Mills, S. L., and Massey, S. C., 1995, Differential properties of two gap junctional pathways made by AII amacrine cells, *Nature* **377:** 734.

Morgan, W. W., and Kamp, C. W., 1980, A GABAergic influence on the light-induced increase in dopamine turnover in the dark-adapted rat retina *in vivo, J. Neurochem.* **34:** 1082.

Morgan, W. W., and Kamp, C. W., 1983, Effect of strychnine and of bicuculline on dopamine synthesis in retinas of dark-maintained rats, *Brain Res.* **278:** 362.

Nguyen-Legros, J., 1988, Morphology and distribution of catecholamine-neurons in mammalian retina, *Progr. Retinal Res.* **7:** 113.

O'Connor, P.M., Zucker, C. L., and Dowling, J. E., 1987, Regulation of dopamine release from interplexiform cell processes in the outer plexiform layer of the carp retina, *J. Neurochem.* **49:** 916.

Pfeiffer-Linn, C. L., and Lasater, E. M., 1998, Multiple second-messenger system modulation of voltage-activated calcium currents in teleost retinal horizontal cells, *J. Neurophysiol.* **80:** 377.

Piccolino, M., Neyton, J., and Gerschenfeld, H. M., 1984, Decrease of gap junction permeability induced by dopamine and cyclic adenosine 3':5'-monophosphate in horizontal cells of turtle retina, *J. Neurosci.* **4:** 2477.

Piccolino, M., Witkovsky, P., and Trimarchi, C., 1987, Dopaminergic mechanisms underlying the reduction of electrical coupling between horizontal cells of the turtle retina induced by *d*-amphetamine, bicuculline, and veratridine, *J. Neurosci.* **7:** 2273.

Pourcho, R.G., 1982, Dopaminergic amacrine cells in the cat retina, *Brain Res.* **252:** 101.

Rice, M. E., Cragg, S. J. and Greenfield, S. A., 1997, Characteristics of electrically evoked somatodendritic dopamine release in substantia nigra and ventral tegmental area *in vitro, J. Neurophysiol.* **77:** 853.

Sandell, J. H., and Masland, R. H., 1986, A system of indoleamine-accumulating neurons in the rabbit retina, *J. Neurosci.* **6:** 3331.

Shulman, L. M., and Fox, D. A., 1996, Dopamine inhibits mammalian photoreceptor Na^+, K^+-ATPase activity via a selective effect on the $\alpha 3$ isozyme. *PNAS USA* **93:** 8034.

Stella, S. L., Jr., and Thoreson, W. B., 2000, Differential modulation of rod and cone calcium currents in tiger salamander retina by D2 dopamine receptors and cAMP, *Eur. J. Neurosci.* **12:** 3537.

Strettoi, E., Raviola, E., and Dacheux, R. F., 1992, Synaptic connections of the narrow-field, bistratified rod amacrine cell (AII) in the rabbit retina, *J. Comp. Neurol.* **325:** 152.

Tamura, N., Yokotani, K., Okuma, M., Ueno, H., and Osumi, Y., 1995, Properties of the voltage-gated calcium channels mediating dopamine and acetylcholine release from the isolated rat retina, *Brain Res.,* **676:**363.

Tauchi, M., Madigan, N. K., and Masland, R. H., 1990, Shapes and distributions of the catecholamine-accumulating neurons in the rabbit retina, *J. Comp. Neurol.* **293:** 178.

Tosini, G., and Menaker, M., 1996, Circadian rhythms in cultured mammalian retina, *Science* **272:** 419.

Ureña, J., Fernández-Chacón, R., Benot, A. R., Alvarez de Toledo, G., and López-Barneo, J., 1994, Hypoxia induces voltage-dependent Ca^{2+} entry and quantal dopamine secretion in carotid body glomus cells, *PNAS USA* **91:** 10208.

Veruki, M. L., 1997, Dopaminergic neurons in the rat retina express dopamine D2/3 receptors, *Eur. J. Neurosci.* **9:** 1096.

Veruki, M. L., and Wässle, H., 1996, Immunohistochemical localization of dopamine D$_1$ receptors in rat retina, *Eur. J. Neurosci.* **8:** 2286.

Voigt, T., and Wässle, H., 1987, Dopaminergic innervation of AII amacrine cells in mammalian retina. *J. Neurosci.* **7:** 4115.

Wässle, H., and Chun, M. H., 1988, Dopaminergic and indoleamine-accumulating amacrine cells express GABA-like immunoreactivity in the cat retina, *J. Neurosci.* **8:** 3383.

Wellis, D. P., and Werblin, F. S., 1995, Dopamine modulates GABA$_c$ receptors mediating inhibition of calcium entry into and transmitter release from bipolar cell terminals in tiger salamander retina, *J. Neurosci.* **15:** 4748.

Wightman, R. M., Jankowski, J. A., Kennedy, R. T., Kawagoe, K. T., Schroeder, T. J., Leszczyszyn, D. J., Near, J. A., Diliberto, E. J., Jr., and Viveros, O. H., 1991, Temporally resolved catecholamine spikes correspond to single vesicle release from individual chromaffin cells, *PNAS USA* **88:** 10754.

Witkovsky, P., 2004, Dopamine and retinal function, *Doc. Ophtholmol.* (in press).

Witkovsky, P., and Dearry, A., 1991, Functional roles of dopamine in the vertebrate retina, *Progr. Retinal Res.* **11:** 247.

Witkovsky, P., Nicholson, C., Rice, M. E., Bohmaker, K., and Meller, E., 1993, Extracellular dopamine concentration in the retina of the clawed frog, *Xenopus laevis. PNAS USA* **90:** 5667.

Witkovsky, P., Veisenberger, E., LeSauter, J., Yan, L., Johnson, M., Zhang, D.-Q., McMahon, D., and Silver, R., 2003, Cellular location and circadian rhythm of expression of the biological clock gene *Period 1* in the mouse retina, *J. Neurosci.* **23:** 7670.

Wulle, I., and Wagner, H.-J., 1990, GABA and tyrosine hydroxylase immunocytochemistry reveal different patterns of colocalization in retinal neurons of various vertebrates, *J. Comp. Neurol.* **296:** 173.

Xin, D., and Bloomfield, S. A., 1999, Dark- and light-induced changes in coupling between horizontal cells in mammalian retina, *J. Comp. Neurol.* **405:** 75.

Xu, H.P., Zhao, J.W., and Yang, X.L., 2003, Cholinergic and dopaminergic amacrine cells differentially express calcium channel subunits in the rat retina, *Neuroscience*, **118:**763.

Zhou, Z., and Misler, S., 1995, Amperometric detection of stimulus-induced quantal release of catecholamines from cultured superior cervical ganglion neurons, *PNAS USA* **92:** 6938.

DETERMINANT CONTROL OF NEURONAL NETWORK ACTIVITY BY VASOPRESSIN AND OXYTOCIN RELEASED FROM DENDRITES IN THE HYPOTHALAMUS

Françoise Moos and Michel Desarménien [1]

1. INTRODUCTION

The hypothalamic magnocellular neurosecretory cells (MNCs) constitute the hypothalamo-neurohypophysial system and secrete oxytocin and vasopressin into the blood circulation. These peptide hormones contribute to the regulation of water and salt homeostasis and to various aspects of reproduction. MNCs are a unique model for the study of various neuronal functions. First, their characteristic modes of action potential firing allow their identification *in vivo* and analysis of the correlations between physiological condition, electrical activity and peptide secretion (Poulain et al., 1977; Renaud and Bourque, 1991; Armstrong, 1995; Armstrong and Stern, 1998; Hatton and Li, 1998). The situation is even simpler if analysis is restricted to the rat supraoptic nucleus, which contains no other neuronal population. Second, the long distance between the somata of MNCs, which populate the supraoptic and paraventricular hypothalamic nuclei, and their axon terminals in the neurohypophysis, has allowed surgical separation of these two cellular elements and has enabled extensive studies of the molecular mechanisms responsible for peptide secretion by axon terminals (e.g. Cazalis et al., 1987; Nordmann et al., 1987; Dayanithi et al., 1992). Third, the discovery that central injection of oxytocin favours the expression of a stereotyped activity by oxytocin neurones at lactation (Freund-Mercier and Richard, 1984) gave rise to the new concept of autocrine-paracrine control by molecules synthesised by the neurone. This retrocontrol differs from all other classical feed-back mechanisms since it is mediated directly by molecules released from the soma and dendrites. However, peptide or neurotransmitter release from soma and dendrites is not unique to MNCs (see other chapters in this book), it is probably a characteristic shared by most neurones, but these cells are the best documented

---

[1] Biologie des Neurones Endocrines CNRS UMR 5101 CCIPE 141 Rue de la Cardonille 34094 Montpellier mdesarmenien@ccipe.cnrs.fr

Dendritic Neurotransmitter Release, edited by M. Ludwig
Springer Science+Business Media, Inc., 2005

example. The literature is abundant on the physiological conditions inducing dendritic release, the mechanisms and sites of action of oxytocin and vasopressin, and the physiological consequences of this release. This review will consider some aspects of the functional consequences of dendritic peptide release in adult rats facing specific physiological conditions and during post-natal development. In both situations, release occurs in conditions of high functional and morphological plasticity, in which autocontrol plays a key role.

2. FUNCTIONAL CONSEQUENCE OF DENDRITIC RELEASE IN ADULTS

Granules containing oxytocin or vasopressin are present in the somata and dendrites in the supraoptic nucleus; exocytosis of these granules can be revealed by electron microscopy (Pow and Morris, 1989), and peptide release can be detected in perfusion fluids *in vitro* (Moos et al., 1984; Wotjak et al., 1994). *In vivo*, dendritic release occurs in case of high hormonal need such as parturition and lactation for oxytocin (Moos et al., 1987; Moos et al., 1989; Landgraf et al., 1992) and dehydration or haemorrhage for vasopressin (Ota et al., 1994). A particularly interesting discovery during the last decade was that vasopressin and oxytocin behave as "optimising factors", fostering their respective neuronal populations to discharge in patterns that are most efficient for hormone secretion. The mechanisms and sites of action of the two neuropeptides are complex, and involve both pre- and post-synaptic control of neuronal activity.

2.1. Vasopressin Optimises the Functioning of the Vasopressin Neuronal Network

In conditions of high and sustained hormonal demand, such as dehydration or haemorrhage, vasopressin neurones adopt a phasic pattern of activity characterised by a succession of active and silent periods (Wakerley et al., 1978), a pattern known to be particularly efficient for vasopressin secretion from pituitary terminals (Cazalis et al., 1985). Because dendritic release of vasopressin occurs in response to hyperosmotic stimulation (Neumann et al., 1993) and haemorrhage (Ota et al., 1994), and can be maintained for several hours (Ludwig et al., 1994), one would expect sustained modulatory effects of dendritic vasopressin on the firing of vasopressin neurones, and this seems to be the case.

In adult rats, the effect of vasopressin on the activity of vasopressin neurones is markedly dependent on the initial firing pattern, since vasopressin excites neurones that display a low activity quotient, triggering or enhancing phasic activity, but inhibits highly active neurones; phasic neurones that initially display an intermediate phasic pattern are not affected. Therefore, the action of vasopressin results in homogenisation of the firing pattern of vasopressin neurones (Gouzenes et al., 1998), ultimately pushing the whole population towards an intermediate activity with active and silent periods of similar duration (10-40 s), as needed for maximally efficient release by the terminals (Cazalis et al., 1985). This optimising effect may be attained through an autocrine mechanism since vasopressin neurones express vasopressin autoreceptors (Hurbin et al., 1998). As these receptors are co-localised with vasopressin in neurosecretory vesicles (Hurbin et al., 2002), simultaneous exposure of both partners at the membrane surface during exocytosis would facilitate binding of the peptide to its receptors. Regulation of the phasic pattern by vasopressin involves at least the V_{1a} receptor (Inenaga and Yamashita, 1986; Ludwig and

Leng, 1997), and probably the V_2-type receptor, although its involvement remains hypothetical (Abe et al., 1983). Nevertheless, these two different autoreceptors may mediate the opposite effects (excitatory or inhibitory) of vasopressin on the phasic pattern of activity as suggested by our recent electrophysiological data (Fig 1). Indeed, juxtamembrane application of a V_{1a} receptor agonist ([Phe^2, Orn^8]vasotocin at 10 or 100nM) depressed the phasic pattern by decreasing the activity quotient, Q, (i.e. the proportion of time in activity; from 0.87 ± 0.05 to 0.67 ± 0.05, n = 9). The V_{1a} receptor antagonists (SR 49059 at 10nM, Serradeil-Le Gal et al., 1993) ; Phaa-(D-Tyr)(Et)-Phe-Gln-Asn-Lys-Pro-Arg at 100nM, Manning et al., 1990) similarly enhanced the phasic pattern (Q increased from 0.46 ± 0.04 to 0.67 ± 0.04, n=17). The V_2 receptor agonist (1-deamino-8-D-AVP at 10nM) increased both Q (from 0.32 ± 0.07 to 0.76 ± 0.08, n=12) and intraburst frequency (from 3.9 ± 1.1 to 6.0 ± 1.1 spikes/s). The V_2 receptor antagonist (desGlyNH$_2$-d(CH$_2$)^5-[D-Ile^2,Ile^4]AVP at 1µM) depressed phasic activity mainly by decreasing the duration of active periods (from 100 ± 37 to 19 ± 2 s, n=7) and intraburst frequency (from 6.8 ± 0.8 to 5.2 ± 1.0 spikes/s).

Microspectrofluorimetric studies on dissociated neurones also indicate the presence of receptors (R) that display the pharmacological properties (Gouzenes et al., 1999) and

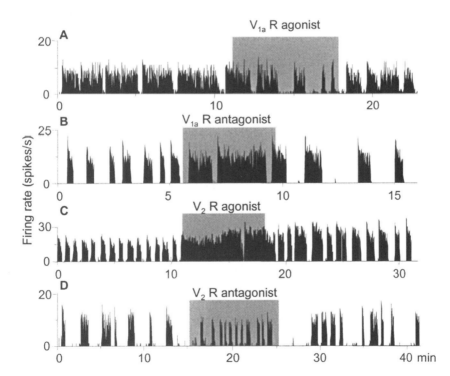

Fig 1: Vasopressin modulates the phasic pattern of vasopressin neurones through V1a and V2-type receptors. Typical examples are illustrated in A-D. A: depression of the phasic pattern by a V1aR agonist. B: enhancement of the phasic pattern by a V1aR antagonist. C: enhancement of the phasic pattern and intraburst frequency by a V2R agonist. D: depression of the phasic activity by a V2R antagonist: note the shortening of active periods and the decrease of intraburst frequency.

activate intracellular transduction signals of the V_{1a} and V_2 types (Sabatier et al., 1998). However, vasopressin neurones express $V_{1a}R$ and $V_{1b}R$ mRNAs but not the V_2R transcript (Hurbin et al., 1998). Although the pharmacological characteristics of the $V_{1b}R$ have not yet been established, a recent study in inner medullary collecting duct suggests they may be candidate V_2-like receptors (Saito et al., 2000). The effect of vasopressin on its autoreceptors could be complemented by a presynaptic action of vasopressin on afferent terminals (Kombian et al., 1997; Hermes et al., 2000; Pittman et al., 2000). Thus vasopressin may induce different effects at pre- and postsynaptic levels. A recurrent action via neighbouring interneurones that project onto magnocellular neurones also cannot be excluded (Leng and Dyball, 1983). As mentioned above, these modulatory actions would be particularly prominent in physiological situations where local vasopressin release is increased. This would allow the expression of a phasic pattern optimal for a sustained systemic vasopressin secretion, in accordance with physiological demand. Interestingly, a former study has reported that dehydration induces most vasopressin neurones to fire phasically (Wakerley et al., 1978) with a pattern similar to that triggered by vasopressin.

2.2. Oxytocin Optimises the Functioning of the Oxytocin Neuronal Network at Lactation

During lactation, oxytocin neurones face a peculiar situation: they are responsible both for triggering milk ejection in response to stimulation of the nipples mechanoreceptors by the pups, and for redressing the changes in water balance brought about by milk synthesis and secretion. During suckling (see review in Leng et al., 1999), oxytocin neurones in the supraoptic and paraventricular nuclei display periodic high-frequency bursts of spikes that are superimposed on low tonic basal activity. The bursts are optimally efficient for stimulus-secretion coupling, and the resulting oxytocin pulses optimise the effectiveness of oxytocin at the mammary gland (during milk-ejection). During dehydration and haemorrhage, oxytocin neurones respond by firing continuously, leading to a continuous release of oxytocin into the peripheral circulation, thus allowing sustained hormonal action of oxytocin (e.g. natriuretic action at the kidneys). The question then arises *'how do oxytocin neurones deal with this complex information and adapt their properties to generate the stereotyped bursting behaviour in response to suckling?'*
Several mechanisms contribute to this ability but the key prerequisite for bursting is oxytocin released from dendrites. As detailed below, oxytocin acts at several levels to favour bursting. Oxytocin first facilitates its own release from soma and dendrites, and it modifies the intrinsic properties of oxytocin neurones to increase the firing irregularity that plays a critical role in the ability to burst. The autocontrol by oxytocin involves mobilisation of calcium stores, putative activators of store-operated channels (SOCs) that may contribute to patterned spiking activity. Oxytocin also enhances correlation of the fluctuations in firing rate between oxytocin neurones, an event strongly correlated with burst generation. It also ensures co-ordination of bursting behaviour in and between all four nuclei (injection of oxytocin into one nucleus not only facilitates bursting of oxytocin cells in the injected nucleus, but also in the other three nuclei). Finally, oxytocin modulates afferent inputs, in particular GABA and glutamate, probably to depress any strong tonic activation that restrains bursting, such as that induced by hyperosmotic stress. Burst synchronisation is ensured by common input pathways that mainly originate

in the medulla oblongata and which project simultaneously to all nuclei. Sectioning this common input reveals local independent rhythmic networks whose functioning is under the dependence of oxytocin.

2.3. Autocrine Control and its Functional Consequences

Oxytocin responsible for burst genesis arises from oxytocin neurones themselves that release oxytocin from their soma and dendrites. Dendritic exocytosis has been visualised by electron microscopy (Pow and Morris, 1989), and quantified by capacitance measurements in dissociated MNCs (de Kock et al., 2003). Exocytosis from soma and dendrites is augmented during lactation since the capacitance response to the same stimulus is greater in neurones from lactating rats than in neurones from virgin rats. In suckled rats, local oxytocin release increases during the milk ejection reflex, even before the occurrence of the first burst at the beginning of suckling (Moos et al., 1989). Since this increase does not occur in rats that fail to milk eject, it was concluded that oxytocin is a prerequisite to bursting activity. Furthermore, oxytocin release could be evoked by stimulating the pituitary stalk, suggesting that antidromic invasion of soma and possibly dendrites by action potentials can increase oxytocin release. Indeed, the entry of calcium and a concomitant increase in membrane capacitance can be evoked by a single action potential, although the response is more pronounced when several action potentials are applied at 20Hz (de Kock et al., 2003). Once released, oxytocin acts on specific autoreceptors localised on soma and dendrites (Freund-Mercier et al., 1994; Freund-Mercier and Stoeckel, 1995) to facilitate its own release (Moos et al., 1984). Since the release is triggered by electrical activity and mediated by the SNARE complex (de Kock et al., 2003), and oxytocin acts on oxytocin receptors to mobilise intracellular calcium stores (Lambert et al., 1994), it can be postulated that the autofacilitatory loop relies on this mobilisation. Ludwig et al. (2002) recently reported that the consequence of store mobilisation is an increase of the readily releasable pool of oxytocin, and suggested that this priming effect involves movement of vesicles closer to the membrane. The functional consequence of this autocontrol could be to modify the excitability of oxytocin neurones, inducing direct excitation of oxytocin neurones (Yamashita et al., 1987) through a depolarising effect as suggested by electrophysiological recordings in hypothalamic slices from adult (Murai et al., 1998) or new-born rats (Chevaleyre et al., 2000).

2.4. Dendritic Oxytocin as a Retrograde Modulator of Afferent Terminals

A few years ago, Kombian et al. (1997) presented evidence that oxytocin released locally acts pre-synaptically to decrease glutamate release from glutamate terminals while Brussaard et al. (1996) suggested that oxytocin both inhibits post-synaptic $GABA_A$ receptors and exerts a pre-synaptic inhibition of GABA release (de Kock et al., 2003). Since application of the adenosine A_1 receptor antagonist, CPT, also significantly reduced the release of GABA evoked by the electrical stimulation of GABA afferents, it was suggested that adenosine was co-released with oxytocin. Very recently, we obtained evidence that supraoptic neurones release ATP in response to oxytocin/vasopressin application. Excised membrane patches from HEK cells transfected with the P_2X_2 ATP receptor, were brought close to an acutely dissociated supraoptic neurone, the release of ATP was recorded in the form of ion channel openings in response to peptide application (Fig 2). So, the process of autocrine-paracrine control is likely much more complex than

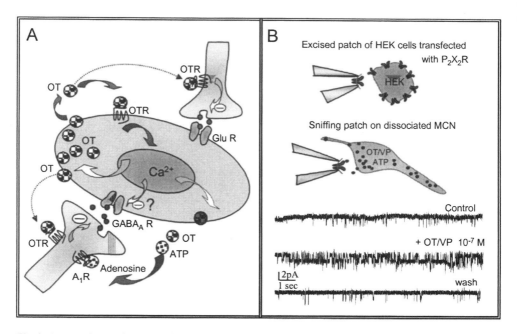

Fig. 2: Auto- and paracrine control by oxytocin and co-released factor.
A: schematic summary of oxytocin actions, once released from soma and dendrites. Activation of oxytocin autoreceptors by oxytocin induces calcium increase via mobilisation of Ca^{2+} from intracellular pools, which favours translocation of dense core vesicles closer to the dendritic membrane. Oxytocin acts pre-synaptically to decrease both glutamate (Glu) and GABA release from their respective terminals. Oxytocin inhibits post-synaptic $GABA_A$ receptors via a mechanism that remains to be determined. Co-released ATP completes the effect of oxytocin on GABA terminals through a A1-type adenosine receptor. B: test of sniffing patch demonstrating the co-release of neuropeptide and ATP. An excised patch of HEK cells transfected with P_2X_2 receptors is approached close to a freshly dissociated supraoptic neurone. Application of both oxytocin and vasopressin at 10^{-7} M, strongly increases ion channel opening as compared to traces before (control) and after application (wash). This response was observed in 3 cases and attests for the co-release of ATP in the facilitatory loop of oxytocin/vasopressin release by their respective neuropeptides.

presently described, and involves co-stored and co-released molecules such as ATP and dynorphin (Brown, C., this volume), which probably modulate the effects of oxytocin and/or vasopressin.

In conclusion, the physiological impact of this auto- and retro-control is to favour bursting behaviour, through an increase in local oxytocin release (which is a prerequisite for bursting), a change in oxytocin cell excitability, and a modulation of the afferent information. The excitatory glutamate and inhibitory GABA inputs need to be finely modulated to favour expression of bursts, and it is necessary to limit the excitatory information unrelated to suckling which could restrain bursting by inducing a strong tonic excitation incompatible with the expression of bursts (Moos, 1995). Let us now try to explain the mechanisms of action of oxytocin on the functionality of the oxytocin neuronal network at lactation, i.e., on burst generation.

2.5. Oxytocin-Dependent Co-ordinated Fluctuations of Firing: A Critical Stage of Burst Generation.

A major step to help understand the role of oxytocin in burst generation was to differentiate the influence of osmotic challenge on oxytocin cell activity from that of suckling. The fact that an acute increase in osmolarity strongly increases basal activity but restrains bursting (Moos, 1995), led us to postulate that, conversely, a decrease in osmotic input might suppress osmotic-related activity and reveal the activity induced by suckling. This hypothesis was tested in water-loaded lactating rats. As expected, water loading strongly depressed both basal and bursting activity, but some lactating rats displayed bursts after oxytocin treatment (Fig. 3). Although oxytocin neurones were almost silent during the interburst period, clusters of spikes were always recorded during

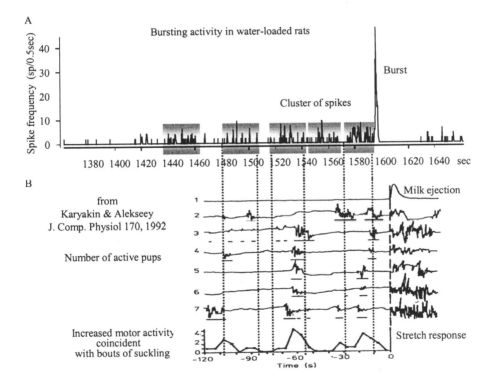

Fig. 3: Basal and bursting activity in water-loaded rats (10ml H$_2$0 i.p.). *A:* In five water-loaded lactating rats, bursting could be triggered by oxytocin (1µl of 1µM solution into the third ventricle). Although the seven oxytocin neurones recorded were almost silent during the inter-burst period, bursts (mean amplitude: 43 ± 8 spikes) were preceded by an increase in firing rate during the last 100 s (up to 0.8 ± 0;1 spikes/s). The increase in firing consisted in periodic clusters of spikes. B: recording from Karyakin and Alekseev (1992), illustrating the motor activity of 6 pups (traces 2 to 7) before the occurrence of milk ejection (trace 1). The motor activity coincided with bouts of suckling. The lower graph representing the averaged motor activity, clearly illustrates the periodicity of the suckling information. Traces in B have been scaled with the electrophysiological recording in A. Note the similarity between periodicity of clusters of spikes with the periods of increased motor activity of pups.

the 100 s preceding bursts. These periodic clusters of spikes might reflect the intermittent nature of the suckling input, since their periodicity appeared similar to that of bouts of suckling exerted by the pups as the milk ejection was approached (Fig 3). Thus, the activity of oxytocin neurones may have two components: (1) the continuous and very regular basal activity that seems to be the response to osmoreceptors, and (2) the periodic clustered activity, revealed by hypo-osmolarity and oxytocin treatment, which might reflect changes in suckling inputs. These periodic fluctuations of firing rate before bursts could play a determinant role in burst generation.

We performed statistical analyses of basal activity during the inter-burst period during facilitation of bursting by oxytocin or when bursting is restrained by hypertonic saline. Our analyses support the hypothesis that changes in basal activity reflects changes in oxytocin cell excitability and that oxytocin neural network plays a determinant role in burst generation. First, there is a narrow window of mean firing rate (1-3 spikes/s) that seems compatible with bursting (Brown and Moos, 1997), and more importantly, bursting is highly correlated with fluctuations in firing rate during the 100-s period before the burst (Brown et al., 2000).

Examining this relationship at a much finer time scale for recorded cell pairs reveals dynamic changes during the period leading to a burst, especially during the last few seconds (Moos et al., 2004a). During the whole inter-burst period there was a progressive increase in the mean firing rate, the variability of firing; and the correlation of firing rate fluctuations between neurones in bilateral nuclei; all these measures strengthened substantially about 10s before the burst (Fig. 4). These changes were high during facilitation of bursting by oxytocin, but when bursting was delayed after hypertonic saline, the mean firing rate was much higher, and less variable. Furthermore, correlation of firing rate between pairs of oxytocin cells was much lower or non-existent. On the other hand, during the inter-burst period, the serial correlation coefficient was negative (spikes at low frequency follow spikes at high frequency and vice-versa). This reflects the irregularity of firing, and indicates that oxytocin neurones may exert negative feedback effects on activity to dampen any strong activation. When bursting is restrained, this coefficient remains negative in a range significantly greater than for oxytocin-treated neurones. The negative serial correlation coefficient abruptly weakens just before a burst, indicating the withdrawal of a restraining influence on the tendency of oxytocin neurones to burst. So, the general process leading to bursting appears to result from a co-ordinated behaviour of the whole population: the progressive increase in mean firing rate and irregularity of firing, as well as the increase in correlation of fluctuations in firing between neurones, would lead more neurones to be recruited in the fluctuating pattern. However, this process is dampened, and, in a sense, controlled, by the factors inducing the negative autocorrelation. This control is maintained until the last few seconds before the burst, when the negative autocorrelation weakens markedly and when fluctuations are more and more synchronised and overcome the restraining effect indicated by the negative serial correlation coefficient. At this time, the fluctuations in firing become strong enough or synchronised enough to overcome the effects of the few remaining neurones that would still exhibit negative autocorrelation, and the cross correlation and mean and variability of firing increase substantially.

Oxytocin neurones have intrinsic properties that might be involved in spike clustering. First, they display depolarising afterpotentials that would increase the probability of subsequent EPSPs to trigger action potentials. They also display a depolarisation-activated sustained outward rectification and a rebound depolarisation

(Stern and Armstrong, 1995, 1996), both being likely to favour the expression of clusters or bursts of spikes. If neurones with a membrane potential close to the spike threshold receive inhibitory inputs, the resulting hyperpolarisation would deactivate the sustained outward rectification, which consequently would trigger a train of high frequency spikes. So, intermittent inhibitory input might also enhance variability of firing during lactation, a phenomenon attested by *in vivo* experiments. Indeed, pulsatile GABA application increases the variability of firing rate (Brown et al., 2000) and facilitates bursting behaviour in suckled rats (Moos, 1995). Irregular firing (clusters of spikes separated by long silent periods) also occurs in oxytocin neurones in organotypic cultures of the supraoptic nucleus, especially after oxytocin application (Israel et al., 2003). Much of this variability was ascribed to the transduction process of the excitatory input into action

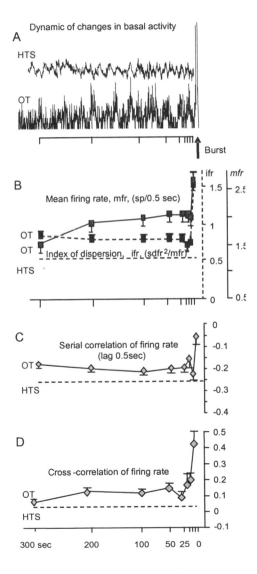

Fig 4: Dynamic of changes in basal activity of oxytocin neurones during a 300-s period preceding the burst. A: examples illustrating the evolution of the mean frequency (in spikes/0.5s) before bursting of two oxytocin neurones submitted to hypertonic saline treatments (HTS; 1 ml NaCl 9% ip) or oxytocin (OT; 1ng icv). B: In oxytocin condition, the mean frequency, *mfr*, and variability of firing (not shown), increase slowly but progressively over the interburst period, and the increase accelerates substantially just before the burst. During HTS treatment, *mfr*, remains elevated (~ 3 spikes/0.5s out of the range of the scale) and very regular. After oxytocin treatment, the index of clustering, *ifr* (= $sdfr^2/mfr$), does not change systematically over most of the interburst period, but increases substantially during the last few seconds before the burst. Under HTS treatment, *ifr* is very low. C: During oxytocin facilitation, the serial correlation of firing rate (at lag of 0.5s) remains negative over the interburst period, which tends to keep the firing rate on a longer time scale quite steady and relatively low. This control is maintained until some time during the last few seconds before the burst, when the negative autocorrelation weakens markedly. During HTS activation, this parameter remains very negative (- 0.25). D: During oxytocin treatment, the average cross correlation between firing of pair recorded neurones, remains low (0.05-0.1), but increases substantially to 0.45 in the last few seconds before the burst, which is not the case after HTS treatment. Adapted from (Moos et al., 2004a)

potential generation: although receiving synchronous EPSPs from glutamate inputs, only one of the pairs of neurones recorded produced an action potential. The clustered activity arising from any of these processes could be enhanced by dendro-dendritic coupling between neurones through oxytocin release. Furthermore, all these processes may occur simultaneously or sequentially, thus enhancing variability of firing and facilitating bursting. Variability of firing might also result from heterosynaptic depression between excitatory and inhibitory inputs (Piet et al., 2003), or from the intermittent bouts of pressure exerted by pups on the nipples (Karyakin and Alekseev, 1992; Fig. 3).

Although the abrupt changes in some characteristics of firing (increase in firing rate and in its irregularity, decrease in negative serial correlation; increase in correlation in fluctuations of firing rate...) were a critical stage of burst generation, it appeared that the key factor was the correlation of activity between neurones (Moos et al., 2004a). Such co-ordination is ensured by oxytocin itself, both within and between nuclei (Moos and Richard, 1989; Lambert et al., 1993a). Within each nucleus, oxytocin-induced interaction and co-ordination of activity between oxytocin neurones may be favoured by the morphological reorganisation that takes place at lactation. Indeed, the retraction of glial cell processes that normally unsheathe oxytocin cell bodies leads to numerous soma/soma and soma/dendrite appositions (Theodosis and Poulain, 1999; Theodosis, 2002) and also to bundles of dendrites (Hatton, 1999) that will favour neurone interaction and co-ordination and also the autocrine/paracrine oxytocin control. Co-ordination between bilateral nuclei by oxytocin, is probably ensured indirectly, through reciprocal interconnections with neurones of the dorso-chiasmatic area (Thellier et al., 1994) or the bed nucleus of the stria terminalis (Lambert et al., 1993b).

2.6. Synchronisation of Bursting in Bilateral Oxytocin Neurones

Within each nucleus, simultaneous activation of several oxytocin neurones is favoured by the multiple synapses between a terminal and several soma-dendrites (Theodosis, 2002). Synchronisation between neurones in the four nuclei requires common inputs, which could be constituted, at least, by the noradrenergic pathways arising from the mesencephalic A1-A2 regions. They have been shown to project simultaneously to the supraoptic and paraventricular nuclei (Sawchenko and Swanson, 1983) and recent electrophysiological data have brought evidence that noradrenaline can act directly on oxytocin neurones or indirectly via a pre-synaptic facilitation of glutamate release (Boudaba et al., 2003). The fact that the pre-synaptic receptors are of the α_1 and α_2 types is of particular relevance since α_2 receptor agonist administration facilitates bursting (Bailey et al., 1997). However, the noradrenergic pathway is probably not exclusive in ensuring possible synchronisation of oxytocin neurones. Indeed, our recent morphological and electrophysiological work brought evidence for a non-aminergic common pathway arising from the medulla-oblongata (Moos et al., 2004b).

The proposed reconstruction of this pathway is as follow: most neurones of the medulla oblongata project unilaterally, some project contralaterally, and some neurones project bilaterally. Branching occurs at the level of the supraoptic decussation from the caudal level of the paraventricular nucleus up to the most rostral part of the supraoptic nucleus. These data suggest that synchronisation of oxytocin neurones in the four nuclei may be facilitated by common pathways arising from the medulla oblongata. These experiments did not exclude involvement of noradrenergic pathways in the synchronisation of bilateral oxytocin neurones. Furthermore, each nucleus with its inputs seems to behave as an oscillator, in which oxytocin neurones exert a positive feed-back through a local release of oxytocin. Although oxytocin neurones in each nucleus may behave as intrinsic oscillators, electrophysiological recordings in lactating rats suggest that neurones in the medial portion of the hypothalamus, posterior to the paraventricular nucleus, relay a rhythmic input to the oxytocin neurones (Takano et al., 1992). On the other hand, electrophysiological recordings in organotypic culture of supraoptic nucleus suggested the existence of a local, glutamatergic intra-hypothalamic network that behaves as a pulse generator, and whose functioning is facilitated by oxytocin (Israel et al., 2003). In these cultures, glutamate neurones govern burst generation and synchronisation in oxytocin neurones, and reciprocally, oxytocin release influences the periodic activation of glutamate neurones. Furthermore, although oxytocin cells are driven by similar input, their bursting behaviour and background activity are modulated by their intrinsic properties. This last observation tallies with the conclusions arising from our previous statistical analyses. Indeed, we found a strong correlation between bursting behaviour and irregularity of firing that reflects changes in intrinsic properties and in suckling inputs. Whether the bursts result from intrinsic oscillatory properties of oxytocin neurones or from other oscillatory neurones through volleys of EPSPs, it is clear that the capacity of oxytocin neurones to generate a burst depends on oxytocin itself.

In conclusion: it appears that, in each of the four magnocellular nuclei, there is a local oscillatory network that is driven by suckling information and which is influenced or even perturbed by osmotic inputs. It is still not clear whether, in these networks, oxytocin neurones are themselves intrinsic oscillators, or whether bursts are driven by other neurones (proximal glutamate neurones or neurones in the medial portion of the hypothalamus posterior to the paraventricular nucleus). If the second possibility is valid, the bursting behaviour and background activity nevertheless appear to be modulated by intrinsic properties. In all cases, functioning of these local oscillatory networks is under the dependence of oxytocin, and probably co-released ATP. Co-ordination in basal and bursting activity between bilateral oxytocin neurones is ensured by oxytocin acting indirectly on inputs connecting the four nuclei, in particular the dorso-chiasmatic areas and the limbic system (bed nucleus of the stria terminalis and ventral lateral septum). Synchronisation within each oscillatory network is under the dependence of oxytocin, and synchronisation of the four oscillatory networks is ensured by common inputs among which pathways from the medulla oblongata seem to play a determining role.

3. DEVELOPMENT OF AUTOCONTROL, AUTOCONTROLLED DEVELOPMENT

To understand the wiring of neural networks, which occurs, for a large part, after birth under the influence of sensory inputs, a major question is to understand the respective contributions of innate and acquired properties. For human characters, this question can be approached by the study of twins, monozygotic or not, nurtured in the same or different environments. The same is true for neurones, and MNCs are, here again, a unique model. In the rat, they are produced together in the same region of the neural tube from embryonic day E13 to E16, and they settle in their final position between E15 and E18 (Altman and Bayer, 1986). At that stage, they constitute a unique population of hypothalamo-neurohypophysial neurones, under the control of the Brn-2 gene (Nakai et al., 1995). Once settled, they further differentiate as oxytocin or vasopressin neurones, and begin to synthesize their hormones a few days (vasopressin) or immediately (oxytocin) before birth (Buijs et al., 1980; Altstein and Gainer, 1988; Jing et al., 1998). This differentiation is not restricted to the identity of the secreted peptide: neurones also differ by their selective expression of oxytocin or vasopressin autoreceptors (Moos et al., 1998) and by some electrophysiological properties (Armstrong and Stern, 1998) supporting either the phasic activity of vasopressin neurones or the bursting pattern displayed by oxytocin neurones. However, this differentiation is neither absolute nor definitive: even in adults, all oxytocin neurones contain a significant amount of vasopressin mRNA and vice versa (Glasgow et al., 1999), and 5% of them express properties of the two populations (Moos and Ingram, 1995; Nakai et al., 1995). This latter proportion varies with the physiological condition (Glasgow et al., 1999), suggesting that vasopressin neurones can acquire properties of oxytocin neurones and vice versa, when one hormone is needed predominantly.

Supraoptic oxytocin and vasopressin neurones acquire their electrophysiological properties in parallel. The beginning of the first postnatal week is a period of chaotic activity: the action potentials are large and of small amplitude, their erratic pattern reflects a very unstable resting potential and synaptic activity is restricted to GABAergic depolarizing potentials. The action potentials mature during the course of this first week, while the resting potential stabilises. The hallmark of the second week is synaptic regulation; GABAergic potentials become hyperpolarizing and inhibitory, whilst glutamatergic excitation appears. The typical patterns of activity, continuous for oxytocin neurones and phasic for vasopressin ones, appear early during the third week (Desarmenien et al., 1994; Widmer et al., 1997; Chevaleyre et al., 2001). Just prior to weaning, the neurones are ready to respond to physiological needs with the appropriate hormone secretion.

MNCs can take their time to become fully functional since the regulation of water balance is not more functional (Rajerison et al., 1976) than reproduction at birth, and the absence of oxytocin and vasopressin neurones by Brn-2 null mutation becomes lethal only 10 days later (Nakai et al., 1995). However, the two peptides are secreted before they can exert their hormonal action; therefore, they might play some developmental role. Indeed, proliferative properties have been reported for vasopressin, together with various developmental actions in several brain structures (Boer et al., 1980). These studies are contradicted by the fact that both the Brattleboro rat (which lacks vasopressin) and the oxytocin knock-out mouse (Young et al., 1996; 1998) do not show major signs of brain dysfunctions. However, only one of the two peptides is absent in these models and compensation by the remaining one may occur.

Apart from the establishment of their electrical activity patterns, MNCs undergo two major changes during the developmental period: a transient extension of the dendritic tree and the establishment of synaptic connection with afferent networks. This, of course, is not specific to MNCS, but the fact that both processes are under the control of dendritic hormone release justifies a mention in this book.

3.1. Development of Autocontrol

Using superfusion of isolated supraoptic nuclei and radioimmunoassay, we have demonstrated that both oxytocin and vasopressin are released in the supraoptic nucleus from the postnatal day 2. Autocontrol of this release is very rapidly functional; an oxytocin agonist specifically increases oxytocin release whilst vasopressin agonists stimulate vasopressin release. This autocontrol is maximal during the second postnatal week and then progressively declines to the values attained in adults (Chevaleyre et al., 2000).

During the same period, oxytocin specifically stimulates firing of action potentials in half of the neurones tested, the others being stimulated by vasopressin. As expected, 5% of the neurones respond to both peptides. Interestingly, oxytocin triggers a continuous activity whilst vasopressin evokes phasic activity. Both excitations are supported by a slight but consistent depolarization, a notable difference with adult neurones. Once again, the proportion of neurones displaying this depolarisation is maximal during the second postnatal week. Electrical activity is not only activated by exogenous peptides, it is entirely supported by endogenous peptides. Indeed, application of an oxytocin antagonist on a neurone responding to oxytocin completely blocks spontaneous firing, whilst vasopressin antagonists block vasopressin-responding neurones. Thus MNCs can release

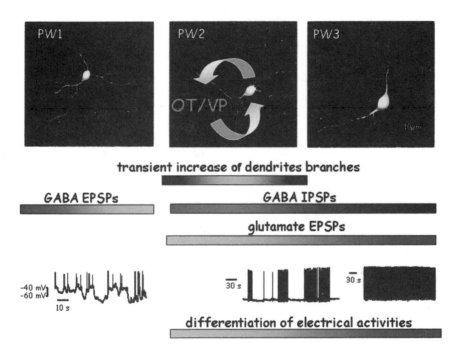

Fig. 5. Morphological plasticity and development of electrophysiological properties of MNCs. At birth and during the first postnatal week (PW1), MNCs possess 2 to 3 primary dendrites, with a small number of proximal branches. The dendritic tree extends transiently during PW2, under the control of locally released oxytocin and vasopressin. During PW3, the dendritic tree retracts to attain its mature shape, typically bipolar with few proximal branches. The synaptic GABA activity, initially excitatory, becomes inhibitory during PW2, when glutamate takes the lead of synaptic excitation. Finally, the initially erratic electrical activity shapes progressively to the typical tonic or phasic firing patterns that respectively characterize oxytocin and vasopressin neurones.

(A color version of this figure appears in the signature between pp. 256 and 257.)

oxytocin and vasopressin from their soma and dendrites during the developmental period, and the peptides auto-stimulate their own release and support electrical activity (Chevaleyre et al., 2000).

3.2. Autocontrol of Development

Patch-clamp recordings from developing supraoptic MNCs has revealed that the membrane resistance and capacitance values at birth are very close to the mature ones. However, during the second postnatal week, a transient resistance decrease and capacitance increase occurs, which reflects a transient increase in the complexity of the dendritic tree (Chevaleyre et al., 2001). The number of initial dendrites (extending from the soma) and of proximal 2nd, 3rd and 4th order branches increases significantly, as if the neurones tried to fill the space around their soma and enhance the probability of coming in contact with afferent axons. The prominent influence of oxytocin and vasopressin in this transient extension was demonstrated by daily injection of antagonists during a seven day period after birth (Chevaleyre et al., 2002). Supraoptic neurones from treated pups possessed significantly fewer dendritic branches. This was confirmed, and further analysed *in vitro*. Incubation of slices from *young* pups (3 – postnatal days) in a mixture of oxytocin and vasopressin increased the number of dendritic branches, whilst treatment with antagonists of slices from *older* pups (7-9 days) decreased it. Thus local release of oxytocin and vasopressin favours transient extension of the dendritic tree, which happens when this release is maximal. But autocontrol is not the only contributor to this morphological change. Glutamate, presumably secreted by incoming afferents, mimics the peptides effect via activation of NMDA type receptors, and NMDA antagonist decreases the dendritic tree extension. In addition, the peptides stimulate the releasing activity of glutamatergic afferents. The third partner is electrical activity; activation by potassium chloride or inhibition by tetrodotoxin respectively increases or decreases the number of dendritic branches. A very peculiar observation is that the three partners act together, any of the antagonists blocking the effect of any of the agonists. In addition, blocking calcium entry through voltage-dependent channels or calcium mobilisation from intracellular stores also blocks agonist-induced extension.

Thus, electrical activity, local oxytocin and vasopressin release as well as endogenous glutamate act in concert to favour a transient extension of the dendritic tree of supraoptic neurones during the second postnatal week. Finally, the fact that daily injection of oxytocin and vasopressin antagonists during the first two postnatal weeks reduces the frequency (but not the size) of spontaneous EPSCs in slices from young adults (Chevaleyre et al., 2002) demonstrates that this transient increase contributes to the establishment of excitatory synapses on MNCs dendrites.

In adults, key roles are played by neuropeptides released from the soma and dendrites of MNCs: control of the morphological plasticity and patterning of electrical activity and hormone secretion. During development, the same release mechanism play equivalent roles for different aims: morphological plasticity then concerns the shape of the dendritic tree and the functional adjustment is a facilitation of synaptogenesis. In both cases, locally released oxytocin and vasopressin act both on MNCs and on their afferents, the pattern of electrical activity resulting from complex interactions between intrinsic MNC properties and network activities.

4. ACKNOWLEDGEMENTS

The developmental part of this study is essentially due to the work of Vivien Chevaleyre and the *in vivo* pharmacological analysis of vasopressin receptors is due to Laurent Gouzènes, both when they were PhD students in our lab. We thank Dr F. Rassendren and S. Chaumont for kindly providing HEK cells expressing P2X2 receptor. Work supported by CNRS, French Research Ministry and Region Languedoc Roussillon.

5. REFERENCES

Abe, H., Inoue, M., Matsuo, T., and Ogata, N., 1983, The effects of vasopressin on electrical activity in the guinea-pig supraoptic nucleus *in vitro*, *J. Physiol.* **337**: 665.

Altman, J., and Bayer, S. A., 1986, The development of the rat hypothalamus, *Adv. Anat. Embryol. Cell Biol.* **100**: 1.

Altstein, M., and Gainer, H., 1988, Differential biosynthesis and post-translational processing of vasopressin and oxytocin in rat brain during embryonic and postnatal development, *J. Neurosci.* **8**: 3967.

Armstrong, W. E., 1995, Morphological and electrophysiological classification of hypothalamic supraoptic neurons, *Prog. Neurobiol.* **47**: 291.

Armstrong, W. E., and Stern, J. E., 1998, Electrophysiological distinctions between oxytocin and vasopressin neurons in the supraoptic nucleus, *Adv. Exp. Med. Biol.* **449**: 67.

Bailey, A., Clarke, G., and Wakerley, J., 1997, The role of alpha-2 adrenoceptors in the regulation of oxytocin neurones in the suckled rat, *Brain Res. Bull.* **44**: 193.

Boer, G. J., Swaab, D. F., Uylings, H. B., Boer, K., Buijs, R. M., and Velis, D. N., 1980, Neuropeptides in rat brain development, *Prog. Brain Res.* **53**: 207.

Boudaba, C., Di, S., and Tasker, J. G., 2003, Presynaptic noradrenergic regulation of glutamate inputs to hypothalamic magnocellular neurones, *J. Neuroendocrinol.* **15**: 803.

Brown, D., and Moos, F., 1997, Onset of bursting in oxytocin cells in suckled rats, *J. Physiol.* **503**: 625.

Brown, D., Fontanaud, P., and Moos, F. C., 2000, The variability of basal action potential firing is positively correlated with bursting in hypothalamic oxytocin neurones, *J. Neuroendocrinol.* **12**: 506.

Brussaard, A. B., Kits, K. S., and de Vlieger, T. A., 1996, Postsynaptic mechanism of depression of GABAergic synapses by oxytocin in the supraoptic nucleus of immature rat, *J. Physiol.* **497**: 495.

Buijs, R. M., Velis, D. N., and Swaab, D. F., 1980, Ontogeny of vasopressin and oxytocin in the fetal rat: early vasopressinergic innervation of the fetal brain, *Peptides* **1**: 315.

Cazalis, M., Dayanithi, G., and Nordmann, J. J., 1985, The role of patterned burst and interburst interval on the excitation- coupling mechanism in the isolated rat neural lobe, *J. Physiol.* **369**: 45.

Cazalis, M., Dayanithi, G., and Nordmann, J. J., 1987, Hormone release from isolated nerve endings of the rat neurohypophysis, *J. Physiol.* **390**: 55.

Chevaleyre, V., Moos, F. C., and Desarmenien, M. G., 2001, Correlation between electrophysiological and morphological characteristics during maturation of rat supraoptic neurons, *Eur. J. Neurosci.* **13**: 1136.

Chevaleyre, V., Moos, F. C., and Desarmenien, M. G., 2002, Interplay between presynaptic and postsynaptic activities is required for dendritic plasticity and synaptogenesis in the supraoptic nucleus, *J. Neurosci.* **22**: 265.

Chevaleyre, V., Dayanithi, G., Moos, F. C., and Desarmenien, M. G., 2000, Developmental regulation of a local positive autocontrol of supraoptic neurons, *J. Neurosci.* **20**: 5813.

Dayanithi, G., Stuenkel, E. L., and Nordmann, J. J., 1992, Intracellular calcium and hormone release from nerve endings of the neurohypophysis in the presence of opioid agonists and antagonists, *Exp. Brain Res.* **90**: 539.

de Kock, C. P., Wierda, K. D., Bosman, L. W., Min, R., Koksma, J. J., Mansvelder, H. D., Verhage, M., and Brussaard, A. B., 2003, Somatodendritic secretion in oxytocin neurons is upregulated during the female reproductive cycle, *J. Neurosci.* **23**: 2726.

Desarmenien, M. G., Dayanithi, G., Tapia-Arancibia, L., and Widmer, H., 1994, Developmental autoregulation of calcium currents in mammalian central neurones, *Neuroreport* **5**: 1953.

Freund-Mercier, M. J., and Richard, P., 1984, Electrophysiological evidence for facilitatory control of oxytocin neurones by oxytocin during suckling in the rat, *J. Physiol.* **352**: 447.

Freund-Mercier, M. J., and Stoeckel, M. E., 1995, Somatodendritic autoreceptors on oxytocin neurones, *Adv. Exp. Med. Biol.* **395**: 185.

Freund-Mercier, M. J., Stoeckel, M. E., and Klein, M. J., 1994, Oxytocin receptors on oxytocin neurones: Histoautoradiographic detection in the lactating rat, *J. Physiol.* **480:** 155.

Glasgow, E., Kusano, K., Chin, H., Mezey, E., Young, W. S., 3rd, and Gainer, H., 1999, Single cell reverse transcription-polymerase chain reaction analysis of rat supraoptic magnocellular neurons: neuropeptide phenotypes and high voltage-gated calcium channel subtypes, *Endocrinology* **140:** 5391.

Gouzenes, L., Desarmenien, M. G., Hussy, N., Richard, P., and Moos, F. C., 1998, Vasopressin regularizes the phasic firing pattern of rat hypothalamic magnocellular vasopressin neurons, *J. Neurosci.* **18:** 1879.

Gouzenes, L., Sabatier, N., Richard, P., Moos, F. C., and Dayanithi, G., 1999, V1a- and V2-type vasopressin receptors mediate vasopressin-induced Ca^{2+} responses in isolated rat supraoptic neurones, *J. Physiol.* **517:** 771.

Hatton, G. I., 1999, Astroglial modulation of neurotransmitter/peptide release from the neurohypophysis: present status, *J. Chem. Neuroanat.* **16:** 203.

Hatton, G. I., and Li, Z., 1998, Mechanisms of neuroendocrine cell excitability, *Adv. Exp. Med. Biol.* **449:** 79.

Hermes, M. L., Ruijter, J. M., Klop, A., Buijs, R. M., and Renaud, L. P., 2000, Vasopressin increases GABAergic inhibition of rat hypothalamic paraventricular nucleus neurons *in vitro*, *J. Neurophysiol.* **83:** 705.

Hurbin, A., Orcel, H., Alonso, G., Moos, F., and Rabie, A., 2002, The vasopressin receptors colocalize with vasopressin in the magnocellular neurons of the rat supraoptic nucleus and are modulated by water balance, *Endocrinology* **143:** 456.

Hurbin, A., Boissin-Agasse, L., Orcel, H., Rabie, A., Joux, N., Desarmenien, M. G., Richard, P., and Moos, F. C., 1998, The V1a and V1b, but not V2, vasopressin receptor genes are expressed in the supraoptic nucleus of the rat hypothalamus, and the transcripts are essentially colocalized in the vasopressinergic magnocellular neurons, *Endocrinology* **139:** 4701.

Inenaga, K., and Yamashita, H., 1986, Excitation of neurones in the rat paraventricular nucleus *in vitro* by vasopressin and oxytocin, *J. Physiol.* **370:** 165.

Israel, J. M., Le Masson, G., Theodosis, D. T., and Poulain, D. A., 2003, Glutamatergic input governs periodicity and synchronization of bursting activity in oxytocin neurons in hypothalamic organotypic cultures, *Eur. J. Neurosci.* **17:** 2619.

Jing, X., Ratty, A. K., and Murphy, D., 1998, Ontogeny of the vasopressin and oxytocin RNAs in the mouse hypothalamus, *Neurosci. Res.* **30:** 343.

Karyakin, M. G., and Alekseev, N. P., 1992, The influence of tactile stimulation of nipples on the milk ejection reflex in the rat, *J. Comp. Physiol.* **170:** 645.

Kombian, S. B., Mouginot, D., and Pittman, Q. J., 1997, Dendritically released peptides act as retrograde modulators of afferent excitation in the supraoptic nucleus *in vitro*, *Neuron* **19:** 903.

Lambert, R. C., Moos, F. C., and Richard, P., 1993a, Action of endogenous oxytocin within the paraventricular or supraoptic nuclei: a powerful link in the regulation of the bursting pattern of oxytocin neurons during the milk-ejection reflex in rats, *Neuroscience* **57:** 1027.

Lambert, R. C., Dayanithi, G., Moos, F. C., and Richard, P., 1994, A rise in the intracellular Ca^{2+} concentration of isolated rat supraoptic cells in response to oxytocin, *J. Physiol.* **478:** 275.

Lambert, R. C., Moos, F. C., Ingram, C. D., Wakerley, J. B., Kremarik, P., Guerne, Y., and Richard, P., 1993b, Electrical activity of neurons in the ventrolateral septum and bed nuclei of the stria terminalis in suckled rats: statistical analysis gives evidence for sensitivity to oxytocin and for relation to the milk-ejection reflex, *Neuroscience* **54:** 361.

Landgraf, R., Neumann, I., Russell, J. A., and Pittman, Q. J., 1992, Push-pull perfusion and microdialysis studies of central oxytocin and vasopressin release in freely moving rats during pregnancy, parturition, and lactation, *Ann. NY Acad. Sci.* **652:** 326.

Leng, G., and Dyball, R. E., 1983, Intercommunication in the rat supraoptic nucleus. *Exp. Physiol.* **68:** 493.

Leng, G., Brown, C. H., and Russell, J. A., 1999, Physiological pathways regulating the activity of magnocellular neurosecretory cells, *Prog. Neurobiol.* **57:** 625.

Ludwig, M., and Leng, G., 1997, Autoinhibition of supraoptic nucleus vasopressin neurons in vivo: a combined retrodialysis/electrophysiological study in rats, *Eur. J. Neurosci.* **9:** 2532.

Ludwig, M., Callahan, M. F., Neumann, I., Landgraf, R., and Morris, M., 1994, Systemic osmotic stimulation increases vasopressin and oxytocin release within the supraoptic nucleus, *J. Neuroendocrinol.* **6:** 369.

Ludwig, M., Sabatier, N., Bull, P. M., Landgraf, R., Dayanithi, G., and Leng, G., 2002, Intracellular calcium stores regulate activity-dependent neuropeptide release from dendrites, *Nature* **418:** 85.

Manning, M., Stoev, S., Kolodziejczyk, A., Klis, W. A., Kruszynski, M., Misicka, A., Olma, A., Wo, N. C., and Sawyer, W. H., 1990, Design of potent and selective linear antagonists of vasopressin (V1-receptor) responses to vasopressin, *J. Med. Chem.* **33:** 3079.

Moos, F., and Richard, P., 1989, Paraventricular and supraoptic bursting oxytocin cells in rat are locally regulated by oxytocin and functionally related, *J. Physiol.* **408:** 1.

Moos, F., Fontanaud, P., Mekaouche, M., and Brown, D., 2004a, Oxytocin neurones are recruited into co-ordinated fluctuations of firing before bursting in the rat, *Neuroscience* **125**: 391.

Moos, F., Marganiec, A., Fontanaud, P., Guillou-Duvoid, A., and Alonso, G., 2004b, Synchronisation of oxytocin neurones in suckled rats : possible role of bilateral innervation of hypothalamic supraoptic nuclei by single medullary neurones, *Eur. J. Neurosci.* (in Press).

Moos, F., Freund-Mercier, M. J., Guerne, Y., Guerne, J. M., Stoeckel, M. E., and Richard, P., 1984, Release of oxytocin and vasopressin by magnocellular nuclei *in vitro*: specific facilitatory effect of oxytocin on its own release, *J. Endocrinol.* **102**: 63.

Moos, F., Poulain, D. A., Rodriguez, F., Guerne, Y., Vincent, J. D., and Richard, P., 1989, Release of oxytocin within the supraoptic nucleus during the milk ejection reflex in rats, *Exp. Brain Res.* **76**: 593.

Moos, F., Strosser, M. T., Poulain, D., Di Scala-Guenot, D., Guerne, Y., Rodriguez, F., Richard, P., and Vincent, J. D., 1987, Evaluation of perfusion technics of the magnocellular nucleus of the hypothalamus *in vitro* and *in vivo*, *Ann. Endocrinol.* **48**: 419.

Moos, F., Gouzenes, L., Brown, D., Dayanithi, G., Sabatier, N., Boissin, L., Rabie, A., and Richard, P., 1998, New aspects of firing pattern autocontrol in oxytocin and vasopressin neurones, *Adv. Exp. Med. Biol.* **449**: 153.

Moos, F. C., 1995, GABA-induced facilitation of the periodic bursting activity of oxytocin neurones in suckled rats, *J. Physiol.* **488**: 103.

Moos, F. C., and Ingram, C. D., 1995, Electrical recordings of magnocellular supraoptic and paraventricular neurons displaying both oxytocin- and vasopressin-related activity, *Brain Res.* **669**: 309.

Murai, Y., Nakashima, T., Miyata, S., and Kiyohara, T., 1998, Different effect of oxytocin on membrane potential of supraoptic oxytocin neurons in virgin female and male rats *in vitro*, *Neurosci. Res.* **30**: 35.

Nakai, S., Kawano, H., Yudate, T., Nishi, M., Kuno, J., Nagata, A., Jishage, K., Hamada, H., Fujii, H., and Kawamura, K., et al., 1995, The POU domain transcription factor Brn-2 is required for the determination of specific neuronal lineages in the hypothalamus of the mouse, *Genes Dev.* **9**: 3109.

Neumann, I., Ludwig, M., Engelmann, M., Pittman, Q. J., and Landgraf, R., 1993, Simultaneous microdialysis in blood and brain: oxytocin and vasopressin release in response to central and peripheral osmotic stimulation and suckling in the rat, *Neuroendocrinology* **58**: 637.

Nordmann, J. J., Dayanithi, G., and Lemos, J. R., 1987, Isolated neurosecretory nerve endings as a tool for studying the mechanism of stimulus-secretion coupling, *Biosci. Rep.* **7**: 411.

Ota, M., Crofton, J. T., and Share, L., 1994, Hemorrhage-induced vasopressin release in the paraventricular nucleus measured by in vivo microdialysis, *Brain Res.* **658**: 49.

Piet, R., Bonhomme, R., Theodosis, D. T., Poulain, D. A., and Oliet, S. H., 2003, Modulation of GABAergic transmission by endogenous glutamate in the rat supraoptic nucleus, *Eur. J. Neurosci.* **17**: 1777.

Pittman, Q. J., Hirasawa, M., Mouginot, D., and Kombian, S. B., 2000, Neurohypophysial peptides as retrograde transmitters in the supraoptic nucleus of the rat, *Exp. Physiol.* **85**: 139S.

Poulain, D. A., Wakerley, J. B., and Dyball, R. E., 1977, Electrophysiological differentiation of oxytocin- and vasopressin- secreting neurons, *Proc. R. Soc. Lond. B. Biol. Sci.* **196**: 367.

Pow, D. V., and Morris, J. F., 1989, Dendrites of hypothalamic magnocellular neurons release neurohypophysial peptides by exocytosis, *Neuroscience* **32**: 435.

Rajerison, R. M., Butlen, D., and Jard, S., 1976, Ontogenic development of antidiuretic hormone receptors in rat kidney: comparison of hormonal binding and adenylate cyclase activation, *Mol. Cell. Endocrinol.* **4**: 271.

Renaud, L. P., and Bourque, C. W., 1991, Neurophysiology and neuropharmacology of hypothalamic magnocellular neurons secreting vasopressin and oxytocin, *Prog. Neurobiol.* **36**: 131.

Sabatier, N., Richard, P., and Dayanithi, G., 1998, Activation of multiple intracellular transduction signals by vasopressin in vasopressin-sensitive neurones of the rat supraoptic nucleus, *J. Physiol.* **513**: 699.

Saito, M., Tahara, A., Sugimoto, T., Abe, K., and Furuichi, K., 2000, Evidence that atypical vasopressin V(2) receptor in inner medulla of kidney is V(1B) receptor, *Eur. J. Pharmacol.* **401**: 289.

Sawchenko, P. E., and Swanson, L. W., 1983, The organization of forebrain afferents to the paraventricular and supraoptic nuclei of the rat, *J. Comp. Neurol.* **218**: 121.

Serradeil-Le Gal, C., Wagnon, J., Garcia, C., Lacour, C., Guiraudou, P., Christophe, B., Villanova, G., Nisato, D., Maffrand, J. P., Le Fur, G., et al., 1993, Biochemical and pharmacological properties of SR 49059, a new, potent, nonpeptide antagonist of rat and human vasopressin V1a receptors, *J. Clin. Invest.* **92**: 224.

Stern, J. E., and Armstrong, W. E., 1995, Electrophysiological differences between oxytocin and vasopressin neurones recorded from female rats *in vitro*, *J. Physiol.* **488**: 701.

Stern, J. E., and Armstrong, W. E., 1996, Changes in the electrical properties of supraoptic nucleus oxytocin and vasopressin neurons during lactation, *J. Neurosci.* **16**: 4861.

Takano, S., Negoro, H., Honda, K., and Higuchi, T., 1992, Lesion and electrophysiological studies on the hypothalamic afferent pathway of the milk ejection reflex in the rat, *Neuroscience* **50**: 877.

Thellier, D., Moos, F., Richard, P., and Stoeckel, M. E., 1994, Evidence for reciprocal connections between the dorsochiasmatic area and the hypothalamo neurohypophyseal system and some related extrahypothalamic structures, *Brain Res. Bull.* **35:** 311.

Theodosis, D. T., 2002, Oxytocin-secreting neurons: A physiological model of morphological neuronal and glial plasticity in the adult hypothalamus, *Front. Neuroendocrinol.* **23:** 101.

Theodosis, D. T., and Poulain, D. A., 1999, Contribution of astrocytes to activity-dependent structural plasticity in the adult brain, *Adv. Exp. Med. Biol.* **468:** 175.

Wakerley, J. B., Poulain, D. A., and Brown, D., 1978, Comparison of firing patterns in oxytocin- and vasopressin-releasing neurones during progressive dehydration, *Brain Res.* **148:** 425.

Widmer, H., Amerdeil, H., Fontanaud, P., and Desarmenien, M. G., 1997, Postnatal maturation of rat hypothalamoneurohypophysial neurons: evidence for a developmental decrease in calcium entry during action potentials, *J. Neurophysiol.* **77:** 260.

Wotjak, C. T., Ludwig, M., and Landgraf, R., 1994, Vasopressin facilitates its own release within the rat supraoptic nucleus in vivo, *Neuroreport* **5:** 1181.

Yamashita, H., Okuya, S., Inenaga, K., Kasai, M., Uesugi, S., Kannan, H., and Kaneko, T., 1987, Oxytocin predominantly excites putative oxytocin neurons in the rat supraoptic nucleus *in vitro*, *Brain Res.* **416:** 364.

Young, W. S. 3rd, Shepard, E., Amico, J., Hennighausen, L., Wagner, K. U., LaMarca, M. E., McKinney, C., and Ginns, E. I., 1996, Deficiency in mouse oxytocin prevents milk ejection, but not fertility or parturition, *J. Neuroendocrinol.* **8:** 847.

Young, W. S. 3rd, Shepard, E., DeVries, A. C., Zimmer, A., LaMarca, M. E., Ginns, E. I., Amico, J., Nelson, R. J., Hennighausen, L., and Wagner, K. U., 1998, Targeted reduction of oxytocin expression provides insights into its physiological roles, *Adv. Exp. Med. Biol.* **449:** 231.

14

CONDITIONAL PRIMING OF DENDRITIC NEUROPEPTIDE RELEASE

Mike Ludwig[*] and Gareth Leng

1. INTRODUCTION

Among the best-established sites of dendritic release are the hypothalamic supraoptic and paraventricular nuclei, where magnocellular neurons synthesise the neuropeptides vasopressin and oxytocin. Studies on dendritic release from these neurons have been facilitated by the fact that their somata and dendrites are tightly grouped, making it possible to measure release *in vivo* using microdialysis probes, and by the fact that these neurons make vast quantities of their peptide products. From their axonal nerve terminals in the posterior pituitary gland, these neurons secrete enough vasopressin and oxytocin into the systemic circulation to regulate peripheral target organs – notably the kidney, uterus and mammary gland. However, within the hypothalamus the axons of vasopressin and oxytocin neurons have few if any collaterals; since axon terminals are, almost exclusively, located distantly in the posterior pituitary, studies of dendritic release can be accomplished in the absence of contaminating axonal release of peptide. Over the last decade, *in vivo* and *in vitro* studies have amplified the data on dendritic vasopressin and oxytocin release and revealed many aspects of its control (Landgraf, 1995; Ludwig, 1998; Ludwig and Pittman, 2003).

Depending upon the stimulus, electrical activity in the cell bodies can release peptides from nerve terminals in the posterior pituitary with little or no release from the dendrites. Furthermore, some stimuli (Engelmann et al., 1999; Ludwig et al., 2002b; Sabatier et al., 2003; Wotjak et al., 1998), induce peptide release from dendrites without increasing the electrical activity of the cell body, and without inducing secretion from nerve terminals. Differential release of neurotransmitters from different compartments of a single neurone requires subtle regulatory mechanisms. Here we describe a novel mechanism of conditional priming of dendritic neurosecretory granules which enables these neurones to respond to appropriate stimulation in an activity-dependent manner for prolonged periods.

[*] School of Biomedical and Clinical Laboratory Sciences, University of Edinburgh, Edinburgh, EH8 9XD, UK

The magnocellular neurons of the paraventricular and supraoptic nucleus are by far the richest source of oxytocin and vasopressin within the brain. Each supraoptic nucleus contains about 2 ng of oxytocin, and rather more vasopressin, and although some of this peptide is destined for transport to the vast storage and secretion site in the posterior pituitary gland, about 90% is present in neurosecretory granules within dendrites (see Russell et al., 2003). Ultrastructural studies using electron microscopy have shown exocytotic profiles of neurosecretory granules at regions that apparently display no specific synaptic specialisations (Pow and Morris, 1989). The estimated vasopressin and oxytocin concentration in the extracellular fluid of the supraoptic nucleus (1,000 to 10,000 pg/ml, Landgraf et al., 1992; Ludwig and Leng, 1997) is 100-1000 fold higher than the plasma concentration. Thus, an effective concentration of these peptides in the brain extracellular fluid can be expected even after substantial diffusion from the nucleus. By carrying specific information not only across synapses, but also to distant brain regions the peptides act as *paracrine* signals and may induce widespread, long-lasting effects within the brain.

2. REGULATION OF DENDRITIC NEUROPEPTIDE RELEASE

Exocytosis of vasopressin and oxytocin from the axonal terminals in the posterior pituitary gland is linked to electrical activity, resulting from Ca^{2+} entry via voltage-gated channels following depolarisation of the terminals by invading action potentials (Leng et al., 1999). However, some chemical signals, notably oxytocin and vasopressin themselves (Moos et al., 1989; Wotjak et al., 1994, Fig. 2), can elicit dendritic peptide release without increasing the electrical activity of the neurons. Oxytocin neurons and vasopressin neurons express oxytocin receptors and vasopressin receptors respectively (Freund-Mercier et al., 1994; Hurbin et al., 2002), and the peptides act at these receptors to produce a cell-type-specific rise in intracellular Ca^{2+} concentration. In acutely-dissociated oxytocin neurons of the supraoptic nucleus, oxytocin mobilises Ca^{2+} from thapsigargin-sensitive intracellular Ca^{2+} stores (Lambert et al., 1994), whereas the response induced by vasopressin in vasopressin cells requires an influx of external Ca^{2+} through voltage-gated Ca^{2+} channels as well as some contribution from mobilisation of thapsigargin-sensitive stores (Lambert et al., 1994; Sabatier et al., 1997). Thus, activation of G-protein coupled receptors on the dendrites or cell body may elevate intracellular Ca^{2+} concentrations sufficiently to trigger exocytosis of neurosecretory granules. Once triggered, dendritic peptide release may be self-sustaining and hence long-lasting.

Part of the function of dendritically-released peptides is to *autoregulate* the electrical activity of the cells of origin. Both oxytocin and vasopressin act on their respective neurons to modify their excitability. Vasopressin inhibits or excites vasopressin neurones, depending on their ongoing electrical activity, so that fast-firing neurones are slowed, and slow-firing neurones are excited (Gouzenes et al., 1998). These effects involve two receptor subtypes expressed by vasopressin cells (V1a and V1b receptors), modulation of several intracellular second messenger pathways (Sabatier et al., 1998); and presynaptic actions on synaptic inputs (Kombian et al., 2000). These actions appear to maintain the vasopressin cell in an activity pattern conducive to maximally efficient secretion from the axon terminals (Gouzenes et al., 1998).

In vitro studies indicate that oxytocin can act both presynaptically and postsynaptically to attenuate the effects of GABA inputs (Brussaard et al., 1996) and to

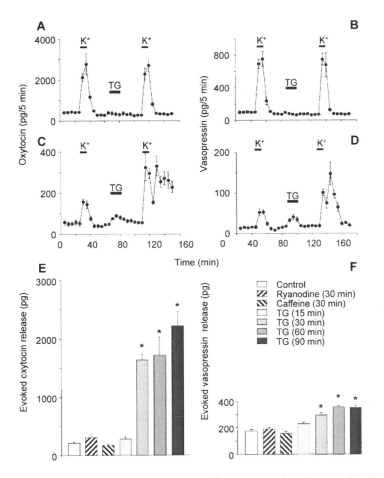

Figure 1. Priming in isolated supraoptic nuclei *in vitro*. K[+] evoked oxytocin and vasopressin release from isolated SON (**C,D**), but not neural lobes (**A,B**), was strongly potentiated after pretreatment with thapsigargin (TG) (**C,D**). The potentiation was specific to thapsigargin (no effect of ryanodine or caffeine), time-dependent (no significant effect within the first 15 min) and long-lasting (at least 90 min after exposure to thapsigargin). Adapted from Ludwig et al 2002a,b.

reduce glutamatergic excitation (Kombian et al., 1997) through an action on voltage-activated Ca^{2+} currents in presynaptic terminals (Hirasawa et al., 2001).

There is a probably more important consequence than simple autoregulation of the electrical activity of the cells of origin. Spike activity in oxytocin neurons or spike activity induced by antidromic stimulation of the neural stalk does *not* always result in dendritic oxytocin release (Ludwig et al., 2002b). For example, systemic osmotic stimulation leads to prompt and long-lasting activation of both oxytocin and vasopressin secretion into the systemic circulation; osmotic stimuli also lead to oxytocin and vasopressin release within the hypothalamus, but much later than secretion into the blood (Ludwig et al., 1994); in these circumstances, dendritic release is apparently delayed by

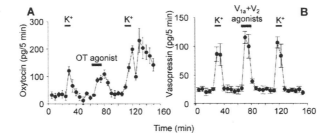

Figure 2. Application of oxytocin (**A**) or vasopressin receptor agonists (**B**) stimulates dendritic oxytocin and vasopressin release respectively, indicating that the peptides themselves are capable of evoking dendritic release (positive feed-back). Furthermore, oxytocin, but not vasopressin, primes dendritic stores of neurosecretory vesicles, indicated by a potentiated release evoked by K^+ similar to that seen after thapsigargin administration. Adapted from Ludwig et al., 2002a,b.

at least an hour compared to axonal secretion, but in other circumstances dendritic release appears to be coincident with or in advance of peripheral secretion (see Ludwig, 1998).

However, after exposure of supraoptic neurons to thapsigargin to mobilise Ca^{2+} from intracellular stores, dendritic peptide release in response to subsequent osmotic stimulation, or in response to depolarisation with high potassium solutions, or in response to antidromically-induced electrical stimulation, is dramatically potentiated *in vivo* and *in vitro* (Fig. 1). This "priming" effect is long-lasting, since a large potentiation was observed in response to high K^+ (50mM) administered 30, 60 or 90 min after thapsigargin (Fig. 1). It seems likely that any signal that mobilises Ca^{2+} from these intracellular stores may prime dendritic secretion; most importantly, oxytocin itself both mobilises intracellular Ca^{2+} and primes its own dendritic release (Ludwig et al., 2002b, Fig. 2)

Thus, we concluded that thapsigargin, and oxytocin itself, *prime* dendritic stores of oxytocin to enable them for subsequent activity-dependent release for a prolonged period – at least 90 min after exposure to thapsigargin (Ludwig et al., 2002b). Interestingly, although vasopressin is effective in inducing dendritic vasopressin release, and does provoke some release of Ca^{2+} from thapsigargin-sensitive stores, it is ineffective in producing priming (Ludwig et al., 2002a, Fig. 2).

3. MECHANISMS OF PRIMING OF DENDRITIC NEUROPEPTIDE RELEASE

Priming in vasopressin and oxytocin cells is not a consequence of long-lasting elevation of intracellular $[Ca^{2+}]$ induced by thapsigargin, since priming outlasts the increase in intracellular $[Ca^{2+}]$ (Lambert et al., 1994; Dayanithi et al., 1996). No potentiation was seen in response to high-K^+ concentrations administered 5 min after exposure to thapsigargin, but a large potentiation was observed in response to high-K^+ administered 30, 60 or 90 min after thapsigargin. Remarkably, this prolonged release had been noted in microdialysis experiments, where dendritic release outlasted that of peripheral secretion into the systemic circulation (Ludwig et al., 1994).

Thapsigargin induces an increase in intracellular $[Ca^{2+}]$ by inhibiting the endoplasmic reticulum Ca^{2+}-ATPase, and thapsigargin targeted stores include the IP_3

sensitive stores (Thastrup et al., 1990). Thapsigargin blocks the refilling of these stores, this may therefore also result in opening of store-operated cation channels, which may affect the strength or effectiveness of subsequent secretory stimuli such as depolarisation. Store-operated cation channels in magnocellular neurons are Ca^{2+} permeable and are activated at membrane potentials more negative than $-40mV$ (Tobin et al., 2002). The irreversible depletion of intracellular Ca^{2+} stores will permanently activate these currents, the resulting small but constant Ca^{2+} current may contribute to the activation of the Ca^{2+}-sensitive exocytotic machinery, or increase voltage-dependent Ca^{2+} influx by increasing the membrane potential range at which Ca^{2+} influx occurs through voltage-gated channels.

Priming may also involve actions on translation/protein processing such as local synthesis of proteins that support exocytosis (Ma and Morris, 2002). Furthermore, it may entail changes in vesicle tethering, maturation of docked vesicles and/or active vesicle transport. Recent studies using electron microscopy determined whether priming in magnocellular neurones involves a translocation of dense-core vesicles closer to the dendritic membrane. Thapsigargin treatment resulted in an increase in the total number of vesicles within 500 nm and a higher frequency of vesicles within the first 200nm of the plasma membrane, indicating that the thapsigargin-induced priming involves a relocation of vesicles closer to sites of secretion (Tobin et al., 2004, Fig. 3).

A recent immunocytochemical study localized vasopressin receptors to secretory

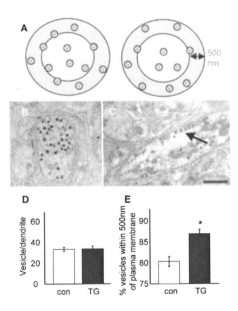

Figure 3. Priming involves a translocation of dense core vesicles closer to the dendritic membrane as indicated in the schematic drawing (**A**). Selected electron micrographs showing dendritic distribution of dense core vesicles from a control (**B**) and thapsigargin (TG)-treated rat (**C**). Note the greater number of vesicles closer to the membrane after TG treatment. Statistical analysis revealed that there were no differences in the total number of dense core vesicles within the dendrites (**D**), but TG treatment significantly increases the number of dense core vesicles within a 500 nm margin to the dendritic plasma membrane (**E**). Scale bar = 1μm. Adapted from Tobin et al., 2004.

vesicles that also contained vasopressin (Hurbin et al., 2002). There is also evidence that other receptors, including opioid receptors are inserted into the cell membrane by the process of exocytosis of neurosecretory vesicles. We should note therefore that, when dendritic secretion is initiated, insertion of additional receptors in the plasma membrane of dendrites may progressively amplify the autoregulatory actions of dendritically released peptides, amplifying further the consequences of priming itself.

Priming has previously been shown in other hormonal systems. In LH-secreting gonadotrophs of the anterior pituitary gland of oestrogen-primed rats, luteinizing hormone-releasing hormone (LHRH) is capable of "self-priming", causing a characteristic and dramatic potentiation of LH secretion in response to successive challenges with LHRH (Aiyer et al., 1974; Scullion et al., 2004). LHRH self-priming is also time-dependent (occurs within 30-40 min of LHRH exposure, Waring and Turgeon, 1983), involves protein synthesis (Curtis et al., 1985) and involves calcium- and ATP-dependent translocation of vesicles to docking sites and maturation of docked dense-cored vesicles at the plasma membrane (Lewis et al., 1986; Pickering and Fink, 1979).

4. CONSEQUENCES OF PRIMING - THE MILK EJECTION REFLEX

Under basal conditions, and in males, oxytocin neurons are continuously active; but in the lactating animal, suckling modifies this pattern. During lactation, in response to suckling, oxytocin cells discharge, approximately synchronously, with brief, intense bursts of action potentials (Lincoln and Wakerley, 1974); these bursts release boluses of oxytocin into the circulation that result in milk let-down from the mammary glands. The bursting activity can be blocked by central administration of oxytocin antagonists (Lambert et al., 1993), thus central as well as peripheral oxytocin is essential for milk let-down. As mentioned above, once triggered, oxytocin release in the supraoptic nucleus is self-sustaining and hence long-lasting, and acts in a positive-feedback manner to evoke bursting (Leng et al., 1999).

To investigate the physiological significance of priming of activity-dependent dendritic release for bursting behaviour, we recorded from single oxytocin neurons in *virgin* female rats while dialysing the supraoptic nucleus with thapsigargin *in vivo*. Thapsigargin had no effect upon the mean electrical discharge rate of cells, but statistical analysis revealed a subtle but important effect on discharge patterning. In oxytocin cells under control conditions, interspike intervals that are much shorter than the average interval tend to be followed by interspike intervals that are longer than average, reflecting the summating effects of a slow post-spike afterhyperpolarisation (Kirkpatrick and Bourque, 1996). However, in some cells after thapsigargin, intervals much shorter than average were followed by intervals that were also significantly shorter than average. This suggested that the normal post-spike afterhyperpolarisation may be being overridden by post-spike depolarisation, and thus that activity-dependent exocytosis from the dendrites may have a positive-feedback effect on electrical activity.

If oxytocin and/or other substances released from dendrites affect the activity of neighbouring cells, then this should be particularly apparent from the effects of "constant collision" stimulation (Leng, 1981, Fig. 4A). This protocol involves activating the neighbours of a recorded cell synchronously in an activity-dependent manner, mimicking the co-ordination of neuronal activity that precedes reflex milk-ejection (Brown and Moos, 1997). Electrical stimulus pulses applied to the neural stalk evoke antidromic

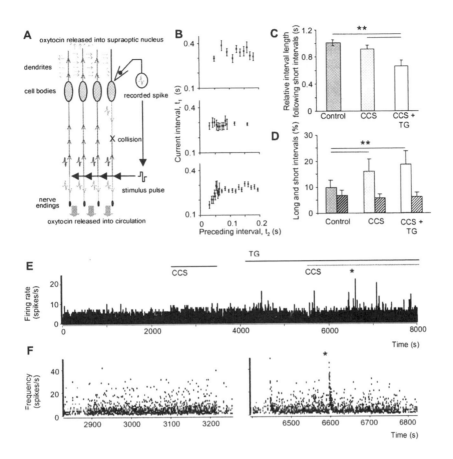

Figure 4. The effects of constant-collision stimulation (CCS, schematic, **A**) and thapsigargin (TG) on oxytocin neurons. After TG (administered by retrodialysis *in vivo*), CCS produced clustering of activity with occasional bursts (asterisks in **E, F**). **E** shows second-by-second firing rate, **F** shows instantaneous frequency, and **B** shows mean t_1 ($\pm$ S.E.) against t_2 for this cell in control conditions (top), during CCS (middle), and during CCS after thapsigargin (bottom). **C** and **D** show means + S.E.. **C** shows the mean t_2 when following short values of t_1, compared to the overall mean t_2 for each cell. The bars show this ratio in control conditions, during CCS, and during CCS after thapsigargin. **D** shows mean relative incidence of short intervals (<50 ms, hatched bars) and long intervals (>700 ms, shaded bars) in the interspike interval distributions recorded in control conditions, during CCS, and during CCS after thapsigargin. ** P<0.01, paired t-tests. Adapted from Ludwig et al., 2002b.

spikes in all supraoptic neurons, except when a spike has occurred in the 10ms before stimulation, in which case the antidromic spike evoked in that cell will be extinguished by collision with the descending spontaneous spike. During recording of each cell, an electrical stimulus pulse was applied to the neural stalk 5 ms after every spontaneous spike. This constant-collision stimulation does not affect the recorded cell directly, as all spikes evoked antidromically in that cell are extinguished, but antidromic spikes cause approximately synchronous activation of neighbouring cells, and hence will cause synchronised activity-dependent exocytosis from neighbouring dendrites. Constant-collision stimulation induces clustered discharge in some oxytocin cells with little effect on mean discharge rate (Leng, 1981). We compared current and preceding interspike intervals to look specifically for positive feedback effects. During constant-collision stimulation, the neurons showed a positive correlation between the current interspike interval and preceding interval lengths for the shortest intervals (Fig. 4). After thapsigargin, positive correlations were stronger in each cell during constant-collision stimulation, and positive correlations were seen in cells that had not displayed them before thapsigargin treatment (Fig. 4B).

Overall, there was no change in the relative incidence of short intervals (< 50 ms) in the cells tested with constant-collision stimulation. On the contrary, there was an increase in the incidence of *long* interspike intervals (>700 ms, Fig. 4D). Thus, during constant-collision stimulation after thapsigargin, there is no increase in firing rate of oxytocin cells, and no generally increased incidence of short intervals, but there is a reorganisation of spike activity reflected by changes in the interspike interval distributions. Very short intervals tend to be followed by short intervals (Fig. 4C) in clusters of spikes that are followed by long intervals, and hence the second-by-second firing rate shows greater variability, as observed before suckling-induced milk-ejection bursts (Brown and Moos, 1997). Some cells displayed clear bursts of activity (Lincoln and Wakerley, 1972, Fig. 4E, F). These bursts were less intense than "classic" milk-ejection bursts that are characterised by extremely short interspike intervals (<10 ms) never normally seen in spontaneous activity (Dyball and Leng, 1986), but were similar to bursts seen at the beginnings of reflex milk ejection (Belin and Moos, 1986).

In lactating rats, microinjections of oxytocin into the cerebral ventricle, or directly into the supraoptic nucleus facilitate the suckling-induced milk ejection reflex for about 30 min (Lambert et al., 1993). When constant-collision stimulation was applied to oxytocin cells before and after intracerbroventricular injection of oxytocin into the lateral ventricle, the effects of oxytocin upon spike patterning during constant-collision stimulation were similar to those observed with thapsigargin. However, after priming, activity-dependent dendritic release facilitates "clustered" discharge patterning in oxytocin cells. Whether this is a direct action on oxytocin cells, or involves modulation of presynaptic neurotransmitter release (Kombian et al., 1997), or actions on interneurons (Jourdain et al., 1998) has not been determined. However, it seems likely that this mechanism underlies the oxytocin-dependent generation of bursting activity in response to suckling.

5. DIGITAL VERSUS ANALOG SIGNALLING

Once released within the brain, the role of neuropeptides in interneuronal communication (including feedback actions on their own neurons) is based on actions as neuromodulators and/or neurotransmitters (Landgraf and Neumann, 2004). As transmitters, neuropeptides contribute to the synaptic mode of information transfer, which refers to fast point-to-point signalling including transient actions, which are limited to postsynaptic sites. Since fast transmitter release at synapses mediates the transfer of information encoded by action potentials relatively faithfully, neurotransmitters may be regarded as mediating *digitally* encoded information transfer between neurons.

Immunohistochemical detection of neuropeptide-containing structures supports the hypothesis of some central release from axonal terminals, in that occasional peptide-containing vesicles are found associated with synapses, although generally, as observed in central oxytocin-containing synapses, the synaptic endings of peptide-containing neurons also contain large numbers of small vesicles that apparently do not contain the peptide (Meeker et al., 1991). This suggests that peptide-containing neurons can also synthesise other conventional neurotransmitters that are packaged separately. In many cases there is direct evidence for this, for instance neuropeptide Y–containing neurons of the arcuate nucleus use GABA as their principal neurotransmitter product (Cowley et al., 2001), and the noradrenergic cell groups of the brainstem co-express a variety of peptides, including enkephalin, galanin, and neuropeptide Y (Kalra et al., 1999).

This raises the question of the functional significance of peptides that co-exist with conventional neurotransmitters. The simplest assumption is that the co-existing peptide is co-released with the transmitter, and that their postsynaptic actions are complementary (Brown et al., 2004). However, if the transmitter and peptide are packaged separately, then this raises the question of whether their release is regulated differentially. Indeed there is evidence from noradrenergic systems of differences in stimulus-secretion coupling for noradrenaline and co-existing peptides (Lundberg and Hokfelt, 1985). Stimulus-secretion coupling at the neurosecretory terminals of oxytocin and vasopressin neurons has been extensively studied, and for this system it is very clear that the patterning of action potentials is of decisive importance for the efficiency of peptide secretion at this site. A plausible hypothesis supported by various lines of circumstantial evidence and by some specific direct examples, is that while neurotransmitter secretion can follow action potential activity relatively faithfully, peptide secretion is coupled very non-linearly to action potential activity, being strongly potentiated at high frequencies of stimulation.

Conventional transmitters mediate highly specific information because their release and actions are tightly restricted by specificity of location (at synapses) and specificity of timing (due to their very short biological half-lives). Accordingly, many different types of information may be encoded by different neurons that use a single transmitter. Indeed, only a relatively small number of conventional fast neurotransmitters are known to be used in the brain. Most central synapses use either glutamate or GABA, and most of the rest use acetylcholine, dopamine, histamine, noradrenaline or serotonin. However, the list of peptides secreted in the hypothalamus alone is enormous: to consider just one small region, neurons in the arcuate nucleus synthesise agouti-related peptide, enkephalin, galanin, ghrelin, growth-hormone releasing hormone, neurotensin, neuropeptide Y, neuromedin U, somatostatin, and peptides derived from pro-opiomelanocortin including β-endorphin and α-melanocyte-stimulating hormone, all of which are released centrally and act on specific, discretely-expressed receptors (Kalra et al., 1999).

Unlike conventional neurotransmitters, peptides generally have very long biological half-lives, and typically they act at G-protein coupled receptors with nanomolar affinity. Accordingly, peptides are unlikely to be restricted in their actions to the precise synaptic localization of their release. In fact, at least for vasopressin and oxytocin, the synapse does not appear to be a particularly favored site of release at all; these peptides can be released from all parts of the neuron: soma, dendrites, axons, axonal swellings and neurosecretory terminals, and they appear to be released at all sites in proportion to the number of vesicles that lie close to the membrane (Morris and Pow, 1991). Receptors for these peptides on the other hand are not conspicuously expressed close to the sites of

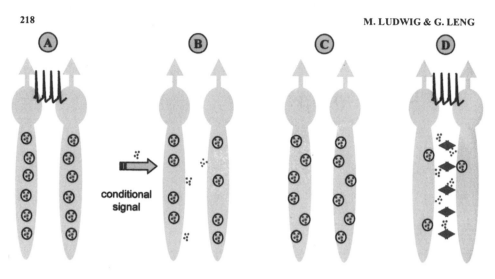

Figure 5. Proposed mechanisms of conditional priming. (A) Under normal conditions dendritic peptide release is not activated by electrical activity. A conditional (probably peptidergic) signal triggers release from dendrites independently of the electrical activity (B) and primes dendritic stores. Priming occurs partially by relocation of dendrite neurosecretory vesicles closer to the membrane (C) where they are available for activity-dependent released for a prolonged period (D).

(A color version of this figure appears in the signature between pp. 256 and 257.)

axonal termination – indeed, on the contrary, gross mismatches of peptide receptor distribution and terminals of peptide-containing neurons are very common throughout the brain (Herkenham, 1987). Thus stimulus-secretion coupling studies indicate that peptide release reflects integrated electrical activity of neurons, not instantaneous electrical activity; peptides generally seem likely to reach their targets by diffusion, possibly over long distances; and peptide actions once initiated are long lasting. Even the slow enzymatic degradation in the extracellular fluid may produce shorter but still active peptide fragments (Reijmers et al., 1998). These features suggest that peptidergic signalling is a sluggish form of communication, where specificity is conferred by the precise distribution of specific receptors rather that the precise distribution of release sites, and where temporal signalling is slow and *analog*, rather than fast and digital.

In some cases, communication via released peptides may span very large distances indeed within the brain. In many areas, neuropeptides can be measured in microdialysates of extracellular fluid at concentrations implausibly high given the local tissue peptide content (Landgraf and Neumann, 2004; Proescholdt et al., 2000)., indicating that these concentrations must arise by diffusion from distant sites of release rather than by local release.

The measured concentration of oxytocin and vasopressin in the extracellular fluid of the supraoptic nucleus is at least 100 fold higher than the concentrations needed to act at specific receptors. In some physiological states, such as lactation or during chronic dehydration, the glial cells of the supraoptic nucleus retract from their neuronal normal appositions, and any role that they play in uptake or degradation of peptides will be diminished (Theodosis, 2002). Thus 'volume transmission' reflecting the overflow diffusion of peptides from their release sites to distant targets may mediate peptide actions on more distant neurons in regions such as the septum, hippocampus, olfactory bulb and the amygdala, all of which are concerned with behaviours that are influenced by oxytocin and vasopressin (Engelmann et al., 1996; Ferguson et al., 2002), but which generally appear to display little or no innervation by oxytocin- or vasopressin-containing nerve fibres.

6. CONCLUSION

Dendritically-released neuropeptides act as autocrine and paracrine signals that modulate synaptic transmission and the electrical activity and, in some cases, cell morphology within the hypothalamic nuclei. The neuropeptides may have a very long half life in the extracellular fluid and thus are able to diffuse considerable distances to extrahypothalamic areas. The slow time scale of diffusion prolongs temporal and spatial summation, and allows intercommunication between dispersed elements. Interactions are also governed by the availability of secretory product for release at particular sites. Peptide-induced priming of these stores alters their availability for activity-dependent release. Priming involves release of a neuromodulator that results in a prolonged change in functional connectivity by its influence on target neurons that express appropriate receptors, without any anatomical communication between neurons, and without a direct effect upon neuron excitability. Priming allows one network to be recruited for different tasks, while feedback from the environment guides adaptation and learning. Thus, priming can contribute to a form of metaplasticity by changing the nature of interactions between neurons and their inputs over a long time period. While this has been shown for the releasable pools of oxytocin and vasopressin in dendrites, similar mechanisms may apply generally, perhaps including priming of release from synapses.

7. ACKNOWLEDGEMENTS

This work was funded by the Wellcome Trust and the BBSRC and conducted in collaboration with a number of excellent postgraduate students, postdoctoral fellows and colleagues in the lab and around the world.

8. REFERENCES

Aiyer, M. S., Chiappa, S. A., and Fink, G., 1974, A priming effect of luteinizing hormone releasing factor on the anterior pituitary gland in the female rat, *J. Endocrinol.* **62**: 573.

Belin, V. and Moos, F., 1986, Paired recordings from supraoptic and paraventricular oxytocin cells in suckled rats: recruitment and synchronization, *J. Physiol.* **377**: 369.

Brown, C. H., Ludwig, M., and Leng, G., 2004, Temporal dissociation of the feedback effects of dendritically co-released peptides on rhythmogenesis in magnocellular neurosecretory cells, *Neuroscience* **124**: 105.

Brown, D., and Moos, F., 1997, Onset of bursting in oxytocin cells in suckled rats, *J. Physiol.* **503**: 625.

Brussaard, A. B., Kits, K. S., and de Vlieger, T. A., 1996, Postsynaptic mechanism of depression of GABAergic synapses by oxytocin in the supraoptic nucleus of immature rat, *J. Physiol.* **497**: 495.

Cowley, M. A., Smart, J. L., Rubinstein, M., Cordan, M. G., Diano, S., Horvath, T. L., Cone, R. D., and Low, M. J., 2001, Leptin activates anorexigenic POMC neurons through a neural network in the arcuate nucleus, *Nature* **411**: 480.

Curtis, A., Lyons, V., and Fink, G., 1985, The priming effect of LH-releasing hormone: effects of cold and involvement of new protein synthesis, *J. Endocrinol.* **105**: 163.

Dayanithi, G., Widmer, H., and Richard, P., 1996, Vasopressin-induced intracellular Ca^{2+} increase in isolated rat supraoptic cells, *J. Physiol.* **490**: 713.

Dyball, R. E. and Leng, G., 1986, Regulation of the milk ejection reflex in the rat, *J. Physiol.* **380**: 239.

Engelmann, M., Ebner, K., Landgraf, R., Holsboer, F., and Wotjak, C. T., 1999, Emotional stress triggers intrahypothalamic but not peripheral release of oxytocin in male rats, *J. Neuroendocrinol.* **11**: 867.

Engelmann, M., Wotjak, C. T., Neumann, I., Ludwig, M., and Landgraf, R., 1996, Behavioral consequences of intracerebral vasopressin and oxytocin: focus on learning and memory, *Neurosci. Biobehav. Rev.* **20**: 341.

Ferguson, J. N., Young, L. J., and Insel, T. R., 2002, The neuroendocrine basis of social recognition, *Front. Neuroendocrinol.* **23**: 200.

Freund-Mercier, M. J., Stoeckel, M. E., and Klein, M. J., 1994, Oxytocin receptors on oxytocin neurones: Histoautoradiographic detection in the lactating rat, *J. Physiol.* **480**: 155.

Gouzenes, L., Desarmenien, M. G., Hussy, N., Richard, P., and Moos, F. C., 1998, Vasopressin regularizes the phasic firing pattern of rat hypothalamic magnocellular neurons, *J. Neurosci.* **18**: 1879.

Herkenham, M., 1987, Mismatches between neurotransmitter and receptor localizations in brain: observations and implications, *Neuroscience* **23**: 1.

Hirasawa, M., Kombian, S. B., and Pittman, Q. J., 2001, Oxytocin retrogradely inhibits evoked, but not miniature, EPSCs in the rat supraoptic nucleus: role of N- and P/Q-type calcium channels, *J. Physiol.* **532**: 595.

Hurbin, A., Orcel, H., Alonso, G., Moos, F., and Rabie, A., 2002, The vasopressin receptors colocalize with vasopressin in the magnocellular neurons of the rat supraoptic nucleus and are modulated by water balance, *Endocrinology* **143**: 456.

Jourdain, P., Israel, J. M., Dupouy, B., Oliet, S. H., Allard, M., Vitiello, S., Theodosis, D. T., and Poulain, D. A., 1998, Evidence for a hypothalamic oxytocin-sensitive pattern-generating network governing oxytocin neurons in vitro, *J. Neurosci.* **18**: 6641.

Kalra, S. P., Dube, M. G., Pu, S. Y., Xu, B., Horvath, T. L., and Kalra, P. S., 1999, Interacting appetite-regulating pathways in the hypothalamic regulation of body weight, *Endocri. Rev.* **20**: 68.

Kirkpatrick, K. and Bourque, C. W., 1996, Activity dependence and functional role of the apamin-sensitive K+ current in the rat supraoptic neurones in vitro, *J. Physiol.* **494**: 389.

Kombian, S. B., Mouginot, D., Hirasawa, M., and Pittman, Q. J., 2000, Vasopressin preferentially depresses excitatory over inhibitory synaptic transmission in the rat supraoptic nucleus in vitro, *J. Neuroendocrinol.* **12**: 361.

Kombian, S. B., Mouginot, D., and Pittman, Q. J., 1997, Dendritic released peptides act as retrograde modulators of afferent excitation in the supraoptic nucleus in vitro, *Neuron* **19**: 903.

Lambert, R. C., Dayanithi, G., Moos, F. C., and Richard, P., 1994, A rise in the intracellular Ca^{2+} concentration of isolated rat supraoptic cells in response to oxytocin, *J. Physiol.* **478**: 275.

Lambert, R. C., Moos, F. C., and Richard, P., 1993, Action of endogenous oxytocin within the paraventricular or supraoptic nuclei: A powerful link in the regulation of the bursting pattern of oxytocin neurons during the milk-ejection reflex in rats, *Neuroscience* **57**: 1027.

Landgraf, R., 1995, Intracerebrally released vasopressin and oxytocin: measurement, mechanisms and behavioural consequences, *J. Neuroendocrinol.* **7**: 243.

Landgraf, R., Neumann, I., Russell, J. A., and Pittman, Q. J., 1992, Push-pull perfusion and microdialysis studies of central oxytocin and vasopressin release in freely moving rats during pregnancy, parturition, and lactation, *Ann. NY Acad. Sci.* **652**: 326.

Landgraf, R. and Neumann, I. D., 2004, Neuropeptide release within the brain: a dynamic concept of multiple and variable modes of communication, *Front. Neuroendocrinol.* (in press).

Leng, G., 1981, The effects of neural stalk stimulation upon firing patterns in rat supraoptic neurones, *Exp. Brain Res.* **41**: 135.

Leng, G., Brown, C. H., and Russell, J. A., 1999, Physiological pathways regulating the activity of magnocellular neurosecretory cells, *Prog. Neurobiol.* **57**: 625.

Lewis, C. E., Morris, J. F., Fink, G., and Johnson, M., 1986, Changes in the granule population of gonadotrophs of hypogonadal (hpg) and normal female mice associated with the priming effect of LH-releasing hormone in vitro, *J. Endocrinol.* **109**: 35.

Lincoln, D. W. and Wakerley, J. B., 1972, Accelerated discharge of paraventricular neurosecretory cells correlated with reflex release of oxytocin during suckling, *J. Physiol.* **111**: 327.

Lincoln, D. W. and Wakerley, J. B., 1974, Electrophysiological evidence for the activation of supraoptic neurones during the release of oxytocin, *J. Physiol.* **242**: 533.

Ludwig, M., 1998, Dendritic release of vasopressin and oxytocin, *J. Neuroendocrinol.* **10**: 881.

Ludwig, M., Callahan, M. F., Neumann, I., Landgraf, R., and Morris, M., 1994, Systemic osmotic stimulation increases vasopressin and oxytocin release within the supraoptic nucleus, *J. Neuroendocrinol.* **6**: 369.

Ludwig, M., and Leng, G., 1997, Autoinhibition of supraoptic nucleus vasopressin neurons in vivo - a combined retrodialysis/electrophysiological study in rats, *Eur. J. Neurosci.* **9**: 2532.

Ludwig, M., and Pittman, Q. J., 2003, Talking back: dendritic neurotransmitter release, *Trends Neurosci.* **26**: 255.

Ludwig, M., Sabatier, N., Bull, P. M., Baxter, S. L., Landgraf, R., Dayanithi, G., and Leng, G., 2002a, Mechanisms of activity-dependent vasopressin and oxytcoin secretion from dendrites, *The 5th International Congress of Neuroendocrinology, Bristol, UK* P260.

Ludwig, M., Sabatier, N., Bull, P. M., Landgraf, R., Dayanithi, G., and Leng, G., 2002b, Intracellular calcium stores regulate activity-dependent neuropeptide release from dendrites, *Nature* **418**: 85.

Lundberg, J. M., and Hokfelt, T., 1985, Coexistence of Peptide and Classical Transmitters, in: *Neurotransmitter in Action*, D. Bousfield, ed., Elsevier, Amsterdam, pp. 104-118.

Ma, D., and Morris, J. F., 2002, Protein synthetic machinery in the dendrites of the magnocellular neurosecretory neurons of wild-type Long-Evans and homozygous Brattleboro rats, *J. Chem. Neuroanat.* **23**: 171.

Meeker, R. B., Swanson, D. J., Greenwood, R. S., and Hayward, J. N., 1991, Ultrastructural distribution of glutamate immunoreactivity within neurosecretory endings and pituicytes of the rat neurohypophysis, *Brain Res.* **564**: 181.

Moos, F., Poulain, D. A., Rodriguez, F., Guerne, Y., Vincent, J. D., and Richard, P., 1989, Release of oxytocin within the supraoptic nucleus during the milk ejection reflex in rats, *Exp. Brain Res.* **76**: 593.

Morris, J. F., and Pow, D. V., 1991, Widespread release of petides in the central nervous system: quantitation of tannic acid-captured exocytosis, *Anat. Record* **231**: 437.

Pickering, A. J., and Fink, G., 1979, Priming effect of luteinizing hormone releasing factor in vitro: role of protein synthesis, contractile elements, Ca^{2+} and cyclic AMP, *J. Endocrinol.* **81**: 223.

Pow, D. V., and Morris, J. F., 1989, Dendrites of hypothalamic magnocellular neurons release neurohypophysial peptides by exocytosis, *Neuroscience* **32**: 435.

Proescholdt, M. G., Hutto, B., Brady, L. S., and Herkenham, M., 2000, Studies of cerebrospinal fluid flow and penetration into brain following lateral ventricle and cisterna magna injections of the tracer [C-14]inulin in rat, *Neuroscience* **95**: 577.

Reijmers, L. G., Van Ree, J. M., Spruijt, B. M., Burbach, J. P., and De Wied, D., 1998, Vasopressin metabolites: a link between vasopressin and memory?, *Prog. Brain Res.* **119**: 523.

Russell, J. A., Leng, G., and Douglas, A. J., 2003, The magnocellular oxytocin system, the fount of maternity: adaptations in pregnancy, *Front. Neuroendocrinol.* **24**: 27.

Sabatier, N., Caquineau, C., Dayanithi, G., Bull, P., Douglas, A. J., Guan, X. M. M., Jiang, M., Van der, P. L., and Leng, G., 2003, alpha-melanocyte-stimulating hormone stimulates oxytocin release from the dendrites of hypothalamic neurons while inhibiting oxytocin release from their terminals in the neurohypophysis, *J. Neurosci.* **23**: 10351.

Sabatier, N., Richard, P., and Dayanithi, G., 1997, L-, N- and T- but neither P- nor Q-type Ca^{2+} channels control vasopressin-induced Ca^{2+} influx in magnocellular vasopressin neurones isolated from the rat supraoptic nucleus, *J. Physiol.* **503**: 253.

Sabatier, N., Richard, P., and Dayanithi, G., 1998, Activation of multiple intracellular transduction signals by vasopressin in vasopressin-sensitive neurones of the rat supraoptic nucleus, *J. Physiol.* **513**: 699.

Scullion, S., Brown, D., and Leng, G., 2004, Modelling the pituitary response to luteinizing hormone-releasing hormone. *J. Neuroendocrinol.* **16**: 265.

Thastrup, O., Cullen, P. J., Drobak, B. K., Hanley, M. R., and Dawson, A. P., 1990, Thapsigargin, a tumor promoter, discharges intracellular Ca^{2+} stores by specific inhibition of the endoplasmic reticulum Ca^{2+}-ATPase, *PNAS USA* **87**: 2466.

Theodosis, D. T., 2002, Oxytocin-secreting neurons: A physiological model of morphological neuronal and glial plasticity in the adult hypothalamus, *Front. Neuroendocrinol.* **23**:101.

Tobin, V. A., Hurst, G., Norrie, L., Dal Rio, F., Bull, P. M., and Ludwig, M., 2004, Thapsigargin-induced mobilisation of dendritic dense-cored vesicles in rat supraoptic neurones, *Eur. J. Neurosci.***19**: 2909.

Tobin, V. A., Moos, F. C., and Desarmenien, M. G., 2002, ICRAC in magnocellular neurones from adult male rats, *The 5th International Congress of Neuroendocrinology, Bristol, UK* P262.

Waring, D. W., and Turgeon, J. L., 1983, LHRH self priming of gonadotrophin secretion: time course of development, *Am. J. Physiol.* **244**: C410.

Wotjak, C. T., Ganster, J., Kohl, G., Holsboer, F., Landgraf, R., and Engelmann, M., 1998, Dissociated central and peripheral release of vasopressin, but not oxytocin, in response to repeated swim stress: new insights into the secretory capacities of peptidergic neurons, *Neuroscience* **85**: 1209.

Wotjak, C. T., Ludwig, M., and Landgraf, R., 1994, Vasopressin facilitates its own release within the rat supraoptic nucleus in vivo, *Neuroreport* **5**: 1181.

15

AUTOCRINE MODULATION OF EXCITABILITY BY DENDRITIC PEPTIDE RELEASE FROM MAGNOCELLULAR NEUROSECRETORY CELLS

Colin H. Brown[*]

1. INTRODUCTION

Neuropeptides are expressed by many neurones in the central nervous system, virtually always co-existing with other transmitters (Hokfelt et al., 2000). While much is known about the effects of exogenously applied neuropeptides, the physiological roles of endogenous neuropeptides released as co-transmitters remain largely unknown. One class of neuropeptide for which endogenous functions have been extensively studied are the opioid peptides. Secretion of the endogenous κ-opioid peptide, dynorphin, by hippocampal granule cells of the dentate gyrus has been shown to block long term potentiation (Chavkin, 2000; Drake and Chavkin, this volume). Granule cells form powerful excitatory glutamatergic synapses on the proximal dendrites of CA3 pyramidal cells via their mossy fibre terminals and dynorphin acts as an autocrine inhibitor of glutamate (and dynorphin) release from these terminals. In addition, granule cell dendrites also release dynorphin following perforant path activation and this dendritic dynorphin acts as a retrograde inhibitor of excitatory amino acid release from perforant path terminals (Chavkin, 2000). Dendritic dynorphin release has now also been implicated in activity patterning in hypothalamic magnocellular neurosecretory cells (MNCs) that release vasopressin into the circulation from the posterior pituitary gland (neurohypophysis). While vasopressin released with dynorphin from dendrites appears to act in a paracrine manner to inhibit glutamate release from excitatory synaptic inputs, co-released dynorphin exerts its action in an autocrine rather than paracrine manner via modulation of intrinsic membrane properties to induce activity-dependent termination of action potential firing and thus generate the rhythmic 'phasic' activity characteristic of these neurones. The mechanisms of dynorphin feedback inhibition involve inactivation of plateau potentials that sustain action potential firing by maintaining membrane potential close to, or above, threshold. Thus, dendritic dynorphin secretion serves as an autocrine

[*]School of Biomedical and Clinical Laboratory Sciences, University of Edinburgh, EDINBURGH, EH8 9XD, UK

Dendritic Neurotransmitter Release, edited by M. Ludwig
Springer Science+Business Media, Inc., 2005

regulator of activity patterning in vasopressin cells. The characterisation of this autocrine modulation of activity patterning in vasopressin cells has recently been investigated using both *in vivo* and *in vitro* electrophysiological techniques and is described in detail below.

In addition to κ-opioid interactions with MNCs, chronic intracerebroventricular (i.c.v.) administration of morphine induces tolerance (desensitisation) to, and dependence upon, this μ-opioid alkaloid in oxytocin MNCs; dependence is revealed by hyper-excitation of oxytocin MNCs after withdrawal of morphine. Oxytocin MNCs also release neuropeptides from their dendrites to modulate their activity. By contrast to vasopressin, oxytocin released from dendrites causes autoexcitation that is involved in the organisation of the response of the oxytocin system to physiological stimuli. As described below, oxytocin released from dendrites also interacts with opioid influences on oxytocin MNCs, contributing to the increased firing rate of oxytocin MNCs evident during the excitation of these neurones upon withdrawal of morphine after chronic administration of this μ-opioid receptor agonist over five days.

2. PHASIC PATTERNING IN VASOPRESSIN MNCS

Vasopressin (antidiuretic hormone) is synthesised by MNCs in the hypothalamic supraoptic nucleus (SON) and paraventricular nucleus (PVN) and is secreted into the circulation to maintain plasma osmolality and blood pressure by promoting antidiuresis and vasoconstriction (Holmes et al., 2001). Since MNC terminals cannot sustain intrinsic repetitive firing (Bourque, 1990), secretion is primarily determined by action potential (spike) activity that is entirely dependent upon synaptic input. While activity *per se* depends upon synaptic inputs (Brown et al., 2004a; Nissen et al., 1994; Nissen et al., 1995) that are essentially randomly patterned, vasopressin MNCs fire spikes in a highly organised rhythmic 'phasic' pattern, alternating between periods of silence and activity that each typically last for tens of seconds (Brimble and Dyball, 1977; Brown et al., 1998a; Harris et al., 1975; Wakerly et al., 1975) (Fig. 1). By contrast to vasopressin MNCs, oxytocin MNCs (that are found alongside vasopressin MNCs in the SON and PVN) fire spikes randomly under basal conditions, presumably reflecting the activity of their afferent inputs. Phasic patterning is principally generated by the intrinsic membrane properties of vasopressin MNCs to potentiate vasopressin release into the circulation (Bourque, 1990; Dutton and Dyball, 1979; Jackson et al., 1991). However, bursts are not co-ordinated between MNCs (Leng and Dyball, 1983) and so vasopressin release into the circulation is continuous rather than pulsatile.

Excitatory amino acid receptor antagonists block spontaneous and evoked bursts in vasopressin neurones recorded from anaesthetised rats (Brown et al., 2004a; Nissen et al., 1994; Nissen et al., 1995), indicating that summation of excitatory postsynaptic potentials (EPSPs) is an essential component of burst initiation. In intracellular recordings *in vitro*, each spike in a vasopressin MNC is followed by a prominent non-synaptic post-spike depolarising after-potential (DAP) that lasts 1 – 3 seconds (Andrew and Dudek, 1983) (Fig. 2). Temporal summation of DAPs that follow consecutive spikes generates a sustained plateau potential that maintains firing during bursts (Andrew and Dudek, 1983; Ghamari-Langroudi and Bourque, 1998). DAPs are attenuated when evoked immediately following the end of a spontaneous burst (Andrew and Dudek, 1983), suggesting that plateau potentials are, at least partially, inactivated at burst termination.

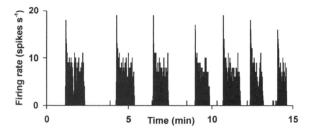

Figure 1. Phasic activity in MNCs. Ratemeter record of an extracellular single-unit recording of spontaneous activity (averaged in 1 s bins) of a vasopressin MNC in the SON of a urethane-anaesthetised rat, showing characteristic periods of activity (bursts with spike frequency adaptation, typical of phasic MNCs) and silence each lasting several tens of seconds.

DAPs and plateau potentials can also be evoked by triggering brief trains of action potentials with depolarising current pulses. Similarly to DAPs evoked soon after a spontaneous burst, DAPs evoked soon after a preceding evoked DAP are also attenuated. The degree of this attenuation is increased by increasing the number of spikes elicited before the evoked DAP (Brown and Bourque, 2004) (Fig. 2B). Thus DAPs, and presumably the plateau potentials that sustain burst firing, undergo pronounced activity-dependent inhibition that reduces the probability of spontaneous spikes firing as bursts progress, and so terminates burst firing. Thus, bursts are both sustained *and* terminated by activity-dependent modulation of DAPs (and associated plateau potentials) in vasopressin MNCs. It has now become apparent that the activity-dependent inhibition of DAPs responsible for burst termination is generated by feedback effects of dendritically released dynorphin.

Activity-dependent inhibition of DAPs can also be tracked indirectly *in vivo*. Since DAPs effectively sustain firing by increasing the probability of a spike firing soon after each spike (over the time-course of the associated DAP), plotting the probability of a spike firing at any time following the preceding spike will reflect the influence of the DAP on spontaneous activity *in vivo* (Leng et al., 1995). The probability of spike firing (hazard) is calculated from the inter-spike interval histogram of individual cells. Construction of inter-spike interval histograms for the periods 0 – 15 s, 16 – 30 s and 31 – 45 s after the start of each burst shows that progressively fewer spikes fire towards the end of bursts relative to the onset of bursts (Fig. 3A), reflecting the spike frequency adaptation that occurs during bursts. However, there are differences not only in the overall *number* of spikes that occur earlier and later in bursts, but also in the *timing* of spikes relative to each other at these times. The differences in spike timing are more clearly exposed by the plots of hazard functions, which reveal that neuronal excitability soon after each spike is greater earlier in bursts than later in bursts (Brown et al., 2004b) (Fig. 3B), presumably reflecting the presence of DAPs after each spike that are progressively inhibited over the course of bursts.

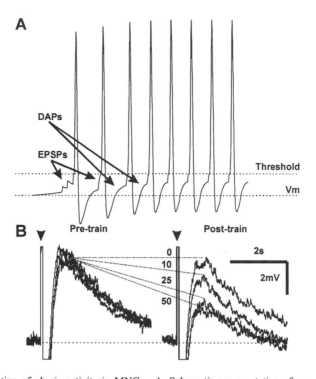

Figure 2. Generation of phasic activity in MNCs. A, Schematic representation of spontaneous burst onset. Summation of EPSPs induces spike firing. After each spike, a long lasting non-synaptic, post-spike DAP is activated that brings the membrane potential (Vm) closer to threshold increasing the probability that on-going EPSPs will again reach threshold. If subsequent spikes fire close enough together, DAPs summate to form a plateau potential that sustains firing throughout the burst. B, DAPs (averages of 5) that each follows 5 spikes (arrowheads; spikes truncated) evoked by a 80 ms depolarising pulse (+150 pA) 4 s before (pre-train) and 4 s after (post-train) a conditioning train of 0, 10, 25 or 50 spikes delivered over 1 s (evoked by a corresponding number of 5 ms, +500 pA dc-current pulses every 60 s). The amplitude of pre-train DAPs was consistent throughout the recording, while there was an activity-dependent reduction in amplitude of the post-train DAPs. It is hypothesised that the inhibition of DAPs over the course of spontaneous phasic bursts reduces the probability of membrane potential reaching threshold and so terminates bursts firing.

3. DIFFERENTIAL EFFECTS OF PEPTIDES CO-RELEASED FROM DENDRITES ON PHASIC ACTIVITY

MNCs possess between one and three dendrites that also release neuropeptides by exocytosis (Pow and Morris, 1989) and it has become clear that the secretion of

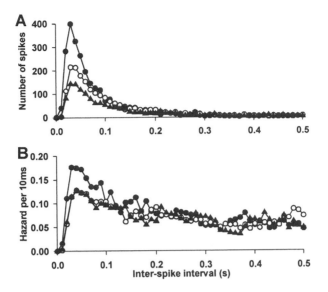

Figure 3. Activity-dependent reduction in post-spike excitability of vasopressin cells during bursts. A. Inter-spike intervals in the periods 0 – 15 s (filled circles), 16 – 30 s (open circles) and 31 – 45 s (filled triangles) after the onset of bursts (n = 17) in a phasic vasopressin MNC under control conditions. B. The hazard function generated from the inter-spike intervals of the cell in A during the periods 0 – 15 s (filled circles), 16 – 30 s (open circles) and 31 – 45 s (filled triangles) after the onset of bursts, calculated using the formula: hazard = n_i / (n_t - n_{i-1}), (where n_i is the number of events per interval, n_{i-1} is the total number of events preceding the current interval and n_t is the total number of events in all intervals) and gives the inferred probability (as a decimal) of a cell firing a subsequent spike in any interval after a spike (at time 0), given that another spike has not occurred earlier. Note that the probability of spike firing varies as a function of time following the preceding spike; this is inferred to reflect the influence of post-spike potentials on the excitability of vasopressin cells; as bursts progress the early increase in hazard (< 0.1 s) decreases, presumably reflecting a reduced influence of the DAP.

neuropeptides, including dynorphin, from these dendrites is of critical importance in the generation of phasic activity.

Vasopressin MNCs synthesise a number of peptides in addition to vasopressin. Principal among these is the κ-opioid peptide, dynorphin that is expressed by vasopressin cells at the highest levels found in the brain (Molineaux et al., 1982; Morsette et al., 1998; Watson et al., 1982). The SON contains a high density of κ-opioid receptors (Sumner et al., 1990) that are localised within the same neurosecretory vesicles that contain vasopressin (Shuster et al., 1999), vasopressin receptors (Hurbin et al., 2002) and dynorphin (Shuster et al., 2000). Thus, dendritic neurosecretion will generate high local concentrations of vasopressin and dynorphin over the cell membrane containing newly-exposed vasopressin and κ-opioid receptors and these may be sufficient to mediate feedback inhibition of vasopressin cell activity over the time-course of bursts.

Both the vasopressin V_1-receptor antagonist, OPC 21268, and the κ-opioid receptor antagonist, *nor*-binaltorphimine (BNI) increase burst duration and intra-burst firing rate in

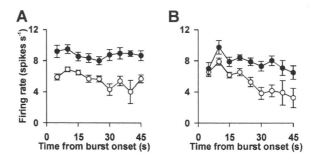

Figure 4. Differential effects of endogenous peptides on phasic bursts. A, The mean intra-burst firing rate analysed in 5 s bins (± SEM) of a phasic MNC recorded before (open circles; n = 9 bursts) and during (filled circles; n = 7 bursts) administration of the V_1 receptor antagonist, OPC 21268 (1 µg µl^{-1} intra-SON through a dialysis probe at 2 µl min^{-1}). OPC 21268 increased the firing rate throughout bursts. B, The mean intra-burst firing rate of a phasic MNC recorded before (open circles; n = 8 bursts) and during (filled circles; n = 11 bursts) administration of the κ-opioid receptor antagonist, BNI (0.2 µg µl^{-1} intra-SON through a dialysis probe at 2 µl min^{-1}). Unlike OPC 21268, the BNI-induced increase in firing rate emerged as the bursts progressed.

phasic MNCs recorded from anaesthetised rats, indicating that both endogenous peptides restrain the activity of phasic MNCs under basal conditions (Brown et al., 1998a; Ludwig and Leng, 1997).

However, there are marked differences between the pattern of this restraint between the two peptides; the increase in firing rate induced by V_1-receptor antagonist administration is evident throughout each burst whereas that induced by κ-opioid receptor antagonist administration is absent at the onset of each burst and emerges as the burst progresses (Brown et al., 2004b) (Fig. 4), indicating that the inhibitory influence of endogenous vasopressin is tonically present, while there must be a progressive activation of κ-opioid receptor mechanisms during bursts.

The temporal dissociation of the feedback inhibition by vasopressin and dynorphin (released from the same vesicles) may have one (or more) of several explanations. The simplest explanation for the observed effects of dynorphin is that dendritic release itself is activity-dependent. But if so, why are the effects of vasopressin itself apparently not activity-dependent? While both vasopressin (Burbach et al., 1993) and dynorphin (Hiranuma et al., 1998) are broken down by extracellular peptidases, vasopressin is more abundant than dynorphin in MNC neurosecretory vesicles and so there are measurable quantities of vasopressin in the extacellular fluid of the SON under basal conditions (Ludwig, 1998). Thus, from a lower starting concentration, co-secreted dynorphin may be cleared from the local environment between each burst while extracellular vasopressin is not completely cleared but rather is 'topped-up' during each burst of spikes.

However, the differential time-course of the feedback effects of the two peptides on phasic activity might arise from the different target of each, since V_1 receptor activation decreases evoked excitatory postsynaptic currents (EPSCs) in vasopressin MNCs (Hirasawa et al., 2003; Kombian et al., 2000) while κ-opioid receptor activation inhibits DAPs (Brown et al., 1999). Such effects identified *in vitro* would be expected to induce different changes in activity patterning. Since synaptic input activity is randomly patterned, a reduction in the strength of synaptic inputs would be expected to induce a general reduction in excitability at any time (such as is seen for vasopressin) while effects mediated by modulation of activity-dependent DAPs would be expected to also be activity-dependent (as is seen for dynorphin).

The differences between the effects of endogenous vasopressin and dynorphin *in vivo* can be further elucidated by analysis of the probability of spike firing. Since κ-opioid receptor activation reduces DAP amplitude (Brown et al., 1999) and DAPs sustain bursts by bringing the membrane potential closer to threshold to transiently increase the probability of spike firing, the effects of endogenous dynorphin release should be reflected in a reduction of post-spike excitability. Generation of the mean hazard function from the inter-spike interval histograms of phasic cells before and during application of the κ-opioid receptor antagonist, BNI, reveals an increase in post-spike excitability during BNI infusion that emerges for shorter inter-spike intervals over the course of bursts (Brown et al., 2004b), indicating that κ-opioid receptor activation by an endogenously-released agonist does indeed specifically reduce post-spike excitability in an activity dependent manner.

Vasopressin, on the other hand, reduces EPSC amplitude (Hirasawa et al., 2003; Kombian et al., 2000) which, since EPSCs are randomly-patterned, should cause a general reduction in excitability. Furthermore, since the effects of endogenous vasopressin on firing rate are of the same magnitude throughout bursts, the effects of vasopressin on excitability should also be the same throughout bursts. Plotting the mean hazard function for the inter-spike interval distributions of phasic cells before and during V_1 receptor antagonist administration reveals a similar increase in neuronal excitability throughout bursts (Brown et al., 2004b), consistent with a tonic inhibitory effect of endogenous vasopressin under basal conditions.

Thus, endogenous activation of both V_1 and κ-opioid receptors decreases the activity of vasopressin cells during burst firing. The actions of endogenous vasopressin appear to be generally inhibitory and tonically present. Unlike vasopressin itself, co-released dynorphin seems to specifically reduce post-spike excitability, and this inhibition is progressively more apparent towards the end of phasic bursts and so probably contributes to burst termination in phasic cells.

4. FEEDBACK MODULATION OF PHASIC ACTIVITY BY DENDRITIC DYNORPHIN RELEASE

Administration of the arylacetamide κ-opioid receptor agonist,U50, 488H, inhibits phasic MNCs *in vivo* and this inhibition is blocked by antagonist administration directly into the SON by microdialysis (Brown et al., 1998a; Ludwig et al., 1997), indicating that the inhibition is due to a local activation of κ-opioid receptors within the SON. The critical involvement of endogenous κ-opioid agonists in the generation of phasic activity is shown by the responses of MNCs to intra-SON infusion of U50,488H, over 5 days. This treatment desensitises SON κ-opioid receptors to acute inhibition by U50,488H (and presumably endogenous dynorphin) since acute administration of U50,488H at doses that consistently silence MNCs in untreated rats does not affect the residual activity that persists in the face of chronic U50,488H administration. This chronic administration of

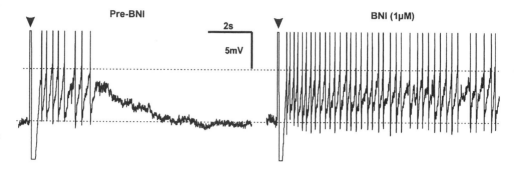

Figure 5. Endogenous κ-opioid peptides prolong phasic bursts by inhibition of plateau potentials. Intracellular recordings of suprathreshold plateau potentials (evoked by a 80 ms spike train: arrowheads) with superimposed spontaneous spikes (truncated) recorded before (Pre-BNI) and during BNI superfusion in a hypothalamic explant. The lower dashed line is aligned with the inter-burst membrane potential and the upper dashed line with spike threshold in the two recordings. Note that the plateau potentials before and during BNI superfusion are of the same amplitude but that in the presence of the κ-opioid receptor antagonist, the plateau potential no longer spontaneously terminated, indicating that an endogenous κ-opioid receptor agonist reduces burst duration in an activity-dependent manner.

U50,488H eliminates spontaneous phasic activity in vasopressin cells *in vivo* even when activity is increased by intense osmotic stimulation (Brown et al., 1998a) and reduces post-spike excitability in MNCs (Brown and Leng, 2000), suggesting that the elimination of phasic activity may be associated with a reduction of DAP amplitude. Similarly, continuous infusion of U50,488H i.c.v. for up to 12 h silences phasic MNCs within ~15 min but these MNCs recover activity (but not *phasic* activity) after several hours of U50,488H administration; this post-recovery activity is also associated with a reduced post-spike excitability as revealed by the generation of the hazard function (Brown and Leng, 2000). Thus, an essential component of the expression of phasic activity by MNCs is activation of SON κ-opioid receptor mechanisms by an endogenous agonist to induce an activity-dependent inhibition of post-spike excitability.

While κ-opioid receptor activation also affects K^+ currents in MNCs (Muller et al., 1999) as well as their synaptic inputs (Inenaga et al., 1994), the modulation of *phasic* activity is probably mediated through inhibition of plateau potentials since U50,488H inhibits DAPs *in vitro* (Brown et al., 1999) and reduces post-spike excitability *in vivo* (Brown and Leng, 2000). More importantly, as outlined above, activation of SON κ-opioid receptors by endogenous dynorphin also induces an activity-dependent inhibition of post-spike excitability *in vivo* (Brown et al., 2004b) and DAP amplitude *in vitro* as well as reducing burst duration *in vivo* (Brown et al., 1998a) and *in vitro* (Brown and Bourque, 2004) (Fig. 5). This κ-opioid inhibition of DAPs (and plateau potentials) requires local transmitter release since depletion of SON neurosecretory vesicles, to remove the source of endogenous dynorphin (by superfusion of the black widow spider venom, α-latrotoxin, Ushkaryov, 2002), reduces activity-dependent DAP inhibition. This autocrine inhibition of DAP amplitude is specific to dynorphin rather than vasopressin since vasopressin receptor antagonists do not alter activity-dependent inhibition of DAPs (Brown and Bourque, 2004).

The mechanisms underlying κ-opioid modulation of phasic activity are not known but almost certainly involve actions on the DAP. DAPs are dependent on both external Ca^{2+} (Andrew, 1987; Bourque, 1986; Li and Hatton, 1997a) and Ca^{2+} release from internal stores (Li and Hatton, 1997a). Calbindin reduces DAP amplitude and converts phasic activity to continuous firing, while immunoneutralisation of calbindin has the opposite effect on firing pattern (Li et al., 1995). The current that underlies the DAP is not known, with evidence in support of a Ca^{2+} induced reduction of a persistent outward K^+ current (Li and Hatton, 1997b) or activation of a Ca^{2+} activated non-selective cation channel (Ghamari-Langroudi and Bourque, 2002). Since κ-opioid receptor activation can activate K^+ currents or inhibit Ca^{2+} currents, as well as exert its actions through intracellular mechanisms, it is possible that endogenous dynorphin could affect DAPs at any of these levels.

What is clear is that the activity-dependent inhibition of DAPs by endogenous dynorphin probably reflects an action on the DAP itself. MNCs also display a prominent activity dependent after-hyperpolarisation (AHP) (Kirkpatrick and Bourque, 1996) that overlaps temporally with evoked DAPs (Ghamari-Langroudi and Bourque, 1998). The AHP has two components. The major (fast) component of the AHP in MNCs is generated by Ca^{2+} dependent K^+ channels of small single channel conductance (SK channel) that is blocked by the bee venom polypeptide, apamin (Bourque and Brown, 1987). The minor (slow) component of the AHP is also generated by Ca^{2+} dependent K^+ channels, but of intermediate single channel conductance (IK channel) that is blocked by the scorpion venom, charybdotoxin (Greffrath et al., 1998). Neither component of the AHP is affected by κ-opioid receptor agonists or antagonists (Brown and Bourque, 2004), even during blockade of DAPs with 3 mM CsCl (Ghamari-Langroudi and Bourque, 1998). Thus, activity-dependent κ-opioid inhibition of DAPs is unlikely to result from a reduction of the amplitude of the coincident AHP. MNCs also display a prominent transient hyperpolarising after-potential (HAP) after individual spikes, that again is a Ca^{2+} dependent K^+ conductance (Bourque et al., 1985). Similarly to post-train AHPs, post-spike HAPs are not affected by κ-opioid receptor antagonism (Brown & Bourque, unpublished observations) suggesting that the Ca^{2+} dependent K^+ conductances underlying HAPs and AHPs are not involved in activity dependent κ-opioid inhibition of DAPs.

MNCs display frequency-dependent spike broadening that is attributed to intracellular Ca^{2+} accumulation during spike trains (Renaud and Bourque, 1991). While exogenous dynorphin decreases the Ca^{2+} component of MNC action potentials that have been pharmacologically prolonged (Inenaga et al., 1994), spike broadening is not altered in recordings that demonstrate activity-dependent DAP inhibition (Brown & Bourque, unpublished observations). Thus, the Ca^{2+} conductances responsible for spike broadening probably do not mediate autocrine dynorphin feedback inhibition of DAPs in MNCs.

5. AUTOCRINE MODULATION OF MORPHINE WITHDRAWAL EXCITATION IN OXYTOCIN NEURONES

Like vasopressin MNCs, oxytocin MNCs are also inhibited by the κ-opioid receptor

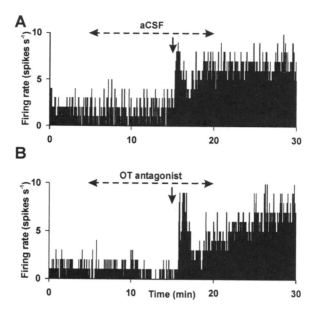

Figure 6. Endogenous oxytocin augments morphine withdrawal excitation of oxytocin MNCs. Ratemeter records of extracellular single-unit recordings of spontaneous activity (averaged in 1 s bins) of oxytocin MNCs in the SON of urethane-anaesthetised morphine-dependent rats, showing characteristic slow continuous activity typical of oxytocin MNCs. Morphine dependence was revealed by i.v. injection of the opioid receptor antagonist, naloxone at 15 min (arrows) which caused a prompt increase in the firing rate of both MNCs. The naloxone-induced increase in firing rate was more marked in the MNC administered i.c.v. (A) aCSF than (B) oxytocin (OT) antagonist (10 μg, followed by infusion of 1.25 μg min^{-1} for 15 min), indicating that endogenous oxytocin contributes to the increased firing rate of oxytocin MNCs during naloxone-precipitated morphine withdrawal excitation.

agonist, U50,488H, and chronic intra-SON administration of U50,488H desensitises oxytocin MNCs to inhibition by acute U50,488H administration (Brown et al., 1998a). However, it is not the interaction of κ-opioid peptides with oxytocin MNCs that is remarkable, rather it is the effects of μ-opioid drugs on these neurones that is of interest since chronic i.c.v. administration of morphine induces both desensitisation to the inhibitory effects of morphine (tolerance) but also induces dependence upon morphine (Bicknell et al., 1988). Morphine dependence involves changes in oxytocin MNCs such that they require the continued presence of morphine to function apparently normally, and is revealed by a rebound hyper-excitation after withdrawal of morphine. Morphine withdrawal can be acutely precipitated by administration of the broad spectrum opioid receptor antagonist, naloxone, which causes an approximate 100-fold increase in oxytocin release (Bicknell et al., 1988). This increased secretion occurs primarily as a result of increased oxytocin cell firing rate (Fig. 6A) that is potentiated by frequency-facilitation of release and naloxone antagonism at κ-opioid receptors in the neurohypophysis. In addition to increased electrical activity, oxytocin heteronuclear RNA and immediate early gene expression in the SON (and PVN) are also increased by morphine withdrawal excitation (Jhamandas et al., 1996; Johnstone et al., 2000).

While afferent inputs to oxytocin MNCs are themselves excited by morphine withdrawal excitation (Murphy et al., 1997), the mechanisms underlying morphine dependence in oxytocin MNCs appear to reside within the oxytocin MNCs themselves since withdrawal excitation of oxytocin MNCs can be induced by naloxone administration into the SON (Ludwig et al., 1997). Furthermore, intra-SON injection of pertussis toxin (Brown et al., 2000a) or Ca^{2+} channels blockers (Blackburn-Munro et al., 2000) attenuates withdrawal excitation in oxytocin MNCs. Also, the proportion of inputs that project to the SON and that are activated by withdrawal is small in comparison to those activated by other less intense stimuli (Onaka et al., 1995) and the sensitivity of oxytocin MNCs to afferent inputs is unchanged in morphine dependence and withdrawal excitation (Brown et al., 1996). Finally, while acute pharmacological interruption of synaptic inputs to oxytocin MNCs attenuates withdrawal excitation (Brown et al., 1998b), lesion of central projections to oxytocin MNCs does not block withdrawal excitation in these cells (Brown et al., 1998b; Russell et al., 1992b). Thus, there is a requirement for active excitatory synaptic inputs to dependent cells for the expression of withdrawal excitation as an increase in firing rate. Nevertheless, the fundamental mechanisms of withdrawal excitation reside in the oxytocin MNCs themselves (Brown et al., 2000b).

By contrast to oxytocin MNCs, there is no systematic change in the firing *rate* of vasopressin MNCs during withdrawal. However, the firing *pattern* within bursts indicates increased excitability (Bicknell et al., 1988) that may reflect a change in DAPs. Indeed, oxytocin MNCs show an increased probability of spike firing 30 – 150 ms after the previous spike during withdrawal excitation (Brown et al., 2000b), which could reflect an increase in DAP amplitude (or decrease in AHP amplitude) in oxytocin cells that contributes to withdrawal excitation.

Dendritic release of oxytocin is critically involved in the physiological functioning of oxytocin MNCs. Under basal conditions, oxytocin MNCs exhibit slow continuous activity, but during lactation (and parturition) these neurones fire in high frequency bursts (of up to 100 Hz) that typically last only 1 – 2 s. These bursts are co-ordinated across the population of oxytocin MNCs and result in milk let-down during suckling. The mechanisms generating these milk-ejection bursts are reviewed in detail elsewhere (Leng et al., 1999; Richard et al., 1997) and involve oxytocin release within the SON since this is increased immediately prior to each milk ejection burst (Moos et al., 1989) and administration of an oxytocin receptor antagonist into the SON inhibits milk ejection bursts (Lambert et al., 1993).

Dendritic oxytocin, but not vasopressin, release is doubled during morphine withdrawal excitation (Russell et al., 1992a) and i.c.v. administration of an oxytocin receptor antagonist reduces the withdrawal-induced increase in firing rate of oxytocin MNCs (Brown et al., 1997) (Fig. 6B), indicating that *endogenous* oxytocin contributes to morphine withdrawal excitation of oxytocin MNCs.

6. CONCLUSIONS

Dendritic neuropeptide release is an important modulator of activity in MNCs. In addition to physiological regulation of oxytocin secretion, autocrine actions of dendritic oxytocin

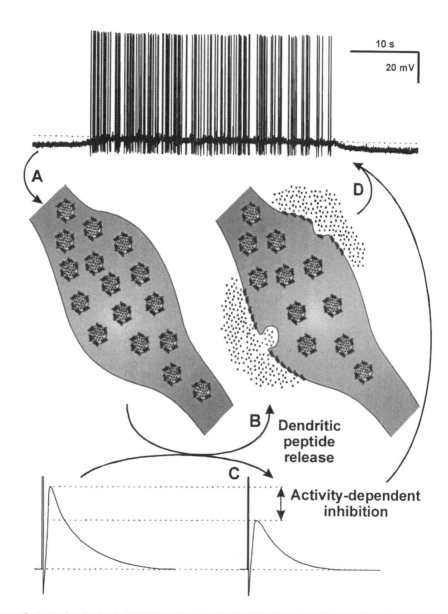

Figure 7. Autocrine feedback inhibition of activity by dendritically-released dynorphin. The top panel shows the membrane potential of a MNC displaying a spontaneous phasic burst superimposed upon an activity-dependent plateau potential. A, it is proposed that before activity starts, dynorphin effects are absent from the system. B, during action potential firing, there is activity-dependent dendritic release of dynorphin and insertion of κ-opioid receptors into the dendritic membrane. C, dendritically-released dynorphin causes inhibition of the plateau potential (illustrated by inactivation of evoked DAPs, see Fig. 2). D, Feedback inhibition of the plateau potential terminates firing; dynorphin is then cleared from the system, which is re-set for the subsequent cycle of activity.

release are involved in the generation of morphine withdrawal excitation of oxytocin MNCs. By contrast, dendritic vasopressin release appears to induce a general inhibition of vasopressin MNC activity while co-released dynorphin specifically modulates phasic activity patterning in vasopressin MNCs. The autocrine κ-opioid modulation of phasic activity is not static but develops as each burst progresses to cause activity-dependent inhibition of plateau potentials and thus terminates firing. During the silent period between bursts, dynorphin is effectively cleared from the local environment and the inactivation of plateau potentials recovers with a time constant of ~ 5 seconds (Brown and Bourque, 2004). At the onset of the subsequent burst, κ-opioid inhibition of firing is again absent but, as before, evolves as the burst progresses to establish a cycle of repetitive activity-dependent autocrine inhibition that generates phasic activity (Fig. 7).

7. ACKNOWLEDGEMENTS

This work was funded by the Wellcome Trust and the Medical Research Council. I am grateful to Profs. Charles W. Bourque, Gareth Leng and John A. Russell for their support in completing the above work.

8. REFERENCES

Andrew, R. D., 1987, Endogenous bursting by rat supraoptic neuroendocrine cells is calcium dependent, *J. Physiol.* **384**: 451.

Andrew, R. D., and Dudek, F. E., 1983, Burst discharge in mammalian neuroendocrine cells involves an intrinsic regenerative mechanism, *Science* **221**: 1050.

Bicknell, R. J., Leng, G., Lincoln, D. W., and Russell, J. A., 1988, Naloxone excites oxytocin neurones in the supraoptic nucleus of lactating rats after chronic morphine treatment, *J. Physiol.* **396**: 297.

Blackburn-Munro, G., Brown, C. H., Neumann, I. D., Landgraf, R., and Russell, J. A., 2000, Verapamil prevents withdrawal excitation of oxytocin neurones in morphine-dependent rats, *Neuropharmacology* **39**: 1596.

Bourque, C. W., 1986, Calcium-dependent spike after-current induces burst firing in magnocellular neurosecretory cells, *Neurosci. Lett.* **70**: 204.

Bourque, C. W., 1990, Intraterminal recordings from the rat neurohypophysis *in vitro*, *J. Physiol.* **421**: 247.

Bourque, C. W., and Brown, D. A., 1987, Apamin and d-tubocurarine block the afterhyperpolarization of rat supraoptic neurosecretory neurons, *Neurosci. Lett.* **82**: 185.

Bourque, C. W., Randle, J. C., and Renaud, L. P., 1985, Calcium-dependent potassium conductance in rat supraoptic nucleus neurosecretory neurons, *J. Neurophysiol.* **54**: 1375.

Brimble, M. J., and Dyball, R. E., 1977, Characterization of the responses of oxytocin- and vasopressin-secreting neurones in the supraoptic nucleus to osmotic stimulation, *J. Physiol.* **271**: 253.

Brown, C. H., and Bourque, C. W., 2004, Autocrine feedback inhibition of plateau potentials terminates phasic bursts in magnocellular neurosecretory cells of the rat supraoptic nucleus, *J. Physiol.* **557**: 949.

Brown, C. H., Bull, P. M., and Bourque, C. W., 2004a, Phasic bursts in rat magnocellular neurosecretory cells are not intrinsically regenerative *in vivo*, *Eur. J. Neurosci.* **19**: 2977.

Brown, C. H., Ghamari-Langroudi, M., Leng, G., and Bourque, C. W., 1999, Kappa-opioid receptor activation inhibits post-spike depolarizing after-potentials in rat supraoptic nucleus neurones *in vitro*, *J. Neuroendocrinol.* **11**: 825.

Brown, C. H., Johnstone, L. E., Murphy, N. P., Leng, G., and Russell, J. A., 2000a, Local injection of pertussis toxin attenuates morphine withdrawal excitation of rat supraoptic nucleus neurones, *Brain Res. Bull.* **52**: 115.

Brown, C. H., and Leng, G., 2000, *In vivo* modulation of post-spike excitability in vasopressin cells by kappa-opioid receptor activation, *J. Neuroendocrinol.* **12**: 711.

Brown, C. H., Ludwig, M., and Leng, G., 1998a, Kappa-opioid regulation of neuronal activity in the rat supraoptic nucleus *in vivo*, *J. Neurosci.* **18**: 9480.

Brown, C. H., Ludwig, M., and Leng, G., 2004b, Temporal dissociation of the feedback effects of dendritically co-released peptides on rhythmogenesis in vasopressin cells, *Neuroscience* **124**: 105.

Brown, C. H., Munro, G., Johnstone, L. E., Robson, A. C., Landgraf, R., and Russell, J. A., 1997, Oxytocin neurone autoexcitation during morphine withdrawal in anaesthetized rats, *Neuroreport* **8**: 951.

Brown, C. H., Munro, G., Murphy, N. P., Leng, G., and Russell, J. A., 1996, Activation of oxytocin neurones by systemic cholecystokinin is unchanged by morphine dependence or withdrawal excitation in the rat, *J. Physiol.* **496**: 787.

Brown, C. H., Murphy, N. P., Munro, G., Ludwig, M., Bull, P. M., Leng, G., and Russell, J. A., 1998b, Interruption of central noradrenergic pathways and morphine withdrawal excitation of oxytocin neurones in the rat, *J. Physiol.* **507**: 831.

Brown, C. H., Russell, J. A., and Leng, G., 2000b, Opioid modulation of magnocellular neurosecretory cell activity, *Neurosci. Res.* **36**: 97.

Burbach, J. P., De Bree, F. M., Terwel, D., Tan, A., Maskova, H. P., and Van Der Kleij, A. A., 1993, Properties of aminopeptidase activity involved in the conversion of vasopressin by rat brain membranes, *Peptides* **14**: 807.

Chavkin, C., 2000, Dynorphins are endogenous opioid peptides released from granule cells to act neurohumorly and inhibit excitatory neurotransmission in the hippocampus, *Prog. Brain Res.* **125**: 363.

Dutton, A., and Dyball, R. E., 1979, Phasic firing enhances vasopressin release from the rat neurohypophysis, *J. Physiol.* **290**: 433.

Ghamari-Langroudi, M., and Bourque, C. W., 1998, Caesium blocks depolarizing after-potentials and phasic firing in rat supraoptic neurones, *J. Physiol.* **510**: 165.

Ghamari-Langroudi, M., and Bourque, C. W., 2002, Flufenamic acid blocks depolarizing afterpotentials and phasic firing in rat supraoptic neurones, *J. Physiol.* **545**: 537.

Greffrath, W., Martin, E., Reuss, S., and Boehmer, G., 1998, Components of after-hyperpolarization in magnocellular neurones of the rat supraoptic nucleus *in vitro*, *J. Physiol.* **513**: 493.

Harris, M. C., Dreifuss, J. J., and Legros, J. J., 1975, Excitation of phasically firing supraoptic neurones during vasopressin release, *Nature* **258**: 80.

Hiranuma, T., Kitamura, K., Taniguchi, T., Kanai, M., Arai, Y., Iwao, K., and Oka, T., 1998, Protection against dynorphin-(1-8) hydrolysis in membrane preparations by the combination of amastatin, captopril and phosphoramidon, *J. Pharmacol. Exp. Ther.* **286**: 863.

Hirasawa, M., Mouginot, D., Kozoriz, M. G., Kombian, S. B., and Pittman, Q. J., 2003, Vasopressin differentially modulates non-NMDA receptors in vasopressin and oxytocin neurons in the supraoptic nucleus, *J. Neurosci.* **23**: 4270.

Hokfelt, T., Broberger, C., Xu, Z. Q., Sergeyev, V., Ubink, R., and Diez, M., 2000, Neuropeptides--an overview, *Neuropharmacology* **39**: 1337.

Holmes, C. L., Patel, B. M., Russell, J. A., and Walley, K. R., 2001, Physiology of vasopressin relevant to management of septic shock, *Chest* **120**: 989.

Hurbin, A., Orcel, H., Alonso, G., Moos, F., and Rabie, A., 2002, The vasopressin receptors colocalize with vasopressin in the magnocellular neurons of the rat supraoptic nucleus and are modulated by water balance, *Endocrinology* **143**: 456.

Inenaga, K., Nagatomo, T., Nakao, K., Yanaihara, N., and Yamashita, H., 1994, Kappa-selective agonists decrease postsynaptic potentials and calcium components of action potentials in the supraoptic nucleus of rat hypothalamus *in vitro*, *Neuroscience* **58**: 331.

Jackson, M. B., Konnerth, A., and Augustine, G. J., 1991, Action potential broadening and frequency-dependent facilitation of calcium signals in pituitary nerve terminals, *Proc. Natl. Acad. Sci. U. S. A.* **88**: 380.

Jhamandas, J. H., Harris, K. H., Petrov, T., and Jhamandas, K. H., 1996, Activation of nitric oxide-synthesizing neurones during precipitated morphine withdrawal, *Neuroreport* **7**: 2843.

Johnstone, L. E., Brown, C. H., Meeren, H. K., Vuijst, C. L., Brooks, P. J., Leng, G., and Russell, J. A., 2000, Local morphine withdrawal increases c-*fos* gene, Fos protein, and oxytocin gene expression in hypothalamic magnocellular neurosecretory cells, *J. Neurosci.* **20**: 1272.

Kirkpatrick, K. and Bourque, C. W., 1996, Activity dependence and functional role of the apamin-sensitive K^+ current in rat supraoptic neurones *in vitro*, *J. Physiol.* **494**: 389.

Kombian, S. B., Mouginot, D., Hirasawa, M., and Pittman, Q. J., 2000, Vasopressin preferentially depresses excitatory over inhibitory synaptic transmission in the rat supraoptic nucleus *in vitro*, *J. Neuroendocrinol.* **12**: 361.

Lambert, R. C., Moos, F. C., and Richard, P., 1993, Action of endogenous oxytocin within the paraventricular or supraoptic nuclei: a powerful link in the regulation of the bursting pattern of oxytocin neurons during the milk-ejection reflex in rats, *Neuroscience* **57**: 1027.

Leng, G., Brown, D., and Murphy, N. P., 1995, Patterning of electrical activity in magnocellular neurones, in: *Neurohypophysis: Recent Progress of Vasopressin and Oxytocin Research*, T. Saito, K. Kurokawa, and S. Yoshida, eds., Elsevier Science B.V., Amsterdam, pp. 225-235.

Leng, G., Brown, C. H., and Russell, J. A., 1999, Physiological pathways regulating the activity of magnocellular neurosecretory cells, *Prog. Neurobiol.* **57**: 625.

Leng, G. and Dyball, R. E., 1983, Intercommunication in the rat supraoptic nucleus, *Q. J. Exp. Physiol.* **68**: 493.

Li, Z., Decavel, C., and Hatton, G. I., 1995, Calbindin-D28k: role in determining intrinsically generated firing patterns in rat supraoptic neurones, *J. Physiol.* **488**: 601.

Li, Z., and Hatton, G. I., 1997a, Ca^{2+} release from internal stores: role in generating depolarizing after-potentials in rat supraoptic neurones, *J. Physiol.* **498**: 339.

Li, Z., and Hatton, G. I., 1997b, Reduced outward K^+ conductances generate depolarizing after-potentials in rat supraoptic nucleus neurones, *J. Physiol.* **505**: 95.

Ludwig, M., 1998, Dendritic release of vasopressin and oxytocin, *J. Neuroendocrinol.* **10**: 881.

Ludwig, M., Brown, C. H., Russell, J. A., and Leng, G., 1997, Local opioid inhibition and morphine dependence of supraoptic nucleus oxytocin neurones in the rat *in vivo*, *J. Physiol.* **505**: 145.

Ludwig, M., and Leng, G., 1997, Autoinhibition of supraoptic nucleus vasopressin neurons *in vivo*: a combined retrodialysis/electrophysiological study in rats, *Eur. J. Neurosci.* **9**: 2532.

Molineaux, C. J., Feuerstein, G., Faden, A. L., and Cox, B. M., 1982, Distribution of immunoreactive dynorphin in discrete brain nuclei; comparison with vasopressin, *Neurosci. Lett.* **33**: 179.

Moos, F., Poulain, D. A., Rodriguez, F., Guerne, Y., Vincent, J. D., and Richard, P., 1989, Release of oxytocin within the supraoptic nucleus during the milk ejection reflex in rats, *Exp. Brain Res.* **76**: 593.

Morsette, D. J., Swenson, K. L., Badre, S. E., and Sladek, C. D., 1998, Regulation of vasopressin release by ionotropic glutamate receptor agonists, *Adv. Exp. Med. Biol.* **449**: 129.

Muller, W., Hallermann, S., and Swandulla, D., 1999, Opioidergic modulation of voltage-activated K^+ currents in magnocellular neurons of the supraoptic nucleus in rat, *J. Neurophysiol.* **81**: 1617.

Murphy, N. P., Onaka, T., Brown, C. H., and Leng, G., 1997, The role of afferent inputs to supraoptic nucleus oxytocin neurons during naloxone-precipitated morphine withdrawal in the rat, *Neuroscience* **80**: 567.

Nissen, R., Hu, B., and Renaud, L. P., 1994, N-methyl-D-aspartate receptor antagonist ketamine selectively attenuates spontaneous phasic activity of supraoptic vasopressin neurons *in vivo*, *Neuroscience* **59**: 115.

Nissen, R., Hu, B., and Renaud, L. P., 1995, Regulation of spontaneous phasic firing of rat supraoptic vasopressin neurones *in vivo* by glutamate receptors, *J. Physiol.* **484**: 415.

Onaka, T., Luckman, S. M., Antonijevic, I., Palmer, J. R., and Leng, G., 1995, Involvement of the noradrenergic afferents from the nucleus tractus solitarii to the supraoptic nucleus in oxytocin release after peripheral cholecystokinin octapeptide in the rat, *Neuroscience* **66**: 403.

Pow, D. V., and Morris, J. F., 1989, Dendrites of hypothalamic magnocellular neurons release neurohypophysial peptides by exocytosis, *Neuroscience* **32**: 435.

Renaud, L. P., and Bourque, C. W., 1991, Neurophysiology and neuropharmacology of hypothalamic magnocellular neurons secreting vasopressin and oxytocin, *Prog. Neurobiol.* **36**: 131.

Richard, P., Moos, F., Dayanithi, G., Gouzenes, L., and Sabatier, N., 1997, Rhythmic activities of hypothalamic magnocellular neurons: autocontrol mechanisms, *Biol. Cell* **89**: 555.

Russell, J. A., Neumann, I., and Landgraf, R., 1992a, Oxytocin and vasopressin release in discrete brain areas after naloxone in morphine-tolerant and -dependent anesthetized rats: push-pull perfusion study, *J. Neurosci.* **12**: 1024.

Russell, J. A., Pumford, K. M., and Bicknell, R. J., 1992b, Contribution of the region anterior and ventral to the third ventricle to opiate withdrawal excitation of oxytocin secretion, *Neuroendocrinology* **55**: 183.

Shuster, S. J., Riedl, M., Li, X., Vulchanova, L., and Elde, R., 1999, Stimulus-dependent translocation of kappa opioid receptors to the plasma membrane, *J. Neurosci.* **19**: 2658.

Shuster, S. J., Riedl, M., Li, X., Vulchanova, L., and Elde, R., 2000, The kappa opioid receptor and dynorphin co-localize in vasopressin magnocellular neurosecretory neurons in guinea-pig hypothalamus, *Neuroscience* **96**: 373.

Sumner, B. E., Coombes, J. E., Pumford, K. M., and Russell, J. A., 1990, Opioid receptor subtypes in the supraoptic nucleus and posterior pituitary gland of morphine-tolerant rats, *Neuroscience* **37**: 635.

Ushkaryov, Y., 2002, Alpha-latrotoxin: from structure to some functions, *Toxicon* **40**: 1.

Wakerly, J. B., Poulain, D. A., Dyball, R. E., and Cross, B. A., 1975, Activity of phasic neurosecretory cells during haemorrhage, *Nature* **258**: 82.

Watson, S. J., Akil, H., Fischli, W., Goldstein, A., Zimmerman, E., Nilaver, G., and Wimersma-Griedanus, T. B., 1982, Dynorphin and vasopressin: common localization in magnocellular neurons, *Science* **216**: 85.

16

GALANIN, A NEW CANDIDATE
FOR SOMATO-DENDRITIC RELEASE

Marc Landry, Zhi-Qing David Xu, André Calas, and Tomas Hökfelt[*]

1. INTRODUCTION

Neuropeptides are widely distributed in the nervous system and are assumed to play manifold functional roles, ranging from synaptic/non-synaptic signalling to trophic actions (Hökfelt et al., 2000). They are synthesized on the endoplasmic reticulum, packaged into secretory granules (SGs) in the Golgi complex and trafficked through the regulated secretory pathway to their sites of release (Kelly, 1985), whereby sorting and packaging is of importance for neuropeptide targeting. In general, as is the case for classic transmitters, release of neuropeptides is assumed to occur at nerve endings. However, there is early evidence that classic transmitters can be released from dendrites, for example dopamine in the substantia nigra zona reticulata (Björklund and Lindvall, 1975; Cheramy et al., 1981). With regard to neuropeptides, dendritic release has mainly been associated with systems where the peptide is the major secretory product, that is in the archetype of neuroendocrine systems, the hypothalamo-posthypophyseal magnocellular neurons (Pow and Morris, 1989; Ludwig, 1998). However, the somato-dendritic release of vasopressin and oxytocin appears to represent an accessory process, complementary to their secretion from the posthypophyseal nerve endings. This is why somato-dendritic release of neuropeptides had not been considered as a general secretion mechanism in neurons in the central nervous system (CNS). In addition, little is known about such 'atypical' secretion of those neuropeptides which are colocalized with the two major magnocellular hormones, for example galanin, whose exact physiological roles still remain to be defined.

Galanin is a 29 amino-acid long, C-terminally amidated neuropeptide first isolated from the porcine gut by Tatemoto et al. (1983). Immunohistochemical analysis has revealed a widespread distribution of galanin throughout the rat central and peripheral nervous system (Rökaeus et al., 1984; Ch'ng et al., 1985; Skofitsch and Jacobowitz, 1985a, b; 1986; Melander et al., 1986a). Galanin coexists with different neurotransmitters

[*] Marc Landry, INSERM E 358, Institut François Magendie, Université Bordeaux 2, Bordeaux, France.
Zhi-Qing David Xu and Tomas Hökfelt, Department of Neuroscience, Karolinska Institutet, Stockholm, Sweden. André Calas, UMR CNRS 7121, Université Paris 6, Paris, France.

Dendritic Neurotransmitter Release, edited by M. Ludwig
Springer Science+Business Media, Inc., 2005

and other neuropeptides, for instance with the vasopressin in hypothalamic magnocellular neurons (Melander et al., 1986b). Galanin has been shown to bind with high affinity to discrete brain regions (Skofitsch et al., 1986; Melander et al., 1988) and exerts its physiological roles through 7-transmembrane, G-protein coupled receptors (GPCRs), the first one being GAL-R1, cloned ten years ago by Habert-Ortoli et al. (1994). Subsequently GAL-R2 and –R3 have been cloned (Branchek et al., 1998; Branchek et al., 2000).

Galanin has been suggested to be involved in numerous neuronal and endocrine functions a s a p utative messenger molecule (Bartfai et a l., 1 993; M erchenthaler et a l., 1993). For example, it has been postulated that endogenous galanin exhibits autocrine/paracrine actions in the supraoptic nucleus (SON) (Landry et al., 1995; Papas and Bourque, 1997) and in locus coeruleus (LC) (Pieribone et al., 1995; Xu et al., 1998; see also Seutin et al., 1989; Sevcik et al., 1993), a noradrenergic cell group in the pons (the A6 group according to Dahlström and Fuxe, 1964). This hypothesis would require galanin-containing SGs to be targeted and released from specific subcellular compartments, such as dendritic processes or axon collaterals.

In the present review, we will first summarize morphological evidence for somato-dendritic release of galanin in the rat both in a neuroendocrine (SON neurons) and a non-neuroendocrine (LC n eurons) system. We w ill then consider two hypotheses regarding possible mechanisms responsible for specific targeting of galanin-containing SGs, (i) specific sorting of galanin into dendritically targeted SGs, and (ii) specific targeting of prepro-galanin transcript leading to local galanin synthesis in dendrites. Finally, the physiological significance of galanin somato-dendritic release will be discussed in the light of electrophysiological and functional data.

2. EVIDENCE FOR SOMATO-DENDRITIC RELEASE OF GALANIN

2.1. Galanin Targeting to Dendrites in the Supraoptic Nucleus

As detailed in previous chapters in the book, the two neurohormones vasopressin and oxytocin are secreted from neurohypophyseal nerve endings into the systemic circulation. However, in addition both oxytocin (Moos et al., 1989; Russell et al., 1992) and vasopressin (Landgraf a nd Ludwig, 1 991; R ussell et a l., 1992) a re r eleased within the magnocellular nuclei themselves upon appropriate stimulation. Early evidence for this mainly arose from ultrastructural studies showing exocytosis of neuropeptides from dendrites and cell bodies of magnocellular neurones (Buma and Nieuwenhuys, 1987; Pow and Morris, 1989). In addition to vasopressin and oxytocin, the neuroendocrine magnocellular nuclei contain numerous other peptides (Brownstein and Mezey, 1986). They are present in much smaller amounts and are less likely to be released in biologically active concentrations into the systemic circulation to exert distant effects, but rather function as paracrine modulators either in the neurohypophysis as demonstrated for cholecystokinin (Jarvis et al., 1995) or at the hypothalamic level. The latter hypothesis would require that coexisting neuropeptides are also released locally within the magnocellular nuclei. Galanin is one of these neuropeptides, and is normally colocalized with vasopressin (Melander et al., 1986b). Moreover, galanin expression and release in vasopressin neurons is regulated under different physiological and experimental conditions (Meister et al., 1990).

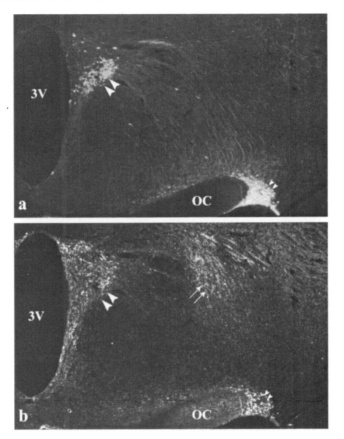

Fig. 1. Light microscopic detection of vasopressin (a) and galanin (b) in rat hypothalamus using double immunohistochemistry.
a: Vasopressin cell bodies are seen in the PVN (large double-arrowhead) and SON (small double-arrowhead). Vasopressin immunoreactivity is also detected in a fiber tract arising from the PVN and running laterally in the hypothalamus.
b: Perikarya appear double-labeled for galanin in the SON (small double-arrowhead). Many fibers are galanin immunoreactive throughout the whole PVN, whereas cell bodies are only weakly labeled. A fiber tract arising from the PVN is also faintly stained. An additional dense network of galanin single-labeled fibers is seen in the dorsal hypothalamus (double-arrow). OC: optic chiasma; 3V: third ventricle.

Double-immunohistochemistry at the light microscope level shows a general overlap of vasopressin (Fig. 1a) - and galanin (Fig 1b)-immunoreactivity at low magnification in the cell bodies in the SON. However, vasopressin-immunoreactive (-IR) projections arching through the lateral hypothalamus do not display a clear signal for galanin. In contrast, a dense network of galanin–IR fibers is seen within SON and PVN, and both peptides are present in nerve endings in the posterior pituitary. The more detailed confocal analysis of SON after double-immunolabeling reveals that galanin and vasopressin signals are partially segregated in positive cell bodies, with the galanin being most prominent in the perinuclear region (Landry et al., 2003).

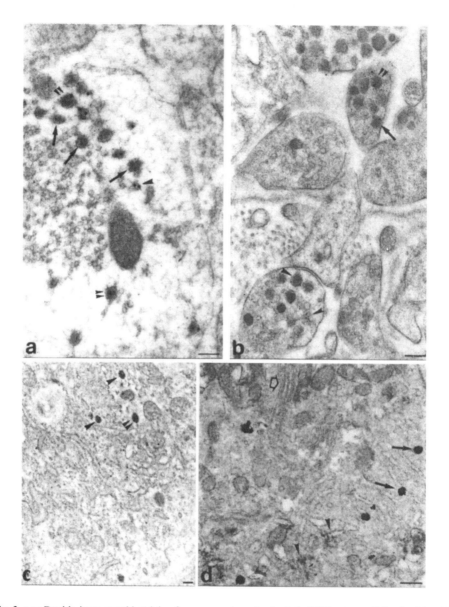

Fig. 2. a-c: Double immunogold staining for vasopressin and galanin in SON magnocellular dendrites (a), neurohypophysis (b) and perikarya (c). The size of gold particles is 10 nm (a) or 20 nm (b and c) for galanin and 20 nm (a) or 10 nm (b and c) for vasopressin. Double-labelled SGs are found in the three subcellular compartments (double-arrowheads in a-c). Many galanin single-labelled SGs are seen in dendritic profiles (arrows in a), whereas they are very rare in the neurohypophysis (arrow in b). Vasopressin single-labelled SGs are more numerous in the perikarya and neurohypophysis than in dendrites (arrowheads in a-c).

d: Double electron microscope detection of galanin (silver-enhanced gold particles, arrows) and vasopressin (dark precipitate, arrowheads) mRNA in SON magnocellular perikarya. The two signals are often segregated, being observed at different locations on the rough endoplasmic reticulum. Large, open arrow indicates the Golgi apparatus. Bars: 0.5 μm

At the ultrastructural level, galanin immunogold labelling has been performed on tissue sections embedded in Lowicryl after progressive lowering of the temperature. Galanin-containing SGs were found in all subcellular compartments of the magnocellular neurons; that is soma, dendrites and neurohypophyseal nerve endings. However, the proportion of galanin- versus vasopressin-immunopositive SGs was higher in the dendrites and lower in the neurohypophysis, as compared to the cell bodies (Fig. 2).

Since the lack of immunogold staining does not necessarily imply lack of neuropeptides, but could very well be due to peptide levels below the detection threshold, a double-face labelling protocol was applied on a grid carrying ultrathin sections (Landry et al., unpublished). This procedure increases the sensitivity of immunogold and confirmed that the proportion of galanin-immunopositive SGs is higher in dendrites than in other compartments. In contrast, no changes in the number of gold particles per SG, reflecting galanin labelling intensity, have been observed in the different subcellular compartments studied. This indicates that galanin immunoreactivity is not masked in any of the compartments. Hence, changes in galanin-positive SG proportions represent true modifications of SG distribution and are not due to methodological hindrance.

Therefore, we propose that galanin-containing SGs are preferentially targeted toward dendrites, where they are likely to be released locally. Our results are further reinforced by detection of galanin in dendritic processes of magnocellular neurons by other authors (Sawchenko and Pfeiffer, 1988). The ensuing somato-dendritic release may represent a general mechanism, also involving other neuropeptides produced in the magnocellular neurons. For instance, local release from magnocellular neurons has been postulated to account for much of the increase in angiotensin II release under osmotic stimulation (Imboden and Felix, 1991). Moreover, whereas vasopressin causes a rapid feedback attenuation of overall excitability of the vasopressin neurons, the colocalized and co-released dynorphin contributes to burst termination of these cells (Brown et al., 2004, chapter 15 this volume). Thus, colocalized peptides, though expressed at comparatively low levels, may act through paracrine/autocrine mechanisms like vasopressin and oxytocin. Since neuropeptides in general have a nano- and even subnanomolar affinity for their receptors, they may have physiological effects even if released in comparatively small amounts.

2.2. Galanin Release in the Locus Coeruleus

An important question is if somato-dendritic release can be considered as a general release mechanism for neuropeptides or if it is restricted to the neurosecretory magnocellular neurons. Therefore, other neuronal, and preferably non-neuroendocrine, systems must be investigated.

The rat noradrenergic LC has the capacity to synthesize galanin (Melander et al., 1986b; Holets et al., 1988; Moore and Gustafson, 1989; Xu et al., 1998). However, for a long time it was difficult to visualize galanin in the LC projection areas, such as cortex and hippocampus (Skofitsch and Jacobowitz, 1985b; Melander et al., 1986a), raising the possibility that galanin is not centrifugally transported. In fact, it had been postulated that galanin synthesized in LC cell bodies is diverted toward axon collaterals or dendrites and released within the LC (Moore and Gustafson, 1989; Pieribone et al., 1995; Xu et al., 1998). Recently however, very sensitive histochemical methodologies have allowed detection of a network of low level galanin-expressing nerve endings in the LC projection areas in the hippocampus which are mainly noradrenergic terminals (Xu et al., 1998).

The putative electron microscopy techniques described above have also been applied to the LC, revealing presence of galanin-containing SGs in LC dendrites (Fig. 3a). Although the number of labelled SGs remained small, these data support the idea of a possible targeting of galanin in dendrites.

The capacity of LC neurons to release neuropeptides from the dendritic compartment has been assessed by the use of tannic acid. This reagent stabilizes the exocytosed proteins, permitting visualization of extruded SGs (Buma and Nieuwenhuys, 1987; Pow and Morris, 1989). Tannic acid does not penetrate the blood-brain barrier and was therefore applied *in vitro* to brainstem slices containing the locus coeruleus. Membrane depolarization and subsequent exocytosis were induced by incubation with 50 mM KCl. The slices were then embedded and processed for electron microscopic analysis. Some figures of SG exocytosis could clearly be observed in soma and dendrites (Fig. 3b) of LC neurons, thus showing that neuropeptides can be released from dendritic processes in a non-neuroendocrine system. Whether these exocytosed SGs indeed contain galanin remains to be investigated. However, this is technically difficult, since the tannic acid procedure reduces immunohistochemical sensitivity, and since galanin levels are comparatively low in LC neurons.

2.3. Galanin Release in Other Systems

Paracrine/autocrine mechanisms have been postulated for galanin in other systems, although not strictly investigated so far. In particular, cancer tumors have drawn some attention. Galanin has been shown to be expressed at highly variable levels in different kinds of tumors, including neuroblastoma (Tuechler et al., 1998; Perel et al., 2002) and pituitary adenomas (Invitti et al., 1999; Leung et al., 2002; Grenbäck et al., 2004). The absence of detectable levels of galanin of pituitary origin in the peripheral circulation points to a local secretion and local effects of galanin secreted from such tumors (Grenbäck et al., 2004). Moreover, the presence of galanin autoreceptors on galanin-expressing tumor cells (Perel et al., 2002) strengthens the hypothesis of a paracrine/autocrine action of galanin in non-neuronal systems, consistent with its somato-dendritic release in healthy, differentiated neuronal systems.

2.4. Regulation of Neuropeptide Somato-Dendritic Release

Interestingly, neuropeptide somato-dendritic release is a regulated phenomenon. This has been unequivocally shown for vasopressin and oxytocin in the SON. Steroids are important players in this control, and neurosteroids have been shown to induce neuropeptide release from SON neurones disconnected from the neurohypohysis. This effect is further controlled by a mechanism partly dependent on $GABA_A$ receptors (Widmer et al., 2003). Moreover, somatodendritic release of oxytocin is increased during lactation (de Kock et al., 2003). In addition, somatodendritic release can be modulated independently from nerve ending secretion by selective mobilization of intracellular calcium stores (Ludwig et al., 2002).

There are also great variations in galanin expression in the SON under physiological and experimental conditions such as salt loading (Meister et al., 1990). In the LC reserpine treatment also increases galanin synthesis (Gundlach et al., 1990). Furthermore, galanin synthesis is regulated in cancer tumors (Perel et al., 2002; Grenbäck et al., 2004).

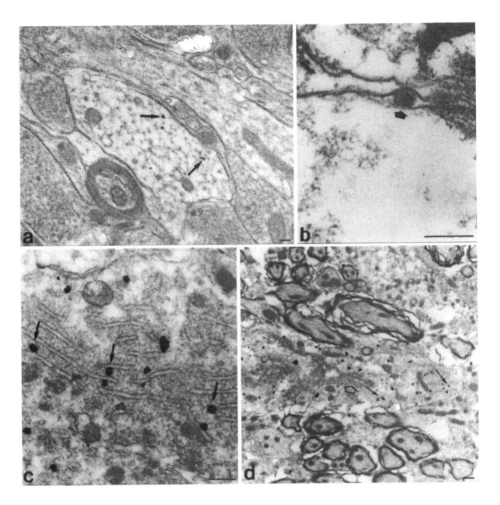

Fig. 3. a. Immunogold detection of galanin immunopositive dendritic profiles (arrows) in the LC. b. SG exocytosis (large arrow) in a dendrite in the LC as seen after the tannic acid procedure. c and d : Electron microscope detection of galanin mRNA (silver-enhanced gold particles, arrows) in the LC perikarya (c) and dendrites (d). The signal is seen on the rough endoplasmic reticulum (c). Galanin mRNA is found in the entire dendritic profile (d) which contains Golgi apparatus (large, open arrow in d).
Bars: 0.5 μm

However, further evidence is needed to demonstrate that changes in galanin are coupled to somato-dendritic release of galanin.

Taken together, it is likely that galanin can be released from the somato-dendritic compartment. This may even represent the principal and most important mechanism of galanin secretion in some neuronal systems. However, in addition to this local release, galanin in SGs is certainly secreted from nerve endings in all the systems investigated so far. Galanin has, for example, been identified in axon terminals originating from the LC and terminating in cortex and hippocampus (Xu et al., 1998), where binding sites exist for a galanin N-terminal fragment (Hedlund et al., 1992). Moreover, reserpine treatment causes depletion of galanin from such hippocampal and cortical nerve terminals (Xu et al., 1998), presumably due to reflex activation of LC neurons.

3. ROUTING OF SOMATO-DENDRITICALLY RELEASED GALANIN

Two different and non-exclusive mechanisms should be considered to account for targeting galanin to dendrites. First, galanin-containing SGs may undergo an active and site-specific sorting process, resulting in their routing to dendrites. Second, galanin could be synthesized in specific subcellular domains, including dendrites themselves, and released through its own secretory pathway.

3.1. Specific Sorting of Galanin-Containing SGs

Both hypotheses, together with the idea that galanin is mainly released from dendrites, would require galanin-containing SGs to be specifically addressed to dendrites, at least partially, independently from other coexpressed neuropeptides. More precisely, if galanin is mainly released from dendrites, and vasopressin mainly from neurohypophyseal nerve endings, then there would be no strict co-packaging of the two peptides into the same SGs in the SON. Such a hypothesis has been tested using electron microscopy and quantitative double-immunogold methods (Landry et al., 2003), focusing on SGs containing detectable galanin and/or vasopressin.

Three subpopulations of immunopositive SGs were observed in perikarya, dendrites and neurohypophyseal nerve endings. Some SGs contained either galanin or vasopressin. The third subpopulation displayed double labelling for both peptides. An objective quantification procedure based on frequency diagrams (Chun et al., 1994) showed a switch toward galanin single-labeled SGs in dendrites (Fig. 2a) and single-labeled vasopressin SGs in the neurohypophysis (Fig. 2b). The three types of immunopositive SGs were seen in SON cell bodies (Fig. 2c). The presence of galanin-single labelled SGs was unexpected in view of the much higher transcription/translation rate for vasopressin than for colocalized neuropeptides (Hekimi et al., 1991; Bean et al., 1994; Merighi, 2002). These data are in agreement with a preferential targeting of galanin to dendrites, and suggest a differential routing of two colocalized neuropeptides in vasopressinergic neurons.

Previous studies with electron microscopy have shown that several neuropeptides are often co-stored within the same SGs (Merighi et al., 1988), including galanin together with substance P and calcitonin gene-related peptide in dorsal root ganglion neurons (Zhang et al., 1993). However, in the cow pituitary, prolactin and growth hormone were mainly found to be stored in separate SGs of the same cells (Fumagalli and Zanini, 1985; Hashimoto et al., 1987). The most convincing evidence for differential neuropeptide packaging stems from studies of the marine mollusc Aplysia. In its bag cell neurons the precursor protein of the egg laying hormone is processed into individual peptides targeted to different SGs (Fisher et al., 1988) together with their respective processing enzymes (Chun et al., 1994; Seidah and Chretien, 1997; Zhou et al., 1999), on a cell-type dependent basis (Klumperman et al., 1996). Moreover, these different subpopulations of SGs can be targeted to different neuronal processes (Sossin et al., 1990).

The specific sorting of neuropeptides in the SON could occur at various levels, including the trans Golgi network (Kelly, 1985), as well as at later (Shennan, 1996) or earlier (Muniz et al., 2001) stages of the regulated secretory pathway. The details on the mechanisms, e.g. aggregation and condensation of cargo in forming SGs (Sossin and Scheller, 1991; Gorr et al., 2001), responsible for galanin and vasopressin sorting into distinct SG subpopulations, remain unknown. However, it is important to note that galanin single-labelled SGs could be observed as soon as they bud off the *trans* Golgi network (Landry et al., 2003), therefore suggesting that galanin could undergo a so-called "sorting by entry" rather than "sorting by retention". Morphological evidence suggests that even early steps of protein synthesis on the endoplasmic reticulum may account for galanin sorting.

3.2. Specific Synthesis Sites for Galanin

3.2.1. Galanin Synthesis Sites in the Soma and Dendrites

Our basic finding is that galanin mRNA has been detected by in situ hybridization in all subcellular compartments studied (Figs. 2d, 3c and d). In the cell body, double in situ experiments with confocal and electron microscopy (Fig. 2d) have revealed topological segregation between galanin and vasopressin mRNA (Landry et al., 2003). Such a spatial dissociation in the translation process would lead to formation of protein aggregates in different subcellular domains, so serving as a sorting mechanism (Sossin and Scheller, 1991). Moreover, van Leeuwen (1992) showed that, in the Brattleboro rat, accumulation of the mutated vasopressin precursor in the endoplasmic reticulum does not prevent galanin mRNA translation, suggesting that galanin and vasopressin mRNA are translated on distinctly separate domains of the endoplasmic reticulum.

Axonal localization of galanin mRNA has been reported after osmotic stimulation (Landry and Hökfelt, 1998). Similarly, vasopressin and oxytocin messengers have been found in axonal swellings of the median eminence (Bloch et al., 1990; Mohr and Richter, 1992; Trembleau et al., 1994). However, protein synthesis in axons has not been demonstrated *in vivo* in mammals, although it cannot be excluded (Alvarez, 2001). Only sparse evidence comes from experiments on cultured neurons (Martin et al., 1997; Aronov et al., 2001). Taken together, the functional significance of axonal neuropeptide mRNAs remains obscure (Mohr, 1999).

Dendritic mRNA is normally found in close association to the endoplasmic reticulum, as is seen in the soma (Fig. 3c). In addition to the cell body, a high amount of galanin mRNA is located in dendrites, in both the SON (Landry and Hökfelt, 1998) and the LC (Fig. 3d).

3.2.2. Regulation of mRNA Localization

Targeted mRNAs reach dendrites through an active sorting mechanism rather than by passive diffusion (St Johnston, 1995). It has been demonstrated that certain neuronal mRNAs contain *cis*-acting dendritic targeting signals in their 3'-untranslated region (3'-UTR) (Kislauskis et al., 1994; Kiebler and DesGroseillers, 2000). Both primary and secondary structures within these localization elements are important for their function and the subsequent dendritic distribution of neuropeptide mRNAs (Prakash et al., 1997; Yaniv and Yisraeli, 2001). The transport of localized mRNAs to their final destination in neuronal processes requires the binding of various *trans*-acting factors to these *cis*-acting elements (Kiebler and DesGroseillers, 2000). An additional level of regulation has emerged from the study of neurotrophin- or synaptic activity-dependent transport of mRNAs (Knowles and Kosik, 1997; Steward et al., 1998; Zhang et al., 1999). So far, the immediate early gene *Arc* mRNA represents the only RNA identified, for which the specific dendritic localization is linked to an event at the cell surface (Steward and Worley, 2001).

Increase in galanin synthesis rate is accompanied by changes in mRNA distribution, and may be an important stimulus for routing galanin mRNA, as for other neuropeptides (Landry and Hökfelt, 1998; Mohr, 1999). Although no data are available, one may suggest that *cis*-acting elements must also be involved in targeting galanin mRNA to axonal and dendritic compartments. Nevertheless, galanin mRNA is found in all subcellular compartments and must then be distinguished from mRNAs specifically localized to the dendrites. It is, therefore, possible that different *trans*-acting factors, interacting with *cis*-acting elements, are responsible for the fate of galanin mRNA. These factors, however, remain to be identified.

3.2.3. Regulation of mRNA Translation

Specific protein synthesis sites have not been considered for the adult neuronal soma, but there are several studies focusing on translation of dendritic mRNAs. One of the main issues to be addressed is whether or not mRNA translation occurs, and is regulated, in dendrites.

Electron microscopy studies have demonstrated that the organelles necessary for protein synthesis, namely ribosomes and rough endoplasmic reticulum harbouring translocation complex, are present in segments of the dendritic tree as distal as dendritic spines in the hippocampus (Pierce et al., 2000). Our own observations revealed the presence of Golgi apparatus in galanin mRNA-containing dendrite profiles in the SON and LC (Fig. 3c).

Several lines of evidence have emerged linking the processes of RNA localization and translational regulation. Translation of localized mRNA can be both activity-dependent and spatially controlled. The former control largely accounts for temporally regulated developmental processes (Kloc et al., 2002) but can also be involved in adult neurons. Thus, synthesis of CaMKII-α protein from dendritic mRNA is stimulated by brain-derived neurotrophic factor (BDNF) upon synaptic stimulation in adult rats (Aakalu et al., 2001).

Translational control elements have been identified in the 3'-UTR of neuropeptide dendritic mRNAs. In most cases, these elements are of importance for determining the polyadenylation status of dendritic mRNAs, and therefore their translational regulation (Mendez and Richter, 2001). One key protein for the polyadenylation mechanism in the cytoplasm is the cytoplasmic polyadenylation element binding protein (CPEB). Studies of dendritically localized mRNAs have identified CPEB in postsynaptic regions of the brain, suggesting polyadenylation-dependent regulation of translation in dendrites (Wu et al., 1998). Moreover, internal ribosome entry sites (IRESs) have been identified for dendritic mRNAs and have proved to be more active in dendrites than elsewhere in the cell, providing an additional level of regulation (Pinkstaff et al., 2001).

Several important issues remain to be solved concerning neuropeptide routing in neurons. In particular, the identity and structural basis of the mechanisms responsible for targeting SGs and/or mRNAs need to be identified. Moreover, the origin of galanin-containing, dendritically released SGs has to be investigated. Do they originate from a common pool of somato-dendritic mRNA or are they synthesized locally within the dendrites ?

4. FUNCTIONAL SIGNIFICANCE OF SOMATO-DENDRITICALLY RELEASED GALANIN

SG targeting has been proposed to represent a critical step in the control of peptide release from neurons (Karhunen et al., 2001). Moreover, neuropeptide release from dendrites and axons may be differentially regulated by intracellular Ca^{2+} mobilization (Ludwig et al., 2002). Therefore, a dissociation between galanin and vasopressin SG routing could lead to their release at distinctly separate sites and upon different stimuli.

In addition to their roles as circulating neurohormones, several groups have proposed specific functions for vasopressin and oxytocin released within the magnocellular nuclei, whereby solid evidence exists for somato-dendritic release of oxytocin in the SON. Intracerebroventricular injections of oxytocin have been shown to induce plastic rearrangements of magnocellular oxytocin neurons and oxytocin release from hypothalamus explants (Moos et al., 1984). Exocytosis of oxytocin from SGs is therefore regarded as a crucial link in a peptidergic feedback mechanism controlling morphological plasticity occurring under physiological conditions such as lactation (Morris et al., 1993). Moreover, oxytocin receptors have been demonstrated on SON neurons (Freund-Mercier et al., 1994).

Since galanin appears to be mainly released from soma and dendrites within the SON, it is more likely to exert an autocrine/paracrine role than a neuroendocrine action through systemic circulation. This local release of galanin would result in a hyperpolarizing effect on magnocellular neurons by enhancing potassium permeability (Papas and Bourque, 1997). Galanin could also regulate *in vivo* gene expression in the magnocellular nuclei (Landry et al., 1995, 2000). As for oxytocin, the possible autocrine/paracrine function of galanin could be mediated by galanin-R1 receptors known to be synthesized in SON neurons (Landry et al., 1998; Burazin et al., 2001), and these receptors thus represent somato-dendritic autoreceptors on SON neurons.

It is, however, important to point out that galanin-positive axon profiles containing numerous SGs have also been observed in the SON and could represent axon collaterals, as already described (Mason et al., 1984). They would constitute an additional, non-dendritic source of galanin. Furthermore, galanin-positive, vasopressin-negative nerve endings have been evidenced in the SON. Since these axonal profiles mainly contain small synaptic vesicles, and only few SGs, it seems likely that they are non-neuroendocrine terminals. Thus, in the SON, magnocellular soma and dendrites are not the only source of active galanin but there are also galanin-positive nerve endings with an intra- and/or extrahypothalamic origin which can release galanin to act on somato-dendritic receptors.

In the LC, galanin may exert an effect at the level of cell bodies as evidenced by electrophysiological studies (Seutin et al., 1989; Sevcik et al., 1993; Pieribone et al., 1995; Xu et al., 1998). Galanin reduces firing rate and induces hyperpolarization/outward current of LC neurons, presumably via an increase in K^+ conductance (Fig. 4a). This action of galanin is mediated by postsynaptic receptors, since the effect persists under

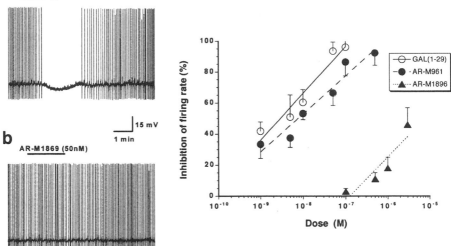

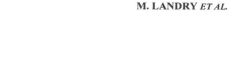

Fig. 4. Effect of galanin and AR-M1896 on the same LC neurons. (a, b). Galanin causes a strong hyperpolarization (a) whereas AR-M1896 hardly has any effect at all (b). The resting potential was -62 mV throughout the experiment. (c) Dose-response curves for LC neuronos to AR-M1896 (solid triangles), AR-M961 (solid circles) and galanin (open circles). The percent inhibition refers to the firing rate measured during last minute of the drug application. A complete cessation of firing/hyperpolarization is considered as 100% inhibition.

conditions in which synaptic input is blocked (TTX and/or plus low Ca^{2+} medium). The transcripts for the GAL-R1 and GAL-R2 receptors have been found in LC neurons, as shown with in situ hybridization (Parker et al., 1995; O'Donnell et al., 1999). Thus, both GAL-R1 and -R2 receptors are potentially postsynaptic receptors in LC neurons. Recently, it has been shown that the selective GAL-R2 agonist Gal(2–11)-NH (AR-M1896) causes inhibition of spike discharge and a slight hyperpolarization only at very high concentrations, on a molar basis being much weaker than galanin itself or AR-M961 (Fig. 4b, c), the latter an agonist at both GAL-R1 and GAL-R2 receptors. This suggests that it is mainly the GAL-R1 receptor that mediates hyperpolarization of LC neurons and acts as somato-dendritic receptor, while GALR2 may be present as a presynaptic receptor on the noradrenergic nerve terminals in the forebrain and other LC projection areas (Ma et al., 2001). Moreover in LC neurons, the noradrenaline-induced outward current is enhanced and prolonged by low concentrations of galanin (0.05-0.1 nM) (Fig. 5a, b), while galanin itself had no detectable effect on the membrane current (Xu et al., 2001). This sensitizing effect may be of physiological importance and could be caused by galanin released from dendrites and soma of galanin/NA neurons and/or from galanin afferents. Thus, besides a direct action, galanin has an indirect, modulating effect on LC neurons.

Taken together, these data suggest that galanin may act extra-synaptically, possibly through a volume transmission mechanism (Fuxe and Agnati, 1991). In this case,

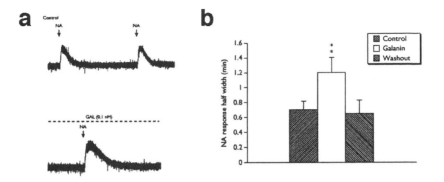

Fig. 5. Effect of galanin on the response of LC neurons to noradrenaline (NA). (a) NA (applied from a pipette at the arrowhead) induces a persistent outward current (upper trace). When galanin (0.1 nM) is present, the NA-induced outward current is enhanced and the duration is prolonged (lower trace). The holding potential was in all cases -62 mV. (b) The duration of the NA current is increased in the presence of galanin.

dendritic release of galanin could contribute to control of the activity of a whole neuronal cell group, such as the SON and the LC. Locally released and diffusible galanin could perhaps also act on the neighbouring Barrington's nucleus, which expresses the galanin-R1 receptor (Xu et al., 1998).

5. CONCLUSIONS

Taken together, our data suggest that neuropeptides are released from the somato-dendritic compartment not only in neuroendocrine systems but also in systems where peptide hormones are not the major messenger molecule. However, in all systems investigated, somato-dendritic targeting and release of galanin coexists with axonal targeting to nerve endings. It is therefore difficult to assess the relative importance of dendritic/autocrine mechanisms versus release from nerve endings. Somato-dendritic release of neuropeptides appears to be a way of increasing their range of functions, intimating their role in non-synaptic signalling. This is also consistent with a role of galanin as a trophic factor involved in neuronal differentiation and regeneration (Zigmond and Sun, 1997; Holmes et al., 2000).

In many neuronal systems, various neuropeptides are colocalized within the same cell bodies but appear to undergo differential sorting and routing processes. These mechanisms would allow their independent regulation within the same cell by two non-mutually exclusive putative mechanisms based on sorting into SGs specifically targeted to soma and dendrites, or on targeting mRNA to dendritic sites of neuropeptide synthesis. In particular, galanin and vasopressin, which are localized to different processes, could be released upon different stimuli, expanding the range of neuronal plasticity. Whether or not this specific routing is regulated upon neuronal stimulation remains an open question. Taken together, the possibility of dendritic release of neuropeptides in general terms expands their potential versatility and further define them as multifunctional signalling molecules.

6. ACKNOWLEDGEMENGS

This work was supported by The Swedish Medical Research Council (04X-2887), The Marianne and Marcus Wallenberg Foundation, an Unrestricted Bristol-Myers Squibb Neuroscience grant and a grant from EC (LSHM-CT-2003-503474).
We are grateful to Dr. E. Vila-Porcile for her very valuable help in the post-embedding immunogold studies.

7. REFERENCES

Aakalu, G., Smith, W. B., Nguyen, N., Jiang, C., and Schuman, E. M., 2001, Dynamic visualization of local protein synthesis in hippocampal neurons, *Neuron* **30**: 489.

Alvarez, J., 2001, The autonomous axon: a model based on local synthesis of proteins, *Biol. Res.* **34**: 103.

Aronov, S., Aranda, G., Behar, L., and Ginzburg, I., 2001, Axonal tau mRNA localization coincides with tau protein in living neuronal cells and depends on axonal targeting signal, *J. Neurosci.* **21**: 6577.

Bartfai, T., Hökfelt, T., and Langel, U., 1993, Galanin--a neuroendocrine peptide. *Crit. Rev. Neurobiol.* **7**: 229.

Bean, A.J., Zhang, X., and Hökfelt, T., 1994, Peptide secretion: what do we know? *Faseb* . **8**: 630.

Björklund, A., and Lindvall, O., 1975, Dopamine in dendrites of substantia nigra neurons: suggestions for a role in dendritic terminals, *Brain Res.* **83**: 531.

Bloch, B., Guitteny, A. F., Normand, E., and Chouham, S., 1990, Presence of neuropeptide messenger RNAs in neuronal processes, *Neurosci. Lett.* **109**: 259.

Branchek, T., Smith, K. E., and Walker, M. W., 1998, Molecular biology and pharmacology of galanin receptors, *Ann. N Y Acad. Sci.* **863**: 94.

Branchek, T. A., Smith, K. E., Gerald, C., and Walker, M. W., 2000, Galanin receptor subtypes, *Trends Pharmacol. Sci.* **21**: 109.

Brown, C. H., Ludwig, M., and Leng, G., 2004, Temporal dissociation of the feedback effects of dendritically co-released peptides on rhythmogenesis in vasopressin cells, *Neuroscience* **124**: 105.

Brownstein, M. J., and Mezey, E., 1986, Multiple chemical messengers in hypothalamic magnocellular neurons, *Prog. Brain Res.* **68**: 161.

Buma, P., and Nieuwenhuys, R., 1987, Ultrastructural demonstration of oxytocin and vasopressin release sites in the neural lobe and median eminence of the rat by tannic acid and immunogold methods, *Neurosci. Lett.* **74**: 151.

Burazin, T. C., Larm, J. A., and Gundlach, A. L., 2001, Regulation by osmotic stimuli of galanin-R1 receptor expression in magnocellular neurones of the paraventricular and supraoptic nuclei of the rat, *J. Neuroendocrinol.* **13**: 358.

Cheramy, A., Leviel, V., and Glowinski, J., 1981, Dendritic release of dopamine in the substantia nigra, *Nature* **289**: 537.

Ch'ng, J. L., Christofides, N. D., Anand, P., Gibson, S. J., Allen, Y. S., Su, H. C., Tatemoto, K., Morrison, J. F., Polak, J. M., and Bloom, S. R., 1985, Distribution of galanin immunoreactivity in the central nervous system and the responses of galanin-containing neuronal pathways to injury, *Neuroscience* **16**: 343.

Chun, J. Y., Korner, J., Kreiner, T., Scheller, R. H., and Axel, R., 1994, The function and differential sorting of a family of aplysia prohormone processing enzymes, *Neuron* **12**: 831.

Dahlström, A., and Fuxe, K., 1964, Evidence for the existence of monoamine neurons in the central nervous system. I. Demonstration of monoamines in the cell bodies of brainstem neurons, *Acta Physiol. Scand.* **62**: S232, 1.

de Kock, C. P., Wierda, K. D., Bosman, L. W., Min, R., Koksma, J. J., Mansvelder, H. D., Verhage, M., and Brussaard, A. B., 2003, Somatodendritic secretion in oxytocin neurons is upregulated during the female reproductive cycle, *J. Neurosci.* **23**: 2726.

Fisher, J. M., Sossin, W., Newcomb, R., and Scheller, R. H., 1988, Multiple neuropeptides derived from a common precursor are differentially packaged and transported, *Cell* **54**: 813.

Freund-Mercier, M. J., Stoeckel, M. E., and Klein, M. J., 1994, Oxytocin receptors on oxytocin neurones: histoautoradiographic detection in the lactating rat, *J Physiol.* **480**: 155.

Fumagalli, G., and Zanini, A., 1985, In cow anterior pituitary, growth hormone and prolactin can be packed in separate granules of the same cell, *J. Cell Biol.* **100**: 2019.

Fuxe, K., and Agnati, L. F., 1991, *Volume Transmission in the Brain,* Raven Press, New York.

Gorr, S. U., Jain, R. K., Kuehn, U., Joyce, P. B., and Cowley, D. J., 2001, Comparative sorting of neuroendocrine secretory proteins: a search for common ground in a mosaic of sorting models and mechanisms, *Mol. Cell. Endocrinol.* **172**: 1.

Grenbäck, E., Bjellerup, P., Wallerman, E., Lundblad, L., Änggård, A., Ericson, K., Åman, K., Landry, M., Schmidt, W. E., Hökfelt, T., and Hulting, A. L., 2004, Galanin in pituitary adenomas, *Regul. Pept.* **117**: 127.

Gundlach, A. L., Rutherfurd, S. D., and Louis, W. J., 1990, Increase in galanin and neuropeptide Y mRNA in locus coeruleus following acute reserpine treatment, *Eur. J. Pharmacol.* **184**: 163.

Habert-Ortoli, E., Amiranoff, B., Loquet, I., Laburthe, M., and Mayaux, J. F., 1994, Molecular cloning of a functional human galanin receptor, *PNAS USA* **91**: 9780.

Hashimoto, S., Fumagalli, G., Zanini, A., and Meldolesi, J., 1987, Sorting of three secretory proteins to distinct secretory granules in acidophilic cells of cow anterior pituitary, *J. Cell. Biol.* **105**: 1579.

Hedlund, P. B., Yanaihara, N., and Fuxe, K., 1992, Evidence for specific N-terminal galanin fragment binding sites in the rat brain, *Eur. J. Pharmacol.* **224**: 203.

Hekimi, S., Fischer-Lougheed, J., and O'Shea, M., 1991, Regulation of neuropeptide stoichiometry in neurosecretory cells, *J. Neurosci.* **11**: 3246.

Hökfelt, T., Broberger, C., Xu, Z. Q., Sergeyev, V., Ubink, R., and Diez, M., 2000, Neuropeptides - an overview, *Neuropharmacology* **39**: 1337.

Holets, V. R., Hökfelt, T., Rökaeus, A., Terenius, L., and Goldstein, M., 1988, Locus coeruleus neurons in the rat containing neuropeptide Y, tyrosine hydroxylase or galanin and their efferent projections to the spinal cord, cerebral cortex and hypothalamus, *Neuroscience* **24**: 893.

Holmes, F. E., Mahoney, S., King, V. R., Bacon, A., Kerr, N. C., Pachnis, V., Curtis, R., Priestley, J. V., and Wynick, D., 2000, Targeted disruption of the galanin gene reduces the number of sensory neurons and their regenerative capacity, *PNAS USA* **97**: 11563.

Imboden, H., and Felix, D., 1991, An immunocytochemical comparison of the angiotensin and vasopressin hypothalamo-neurohypophysial systems in normotensive rats, *Regul. Pept.* **36**: 197.

Invitti, C., Giraldi, F. P., Dubini, A., Moroni, P., Losa, M., Piccoletti, R., and Cavagnini, F., 1999, Galanin is released by adrenocorticotropin-secreting pituitary adenomas *in vivo* and *in vitro*, *J. Clin. Endocrinol. Metab.* **84**: 1351.

Jarvis, C. R., Van de Heijning, B. J., and Renaud, L. P., 1995, Cholecystokinin evokes vasopressin release from perfused hypothalamic-neurohypopheal explants, *Regul. Pept.* **56**: 131.

Karhunen, T., Vilim, F. S., Alexeeva, V., Weiss, K. R., and Church, P. J., 2001, Targeting of peptidergic vesicles in cotransmitting terminals, *J. Neurosci.* **21**: RC127.

Kelly, R. B., 1985, Pathways of protein secretion in eukaryotes, *Science* **230**: 25.

Kiebler, M. A., and DesGroseillers, L., 2000, Molecular insights into mRNA transport and local translation in the mammalian nervous system, *Neuron* **25**: 19.

Kislauskis, E. H., Zhu, X., and Singer, R. H., 1994, Sequences responsible for intracellular localization of beta-actin messenger RNA also affect cell phenotype, *J. Cell. Biol.* **127**: 441.

Kloc, M., Zearfoss, N. R., and Etkin, L. D., 2002, Mechanisms of subcellular mRNA localization, *Cell* **108**: 533.

Klumperman, J., Spijker, S., van Minnen, J., Sharp-Baker, H., Smit, A. B., and Geraerts, W. P., 1996, Cell type-specific sorting of neuropeptides: a mechanism to modulate peptide composition of large dense-core vesicles, *J. Neurosci.* **16**: 7930.

Knowles, R. B., and Kosik, K. S., 1997, Neurotrophin-3 signals redistribute RNA in neurons, *PNAS USA* **94**: 14804.

Landgraf, R., and Ludwig, M., 1991, Vasopressin release within the supraoptic and paraventricular nuclei of the rat brain: osmotic stimulation via microdialysis, *Brain Res.* **558**: 191.

Landry, M., Åman, K., and Hökfelt, T., 1998, Galanin-R1 receptor in anterior and mid-hypothalamus: distribution and regulation, *J. Comp. Neurol.* **399**: 321.

Landry, M., and Hökfelt, T., 1998, Subcellular localization of preprogalanin messenger RNA in perikarya and axons of hypothalamo-posthypophyseal magnocellular neurons: an in situ hybridization study, *Neuroscience* **84**: 897.

Landry, M., Roche, D., and Calas, A., 1995, Short-term effects of centrally administered galanin on the hyperosmotically stimulated expression of vasopressin in the rat hypothalamus. An in situ hybridization and immunohistochemistry study, *Neuroendocrinology* **61**: 393.

Landry, M., Roche, D., Vila-Porcile, E., and Calas, A., 2000, Effects of centrally administered galanin (1-16) on galanin expression in the rat hypothalamus, *Peptides* **21**: 1725.

Landry, M., Vila-Porcile, E., Hökfelt, T., and Calas, A., 2003, Differential routing of coexisting neuropeptides in vasopressin neurons, *Eur. J. Neurosci.* **17**: 579.

Leung, B., Iisma, T. P., Leung, K. C., Hort, Y. J., Turner, J., Sheehy, J. P., and Ho, K. K., 2002, Galanin in human pituitary adenomas: frequency and clinical significance, *Clin. Endocrinol.* **56**: 397.

Ludwig, M., 1998, Dendritic release of vasopressin and oxytocin, *J. Neuroendocrinol.* **10**: 881.

Ludwig, M., Sabatier, N., Bull, P. M., Landgraf, R., Dayanithi, G., and Leng, G., 2002, Intracellular calcium stores regulate activity-dependent neuropeptide release from dendrites, *Nature* **418**: 85.

Ma, X., Tong, Y. G., Schmidt, R., Brown, W., Payza, K., Hodzic, L., Pou, C., Godbout, C., Hokfelt, T., and Xu, Z. Q., 2001, Effects of galanin receptor agonists on locus coeruleus neurons, *Brain Res.* **919**: 169.

Martin, K. C., Casadio, A., Zhu, H., Yaping, E., Rose, J. C., Chen, M., Bailey, C. H., and Kandel, E. R., 1997, Synapse-specific, long-term facilitation of aplysia sensory to motor synapses: a function for local protein synthesis in memory storage, *Cell* **91**: 927.

Mason, W. T., Ho, Y. W., and Hatton, G. I., 1984, Axon collaterals of supraoptic neurones: anatomical and electrophysiological evidence for their existence in the lateral hypothalamus, *Neuroscience* **11**: 169.

Meister, B., Cortes, R., Villar, M. J., Schalling, M., and Hökfelt, T., 1990, Peptides and transmitter enzymes in hypothalamic magnocellular neurons after administration of hyperosmotic stimuli: comparison between messenger RNA and peptide/protein levels, *Cell Tissue Res.* **260**: 279.

Melander, T., Hökfelt, T., and Rokaeus, A., 1986a, Distribution of galaninlike immunoreactivity in the rat central nervous system, *J. Comp. Neurol.* **248**: 475.

Melander, T., Hökfelt, T., Rökaeus, A., Cuello, A. C., Oertel, W. H., Verhofstad, A., and Goldstein, M., 1986b, Coexistence of galanin-like immunoreactivity with catecholamines, 5-hydroxytryptamine, GABA and neuropeptides in the rat CNS, *J. Neurosci.* **6**: 3640.

Melander, T., Köhler, C., Nilsson, S., Hökfelt, T., Brodin, E., Theodorsson, E., and Bartfai, T., 1988, Autoradiographic quantitation and anatomical mapping of 125I-galanin binding sites in the rat central nervous system, *J. Chem. Neuroanat.* **1**: 213.

Mendez, R., and Richter, J. D., 2001, Translational control by CPEB: a means to the end, *Nat. Rev. Mol. Cell. Biol.* **2**: 521.

Merchenthaler, I., Lopez, F. J., and Negro-Vilar, A., 1993, Anatomy and physiology of central galanin-containing pathways, *Prog. Neurobiol.* **40**: 711.

Merighi, A., 2002, Costorage and coexistence of neuropeptides in the mammalian CNS, *Prog. Neurobiol.* **66**: 161.

Merighi, A., Polak, J. M., Gibson, S. J., Gulbenkian, S., Valentino, K. L., and Peirone, S. M., 1988, Ultrastructural studies on calcitonin gene-related peptide-, tachykinins- and somatostatin-immunoreactive neurones in rat dorsal root ganglia: evidence for the colocalization of different peptides in single secretory granules, *Cell Tissue Res.* **254**: 101.

Mohr, E., 1999, Subcellular RNA compartmentalization, *Prog. Neurobiol.* **57**: 507.

Mohr, E., and Richter, D., 1992, Diversity of mRNAs in the axonal compartment of peptidergic neurons in the rat, *Eur. J. Neurosci.* **4**: 870.

Moore, R. Y., and Gustafson, E. L., 1989, The distribution of dopamine-beta-hydroxylase, neuropeptide Y and galanin in locus coeruleus neurons, *J. Chem. Neuroanat.* **2**: 95.

Moos, F., Freund-Mercier, M. J., Guerne, Y., Guerne, J. M., Stoeckel, M. E., and Richard, P., 1984, Release of oxytocin and vasopressin by magnocellular nuclei *in vitro*: specific facilitatory effect of oxytocin on its own release, *J. Endocrinol.* **102**: 63.

Moos, F., Poulain, D. A., Rodriguez, F., Guerne, Y., Vincent, J. D., and Richard, P., 1989, Release of oxytocin within the supraoptic nucleus during the milk ejection reflex in rats, *Exp. Brain Res.* **76**: 593.

Morris, J. F., Pow, D. V., Sokol, H. W., and Ward, A., 1993, Dendritic release of peptides from magnocellular neurons in normal rats, Brattleboro rats and mice with hereditary nephrogenic diabetes insipidus, in: P. Gross, D. Richter, G. L. Robertson, eds., *Vasopressin*. John Libbey Eurotext, Paris, pp. 171-182.

Muniz, M., Morsomme, P., and Riezman, H., 2001, Protein sorting upon exit from the endoplasmic reticulum, *Cell* **104**: 313.

O'Donnell, D., Ahmad, S., Wahlestedt, C., and Walker, P., 1999, Expression of the novel galanin receptor subtype GALR2 in the adult rat CNS: distinct distribution from GALR1, *J. Comp. Neurol.* **409**: 469.

Papas, S., and Bourque, C. W., 1997, Galanin inhibits continuous and phasic firing in rat hypothalamic magnocellular neurosecretory cells, *J. Neurosci.* **17**: 6048.

Parker, E. M., Izzarelli, D. G., Nowak, H. P., Mahle, C. D., Iben, L. G., Wang, J., and Goldstein, M. E., 1995, Cloning and characterization of the rat GALR1 galanin receptor from Rin14B insulinoma cells, *Mol. Brain Res.* **34**: 179.

Perel, Y., Amrein, L., Dobremez, E., Rivel, J., Daniel, J.Y., and Landry, M., 2002, Galanin and galanin receptor expression in neuroblastic tumours: correlation with their differentiation status, *Br. J. Cancer* **86**: 117.

Pierce, J. P., van Leyen, K., and McCarthy, J. B., 2000, Translocation machinery for synthesis of integral membrane and secretory proteins in dendritic spines, *Nat. Neurosci.* **3**: 311.

Pieribone, V. A., Xu, Z. Q., Zhang, X., Grillner, S., Bartfai, T., and Hokfelt, T., 1995, Galanin induces a hyperpolarization of norepinephrine-containing locus coeruleus neurons in the brainstem slice, *Neuroscience* **64**: 861.

Pinkstaff, J. K., Chappell, S. A., Mauro, V. P., Edelman, G. M., and Krushel, L. A., 2001, Internal initiation of translation of five dendritically localized neuronal mRNAs, *PNAS USA* **98**: 2770.

Pow, D. V., and Morris, J. F., 1989, Dendrites of hypothalamic magnocellular neurons release neurohypophysial peptides by exocytosis, *Neuroscience* **32**: 435.

Prakash, N., Fehr, S., Mohr, E., and Richter, D., 1997, Dendritic localization of rat vasopressin mRNA: ultrastructural analysis and mapping of targeting elements, *Eur. J. Neurosci.* **9**: 523.

Rökaeus, A., Melander, T., Hökfelt, T., Lundberg, J. M., Tatemoto, K., Carlquist, M., and Mutt, V., 1984, A galanin-like peptide in the central nervous system and intestine of the rat, *Neurosci. Lett.* **47**: 161.

Russell, J. A., Neumann, I., and Landgraf, R., 1992, Oxytocin and vasopressin release in discrete brain areas after naloxone in morphine-tolerant and -dependent anesthetized rats: push-pull perfusion study, *J. Neurosci.* **12**: 1024.

Sawchenko, P. E., and Pfeiffer, S. W., 1988, Ultrastructural localization of neuropeptide Y and galanin immunoreactivity in the paraventricular nucleus of the hypothalamus in the rat, *Brain Res.* **474**: 231.

Seidah, N. G., and Chretien, M., 1997, Eukaryotic protein processing: endoproteolysis of precursor proteins, *Curr. Opin. Biotechnol.* **8**: 602.

Seutin, V., Verbanck, P., Massotte, L., and Dresse, A., 1989, Galanin decreases the activity of locus coeruleus neurons *in vitro*, *Eur. J. Pharmacol.* **164**: 373.

Sevcik, J., Finta, E. P., and Illes, P., 1993, Galanin receptors inhibit the spontaneous firing of locus coeruleus neurones and interact with mu-opioid receptors, *Eur. J. Pharmacol.* **230**: 223.

Shennan, K. I., 1996, Intracellular targeting of secretory proteins in neuroendocrine cells, *Biochem. So.c Trans.* **24**: 535.

Skofitsch, G., and Jacobowitz, D. M., 1985a, Galanin-like immunoreactivity in capsaicin sensitive sensory neurons and ganglia, *Brain Res. Bull.* **15**: 191.

Skofitsch, G., and Jacobowitz, D. M., 1985b, Immunohistochemical mapping of galanin-like neurons in the rat central nervous system, *Peptides* **6**: 509.

Skofitsch, G., and Jacobowitz, D. M., 1986, Quantitative distribution of galanin-like immunoreactivity in the rat central nervous system, *Peptides* **7**: 609.

Skofitsch, G., Sills, M. A., and Jacobowitz, D. M., 1986, Autoradiographic distribution of 125I-galanin binding sites in the rat central nervous system, *Peptides* **7**: 1029.

Sossin, W. S., and Scheller, R. H., 1991, Biosynthesis and sorting of neuropeptides, *Curr. Opin. Neurobiol.* **1**: 79.

Sossin, W. S., Sweet-Cordero, A., and Scheller, R. H., 1990, Dale's hypothesis revisited: different neuropeptides derived from a common prohormone are targeted to different processes, *PNAS USA* **87**: 4845.

St Johnston, D., 1995, The intracellular localization of messenger RNAs, *Cell* **81**: 161.

Steward, O., Wallace, C. S., Lyford, G. L., and Worley, P. F., 1998, Synaptic activation causes the mRNA for the IEG Arc to localize selectively near activated postsynaptic sites on dendrites, *Neuron* **21**: 741.

Steward, O., and Worley, P. F., 2001, Selective targeting of newly synthesized Arc mRNA to active synapses requires NMDA receptor activation, *Neuron* **30**: 227.

Tatemoto, K., Rokaeus, A., Jörnvall, H., McDonald, T. J., and Mutt, V., 1983, Galanin - a novel biologically active peptide from porcine intestine, *FEBS Lett.* **164**: 124.

Trembleau, A., Morales, M., and Bloom, F. E., 1994, Aggregation of vasopressin mRNA in a subset of axonal swellings of the median eminence and posterior pituitary: light and electron microscopic evidence, *J. Neurosci.* **14**: 39.

Tuechler, C., Hametner, R., Jones, N., Jones, R., Iismaa, T. P., Sperl, W., and Kofler, B., 1998, Galanin and galanin receptor expression in neuroblastoma, *Ann. NY Acad. Sci.* **863**: 438.

van Leeuwen, F., 1992, Mutant vasopressin precursor producing cells of the homozygous Brattleboro rat as a model for co-expression of neuropeptides, *Prog. Brain Res.* **92**: 149.

Widmer, H., Ludwig, M., Bancel, F., Leng, G., and Dayanithi, G., 2003, Neurosteroid regulation of oxytocin and vasopressin release from the rat supraoptic nucleus, *J. Physiol.* **548**: 233.

Wu, L., Wells, D., Tay, J., Mendis, D., Abbott, M. A., Barnitt, A., Quinlan, E., Heynen, A., Fallon, J. R., and Richter, J. D., 1998, CPEB-mediated cytoplasmic polyadenylation and the regulation of experience-dependent translation of alpha-CaMKII mRNA at synapses, *Neuron* **21**: 1129.

Xu, Z. Q., Shi, T. J., and Hökfelt, T., 1998, Galanin/GMAP- and NPY-like immunoreactivities in locus coeruleus and noradrenergic nerve terminals in the hippocampal formation and cortex with notes on the galanin-R1 and -R2 receptors, *J. Comp. Neurol.* **392**: 227.

Xu, Z. Q., Tong, Y. G., and Hökfelt, T., 2001, Galanin enhances noradrenaline-induced outward current on locus coeruleus noradrenergic neurons, *Neuroreport* **12**: 1779.

Yaniv, K., and Yisraeli, J. K., 2001, Defining cis-acting elements and trans-acting factors in RNA localization, *Int. Rev. Cytol.* **203:** 521.

Zhang, X., Nicholas, A. P., and Hökfelt, T., 1993, Ultrastructural studies on peptides in the dorsal horn of the spinal cord. I. Co-existence of galanin with other peptides in primary afferents in normal rats, *Neuroscience* **57:** 365.

Zhang, H. L., Singer, R. H., and Bassell, G. J., 1999, Neurotrophin regulation of beta-actin mRNA and protein localization within growth cones, *J. Cell. Biol.* **147:** 59.

Zhou, A., Webb, G., Zhu, X., and Steiner, D. F., 1999, Proteolytic processing in the secretory pathway, *J. Biol. Chem.* **274:** 20745.

Zigmond, R. E., and Sun, Y., 1997, Regulation of neuropeptide expression in sympathetic neurons. Paracrine and retrograde influences, *Ann. NY Acad. Sci.* **814:** 181.

Chapter 3

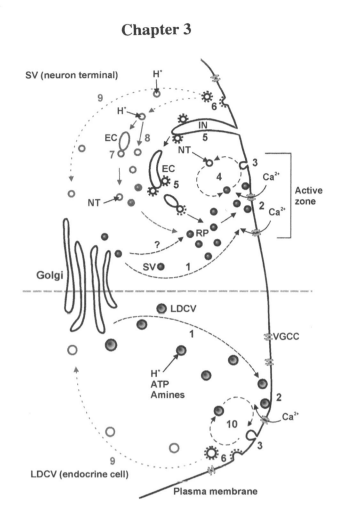

Figure 1. Vesicle lifecycle and membrane trafficking in neuronal and endocrine cells.
Synaptic vesicles (SV) in neurons (top) and large dense cored vesicles (LDCV) in endocrine cells (bottom) bud from the trans-Golgi network of the Golgi apparatus filled with cargo proteins (see text, including peptide transmitters for LDCVs). During their maturation, vesicles are acidified, take up ATP and neurotransmitter (NT, in SVs) or biogenic amines (in LDCVs) and are translocated to the plasma membrane (see step [1]). At the active zone, they dock (see also Figure 2) and are primed [2]. Influx of calcium ions through voltage-gated calcium channels (VGCC) triggers fusion of vesicle and plasma membranes and subsequent release of the vesicle cargo [3]. The vesicle membrane can completely collapse into the plasma membrane and can be retrieved by clathrin-mediated endocytosis [6]. Alternatively, membrane fusion may generate a temporary fusion pore after which the vesicle is quickly retrieved, a mechanism called 'kiss-and-run' exocytosis [4, 10]. In neurons, this mechanism allows quick re-filling of SVs with neurotransmitter (NT) for fast, repetitive exocytosis [4]. Bulk endocytosis provides another mechanism for membrane retrieval in neurons by the invagination of larger membrane areas, which enter endosomal compartments [ECs], from which SVs can bud off [5]. Vesicles retrieved from the plasma membrane by clathrin-mediated endocytosis [6] can either be translocated back to the Golgi for recycling [9] or, in neurons, can fuse to form early endosomal compartments from which new vesicles bud off, ready for re-filling with neurotransmitter [7]. Finally, clathrin-mediated endocytosis can directly deliver vesicles for protonation, refilling and recycling [8].
'Kiss-and-run' exocytosis in neurons provides a mechanism by which SVs can be rapidly retrieved and recycled locally in order to deliver swift, highly repetitive exocytotic response to prolonged stimulation, a long distance away from the cell body. In contrast, because LDCVs in endocrine cells are partially loaded with protein cargo, retrieved vesicles have to return to the Golgi for recycling [9]. However, in endocrine cells LDCVs have also been reported to undergo 'kiss-and-run' exocytosis [10]. This could provide a mechanism to modulate the amount of vesicle cargo released in a single exocytotic event (quantal release).

Chapter 3

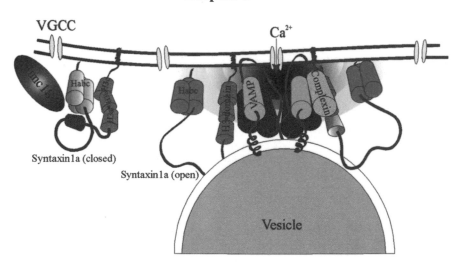

Figure 2. Schematic illustrating the hypothesized conformations of some of the proteins involved in vesicle fusion. The essential t-SNARE protein syntaxin1a exists in the "closed" conformation when bound to the SM protein Munc-18 (also known as nSec-1). In this conformation, the C-terminal SNARE motif of syntaxin1a is occluded by its N-terminal Habc domain thus syntaxin is unavailable to participate with the other SNARE proteins in the formation of the core complex. The subsequent dissociation of Munc-18 results in a large conformational rearrangement of syntaxin1a, "opening" the molecule, and allowing the interaction of the syntaxin1a SNARE motif with the other t-SNARE, SNAP-25, and the v-SNARE, VAMP (also known as synaptobrevin). SNAP-25 is associated with the plasma membrane through a palmitoylated loop, whereas VAMP is an integral vesicle membrane protein. The formation of the tertiary core complex creates a 3-dimensional conformation recognized by complexin, with crystallographic studies revealing its binding to the groove created by syntaxin and VAMP helices, apparently stabilizing the complex. Calcium influx through voltage gated calcium channels (VGCC) creates calcium micro-domains in the vicinity of fusion sites (represented here as purple and blue shading of decreasing intensity to indicate decreasing Ca^{2+} concentration). The precise nature of the "fusion pore" remains undefined, as does the number of core complexes involved. Munc18-mediated sequestering of syntaxin1a has been hypothesized to be an essential modulatory step, preventing ectopic fusion from occurring, however, the regulatory step leading to the dissociation of the two proteins in a cellular environment remains unclear. Furthermore, Munc18 also appears to have other functions important for vesicle docking, as in its absence, docking is reduced.

Chapter 5

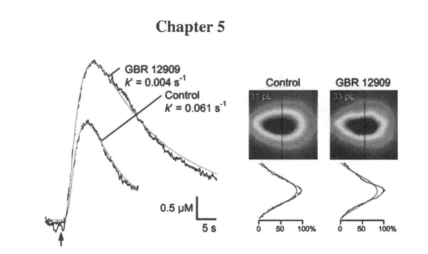

Figure 2. Regulation of diffusing DA in VTA by DATs. Left, Example DA diffusion profiles and right, simultaneous images of co-diffused Texas Red-labelled dextran after pressure ejection with and without the DAT inhibitor, GBR-12909 (2 μM). DA diffusion profiles ~100 μm from site of ejection show the decrease in uptake constant (k') and the increase in $[DA]_o$ when DATs are inhibited, despite similar ejection volumes (31-33 pL). Modified from Cragg et al., 2001.

Chapter 8

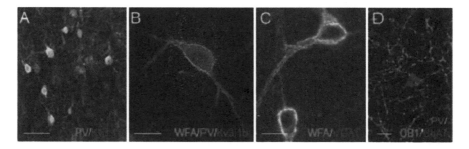

Figure 1. Some distinctive neurochemical hallmarks of cortical fast-spiking (FS) interneurons. (A) FS cells frequently contain the Ca^{2+}-binding protein parvalbumin (PV) and express unique combinations of ion channels that provide anatomical substrates for the maintenance of high-frequency firing. For instance, cortical FS cells show selective expression of the potassium channel 3.1b (Kv3.1b) subunit. (B) In addition, FS cells are often surrounded by perineuronal nets, as revealed by *Wisteria floribunda* agglutinin (WFA) labelling, that likely function as polyanionic buffers for rapid imbalances in cation concentrations during electrical activity (Hartig et al., 1999). (C) FS cell terminals contain high concentrations of vesicular GABA transporter (VGAT) and are embedded in a dense GABAergic interneuron network. (D) FS cells are devoid of CB_1 receptor immunoreactivity but receive modulatory input from various sources including CB_1 receptor-positive and cholinergic (choline-acetyltransferase (ChAT)-immunoreactive) origin. *Scale bars* = 85 μm (A), 20 μm (B,C), 16 μm (D).

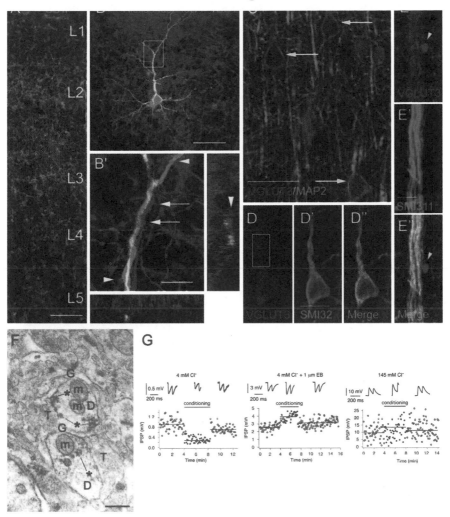

Figure 3. Retrograde signalling at FSN-pyramidal cell synapses is mediated via dendritic vesicular release of glutamate. (A) CB_1 receptor immunocytochemistry revealed highest density of CB_1 receptor-immunoreactive fibres in layer 2/3 of the neocortex, whereas CB_1 receptor-positive cells were predominantly found in deep cortical layers. A neurobiotin (NB)-labelled FSN interneuron-pyramidal cell pair appears in red, and lack CB_1 receptor immunolabelling. L1 – L5 denote cortical layers. (B) Co-localization of CB_1 receptors with identified neuron pairs demonstrated a lack of CB_1 immunoreactivity at inhibitory connections (*open square*). (B') High-power reconstruction of particular inhibitory contacts along the z-axis revealed CB_1 receptor-negative contacts between FSN interneurons and pyramidal cells. Arrows denote CB_1 receptor-negative synaptic contacts between identified cells, whereas arrowheads indicate separate CB_1 receptor-immunoreactive synapses of as yet unidentified origin. AF-488: Alexa Fluor 488. (C-F) VGLUT3-like immunoreactivity is present in dendrites of pyramidal cells in L2/3 of neocortex. (C) Laser-scanning microscopy demonstrated the presence of VGLUT3-like immunoreactivity in populations of putative pyramidal cells immunoreactive for microtubule-associated protein 2 (MAP2, *arrows*). (D-D'') High-power image of a pyramidal cell in layer 2/3 of the neocortex with perisomatic and dendritic VGLUT3-like immunoreactivity. Open square indicates the general localization of (E-E''). (E-E'') VGLUT3-like immunoreactivity in the apical dendritic shaft of a pyramidal cell. Arrowhead denotes VGLUT3-like immunoreactive terminal with a presumed origin in interneurons. (F) Immuno-electron microscopy revealed VGLUT3-like immunoreactivity in vesicle-like structures (*) of pyramidal cell dendrites in rat neocortex. VGLUT3-immunoreactive vesicle-like structures frequently appeared post-synaptically in the vicinity of terminals. *Abbreviations*: D, dendrite; G, glia; m, mitochondria; T, nerve terminal. (G) Evans blue (EB) and high chloride concentrations inhibit glutamate uptake by VGLUT3 also prevent retrograde signalling during conditioning by bAPs. *Scale bars* = 120 μm (A), 65 μm (B), 12 μm (B'), 28 μm (C), 15 μm (D'), 3 μm (E'), 500 nm (F).

Chapter 12

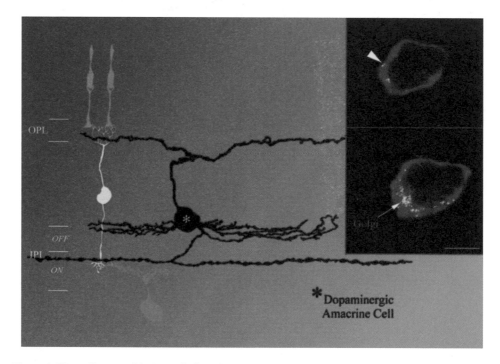

Figure 1. The perikaryon of the DA cell gives rise to a dense dentritic plexus in the scleral stratum of the inner plexiform layer (IPL). From this plexus arise long processes that descend into the middle stratum of the IPL. In addition, a process ascends toward the outer plexiform layer (OPL), where it forms a much looser plexus. Inset: two confocal optical sections of a DA cell double-immunostained for VMAT2 (green) and TH (red). A small number of organelles of varying diameter are scattered throughout the cytoplasm and, occasionally, they are very close to the cell surface (arrowhead). Scale bar: 1μm.

Chapter 13

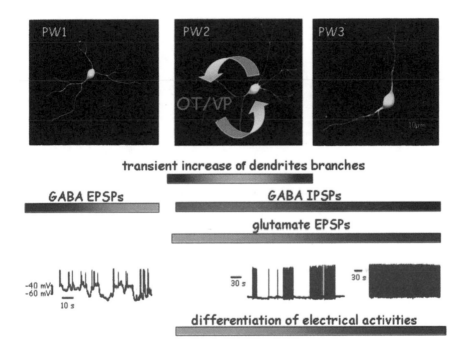

Fig. 5. Morphological plasticity and development of electrophysiological properties of MNCs. At birth and during the first postnatal week (PW1), MNCs possess 2 to 3 primary dendrites, with a small number of proximal branches. The dendritic tree extends transiently during PW2, under the control of locally released oxytocin and vasopressin. During PW3, the dendritic tree retracts to attain its mature shape, typically bipolar with few proximal branches. The synaptic GABA activity, initially excitatory, becomes inhibitory during PW2, when glutamate takes the lead of synaptic excitation. Finally, the initially erratic electrical activity shapes progressively to the typical tonic or phasic firing patterns that respectively characterize oxytocin and vasopressin neurones

Chapter 14

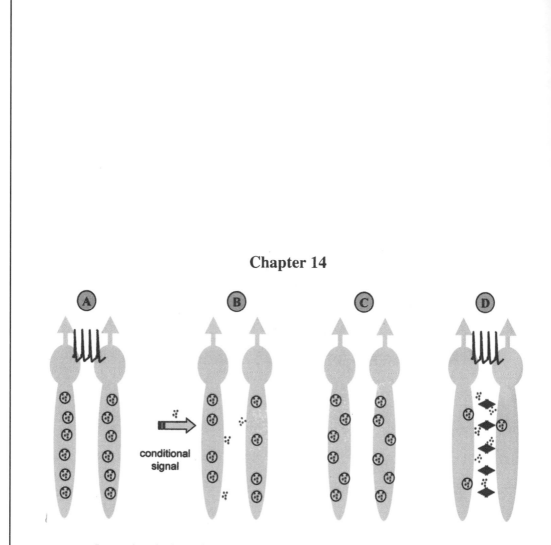

Figure 5. Proposed mechanisms of conditional priming. (A) Under normal conditions dendritic peptide release is not activated by electrical activity. A conditional (probably peptidergic) signal triggers release from dendrites independently of the electrical activity (B) and primes dendritic stores. Priming occurs partially by relocation of dendrite neurosecretory vesicles closer to the membrane (C) where they are available for activity-dependent released for a prolonged period (D).

Chapter 20

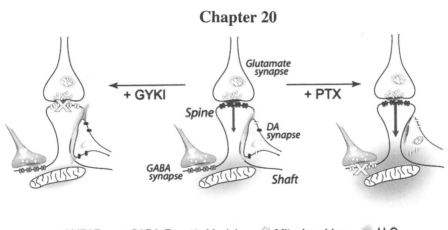

Figure 4. Triad of DA, glutamate and GABA synapses bound together *functionally* by diffusible H_2O_2. Expected H_2O_2 generation in the dendritic shaft of a medium spiny neuron is indicated for control conditions (*center*), $GABA_A$-receptor ($GABA_AR$) blockade by picrotoxin (+ PTX, *right*), and AMPA-receptor (AMPAR) blockade by GYKI-52466 (+ GYKI, *left*). In this model, AMPA-receptor-dependent excitation (red arrows) increases H_2O_2 generation, which opens presynaptic K_{ATP} channels on DA axons; activity-dependent H_2O_2 generation is opposed by $GABA_A$-receptor mediated inhibition. When AMPA-receptors are blocked (+ GYKI), local stimulation does not produce sufficient H_2O_2 to open K_{ATP} channels and alter DA release; opposition by GABA is revealed when $GABA_A$-receptors are blocked (+ PTX), which increase H_2O_2 levels, enhances K_{ATP}-channel opening and further suppresses DA release (circuitry and locations of receptors and mitochondria based on data in Smith and Bolam, 1990; Bernard and Bolam, 1998; Chen et al., 1998; Fujiyama et al., 2000).

Chapter 20

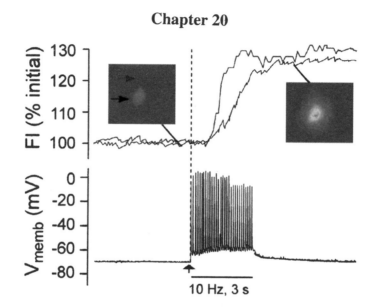

Fig. 5. Generation of H_2O_2 in a striatal medium spiny neuron during local electrical stimulation. Upper panel: fluorescence intensity (FI) in a cell body (black line) and proximal dendrite (red-orange line) in a single, DCF-filled medium spiny neuron under control conditions and during local stimulation (10 Hz, 30 pulses); the apparent plateau is a consequence of irreversible DCF activation. Inset images indicate basal and stimulated fluorescence at monitored sites in the cell body (black arrow) and proximal dendrite (red-orange arrow); DCF (7 μM) was included in the backfill solution of the patch-pipette used for whole-cell recording. DCF images were acquired using a PTI imaging system and Olympus BX-51WI microscope; excitation wavelength was 488 nm with fluorescence emission at 535 nm. Lower panel: simultaneously recorded membrane voltage (V_{memb}) during train stimulation. Dashed vertical line indicates onset of stimulus train.

17

DENDRITIC DYNORPHIN RELEASE IN THE HIPPOCAMPAL FORMATION

Carrie T. Drake[1] and Charles Chavkin*

1. INTRODUCTION

Key questions in neuropeptide research concern how these molecules function as neurotransmitters in the mammalian brain and how the neuropeptide synapse is structurally and spatially organized. Some insights to these questions have been provided by an analysis of the dynorphinergic system in the hippocampus. The endogenous dynorphin opioids are a family of more than five structurally related peptides that are synthesized from a common precursor and coordinately released in the hippocampus (Chavkin et al., 1985). Dynorphins are potent agonists at κ opioid receptors (Chavkin et al., 1982; James et al., 1984; Nock et al., 1990), which are members of the Gi/o family of 7-transmembrane, G-protein coupled receptors (Kieffer, 1995). The conserved presence of dynorphin and κ opioid receptors in the hippocampal formation of species ranging from rats and guinea pigs to humans suggests an important role for this system in hippocampal function. Indeed, κ opioid receptor agonists inhibit hippocampal LTP, a model of learning (Wagner et al., 1992; Wagner et al., 1993; Weisskopf et al., 1993). Hippocampal injection of dynorphin impairs spatial learning in a κ opioid receptor-dependent fashion (Sandin et al., 1998), and elevated dynorphin levels in aged rats are correlated with decreased learning ability (Zhang et al., 1991). Additionally, κ agonists have been reported to inhibit seizure activity in several animal models (Tortella et al., 1986; Tortella, 1988; Tortella and De Coster, 1994); reviewed by (Simonato and Romualdi, 1996). Hippocampal dynorphin expression levels are significantly increased in human temporal lobe epileptics (Houser et al., 1990). Moreover, endogenous dynorphins are protective against temporal lobe epilepsy induced by pilocarpine treatment of rats or mice (Przewlocka et al., 1994; Bausch et al., 1998). Understanding the functional roles of dynorphin has historically been hampered by the apparent spatial mismatch between dynorphin and κ opioid receptors. However, in recent years, the roles played by this neuropeptide family have been significantly clarified using high-resolution anatomical

[1] Department of Neurology and Neurobiology, Weill Medical College of Cornell University, New York, NY 10021 and *Department of Pharmacology, University of Washington, Seattle, WA 98195.

Dendritic Neurotransmitter Release, edited by M. Ludwig
Springer Science+Business Media, Inc., 2005

and physiological techniques. As described below, these studies suggest that granule cells use dynorphin to modulate their excitability via both retrograde and auto-inhibitory mechanisms, and that this modulation critically involves dendritic dynorphin release.

2. DYNORPHIN IS IN GRANULE CELL DENDRITES AND AXON TERMINALS

Although a small amount of dynorphin may be contributed by external afferents (Drake et al., 1994), the vast majority of dynorphin in the hippocampal formation is made by the dentate granule cells (McGinty et al., 1983; Khachaturian et al., 1993). Granule cells extend dendrites throughout the molecular layer, and the outer two-thirds of their dendritic trees receive a massive excitatory input, the perforant path, from entorhinal cortex (Fig. 1). Granule cell axons (the mossy fibers) form numerous collaterals in the hilus, innervating excitatory and inhibitory neurons there (Claiborne et al., 1986). A very small number of axon collaterals penetrate the granule cell layer and extend into the inner portion of the molecular layer, most commonly at the extreme ventral hippocampal pole (Cavazos et al., 1992). Mossy fibers project to stratum lucidum of CA3, where they innervate the proximal portions of CA3 pyramidal cell apical dendrites (Blackstad, 1963). As shown in Fig. 1, the almost complete separation of granule cell axons and dendrites into different fields (hilus and molecular layer, respectively) has rendered this system particularly convenient for studying dendritic versus axonal release of dynorphin.

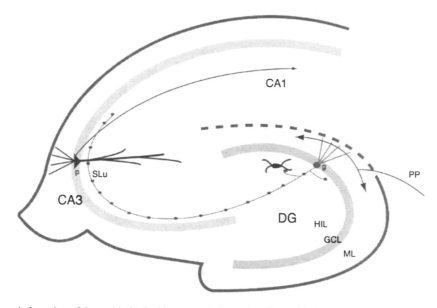

Figure 1. Location of dynorphin in the hippocampal formation. Dynorphin is made by the granule cells (g) of the dentate gyrus (DG). Granule cell somata are tightly packed in the granule cell layer (GCL). The dendrites of granule cells extend into the molecular layer (ML) and receive a massive glutamatergic input from the perforant path (PP). Granule cell axons, known as mossy fibers, give off local collaterals in the hilus (HIL) that innervate excitatory and inhibitory neurons. Granule cells project to stratum lucidum (SLu) in hippocampal region CA3, where they innervate the dendrites of pyramidal cells (p).

Like many other neuropeptides, dynorphin is stored in distinctive organelles known as dense-core vesicles (DCVs; Fig. 2). DCVs consist of an outer membrane, a lumen, and an electron-dense core. In the hippocampus, DCVs are relatively large (80-120 nm in diameter). DCVs are synthesized, transported, and released differently from storage vesicles for fast transmitters such as GABA or glutamate (reviewed in de Camilli and Jahn, 1990). One of the differences of particular relevance for this review is the distinctive stimulation required to trigger release of DCVs. Unlike small synaptic vesicles, such as those containing GABA or glutamate, DCVs require a more prolonged, intense stimulation of the neuron in order to fuse with the membrane and release their contents (Seward et al., 1995). This is linked to differential Ca^{2+} coupling: DCV release is facilitated by a small, widespread elevation of Ca^{2+} while small synaptic vesicle release requires robust, localized Ca^{2+} elevation (Verhage et al., 1991).

DCVs are generally thought to fuse with, and release their contents from, portions of the membrane that are distant from classical synaptic specializations (Thureson-Klein and Klein, 1990). Such nonsynaptic or extrasynaptic sites of release can be the preferential location of DCV fusion, as in *Aplysia californica* (Karhunen et al., 2001). Similarly, in the rat hippocampal mossy fiber terminals, dynorphin-containing DCVs are distributed along both extrasynaptic and synaptic portions of the membrane (Pierce et al., 1999). The majority of these DCVs are extrasynaptic, and changes in the distribution of membrane-associated DCVs following seizure suggest that dynorphin-containing DCVs are

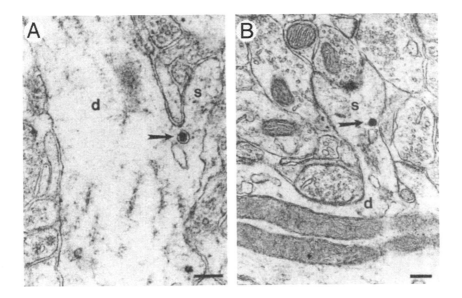

Figure 2. Electron microscopic evidence for dynorphin in granule cell dendrites. **A.** A granule cell dendrite contains a dense-core vesicle (arrow) at the base of a spine (s) extending off a dendritic shaft (d). In this section, which was not processed for immunocytochemistry, the vesicle's dense core can easily be distinguished from its lumen and outer membrane. **B.** A spiny dendrite in the outer molecular layer contains a dynorphin-labeled dense-core vesicle (arrow). The labeled dense-core vesicle is located near the plasma membrane in the spine head (s), but is relatively distant from the excitatory-type synapse. Scale bars: 200 nm. Adapted with permission from Drake et al, 1994 (copyright 1994 by the Society for Neuroscience).

specifically targeted to particular portions of the terminals (Pierce et al., 1999). Thus the shape and dimensions of the neuropeptide "synapse" are defined by the sites of release, the distance to the sites of action and the location of the specific neuropeptide receptors, and these dimensions are strikingly different from those of the fast acting transmitters, glutamate and GABA. The high affinity (usually nanomolar) of neuropeptides for their cognate receptors means that the dilution in the extracellular space following vesicle fusion will not limit the sphere of action to the same extent as it would the lower affinity, fast acting transmitters.

Electron microscopy has revealed dynorphin-containing DCVs in the dendrites (Fig. 2B) as well as the axon terminals of granule cells in both guinea pigs (Drake et al., 1994) and humans (Zhang and Houser, 1999). In the molecular layer of the guinea pig dentate gyrus, the majority (74%) of dynorphin-labeled DCVs are in dendrites. Dynorphin-containing dendrites have the characteristic ultrastructural morphology of granule cells (Claiborne et al., 1990); most notably, the dendrites are spiny, with the complexity of the spines increasing with distance from the granule cell layer. Interestingly, the dendritic spines are themselves a common location of dynorphin-containing DCVs (Drake et al., 1994). In the outer two-thirds of the molecular layer, spines receive virtually all of the excitatory input to granule cells. The large increase in Ca^{2+} concentration within an activated spine provides a depolarization that is transmitted to the dendritic shaft and activates chemical events that are restricted to the spine. In some cases, individual spines behave as discrete functional units, with different Ca^{2+} dynamics from the adjacent dendritic shaft (reviewed by Yuste and Tank, 1996). Such compartmentalization is thought to be critical for synapse-specific plasticity (e.g. LTP) (Yuste and Tank, 1996), which is known to occur in the dentate molecular layer (Levy and Steward, 1979). Our ultrastructural studies of the molecular layer showed that, in dendritic spines, dynorphin-containing DCVs are usually near the extrasynaptic plasma membrane, and are seldom near synapses. However, dynorphin-containing DCVs in dendritic shafts are often in the center of the profile, appearing to be in transit.

A minority (26%) of dynorphin-labeled DCVs in the molecular layer are in axons or axon terminals, particularly in small terminals that may have originated in entorhinal cortex or are rare collaterals of granule cells. Consistent with the latter possibility, the inner molecular layer (which is almost completely devoid of entorhinal afferents) contains the largest proportion of axonal DCVs. Interestingly, at the extreme ventral pole of the hippocampal formation, the number of axonal dynorphin-labeled DCVs was higher than at other levels, with no change in the number of dendritic DCVs, suggesting that some innervation differences are present. The most likely possibility seems to be the mossy fiber collaterals, which as described above are more common at the ventral pole. This dorsal-ventral difference in dynorphin distribution has not been further studied, but is intriguing in light of other functional and structural differences in dynorphin systems of the dorsal compared to the ventral hippocampal formation (McDaniel et al., 1990; Pierce et al., 1994).

3. EVIDENCE FOR DENDRITIC DYNORPHIN RELEASE

Neurotransmission between glutamatergic perforant path afferents and the granule cell dendrites can be studied by locally stimulating the perforant path axons in the molecular layer and measuring the evoked responses in granule cell somata or dendrites.

Applying dynorphin or synthetic ligands of its selective receptor, the κ opioid receptor, depresses perforant path-to-granule cell transmission (Wagner et al., 1992). Pharmacological antagonists have shown that these effects are specifically mediated through κ opioid receptors (Wagner et al., 1992). The basal properties of granule cells are not directly affected by κ receptor agonists. Rather, the κ receptor agonists selectively depress excitatory neurotransmission, indicating a presynaptic locus of action on the perforant path terminals. This is consistent with binding data showing that κ opioid receptors are abundant in the perforant path terminal zone (but not the entire molecular layer) (McLean et al., 1987; Wagner et al., 1991), and that these κ receptors disappear when the entorhinal cortex is lesioned (Simmons et al., 1994).

Other evidence indicates that endogenously released dynorphin can interact with κ receptors in the molecular layer. Using a radioligand displacement assay in hippocampal slices, our group found that stimulating the mossy fibers in the hilus led to the release of an endogenous κ opioid receptor ligand that bound to molecular layer κ receptors, displacing binding of subsequently-added radiolabeled κ receptor agonists (Wagner et al., 1991). Stimulation of perforant path fibers also evoked dynorphin release that displaced radioligand binding. However, this latter method of releasing dynorphin could be blocked by the glutamate receptor antagonist CNQX, indicating the dynorphin was released from a source postsynaptic to the perforant path terminals (Wagner et al., 1991). In physiological studies, high-frequency stimulation in the hilus (which contains the axons of granule cells) led to a depression of the evoked response of granule cells to perforant path stimulation, identical to the effects of applying exogenous κ receptor ligands. This depression was caused by the release of dynorphin, as evidenced by its sensitivity to pretreatment with anti-dynorphin antibodies or the κ receptor antagonist *nor-binaltorphimine* (NBNI)(Wagner et al., 1993; Drake et al., 1994; Terman et al., 1994; Simmons et al., 1995).

Potential sources of dynorphin involved in modulation of the perforant path - granule cell transmission include the dynorphin-rich mossy fibers and sparser, but closer, sources within the molecular layer. Both sources are theoretically plausible. In favor of a hilar source diffusing to the molecular layer, high-frequency stimulation in the hilus directly depolarizes granule cell axons and terminals, and perforant path activity synaptically stimulates them; either type of intense stimulation can trigger release of mossy fiber dynorphin. Although dynorphin released in the hilus would have to diffuse 200-300 μm to the middle and outer molecular layer, dynorphin is relatively resistant to endopeptidase activity (Wagner et al., 1991) suggesting a fairly long diffusion is possible. On the other hand, granule cell somata and dendrites are synaptically depolarized by perforant path stimulation and antidromically depolarized by robust hilar stimulation, so the released dynorphin could come from this source. Moreover, the dynorphin stored in granule cell dendrites, although less abundant than in the mossy fibers, is immediately adjacent to the perforant path terminals.

To establish whether dynorphin acting in the molecular layer was released locally or in the hilus, two approaches were used. First, to test the feasibility of diffusion across various distances, dynorphin was focally applied to various regions of hippocampal slices, simulating local release. To ensure that sufficient dynorphin was applied, the amount of dynorphin delivered in each application (10-100 fmol) was calculated to be greater than the releasable pool of endogenous dynorphin B present in the dentate gyrus (Chavkin et al., 1983; Chavkin et al., 1985). Focused application of dynorphin into the molecular layer reduced excitatory synaptic transmission from the perforant path to the granule cell dendrites, consistent with local dynorphin release in the molecular layer. This effect was sensitive to the κ receptor antagonist NBNI, and was not observed when dynorphin was replaced with saline in the micropipette. In contrast, even the largest applications of dynorphin into the hilus had no effect on molecular layer transmission (Drake et al., 1994). This suggests that diffusion of hilar dynorphin to the molecular layer is effectively limited by dilution and/or degradation, and supports the importance of a local source of dynorphin in the molecular layer.

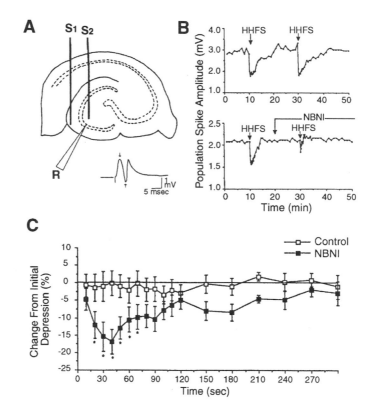

Figure 3. Granule cell responses to perforant path stimulation are depressed following dynorphin-releasing stimulation. **A.** Location of stimulating electrodes (S_1, S_2) and recording electrode (R) in the guinea pig hippocampal slice. The population spike amplitude (arrows) was recorded from the granule cell layer following stimulation of perforant path afferents (with electrode S_1) in the outer molecular layer. **B.** *Top,* Hilar high-frequency stimulation (HHFS) with electrode S_2 produced a transient, reproducible depression of the population spike. *Bottom,* Application of the κ receptor antagonist NBNI before the second HHFS revealed an NBNI-sensitive component of the depression, indicating the involvement of endogenous dynorphin. **C.** Time course of the NBNI-sensitive depression. Data shown are the differences between the first and second responses to HHFS. The second response to HHFS was measured either in control medium (open squares) or in the presence of 100 nM NBNI (solid squares). On the horizontal axis, times listed are seconds following the end of the HHFS; thus the 20-sec measure is actually 31 sec following the start of HHFS. The NBNI-sensitive depression was significantly different (*, $p<0.05$) from control depression for 31-81 sec after beginning HHFS. Adapted with permission from Drake et al, 1994 (copyright 1994 Society for Neuroscience).

Secondly, the kinetics of endogenous dynorphin actions were analyzed to determine whether excitatory activity in the molecular layer was likely to be modulated by local or distant sources of endogenous dynorphin (see Fig. 3). The experimental paradigm consisted of an 11-second high-frequency stimulus (HFS) in the hilus: this stimulation is known to release endogenous dynorphins (Wagner et al., 1991). Excitatory postsynaptic potentials evoked in the molecular layer were measured before and after the HFS. As expected, the HFS led to a NBNI-sensitive depression of the response. Endogenous dynorphin action in the molecular layer was relatively rapid, occurring within 20 seconds of beginning the stimulation, and reached statistical significance by 30 seconds (Drake et al., 1994). Using an estimated diffusion coefficient for dynorphin, we calculated how far dynorphin was likely to diffuse in 20 seconds. Other peptides of the same size have been shown to diffuse in water at 4×10^{-6} cm/sec (Schenk et al., 1991). Adjusting this for the increased tortuosity of the extracellular space and charge interactions between diffusing cations and polyanionic glycosaminoglycans in the extracellular matrix (Rice et al., 1985) yields an effective diffusion coefficient of about 4×10^{-7} cm/sec. Using Fick's law of diffusion ($r^2 = 6D^*t$, where r is the radius of diffusion in the three-dimensional space, D^* is the effective diffusion coefficient, and t is time) (Hille, 1992), the peptide would be expected to move about 70 µm in 20 seconds. Because dynorphins are highly charged and exceptionally "sticky" (Ho et al., 1980), the diffusion coefficient is likely to be somewhat less than estimated, making the actual likely distance even shorter. Nonetheless, these calculations indicate that, to act this quickly, dynorphin should be released within 70 µm of the site of action. Thus, the kinetics of the effect of endogenously released dynorphins are consistent with a local source of dynorphin in the molecular layer.

Further support for the dendrites as an important source of molecular layer dynorphin was obtained when analyzing the Ca^{2+} channels responsible for dynorphin release (Simmons et al., 1995). Antagonists at either the L-type or N-type Ca^{2+} channels were able to block dynorphin release and actions in the molecular layer. However, only N-type Ca^{2+} channel antagonists blocked dynorphin release from the mossy fiber terminals in CA3 (Simmons et al., 1995) or the hilus (Terman et al., 2000). This suggests that the sources of dynorphin acting in the molecular layer are distinct from those contributing dynorphin to the hilus and CA3. Moreover, the differences in Ca^{2+} channel requirements suggest that release of a neurotransmitter from different points of the same neuron type is compartment-specific, consistent with dendritic and axonal differences in receptor and ion channel distribution (Westenbroek et al., 1992).

4. CONSEQUENCES OF DENDRITIC DYNORPHIN RELEASE

Functionally, releasing dynorphin from granule cell dendrites retrogradely inhibits glutamate release from the perforant path afferents (Fig. 4). This synapse in the molecular layer is the first in the classical "trisynaptic pathway" in which cortical information is processed sequentially through the dentate gyrus, hippocampal region CA3, and region CA1. Dynorphins produce inhibition that has a much longer duration than that of the fast transmitters, the released dynorphin produces inhibition that is sustained for many seconds to minutes. The long duration of the dynorphin signal has clear advantages for temporal integration of information.

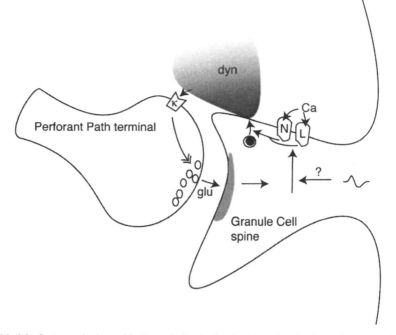

Figure 4. Model of retrograde dynorphin transmission in the dentate molecular layer. Glutamate is released from perforant path terminals onto the dendritic spines of granule cells. When the spine is strongly depolarized (either by strong synaptic stimulation or possibly by antidromic spiking), Ca^{2+} enters the spine through N- and L-type Ca^{2+} channels (N, L) and triggers the release of dynorphin. Released dynorphin diffuses to high-affinity presynaptic κ receptors. Activated κ receptors in turn hyperpolarize the perforant path terminal, decreasing the probability of glutamate release.

There are several potential advantages of retrograde transmission in the molecular layer. First, it provides direct feedback regulation of an excitatory input from its dendritic targets, which limits further inputs. This type of retrograde feedback is an important component of learning models (Medina and Izquierdo, 1995). Second, providing the feedback from the dendrite rather than an axon allows a spatially precise regulation of the active input. It would not be efficient to require that the postsynaptic granule cell grow an axon to deliver dynorphin to each perforant path terminal, especially considering that perforant path terminals form at least 85% of the synapses in the outer and middle molecular layers (Amaral and Witter, 1995). Third, the diffusion of the dynorphin peptides from the sites of release to the targets enables the peptide to activate κ opioid receptors within approximately 70 µm (and possibly more) of the dendritic source. This large sphere of influence allows dynorphin to inhibit surrounding inputs as well as the primary excitatory afferent, while keeping some regional specificity. In the retina, this form of surround inhibition sharpens the spatial boundaries of the input and helps in detecting edges to the signal; surround inhibition by dynorphin may have a similar function in the dentate gyrus. Consistent with this possibility, the perforant path input to the molecular layer is topographically organized (Amaral and Witter, 1995).

Interestingly, κ opioid receptors have also been localized to dense-core vesicles in the dentate gyrus (Drake et al., 1996; Drake et al., 1997) and other CNS regions (Shuster et al., 1999; Svingos et al., 1999; Harris et al., 2004). Many of the κ-receptor-labeled DCVs have been observed in axon terminals, indicating that presynaptic inhibition of transmitter release is a common mechanism. K receptors have also been observed in DCVs in somata of the supraoptic nucleus, a region in which they modulate dendritic transmitter release (see Brown, this volume, and Shuster et al, 1999), and there have been suggestions from confocal microscopy that κ opioid receptors may be in DCVs in dendrites in other regions (Arvidsson et al., 1995). Kappa receptors can be incorporated into the plasma membrane with high-intensity depolarization (Shuster et al., 1999). This further strengthens the argument for an activity-dependent role of the dynorphin-κ receptor system in limiting excitability.

5. COMPARISON WITH AXONAL DYNORPHIN RELEASE

The mossy fibers form dense concentrations of terminals in two zones; the hilus and stratum lucidum of region CA3. In both zones, dynorphin has been shown to modulate excitatory responses and undergo LTP (Weisskopf et al., 1993; Terman et al., 2000). In the hilus, electrophysiological studies have identified an excitatory synaptic input to granule cells that produces a glutamatergic postsynaptic response, including LTP. Both the excitatory postsynaptic response and LTP are diminished by the selective κ receptor agonist U69,593 (Terman et al., 2000). Evoking the release of endogenous dynorphin in the hilus through focal stimulation also reduces the ability of this hilar pathway to excite granule cells. Functionally, this κ-receptor-specific-dynorphin effect is thought to provide endogenous regulation of recurrent excitatory stimulation of granule cells. In the CA3 region, dynorphin released from mossy fiber terminals decreases excitatory synaptic transmission to the pyramidal cells and LTP of this synapse (Weisskopf et al., 1993; Simmons et al., 1995). As described above, axonal dynorphin release in CA3 or the hilar pathway requires only N-type Ca^{2+} channels (Castillo et al., 1994; Simmons et al., 1995; Terman et al., 2000), in contrast to the requirement for both N- and L-type Ca^{2+} channels for dendritic dynorphin release. In contrast to retrograde feedback to afferents in the molecular layer, the mechanism in CA3 is auto-inhibition of glutamate release from the mossy fibers. Dynorphin released from terminals thus acts as a self-limiting modulator of granule cell activity, preventing the transmission of integrated information from the dentate gyrus to the hippocampus.

6. SUMMARY

Endogenous dynorphin opioids are important modulators of excitability in the hippocampal formation. Dynorphin is synthesized by dentate granule cells, which are polarized neurons extending their axons and dendrites into separate fields. Granule cell dendrites and axons release dynorphin by different mechanisms and these two pools of dynorphin affect different cellular targets. Dendritic dynorphin release is controlled by a distinct transmitter release mechanism involving both L- and N-type Ca^{2+} channels, and the released dynorphin retrogradely inhibits glutamate release from cortical afferents. Axon terminal dynorphin release is L-channel independent, and the released dynorphin auto-inhibits glutamate release from mossy fiber terminals. In spite of these differences, dendritic and axonal release of dynorphin may cooperate functionally to yield tight control of granule cell activity. Thus granule cells, by releasing dynorphin into their dendritic field to limit excitatory input and their terminal fields to limit output, can provide an effective brake on their own activity. The requirements for intense stimulation to release dynorphin indicate that this brake would only be applied when the system is highly excited. In the normal hippocampal formation, the role of dynorphin released from both dendritic and axonal sites may be to modulate hippocampal activity between adaptive levels of excitation involved in learning and pathological levels of excitation that promote seizures and cell death.

7. REFERENCES

Amaral D., Witter M., 1995, Hippocampal Formation, in: *The Rat Nervous System*, 2nd Edition (Paxinos G, ed), pp 443-493. San Diego: Academic Press.

Arvidsson U., Riedl M., Chakrabarti S., Vulchanova L., Lee J.-H., Nakano A. H., Lin X., Loh H. H., Law P.-Y., Wessendorf M. W., and Elde R., 1995, The κ-opioid receptor is primarily postsynaptic: Combined immunohistochemical localization of the receptor and endogenous opioids, *PNAS USA* **92**: 5062.

Bausch S. B., Esteb T. M., Terman G. W., and Chavkin C., 1998, Administered and endogenously released kappa opioids decrease pilocarpine-induced seizures and seizure-induced histopathology, *J. Pharmacol. Exp. Ther.* **284**: 1147.

Blackstad T. W., 1963, Ultrastructural studies on the hippocampal region, in: *Progress in Brain Research* (Bargmann W, Schade JP, eds), pp 122-148. New York: Elsevier.

Castillo P. E., Weisskopf M. G., and Nicoll R. A., 1994, The role of Ca^{2+} channels in hippocampal mossy fiber synaptic transmission and long-term potentiation, *Neuron* **12**: 261.

Cavazos J., Golarai G., and Sutula T., 1992, Septohippocampal variation of the supragranular projection of the mossy fiber pathway in the dentate gyrus of normal and kindled rats, *Hippocampus* **2**: 363.

Chavkin C., James I. F., and Goldstein A., 1982, Dynorphin is a specific endogenous ligand of the kappa opioid receptor, *Science* **215**: 413.

Chavkin C., Bakhit C., Weber E., and Bloom F. E., 1983, Relative contents and concomitant release of prodynorphin/neoendorphin-derived peptides in rat hippocampus, *PNAS USA* **80**:7669-7673.

Chavkin C., Shoemaker W. J., McGinty J. F., Bayon A., and Bloom F. E., 1985, Characterization of the prodynorphin and proenkephalin neuropeptide systems in rat hippocampus, *J. Neurosci.* **5**: 808.

Claiborne B. J., Amaral D. G., and Cowan W. M., 1986, A light and electron microscopic analysis of the mossy fibers of the rat dentate gyrus, *J. Comp. Neurol.* **246**: 435.

Claiborne B. J., Amaral D. G., and Cowan W. M., 1990, Quantitative, three-dimensional analysis of granule cell dendrites in the rat dentate gyrus, *J. Comp. Neurol.* **302**: 206.

de Camilli P., and Jahn R., 1990, Pathways to regulated exocytosis in neurons, *Annu. Rev. Physiol.* **52**: 625.

Drake C. T., Chavkin C., and Milner T. A., 1997, Kappa opioid receptor-like immunoreactivity is present in substance P-containing subcortical afferents in guinea pig dentate gyrus, *Hippocampus* **7**: 36.

Drake C. T., Patterson T. A., Simmons M. L., Chavkin C., and Milner T. A., 1996, Kappa opioid receptor-like immunoreactivity in guinea pig brain: Ultrastructural localization in presynaptic terminals in hippocampal formation, *J. Comp. Neurol.* **370**: 377.

Drake C. T., Terman G. W., Simmons M. L., Milner T. A., Kunkel D. D., Schwartzkroin P. A., and Chavkin C. C., 1994, Dynorphin opioids present in dentate gyrus cells may function as retrograde inhibitory neurotransmitters, *J. Neurosci.* **14**: 3736.

Harris J., Chang P., and Drake C., 2004, Kappa opioid receptors in rat spinal cord: sex-linked distribution differences, *Neuroscience* **124**: 879.

Hille B., 1992, *Ionic channels of excitable membranes*. Sunderland, MA: Sinauer Associates.

Ho W., Cox B., Chavkin C., and Goldstein A., 1980, Opioid peptide dynorphin-(1-13): adsorptive losses and potency estimates, *Neuropeptides* **1**: 143.

Houser C. R., Miyashiro J. E., Swartz B. E., Walsh G. O., Rich J. R., and Delgado-Escueta A. V., 1990, Altered patterns of dynorphin immunoreactivity suggest mossy fiber reorganization in human hippocampal epilepsy, *J. Neurosci.* **10**: 267.

James I. F., Fischli W., and Goldstein A., 1984, Opioid receptor selectivity of dynorphin gene products, *J. Pharmacol. Exp. Ther.* **228**: 88.

Karhunen T., Vilim F. S., Alexeeva V., Weiss K. R., and Church P. J., 2001, Targeting of peptidergic vesicles in cotransmitting terminals, *J. Neurosci.* **21**:RC127 (121-125).

Khachaturian H., Schaefer M., and Lewis M., 1993, Anatomy and function of the endogenous opioid system, in: *Opioids I* (Herz A, ed), pp 471-497. Berlin: Springer-Verlag.

Kieffer B., 1995, Recent advances in molecular recognition and signal transduction of active peptides: receptors for opioid peptides, *Cell. Mol. Neurobiol.* **15**: 615.

Levy W., and Steward O., 1979, Synapses as associative memory elements in the hippocampal formation., *Brain Res.* **175**: 233.

McDaniel K. L., Mundy W. R., and Tilson H. A., 1990, Microinjection of dynorphin into the hippocampus impairs spatial learning in rats, *Pharmacol. Biochem. Behav.* **35**: 429.

McGinty J. F., Henriksen S. J., Goldstein A., Terenius L., and Bloom F. E., 1983, Dynorphin is contained within hippocampal mossy fibers: immunochemical alterations after kainic acid administration and colchicine-induced neurotoxicity, *PNAS U S A* **80**: 589.

McLean S., Rothman R. B., Jacobson A. E., Rice K. C., and Herkenham M., 1987, Distribution of opiate receptor subtypes and enkephalin and dynorphin immunoreactivity in the hippocampus of squirrel, guinea pig, rat, and hamster, *J. Comp. Neurol.* **255**: 497.

Medina J., and Izquierdo I., 1995, Retrograde messengers, long-term potentiation and memory, *Brain Res. Rev.* **21**: 185.

Nock B., Giordano A. L., Cicero T. J., and O'Connor L. H., 1990, Affinity of drugs and peptides for U-69,593-sensitive and -insensitive kappa opiate binding sites: the U-69,593-insensitive site appears to be the beta endorphin-specific epsilon receptor, *J. Pharmacol. Exp. Ther.* **254**: 412.

Pierce J. P., Kurucz O. S., and Milner T. A., 1994, An ultrastructural analysis of seizure-induced leu-enkephalin changes in the ventral dentate gyrus of rats, *Regul. Peptides* **54**: 225.

Pierce J. P., Kurucz O. S., and Milner T. A., 1999, Morphometry of a peptidergic transmitter system: dynorphin B-like immunoreactivity in the rat hippocampal mossy fiber pathway before and after seizures, *Hippocampus* **9**: 255.

Przewlocka B., Machelska H., and Lason W., 1994, Kappa opioid receptor agonists inhibit the pilocarpine-induced seizures and toxicity in the mouse, *Eur. Neuropsychopharmacol.* **4**: 527-.

Rice M., Gerhardt G., Hierl P., Nagy G., and Adams R., 1985, Diffusion coefficients of neurotransmitters and their metabolites in brain extracellular fluid space, *Neuroscience* **15**: 891.

Sandin J., Nylander I., Georgieva J., Schott P. A., Ogren S. O., and Terenius L., 1998, Hippocampal dynorphin B injections impair spatial learning in rats: a kappa-opioid receptor-mediated effect, *Neuroscience* **85**: 375.

Schenk J., Morocco M., and Ziemba V., 1991, Interactions between argininyl moieties of neurotensin and catechol protons of dopamine, *J. Neurochem.* **57**: 1787.

Seward E., Chernevskaya N., and Nowycky M., 1995, Exocytosis in peptidergic nerve terminals exhibits two calcium-sensitive phases during pulsatile calcium entry, *J. Neurosci.* **15**: 3390.

Shuster S. J., Riedl M., Li X., Vulchanova L., and Elde R., 1999, Stimulus-dependent translocation of kappa opioid receptors to the plasma membrane, *J. Neurosci.* **19**: 2658.

Simmons M. L., Terman G. W., Drake C. T., and Chavkin C., 1994, Inhibition of glutamate release by presynaptic kappa 1-opioid receptors in the guinea pig dentate gyrus, *J. Neurophysiol.* **72**: 1697.

Simmons M. L., Terman G.W., Gibbs S. M., and Chavkin C., 1995, L-type calcium channels mediate dynorphin neuropeptide release from dendrites but not axons of hippocampal granule cells, *Neuron* **14**: 1265.

Simonato M., and Romualdi P., 1996, Dynorphin and epilepsy, *Prog. Neurobiol.* **50**: 557.

Svingos A. L., Colago E. E., and Pickel V. M., 1999, Cellular sites for dynorphin activation of kappa-opioid receptors in the rat nucleus accumbens shell, *J. Neurosci.* **19**: 1804.

Terman G., Drake C., Simmons M., Milner T., and Chavkin C., 2000, Opioid modulation of recurrent excitation in the hippocampal dentate gyrus, *J. Neurosci.* **20**: 4379.

Terman G. W., Wagner J. J., and Chavkin C., 1994, Kappa opioids inhibit induction of long-term potentiation in the dentate gyrus of the guinea pig hippocampus, *J. Neurosci.* **14**: 4740.

Thureson-Klein A. K., and Klein R. L., 1990, Exocytosis from neuronal large dense-core vesicles, *Int. Rev. Cytol.* **121**: 67.

Tortella F. C., 1988, Endogenous opioid peptides and epilepsy: quieting the seizing brain? *Trends Pharmacol. Sci.* **9**: 366.

Tortella F. C., and De Coster M. A., 1994, Kappa opioids: therapeutic considerations in epilepsy and CNS injury, *Clin. Neuropharmacol.* **17**: 403.

Tortella F. C., Robles L., and Holaday J. W., 1986, U50,488, a highly selective kappa opioid: anticonvulsant profile in rats, *J. Pharmacol. Exp. Ther.* **237**: 49.

Verhage M., McMahaon H., Ghijsen W., Boomsma F., Scholten G., Wiegant V., and Nicholls D., 1991, Differential release of amino acids, neuropeptides, and catecholamines from isolated nerve terminals, *Neuron* **6**: 517.

Wagner J. J., Evans C. J., and Chavkin C., 1991, Focal stimulation of the mossy fibers releases endogenous dynorphins that bind kappa1-opioid receptors in guinea pig hippocampus, *J. Neurochem.* **57**: 333.

Wagner J. J., Caudle R. M., and Chavkin C., 1992, Kappa-opioids decrease excitatory transmission in the dentate gyrus of the guinea pig hippocampus, *J. Neurosci.* **12**: 132.

Wagner J. J., Terman G. W., and Chavkin C., 1993, Endogenous dynorphins inhibit excitatory neurotransmission and block LTP induction in the hippocampus, *Nature* **363**: 451.

Weisskopf M. G., Zalutsky R. A., and Nicoll R. A., 1993, The opioid peptide dynorphin mediates heterosynaptic depression of hippocampal mossy fibre synapses and modulates long-term potentiation, *Nature* **362**: 423.

Westenbroek R. E., Hell J. W., Warner C., Dubel S. J., Snutch T. P., and Catterall W. A., 1992, Biochemical properties and subcellular distribution of an N-type calcium channel alpha 1 subunit, *Neuron* **9**: 1099.

Yuste R., and Tank D. W., 1996, Dendritic integration in mammalian neurons, a century after Cajal, *Neuron* **16**: 701.

Zhang N., and Houser C. R., 1999, Ultrastructural localization of dynorphin in the dentate gyrus in human temporal lobe epilepsy: a study of reorganized mossy fiber synapses, *J. Comp. Neurol.* **405**: 472.

Zhang W. Q., Mundy W. R., Thai L., Hudson P. M., Gallagher M., Tilson H. A., and Hong J. S., 1991, Decreased glutamate release correlates with elevated dynorphin content in the hippocampus of aged rats with spatial learning deficits, *Hippocampus* **1**: 391.

18

CONTROL OF SYNAPTIC TRANSMISSION IN THE CNS THROUGH ENDOCANNABINOID-MEDIATED RETROGRADE SIGNALING

Takako Ohno-Shosaku, Takashi Maejima, Takayuki Yoshida, Kouichi Hashimoto, Yuko Fukudome and Masanobu Kano[*]

1. INTRODUCTION

Psychological and physiological effects of marijuana are caused by binding of its active component (Δ^9-tetrahydrocannabinol) to cannabinoid receptors. The cannabinoid receptors belong to a family of G protein-coupled seven-transmembrane-domain receptors, and consist of type 1 (CB1) and type 2 (CB2) receptors with different distributions (Matsuda et al., 1990; Munro et al., 1993; Felder and Glass, 1998). The CB1 receptor is expressed in the CNS, whereas the CB2 receptor is found in the immune system of the periphery (Klein et al., 1998). Activation of the CB1 receptor induces various effects on neural functions (Di Marzo et al., 1998; Felder and Glass, 1998), including suppression of neurotransmitter release (Gifford and Ashby, 1996; Ishac et al., 1996; Shen et al., 1996; Katona et al., 1999; Hoffman and Lupica, 2000). Several molecules are identified as candidate endogenous ligands for cannabinoid receptors (endocannabinoids). Arachidonylethanolamide (anandamide) and 2-arachidonoylglycerol (2-AG), two major endocannabinoids, are reported to be synthesized from membrane phospholipids in an activity- and a Ca^{2+}-dependent manner (Cadas et al., 1996; Stella et al., 1997; Di Marzo et al., 1998; Bisogno et al., 1999; Piomelli et al., 2000). It is thought that they can diffuse out across the cell membrane. The released endocanabinoids are removed from the extracellular space through uptake and enzymatic degradation (Mechoulam et al., 1998). All these findings suggest that endocannabinoids can work as a diffusible and short-lived mediator that is released from activated neurons, binds to cannabinoid receptors on neighboring neurons to modulate their functions.

Recent electrophysiological studies have revealed that endocannabinoids play an important role in retrograde modulation of synaptic transmission in the CNS (Kreitzer and Regehr, 2001b; Maejima et al., 2001a; Ohno-Shosaku et al., 2001; Wilson and

*T. Ohno-Shosaku, T. Maejima, T.Yoshida, K. Hashimoto, Y. Fukudome and M. Kano, Department of Cellular Neurophysiology, Graduate School of Medical Science, Kanazawa University, Kanazawa 920-8640, Japan

Dendritic Neurotransmitter Release, edited by M. Ludwig
Springer Science+Business Media, Inc., 2005

Nicoll, 2001). Endocannabinoids are released from postsynaptic neurons in response to either depolarization or activation of $G_{q/11}$-coupled receptors such as group I metabotropic glutamate receptors (mGluRs) and M_1/M_3 muscarinic acetylcholine receptors. The released endocannabinoids then activate presynaptic cannabinoid receptors and suppress transmitter release (Maejima et al., 2001b; Alger, 2002; Kano et al., 2002; Kreitzer and Regehr, 2002; Wilson and Nicoll, 2002; Freund et al., 2003; Kano et al., 2003; Piomelli, 2003). Thus, the endocannabinoid signaling is an important mechanism by which postsynaptic neuronal activity can retrogradely influence presynaptic functions. In this review, we introduce recent electrophysiological studies on endocannabinoid-mediated retrograde modulation and discuss its possible physiological roles in the CNS.

2. ENDOCANNABINOID-MEDIATED RETROGRADE MODULATION OF SYNAPTIC TRANSMISSION

2.1. Depolarization-Induced Suppression of Inhibition (DSI)

More than ten years ago, depolarization of a postsynaptic neuron was found to induce a transient suppression of inhibitory postsynaptic currents (IPSCs) in neurons of both the cerebellum (Llano et al., 1991) and the hippocampus (Pitler and Alger, 1992; Ohno-Shosaku et al., 1998). This phenomenon was termed depolarization-induced suppression of inhibition (DSI). DSI is induced by a depolarizing voltage pulse with a long duration (several seconds). However, a train of action potentials can also induce DSI (Pitler and Alger, 1992; Morishita and Alger, 2001; Ohno-Shosaku et al., 2001), suggesting that DSI may occur physiologically *in vivo*. DSI lasts for tens of seconds and the magnitude of DSI depends on the duration of depolarization (Ohno-Shosaku et al., 1998; Lenz and Alger, 1999). Injection of a fast Ca^{2+} chelator, BAPTA, into the postsynaptic neuron inhibits DSI (Pitler and Alger, 1992; Vincent and Marty, 1993; Ohno-Shosaku et al., 2001), indicating that elevation of postsynaptic Ca^{2+} concentration is required for the induction of DSI. DSI accompanies a clear increase in the paired-pulse ratio (Ohno-Shosaku et al., 1998; Wilson and Nicoll, 2001), a widely used indicator of presynaptic modulation (Zucker and Regehr, 2002), but not a comparable decrease in the postsynaptic sensitivity to the inhibitory transmitter GABA (Pitler and Alger, 1992; Ohno-Shosaku et al., 2001). These data indicate that the depolarization-induced elevation of intracellular Ca^{2+} concentration in the postsynaptic neuron induces suppression of GABA release from presynaptic terminals. Thus, it is evident that DSI requires a retrograde signal from depolarized postsynaptic neurons to presynaptic terminals.

Recent electrophysiological studies have revealed that endocannabinoids mediate the retrograde signal of DSI, by using hippocampal cultures (Ohno-Shosaku et al., 2001), hippocampal slices (Wilson and Nicoll, 2001) and cerebellar slices (Kreitzer and Regehr, 2001a; Diana et al., 2002; Yoshida et al., 2002). Our study on hippocampal cultures clearly demonstrates that hippocampal inhibitory synapses are heterogeneous in cannabinoid sensitivity (Ohno-Shosaku et al., 2001). The cannabinoid agonist WIN55,212-2 is effective only at a subset of inhibitory synapses, and suppresses GABA release through activation of the CB1 receptor (Ohno-Shosaku et al., 2001). DSI can be induced only at cannabinoid-sensitive synapses, and blocked by CB1 antagonists (AM 281 and SR141716A) (Fig. 1). These results clearly indicate that DSI is mediated by endocannabinoids (Ohno-Shosaku et al., 2001). Another study on hippocampal slices

(Wilson and Nicoll, 2001) shows essentially the same results as those in cultures. In addition, DSI is not affected by postsynaptically-applied botulinum toxin (Wilson and Nicoll, 2001), suggesting that vesicular release is not involved in DSI. In cerebellar slices, it has been demonstrated that DSI of Purkinje cells is completely occluded by WIN55,212-2-induced suppression of inhibitory transmission, blocked by CB1 antagonists and deficient in CB1-knockout mice (Kreitzer and Regehr, 2001a; Diana et al., 2002; Yoshida et al., 2002). Involvement of endocannabinoids in DSI has also been reported in other regions of the CNS including the basal ganglia (Wallmichrath and Szabo, 2002) and the neocortex (Trettel and Levine, 2003; Trettel et al., 2004). These findings strongly suggest that endocannabinoids mediate DSI widely in the CNS.

2.2. Depolarization-Induced Suppression of Excitation (DSE)

Endocannabinoids, released from depolarized postsynaptic neurons, can also suppress transmitter release from excitatory presynaptic terminals. Depolarization of cerebellar Purkinje cells induces transient suppression of excitatory postsynaptic currents

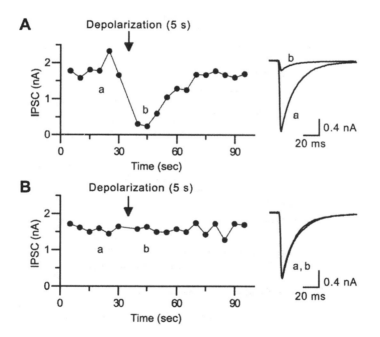

Figure 1. Retrograde modulation induced by depolarization at hippocampal inhibitory synapses. In the experiments shown in Figs. 1-4, inhibitory postsynaptic currents (IPSCs) were recorded from cultured hippocampal neurons prepared from rats, as described previously (Ohno-Shosaku *et al.*, 2001). A. An example of DSI. When a postsynaptic neuron was depolarized from −80 mV to 0 mV for 5 sec, the amplitude of IPSCs was transiently decreased. B: After the treatment with the cannabinoid antagonist AM281 (0.3 μM), the depolarization failed to induce DSI. IPSC traces acquired at the indicated time points are shown on the right.

(EPSCs) elicited by stimulation of climbing fibers or parallel fibers (Kreitzer and Regehr, 2001b; Maejima et al., 2001a). This depolarization-induced suppression of excitation (DSE) lasts for tens of seconds, is blocked by postsynaptic injection of BAPTA, and accompanies a clear change in the paired-pulse ratio (Kreitzer and Regehr, 2001b). DSE is blocked by CB1 antagonists (Kreitzer and Regehr, 2001b), and deficient in CB1-knockout mice (our unpublished data). All these properties are quite similar to those of DSI. These results indicate that endocannabinoids are released from depolarized Purkinje cells, and suppress the release of the excitatory transmitter glutamate through activation of the presynaptic CB1 receptor.

DSE can also be induced in other brain regions including the hippocampus (Ohno-Shosaku et al., 2002b) and ventral tegmental area (Melis et al., 2004). In the hippocampus, however, DSE is less prominent than DSI. The duration of depolarization required for DSE induction is longer than that for DSI, and the magnitude of DSE is much smaller than that of DSI when the neuron is fully depolarized. It is demonstrated that EPSCs are less sensitive to WIN55,212-2 than the cannabinoid-sensitive IPSCs. Thus, presynaptic cannabinoid sensitivity appears to be a major factor that determines the extent of endocannabinoid-mediated suppression.

While the cerebellar DSE is clearly CB1-dependent, identity of the cannabinoid receptor involved in the hippocampal DSE is controversial. It is reported that the third as

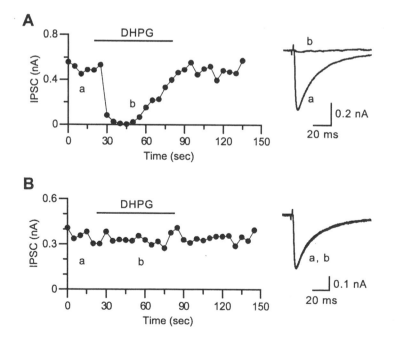

Figure 2. Retrograde modulation induced by activation of group I mGluRs at hippocampal inhibitory synapses. A. Application of a group I mGluR agonist, DHPG (50 µM), induced a suppression of IPSCs. B. This DHPG-induced suppression was blocked by the AM281 treatment.

yet unidentified subtype of cannabinoid receptor (CB3) exists at excitatory synapses in adult hippocampal slices (Hajos et al., 2001; Kofalvi et al., 2003). This is based on the experimental results that the suppression of excitatory synaptic transmission (Hajos et al., 2001) or glutamate release (Kofalvi *et al.*, 2003) by WIN55,212-2 is present in CB1-knockout mice, but the effect is sensitive to the cannabinoid antagonist SR141716A. However, our results on cultured hippocampal neurons indicate that both DSE and the suppression of EPSCs by WIN55,212-2 are totally deficient in CB1-knockout mice (Ohno-Shosaku et al., 2002b). The reason for this discrepancy is unclear. It is possible that the putative CB3 receptor is not expressed in immature neurons, but appears later in adult animals. The molecular identity and the expression pattern of the putative CB3 receptor, if it exists, remain to be elucidated.

2.3. Retrograde Modulation Induced by mGluR Activation

In addition to depolarization, postsynaptic activation of certain types of receptors can induce endocannabinoid-mediated retrograde suppression. This type of modulation was originally found in cerebellar Purkinje cells (Maejima et al., 2001a). In mouse cerebellar slices, application of a group I mGluR agonist, (RS) 3,5 dihydroxyphenylglycine (DHPG), induces suppression of climbing fiber-mediated EPSCs (CF-EPSCs) recorded from Purkinje cells. Several lines of evidence indicate that this suppression is caused by activation of mGluR subtype 1 (mGluR1), a member of group I mGluRs, at postsynaptic Purkinje cells. Firstly, the DHPG-induced suppression is blocked by postsynaptic injection of GDP-β-S or GTP-γ-S. Secondly, the suppression is absent in the mGluR1-knockout mouse. Finally, this defect is restored in the mGluR1-rescue mouse in which rat mGluR1α is introduced into the mGluR1-knockout mouse by using a Purkinje cell specific promotor. Although the induction is postsynaptic, the expression of this suppression is clearly presynaptic. The DHPG-induced suppression is associated with a clear increase in the paired-pulse ratio, indicating that the suppression is caused by the reduction of glutamate release from presynaptic terminals. Therefore, some retrograde signal must exist for this synaptic modulation. It has been demonstrated that the retrograde signal is also mediated by endocannabinoids. The DHPG-induced suppression of CF-EPSCs is mimicked and occluded by the cannabinoid agonist WIN55,212-2, and blocked by CB1 antagonists. Thus, activation of mGluR1 on postsynaptic Purkinje cells stimulates production and release of endocannabinoids, which retrogradely act on the presynaptic CB1 receptor and suppress glutamate release from climbing fiber terminals. Importantly, this mGluR-induced endocannabinoid release does not require elevation of intracellular Ca^{2+} concentration in postsynaptic Purkinje cells, forming a striking contrast to the depolarization-induced release of endocannabinoids. Thus, the endocannabinoid-mediated retrograde modulation can be initiated by two distinct stimuli, namely the depolarization-induced Ca^{2+} elevation and the mGluR1 activation. Furthermore, it should be noted that the activation of mGluR1 by synaptically released glutamate after repetitive stimulation of parallel fibers, which are the other excitatory inputs to Purkinje cells, causes transient cannabinoid-dependent suppression of CF-EPSCs (Maejima et al., 2001a). This result suggests that the mGluR1-mediated generation of the endocannabinoid signal is functional *in vivo* and may contribute to the control of presynaptic functions.

The endocannabinoid-mediated retrograde modulation triggered by activation of group I mGluRs was also found at hippocampal inhibitory synapses (Varma et al., 2001;

Ohno-Shosaku et al., 2002a). A study on cultured hippocampal neurons demonstrates that DHPG suppresses cannabinoid-sensitive IPSCs (Fig. 2A), but not cannabinoid-insensitive IPSCs (Ohno-Shosaku et al., 2002a). This effect is largely blocked by an antagonist specific to mGluR subtype 5 (mGluR5), the other member of group I mGluRs. The DHPG-induced suppression is associated with an increase in the paired-pulse ratio, and blocked by CB1 antagonists (Fig. 2B) (Ohno-Shosaku et al., 2002a), indicating that the activation of group I mGluRs (mostly mGluR5) produces endocannabinoids and suppresses GABA release from cannabinoid-sensitive presynaptic terminals through activation of the CB1 receptor.

2.4. Retrograde Modulation Induced by Muscarinic Activation

Activation of muscarinic receptors produces various effects on neuronal functions in the CNS. Application of cholinergic or muscarinic agonists depresses synaptic transmission in various regions of the brain, primarily by suppressing transmitter release. Recently, it was demonstrated that muscarinic suppression of hippocampal inhibitory transmission is partly cannabinoid-dependent in slices (Kim et al., 2002) and cultured neurons (Fukudome et al., 2004). In the latter study, it was shown that two distinct mechanisms mediate the muscarinic suppression. At a subset of synapses, activation of

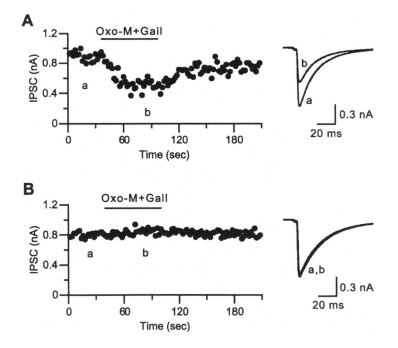

Figure 3. Retrograde modulation induced by activation of muscarinic receptors at hippocampal inhibitory synapses. A. Application of a muscarinic agonist, oxotremoline-M (3 μM), induced suppression of IPSCs under the blockade of the presynaptic M2 muscarinic receptor by gallamine (100 μM). B. This suppression was blocked by the AM281 treatment.

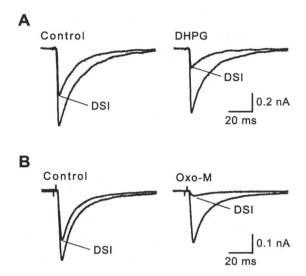

Figure 4. Synergistic effect of depolarization and receptor activation on retrograde suppression. A. DSI induced by a brief depolarization (1 sec) was enhanced by 5 μM DHPG. B. DSI induced by a brief depolarization (1 sec) was enhanced by 0.3 μM oxotremoline-M.

the presynaptic M_2 receptor suppresses GABA release directly. In contrast, at a different subset of synapses, activation of the postsynaptic M_1/M_3 receptors causes endocannabinoid release and subsequently suppresses GABA release by activating the presynaptic CB1 receptor (Fig. 3).

Since both group I mGluRs and M_1/M_3 receptors are coupled to $G_{q/11}$ proteins and capable to trigger endocannabinoid production, it is possible that other $G_{q/11}$-coupled receptors can also be involved in the retrograde modulation by triggering endocannabinoid signaling in the CNS. In this context, it is interesting that the activation of G-protein-coupled corticosteroid receptor triggers the endocannabinoid release and suppresses the glutamate release retrogradely in the hypothalamus (Di et al., 2003), although the molecular identity of this receptor is not clear.

2.5. Interaction between the Depolarization and the Activation of $G_{q/11}$-coupled Receptors

As described above, endocannabinoid release can be triggered by the two distinct stimuli, namely the depolarization-induced Ca^{2+} elevation and the activation of $G_{q/11}$-coupled receptors. These two pathways may interact with each other, because simultaneous application of these two stimuli produces a synergistic effect (Varma et al., 2001; Kim et al., 2002; Ohno-Shosaku et al., 2002a; Ohno-Shosaku et al., 2003).

The activation of group I mGluRs by DHPG enhances DSI in hippocampal slices (Varma et al., 2001; Ohno-Shosaku et al., 2002a) and cultures (Fig. 4A) (Ohno-Shosaku et al., 2002a). This enhancement is much more prominent (about 7-times higher) than what is expected from the simple summation of the depolarization-induced and the group I mGluR-induced endocannabinoid release (Ohno-Shosaku et al., 2002a). DHPG causes no change in depolarization-induced Ca^{2+} transients, indicating that the enhanced DSI by DHPG is not due to the augmentation of Ca^{2+} influx. These results suggest that the two pathways work in a cooperative manner to release endocannabinoids, at least partly, through a common intracellular cascade.

Activation of muscarinic receptors also enhances DSI in hippocampal slices (Kim et al., 2002) and cultures (Ohno-Shosaku et al., 2003). The cholinergic agonist carbachol (CCh) markedly enhances DSI at 0.01-0.3 μM without changing the presynaptic cannabinoid sensitivity (Ohno-Shosaku et al., 2003). The facilitating effect of CCh on DSI is mimicked by a muscarinic agonist (Fig. 4B), and blocked by a muscarinic antagonist. It is also blocked by intracellular injection of GDP-β-S to postsynaptic neurons. By using knockout mice lacking one or two subtypes of muscarinic receptors, it has been shown that both M_1 and M_3 receptors are involved in the muscarinic enhancement of DSI (Ohno-Shosaku et al., 2003). These results indicate that the activation of the postsynaptic M_1/M_3 receptors facilitates the depolarization-induced release of endocannabinoids in the hippocampus.

In addition to these $G_{q/11}$-coupled receptors, D_2 dopamine receptors might also be involved in the modulation of endocannabinoid signalling. In ventral tegmental area dopamine neurons, DSE is partially blocked by a D_2 antagonist, and enhanced by a D_2 agonist without changing the presynaptic cannabinoid sensitivity (Melis et al., 2004). Although the D_2 receptor is $G_{i/o}$-coupled, the authors suggest that its enhancing effect on DSE might be caused by the stimulation of PLC, probably through βγ subunits.

2.6. Limited Diffusion of Endocannabinoids

The endocannabinoids released to the extracellular space are thought to be rapidly removed by uptake and enzymatic degradation (Di Marzo et al., 1998; Piomelli et al., 2000). The limited diffusion of endocannabinoids has been reported in several electrophysiolgial studies on hippocampal and cerebellar slices. In hippocampal slices, it was shown that endocannabinoids released by depolarization can affect only the synapses on neighboring neurons located within 20 μm from the depolarized neurons (Wilson and Nicoll, 2001). In cerebellar slices, it has been demonstrated that endocannabinoids released from a Purkinje cell can not spread to the synaptic terminals onto neighboring Purkinje cells (Maejima et al., 2001a) but can spread to closely located interneurons (Kreitzer et al., 2002). These results indicate that the diffusion of endocannabinoids is rather limited in brain tissues. Thus, endocannabinoids can function as a spatially limited local signal in the CNS.

2.7. Current Model of Endocannabinoid-Mediated Retrograde Modulations

Figure 5 shows a current model for the mechanisms of endocannabinoid-mediated retrograde modulations. Postsynaptic depolarization triggers Ca^{2+} influx by activating voltage-gated Ca^{2+} channels, and causes transient elevation of intracellular Ca^{2+} concentration. This Ca^{2+} elevation and the activation of $G_{q/11}$-coupled receptors such as group I mGluRs and M_1/M_3 muscarinic receptors work in a cooperative manner to produce and release endocannabinoids. The released endocannabinoids diffuse retrogradely and bind to the CB1 receptor on excitatory or inhibitory presynaptic terminals. The binding of endocannabinoids to the CB1 receptor activates $G_{i/o}$ proteins and suppresses the release of glutamate or GABA, presumably by inhibiting voltage-

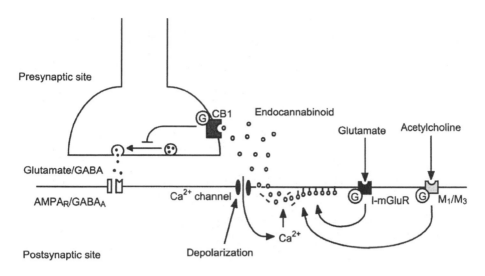

Figure 5. A current model of endocannabinoid-mediated retrograde modulations at central synapses.

gated Ca^{2+} channels (Hoffman and Lupica, 2000; Guo and Ikeda, 2004), activating K^+ channels (Daniel and Crepel, 2001; Guo and Ikeda, 2004) or inhibiting vesicular release machinery directly (Takahashi and Linden, 2000).

2.8. Wide Distribution of the Presynaptic CB1 and Postsynaptic G_q/G_{11}-coupled Receptors in the CNS

Elevation of intracellular Ca^{2+} concentration that triggers endocannabinoid release can be induced by a variety of neural activity through voltage-gated Ca^{2+} channels, NMDA receptors and Ca^{2+} release channels on intracellular stores. Most neurons in the CNS express group I mGluRs on their postsynaptic sites (Abe et al., 1992; Shigemoto et al., 1992; Shigemoto et al., 1997; Thomas et al., 1997). M_1/M_3 muscarinic receptors are also widely distributed in the CNS. In addition, the CB1 receptor is expressed at presynaptic terminals in various regions of the brain (Herkenham et al., 1990; Matsuda et al., 1993; Egertova and Elphick, 2000). Therefore, the endocannabinoid-mediated retrograde modulation observed in the hippocampus and the cerebellum may be a general mechanism, through which postsynaptic Ca^{2+} elevation and/or activation of G_q/G_{11}-coupled receptors can influence presynaptic functions.

3. FUNCTIONAL SIGNIFICANCE OF THE ENDOCANNABINOID-MEDIATED RETROGRADE MODULATION

What could be a functional role of the endocannabinoid-mediated retrograde modulation in the CNS? If endocannabinoids released from an activated neuron suppress its excitatory and inhibitory inputs to the same extent, the excitatory-inhibitory balance of the activated neuron will remain unchanged. In the hippocampus, however, there is a substantial difference in cannabinoid sensitivity between excitatory and inhibitory inputs. When a small amount of endocannabinoid is released from the activated postsynaptic neuron, the cannabinoid-sensitive inhibitory input will be selectively suppressed because of its high cannabinoid sensitivity. Under this condition, the endocannabinoid signal will exert net excitatory action on the postsynaptic neuron. When a large amount of endocannabinoid is released, both the excitatory and the cannabinoid-sensitive inhibitory inputs will be suppressed, but the cannabinoid-insensitive inhibitory input is intact. Under this condition, the endocannabinoid signal will exert net inhibitory action. Thus, endocannabinoids can control the excitatory-inhibitory balance depending on the postsynaptic activity in the hippocampus.

In addition to the short-term modulation, several recent studies suggest that the endocannabinoid system might also be involved in long-term synaptic plasticity, learning and memory. The induction of long-term potentiation (LTP) of the excitatory transmission in the hippocampus is facilitated during DSI (Carlson et al., 2002). Long-term depression (LTD) of the excitatory transmission in the striatum (Gerdeman et al., 2002; Robbe et al., 2002; Ronesi et al., 2004) and the neocortex (Sjöström et al., 2003) is suppressed by cannabinoid antagonists, and the striatal LTP is absent in CB1-knockout mice. LTD of the inhibitory transmission in the basolateral amygdala (Marsicano et al., 2002) and the hippocampus (Chevaleyre and Castillo, 2003) is also suppressed by cannabinoid antagonists. Furthermore, the extinction of certain forms of memory (Marsicano et al., 2002; Varvel and Lichtman, 2002) is reported to be impaired in CB1-knockout mice. Precise mechanisms for the roles of the endocannabinoid system in long-term synaptic plasticity, learning and memory remain to be investigated.

4. UNRESOLVED QUESTIONS

Several important questions remain unanswered. (1) What molecule(s) (anandamide, 2-AG or some other endocannabinoids) contributes to the retrograde modulation induced by the depolarization and/or the receptor activation? (2) What biochemical pathways are involved and which enzyme(s) is rate-limiting in the generation of endocannabinoids? (3) How do the depolarization and the receptor activation interact in the production and release of endocannabinoid? (4) How are endocannabinoids released across the cell membrane to the extracellular space and taken up into the pre-synaptic cells? (5) How does the activation of the presynaptic CB1 receptor lead to the suppression of transmitter release? (6) How are endocannabinoids involved in long-term synaptic plasticity, learning and memory? (7) How does the endocannabinoid signaling contribute to higher brain functions? Future studies should answer these important questions from the molecular/cellular level to the system level by combining various methodologies.

5. REFERENCES

Abe, T., Sugihara, H., Nawa, H., Shigemoto, R., Mizuno, N., and Nakanishi, S., 1992, Molecular characterization of a novel metabotropic glutamate receptor mGluR5 coupled to inositol phosphate/Ca^{2+} signal transduction, *J. Biol. Chem.* **267**: 13361.
Alger, B. E., 2002, Retrograde signaling in the regulation of synaptic transmission: focus on endocannabinoids, *Prog. Neurobiol.* **68**: 247.

Bisogno, T., Melck, D., De Petrocellis, L., and Di Marzo, V., 1999, Phosphatidic acid as the biosynthetic precursor of the endocannabinoid 2-arachidonoylglycerol in intact mouse neuroblastoma cells stimulated with ionomycin, *J. Neurochem.* **72**: 2113.

Cadas, H., Gaillet, S., Beltramo, M., Venance, L., and Piomelli, D., 1996, Biosynthesis of an endogenous cannabinoid precursor in neurons and its control by calcium and cAMP, *J. Neurosci.* **16**: 3934.

Carlson, G., Wang, Y., and Alger, B. E., 2002, Endocannabinoids facilitate the induction of LTP in the hippocampus, *Nat. Neurosci.* **5**: 723.

Chevaleyre, V., and Castillo, P. E., 2003, Heterosynaptic LTD of hippocampal GABAergic synapses: a novel role of endocannabinoids in regulating excitability, *Neuron* **38**: 461.

Daniel, H., and Crepel, F., 2001, Control of Ca^{2+} influx by cannabinoid and metabotropic glutamate receptors in rat cerebellar cortex requires K^+ channels, *J. Physiol.* **537**: 793.

Di Marzo, V., Melck, D., Bisogno, T. and De Petrocellis, L., 1998, Endocannabinoids: endogenous cannabinoid receptor ligands with neuromodulatory action, *Trends Neurosci.* **21**: 521.

Di, S., Malcher-Lopes, R., Halmos, K. C., and Tasker, J. G., 2003, Nongenomic glucocorticoid inhibition via endocannabinoid release in the hypothalamus: a fast feedback mechanism, *J. Neurosci.* **23**: 4850.

Diana, M. A., Levenes, C., Mackie, K., and Marty, A., 2002, Short-term retrograde inhibition of GABAergic synaptic currents in rat Purkinje cells is mediated by endogenous cannabinoids, *J. Neurosci.* **22**: 200.

Egertova, M., and Elphick, M. R., 2000, Localisation of cannabinoid receptors in the rat brain using antibodies to the intracellular C-terminal tail of CB, *J. Comp. Neurol.* **422**: 159.

Felder, C. C., and Glass, M., 1998, Cannabinoid receptors and their endogenous agonists, *Annu. Rev. Pharmacol. Toxicol.* **38**: 179.

Freund, T. F., Katona, I., and Piomelli, D., 2003, Role of endogenous cannabinoids in synaptic signaling, *Physiol. Rev.* **83**: 1017.

Fukudome, Y., Ohno-Shosaku, T., Matsui, M., Omori, Y., Fukaya, M., Tsubokawa, H., Taketo, M. M., Watanabe, M., Manabe, T., and Kano, M., 2004, Two distinct classes of muscarinic action on hippocampal inhibitory synapses: M_2-mediated direct suppression and M_1/M_3-mediated indirect suppression through endocannabinoid signaling, *Eur. J. Neurosci.* in press.

Gerdeman, G. L., Ronesi, J., and Lovinger, D. M., 2002, Postsynaptic endocannabinoid release is critical to long-term depression in the striatum, *Nat. Neurosci.* **5**: 446.

Gifford, A. N., and Ashby, C. R., Jr., 1996, Electrically evoked acetylcholine release from hippocampal slices is inhibited by the cannabinoid receptor agonist, WIN 55212-2, and is potentiated by the cannabinoid antagonist, SR 141716A, *J. Pharmacol. Exp. Ther.* **277**: 1431.

Guo, J., and Ikeda, S. R., 2004, Endocannabinoids modulate N-type calcium channels and G-protein-coupled inwardly rectifying potassium channels via CB1 cannabinoid receptors heterologously expressed in mammalian neurons, *Mol. Pharmacol.* **65**: 665.

Hajos, N., Ledent, C., and Freund, T. F., 2001, Novel cannabinoid-sensitive receptor mediates inhibition of glutamatergic synaptic transmission in the hippocampus, *Neuroscience* **106**: 1.

Herkenham, M., Lynn, A. B., Little, M. D., Johnson, M. R., Melvin, L. S., de Costa, B. R., and Rice, K. C., 1990, Cannabinoid receptor localization in brain, *PNAS USA* **87**: 1932.

Hoffman, A. F., and Lupica, C. R., 2000, Mechanisms of cannabinoid inhibition of GABA_A synaptic transmission in the hippocampus, *J. Neurosci.* **20**: 2470.

Ishac, E. J., Jiang, L., Lake, K. D., Varga, K., Abood, M. E., and Kunos, G., 1996, Inhibition of exocytotic noradrenaline release by presynaptic cannabinoid CB1 receptors on peripheral sympathetic nerves, *Br. J. Pharmacol.* **118**: 2023.

Kano, M., Ohno-Shosaku, T., and Maejima, T., 2002, Retrograde signaling at central synapses via endogenous cannabinoids, *Mol. Psychiatry.* **7**: 234.

Kano, M., Ohno-Shosaku, T., Maejima, T., and Yoshida, T., 2003, Endocannabinoid-mediated modulation of excitatory and inhibitory synaptic transmission, in: *Excitatory-Inhibitory Balance: Synapses, Circuits, System Plasticity,* T. K. Hensch, M. Fagiolini, eds., Kluwer Academic/Plenum Publishers, New York, pp. 99-109.

Katona, I., Sperlagh, B., Sik, A., Kafalvi, A., Vizi, E. S., Mackie, K., and Freund, T. F., 1999, Presynaptically located CB1 cannabinoid receptors regulate GABA release from axon terminals of specific hippocampal interneurons, *J. Neurosci.* **19**: 4544.

Kim, J., Isokawa, M., Ledent, C., and Alger, B. E., 2002, Activation of muscarinic acetylcholine receptors enhances the release of endogenous cannabinoids in the hippocampus, *J. Neurosci.* **22**: 10182.

Klein, T. W., Newton, C., and Friedman, H., 1998, Cannabinoid receptors and immunity, *Immunol. Today* **19**: 373.

Kofalvi, A., Vizi, E. S., Ledent, C., and Sperlagh, B., 2003, Cannabinoids inhibit the release of [^{3}H]glutamate from rodent hippocampal synaptosomes via a novel CB1 receptor-independent action, *Eur. J. Neurosci.* **18**: 1973.

Kreitzer, A. C., Carter, A. G., and Regehr, W. G., 2002, Inhibition of interneuron firing extends the spread of endocannabinoid signaling in the cerebellum, *Neuron* **34**: 787.

Kreitzer, A. C., and Regehr, W. G., 2001a, Cerebellar depolarization-induced suppression of inhibition is mediated by endogenous cannabinoids, *J. Neurosci.* **21**: RC174.

Kreitzer, A. C., and Regehr, W. G., 2001b, Retrograde inhibition of presynaptic calcium influx by endogenous cannabinoids at excitatory synapses onto Purkinje cells, *Neuron* **29**: 717.

Kreitzer, A. C., and Regehr, W. G., 2002, Retrograde signaling by endocannabinoids, *Curr. Opin. Neurobiol.* **12**: 324.

Lenz, R. A., and Alger, B. E., 1999, Calcium dependence of depolarization-induced suppression of inhibition in rat hippocampal CA1 pyramidal neurons, *J. Physiol.* **521**: 147.

Llano, I., Leresche, N., and Marty, A., 1991, Calcium entry increases the sensitivity of cerebellar Purkinje cells to applied GABA and decreases inhibitory synaptic currents, *Neuron* **6**: 565.

Maejima, T., Hashimoto, K., Yoshida, T., Aiba, A., and Kano, M., 2001a, Presynaptic inhibition caused by retrograde signal from metabotropic glutamate to cannabinoid receptors, *Neuron* **31**: 463.

Maejima, T., Ohno-Shosaku, T., and Kano, M., 2001b, Endogenous cannabinoid as a retrograde messenger from depolarized postsynaptic neurons to presynaptic terminals, *Neurosci. Res.* **40**: 205.

Marsicano, G., Wotjak, C. T., Azad, S. C., Bisogno, T., Rammes, G., Cascio, M. G., Hermann, H., Tang, J., Hofmann, C., Zieglgansberger, W., Di Marzo, V., and Lutz, B., 2002, The endogenous cannabinoid system controls extinction of aversive memories, *Nature* **418**: 530.

Matsuda, L. A., Bonner, T. I., and Lolait, S. J., 1993, Localization of cannabinoid receptor mRNA in rat brain, *J. Comp. Neurol.* **327**: 535.

Matsuda, L. A., Lolait, S. J., Brownstein, M. J., Young, A. C., and Bonner, T. I., 1990, Structure of a cannabinoid receptor and functional expression of the cloned cDNA, *Nature* **346**: 561.

Mechoulam, R., Fride, E., and Di Marzo, V., 1998, Endocannabinoids, *Eur. J. Pharmacol.* **359**: 1.

Melis, M., Pistis, M., Perra, S., Muntoni, A. L., Pillolla, G., and Gessa, G. L., 2004, Endocannabinoids mediate presynaptic inhibition of glutamatergic transmission in rat ventral tegmental area dopamine neurons through activation of CB1 receptors, *J. Neurosci.* **24**: 53.

Morishita, W., and Alger, B. E., 2001, Direct depolarization and antidromic action potentials transiently suppress dendritic IPSPs in hippocampal CA1 pyramidal cells, *J. Neurophysiol.* **85**: 480.

Munro, S., Thomas, K. L., and Abu-Shaar, M., 1993, Molecular characterization of a peripheral receptor for cannabinoids, *Nature* **365**: 61.

Ohno-Shosaku, T., Maejima, T., and Kano, M., 2001, Endogenous cannabinoids mediate retrograde signals from depolarized postsynaptic neurons to presynaptic terminals, *Neuron* **29**: 729.

Ohno-Shosaku, T., Matsui, M., Fukudome, Y., Shosaku, J., Tsubokawa, H., Taketo, M. M., Manabe, T., and Kano, M., 2003, Postsynaptic M1 and M3 receptors are responsible for the muscarinic enhancement of retrograde endocannabinoid signalling in the hippocampus, *Eur. J. Neurosci.* **18**: 109.

Ohno-Shosaku, T., Sawada, S., and Yamamoto, C., 1998, Properties of depolarization-induced suppression of inhibitory transmission in cultured rat hippocampal neurons, *Pflugers Arch.* **435**: 273.

Ohno-Shosaku, T., Shosaku, J., Tsubokawa, H., and Kano, M., 2002a, Cooperative endocannabinoid production by neuronal depolarization and group I metabotropic glutamate receptor activation, *Eur. J. Neurosci.* **15**: 953.

Ohno-Shosaku, T., Tsubokawa, H., Mizushima, I., Yoneda, N., Zimmer, A., and Kano, M., 2002b, Presynaptic cannabinoid sensitivity is a major determinant of depolarization-induced retrograde suppression at hippocampal synapses, *J. Neurosci.* **22**: 3864.

Piomelli, D., 2003, The molecular logic of endocannabinoid signalling, *Nat. Rev. Neurosci.* **4**: 873.

Piomelli, D., Giuffrida, A., Calignano, A., and Rodriguez de Fonseca, F., 2000, The endocannabinoid system as a target for therapeutic drugs, *Trends Pharmacol. Sci.* **21**: 218.

Pitler, T. A., and Alger, B. E., 1992, Postsynaptic spike firing reduces synaptic GABA$_A$ responses in hippocampal pyramidal cells, *J. Neurosci.* **12**: 4122.

Robbe, D., Kopf, M., Remaury, A., Bockaert, J., and Manzoni, O. J., 2002, Endogenous cannabinoids mediate long-term synaptic depression in the nucleus accumbens, *PNAS USA* **99**: 8384.

Ronesi, J., Gerdeman, G. L., and Lovinger, D. M., 2004, Disruption of endocannabinoid release and striatal long-term depression by postsynaptic blockade of endocannabinoid membrane transport, *J. Neurosci.* **24**: 1673.

Shen, M., Piser, T. M., Seybold, V. S., and Thayer, S. A., 1996, Cannabinoid receptor agonists inhibit glutamatergic synaptic transmission in rat hippocampal cultures, *J. Neurosci.* **16**: 4322.

Shigemoto, R., Kinoshita, A., Wada, E., Nomura, S., Ohishi, H., Takada, M., Flor, P.J., Neki, A., Abe, T., Nakanishi, S., and Mizuno, N., 1997, Differential presynaptic localization of metabotropic glutamate receptor subtypes in the rat hippocampus, *J. Neurosci.* **17**: 7503.

Shigemoto, R., Nakanishi, S., and Mizuno, N., 1992, Distribution of the mRNA for a metabotropic glutamate receptor (mGluR1) in the central nervous system: an in situ hybridization study in adult and developing rat, *J. Comp. Neurol.* **322**: 121.

Sjöström, P. J., Turrigiano, G. G., and Nelson, S. B., 2003, Neocortical LTD via coincident activation of presynaptic NMDA and cannabinoid receptors, *Neuron* **39**: 641.

Stella, N., Schweitzer, P., and Piomelli, D., 1997, A second endogenous cannabinoid that modulates long-term potentiation, *Nature* **388**: 773.

Takahashi, K. A., and Linden, D. J., 2000, Cannabinoid receptor modulation of synapses received by cerebellar Purkinje cells, *J. Neurophysiol.* **83**: 1167.

Thomas, E. A., Cravatt, B. F., Danielson, P. E., Gilula, N. B., and Sutcliffe, J. G., 1997, Fatty acid amide hydrolase, the degradative enzyme for anandamide and oleamide, has selective distribution in neurons within the rat central nervous system, *J. Neurosci. Res.* **50**: 1047.

Trettel, J., Fortin, D. A., and Levine, E. S., 2004, Endocannabinoid signalling selectively targets perisomatic inhibitory inputs to pyramidal neurones in juvenile mouse neocortex, *J. Physiol.* **556**: 95.

Trettel, J., and Levine, E. S., 2003, Endocannabinoids mediate rapid retrograde signaling at interneuron - pyramidal neuron synapses of the neocortex, *J. Neurophysiol.* **89**: 2334.

Varma, N., Carlson, G. C., Ledent, C., and Alger, B. E., 2001, Metabotropic glutamate receptors drive the endocannabinoid system in hippocampus, *J. Neurosci.* **21**: RC188.

Varvel, S. A., and Lichtman, A. H., 2002, Evaluation of CB1 receptor knockout mice in the Morris water maze, *J. Pharmacol. Exp. Ther.* **301**: 915.

Vincent, P., and Marty, A., 1993, Neighboring cerebellar Purkinje cells communicate via retrograde inhibition of common presynaptic interneurons, *Neuron* **11**: 885.

Wallmichrath, I., and Szabo, B., 2002, Cannabinoids inhibit striatonigral GABAergic neurotransmission in the mouse, *Neuroscience* **113**: 671.

Wilson, R. I., and Nicoll, R. A., 2001, Endogenous cannabinoids mediate retrograde signalling at hippocampal synapses, *Nature* **410**: 588.

Wilson, R. I., and Nicoll, R. A., 2002, Endocannabinoid signaling in the brain, *Science* **296**: 678.

Yoshida, T., Hashimoto, K., Zimmer, A., Maejima, T., Araishi, K., and Kano, M., 2002, The cannabinoid CB1 receptor mediates retrograde signals for depolarization-induced suppression of inhibition in cerebellar Purkinje cells, *J. Neurosci.* **22**: 1690.

Zucker, R. S., and Regehr, W. G., 2002, Short-term synaptic plasticity, *Annu. Rev. Physiol.* **64**: 355.

19

TRANS-SYNAPTIC SIGNALLING BY NITRIC OXIDE

Catherine N. Hall and John Garthwaite[1]

1. INTRODUCTION

The signalling molecule, nitric oxide (NO) modulates synaptic function in numerous networks throughout the nervous system. NO is typically generated in neurons following increased intracellular Ca^{2+} levels, often as a result of the activation of synaptic NMDA receptors, and it can then readily diffuse through membranes to access neighbouring cells. Here, many of its physiological effects are transduced by binding to proteins that have historically been called soluble guanylyl (or guanylate) cyclases. Considering that these proteins are now known to be *bona fide* NO receptors with a ligand binding site and transduction unit, it is preferable to call them guanylyl cyclase-coupled NO receptors, or NO(GC) receptors. On binding NO, the cyclase domain synthesises the second messenger cGMP from GTP. This NO-cGMP pathway is involved in the acute and long-term regulation of synaptic function, and also in the development of the central nervous system. Many aspects of the NO signalling pathway, including its discovery and the identification of the key participants, have already been reviewed extensively (Garthwaite and Boulton, 1995; Moncada et al., 1991). Here, we discuss some recent developments in the understanding of the role of NO in vertebrate neurophysiology, and identify areas in which further clarification is required. Neural NO signalling in invertebrates has already been the subject of a number of reviews (Bicker, 2001; Stefano and Ottaviani, 2002) and will not be covered here.

2. THE NO SIGNAL

Throughout the body, NO is synthesised by three major subtypes of NO synthase (NOS): the endothelial (eNOS), neuronal (nNOS) and inducible (iNOS) isoforms, all of which catalyse the reaction of L-arginine, NADPH and O_2 to produce NO, L-citrulline and NADP. The activity of two of the isoforms, eNOS and nNOS, is dependent on

---

[1] Catherine Hall and John Garthwaite, Wolfson Institute for Biomedical Research, University College London, London, WC1E 6BT

Dendritic Neurotransmitter Release, edited by M. Ludwig
Springer Science+Business Media, Inc., 2005

intracellular Ca^{2+} (Alderton et al., 2001). eNOS is expressed in cerebral blood vessels and iNOS can be expressed in glial cells and/or neurons following damage or inflammatory stimuli, but nNOS is the major isoform expressed physiologically in neurons (Blackshaw et al., 2003). The trigger for NO synthesis is an increase of intracellular Ca^{2+}, which occurs on the activation of NMDA receptors by glutamate, entry through voltage-gated channels, or release from intracellular stores.

nNOS is present throughout the brain, often in discrete neuronal subpopulations having few other phenotypic characteristics in common. For example the enzyme is found in both excitatory and inhibitory cell-types and in both projection neurons and interneurons in different brain regions (de Vente et al., 1998). Recently, the use of weaker fixation protocols and electron microscopy has allowed the subcellular loci of nNOS to be determined. Strongly-stained cells often demonstrate uniform cytosolic labelling, while cells that are stained more weakly, or are even unstained at the light microscopic level, can be seen using the electron microscope to contain NOS in association with membranes or postsynaptic densities (Burette et al., 2002). Such membrane localisation is due to the interaction of the full-length nNOS (nNOSα) with the postsynaptic density protein, PSD-95 (Brenman et al., 1996; Huang et al., 1993). NMDA receptors also bind PSD-95 (Kornau et al., 1995) resulting in a ternary complex with nNOS (Christopherson et al., 1999). This localisation of NMDA receptors with nNOS is functionally significant for NMDA receptor-mediated production of NO, because suppression of PSD-95 reduces NMDA-induced cGMP formation but not that resulting from non-specific depolarisation (Sattler et al., 1999). Association of nNOS with the membrane and NMDA receptors thus enables synaptic activation of NO synthesis. The splice variants nNOSβ/γ lack the domain that binds PSD-95 and, hence, are located in the cytosol. The nNOSβ variant is catalytically active (Eliasson et al., 1997) and so could generate NO in response to Ca^{2+} entry from other sources, such as voltage-gated Ca^{2+} channels.

The pattern of nNOS activation will be instrumental in shaping the NO signal. Repeated activity of groups of neurons may be expected to generate a diffuse "cloud" of NO throughout the neuronal bundle, while sparse activity could lead to an NO signal being localised to individual neurons and synapses (Wood and Garthwaite, 1994). Direct measurement of the temporal and spatial NO profiles with different forms of stimulation would be highly desirable. Unfortunately, adequate methods are currently unavailable. Fluorescent indicators for NO, such as 4,5-diaminofluorescein and its cell-permeable derivative have been used experimentally (Brown et al., 1999; Kojima et al., 1998), but the interpretation of the fluorescence changes is fraught with difficulty due to synergy with Ca^{2+}or Mg^{2+} (Broillet et al., 2001) and cross-reactivity with ascorbic acid (Zhang et al., 2002), and peroxynitrite (Roychowdhury et al., 2002). Electrochemical detection of NO is possible (Shibuki and Kimura, 1997) but the electrodes do not have sufficient spatial or temporal resolution to record endogenous NO profiles accurately.

The way in which the NO is inactivated is probably also an important determinant of the spatiotemporal properties of NO signals. The diffusion of NO is rapid (Malinski et al., 1993) such that the amplitude of a very local signal is likely to be determined solely by NO diffusion kinetics. When more distributed synthesis occurs, however, any biological inactivation is likely to become much more influential (Wood and Garthwaite, 1994). One possible inactivation mechanism is its reaction with haemoglobin in circulating red blood cells but there may well be others. Enzymatic breakdown of NO has been identified in bacteria and mammalian cell lines (Abu-Soud and Hazen, 2000; Coffey et al., 2001; O'Donnell et al., 1999) and, recently, a novel process has been reported in cerebellar cells and in whole-brain homogenate (Griffiths and Garthwaite, 2001). The relative physiological contributions of these potential routes of NO consumption remain to be analysed, however, and more work will be required to elucidate which, if any, of the above constitute a specialised physiological inactivation pathway, and to evaluate how such a mechanism might shape the endogenous NO profile.

While the dynamics of NO signals are currently uncertain, the physiological concentrations of NO reached in the brain are likely to be in the low nanomolar range, rather than the micromolar levels previously thought. In cerebellar slices, for example, electrochemical detection showed the maximum NO concentration achieved following stimulation of the nNOS-containing parallel fibres (20 Hz for 5 s) to be ~ 4 nM (Shibuki and Kimura, 1997). Physiologically relevant NO levels can be at least an order of magnitude lower, as indicated by the finding that afferent fibre stimulation in cortical slices resulted in NO-dependent synaptic plasticity associated with a peak NO concentration in layer V of only 0.4 nM (Wakatsuki et al., 1998). Additionally, NMDA-induced NO synthesis produces sub-maximal activation of NO(GC) receptors (Griffiths and Garthwaite, 2001), the operating range of which is 0.5-10 nM NO at steady-state (Bellamy et al., 2002; Gibb et al., 2003). Both measurement of NO itself and of NO(GC) receptor function, therefore, suggest that physiological NO signals are likely to be in the low nM range or below.

3. NO SIGNAL TRANSDUCTION

The only known physiological transduction pathway for NO is through NO(GC) receptors. Given evidence that endogenously produced NO has biological effects that are independent of these receptors (Jacoby et al., 2001; Lev-Ram et al., 2002) other, as yet unidentified, targets appear to exist. The receptor exists as a heterodimer of α and β subunits. The $\alpha_1\beta_1$ and $\alpha_2\beta_1$ isoforms have been detected at the protein level (Russwurm et al., 2001) and show differential expression across the brain at the mRNA level (Gibb and Garthwaite, 2001). Binding of NO to a prosthetic haem group triggers a conformational change in the protein that is transmitted to the catalytic site, resulting in the formation of cGMP from GTP (Hobbs, 1997; Koesling, 1999; Lucas et al., 2000).

Both isoforms share similar kinetic properties, with EC_{50} values for NO of 0.5-2 nM under steady-state conditions in cells (Bellamy et al., 2002; Gibb et al., 2003). In a cellular environment, the receptor is negatively-regulated by at least two mechanisms. On stimulation by NO, the enzyme activity desensitises within a few seconds (Bellamy et al., 2000; Bellamy and Garthwaite, 2001b) in a process that appears to depend on the level of NO and cGMP (Bellamy and Garthwaite, 2001b; Wykes et al., 2002). This rapid desensitisation is characteristic of the dynamic nature of cellular NO(GC) receptor activity, as is also evidenced by the fast activation (within 20 ms) and deactivation (half-time of 200 ms) on addition or removal of NO (Bellamy and Garthwaite, 2001b). When exposure to NO is prolonged (hours), NO(GC) receptor protein is reduced as a result of mRNA destabilisation (Kloss et al., 2003).

The relationship between the sites of synthesis and action of NO is revealing about the functioning of NO as an intercellular signalling molecule. NO(GC) receptors and cGMP are found in discrete neuronal sub-populations that are usually complementary to that of nNOS (de Vente et al., 1998; Southam and Garthwaite, 1993), a juxtaposition of source and target which is yet more apparent at a synaptic level. In the hippocampus, for example, nNOS is present at the postsynaptic membrane and the NO receptors are found in apposing presynaptic terminals (Burette et al., 2002). The identity of the receptor isoform in this case is unknown but the $\alpha_2\beta_1$ isoform, which is abundant in the hippocampus (Gibb and Garthwaite, 2001) can associate with membranes via PSD-95 and/or presynaptic proteins (Russwurm et al., 2001) while, in rat heart at least, the $\alpha_1\beta_1$ isoform can translocate to the membrane in a Ca^{2+}-dependent manner (Zabel et al., 2002).

The distribution of the sources and targets of NO, therefore, support its role as a transsynaptic signalling molecule. It can be synthesised presynaptically in axon terminals, or postsynaptically in dendrites and spines, permitting signalling in either the anterograde or retrograde directions in different neural systems. In addition, there is evidence that neuronally-derived NO can also signal to astrocytes (Matyash et al., 2001) and blood vessels (Yang and Iadecola, 1998).

4. THE cGMP SIGNAL

As for the NO signal, the cGMP profile is shaped by the balance between its synthesis *via* NO(GC) receptor activation and its breakdown by phosphodiesterases (PDEs). PDE5 and 9 which are selective for cGMP over cAMP, and PDE1, 2, 3, 10, 11 and, to some extent PDE4 (Bellamy and Garthwaite, 2001a), which all also hydrolyse cAMP, are potentially responsible for the breakdown of NO-evoked cGMP accumulation in the brain (Beavo, 1995; Fawcett et al., 2000; Soderling and Beavo, 2000). While the functional properties of NO(GC) receptors are similar between isoforms, those of the various PDEs are much more variable. Several contain modulatory binding sites for cGMP, termed GAF domains (Martinez et al., 2002), the effects of which have been well characterised in PDE2 and 5. Binding of cGMP increases the rate of cyclic nucleotide hydrolysis (Beavo, 1995) and, in the case of PDE5, also allows phosphorylation of the enzyme by cGMP-dependent protein kinase I (Thomas et al., 1990), which may contribute to the enhanced activity of PDE5 (Mullershausen et al., 2003).

With such functional heterogeneity, the type and level of PDE expression is critical in setting the profile of the cGMP signal in different subpopulations of cells. The rapid activation and desensitisation kinetics of NO(GC) receptors combined with low PDE activity produce a rapid rise of cGMP to a prolonged plateau in cerebellar astrocytes (Bellamy et al., 2000) while, in Purkinje cells, high PDE activity results in only a transient cGMP signal on NO stimulation (Honda et al., 2001). An intermediate, triphasic, profile is seen in a population of striatal neurons (Wykes et al., 2002). Presumably the diversity of cGMP responses impacts on the selection of downstream targets, but information on this issue is scant at present.

5. TARGETS FOR cGMP

5.1. Protein Kinases

cGMP-dependent protein kinases (cGKs) are established receptors for cGMP and are sensitive to submicromolar concentrations (Hofmann et al., 2000). There are two

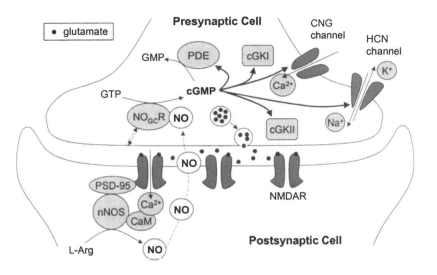

Figure 1. The NO-cGMP signalling pathway at the synapse.

major subtypes: cGKI exists in two cytosolic alternatively-spliced isoforms, cGKα and cGKβ. Conversely, cGKII is membrane-bound, due to N-terminal myristoylation. The expression of cGKI is restricted to Purkinje cells of the cerebellum, the hippocampus, and the dorsomedial thalamus (El Husseini et al., 1999), while cGKII is widely expressed in distinct cell populations across the brain, especially in the olfactory bulb, cerebral cortex and thalamus (de Vente et al., 2001).

The physiological substrates for cGKIα/β and cGKII remain unclear. The best-characterised is the phosphatase inhibitor known as G-substrate, which is located primarily in cerebellar Purkinje cells (Endo et al., 1999). Both cGKI and II also undergo autophosphorylation which, in cGKI, increases cGMP binding (Lohmann et al., 1997). Inositol trisphosphate receptors (IP$_3$Rs) are phosphorylated by cGKI in cerebellum (Haug et al., 1999). Other candidate substrates include DARPP-32, a phosphatase inhibitor sharing homology with G-substrate (Tsou et al., 1993), and several additional cGK-specific substrates which have been detected in brain, though not identified (Wang and Robinson, 1995). cGK activity has been implicated in numerous neurophysiological processes, including synaptic plasticity (Kleppisch et al., 2003) and membrane excitability (Klyachko et al., 2001), but which, if any, of these substrates are responsible for these functional changes has yet to be investigated.

5.2. Cyclic Nucleotide-Gated Channels

Cyclic nucleotide-gated (CNG) channels are cationic channels, which open on binding of either cGMP or cAMP and predominantly conduct Ca^{2+} under physiological conditions (Kaupp and Seifert, 2002). CNGs are tetramers of α and β subunits. Only α$_{1-3}$ subunits form functional channels when expressed alone, but co-expression of β subunits imparts properties that are more like those seen in the native channels, such as flickering between open and closed states. The major difference between the sub-types of CNG channel is their cyclic nucleotide sensitivity and selectivity. Rod and cone-type channels are highly selective for cGMP over cAMP but show relatively low sensitivity, with half-maximal activation occurring at 50 μM cGMP. Olfactory channels, however, respond to both ligands and are activated at low micromolar concentrations (Kaupp and Seifert, 2002).

The subunits comprising olfactory-type channels appear to be expressed in several brain regions, including the hippocampus, cerebellar Purkinje cells and cortex (Kingston et al., 1999). Those of rod- and cone-type CNGs are traditionally thought to be limited to retinal cells (Bradley et al., 1997) but they may be expressed more widely, as mRNA of rod-type channels has been detected in hippocampus (Kingston et al., 1996). There is, therefore, scope for modulation of membrane potential and Ca^{2+}-influx by NO/cGMP-mediated activation of CNGs, but direct physiological evidence for this is scarce. NO/cGMP modulation of transmitter release has been demonstrated in salamander and lizard retina (Savchenko et al., 1997) and NO(GC) receptor-dependent depolarisation of the membrane potential has been observed in rat hippocampal pyramidal cells (Kuzmiski and MacVicar, 2001). Further work is evidently required in order to clarify the precise role of CNGs as a target for the NO/cGMP signalling pathway.

5.3. HCN Channels

Hyperpolarisation-activated cyclic nucleotide-regulated non-selective cation (HCN) channels represent a potential, but unconfirmed, target for NO/cGMP signalling. Activation of HCN channels occurs on hyperpolarisation of the membrane and the activity is potentiated by cyclic nucleotide binding, leading to the slow onset of an inward non-selective cation current, (I_h) and membrane depolarisation. I_h can act as a pacemaker current in some cell types (e.g. thalamocortical relay neurons), being activated by hyperpolarisation following an action potential and then depolarising the membrane towards the threshold for firing a subsequent action potential. It also helps regulate the resting membrane potential and the cable properties of dendrites (Robinson and Siegelbaum, 2003).

There are four HCN subunits which exhibit differential expression across the brain. HCN1 is predominantly cortical, HCN2 is widespread and HCN3 and 4 are found subcortically (Franz et al., 2000). The subtypes exhibit substantial differences in terms of kinetics of activation, gating, and of cyclic nucleotide sensitivity (Robinson and Siegelbaum, 2003). HCN1 homotetramers show fast activation kinetics but low cyclic nucleotide sensitivity; HCN2 are slower, but have greater cyclic nucleotide sensitivity, while HCN4 have slowest activation kinetics and are also strongly regulated by cyclic nucleotides. Native channels are presumably heteromeric and their properties are best described by a combination of the properties of homomeric channels. For example, cortical layer V pyramidal neurons express HCN1 and 2 and the I_h current in these neurons exhibits activation kinetics intermediate between those exhibited by HCN1 or 2 homomers (Santoro et al., 2000).

Regulation of HCN channels by cAMP has been described in several neuronal systems and has been implicated in a number of functions, from LTP to modulation of membrane conductance (Cuttle et al., 2001; Huang and Hsu, 2003). The channels can also be activated by cGMP (Zagotta et al., 2003) making them putative targets for NO-cGMP signalling, though direct evidence for this possibility is lacking. Exogenously-applied cGMP increases I_h in nodose and trigeminal ganglion neurons (Ingram and Williams, 1996) and in the calyx of Held (Cuttle et al., 2001), while exogenous NO modulates I_h in thalamocortical relay neurons in a manner that is mimicked by applied 8-Br-cGMP (Pape and Mager, 1992), but a physiological link between NO synthesis and subsequent cGMP-mediated effects on I_h remains elusive. Endogenous NO/cGMP-induced increases in membrane excitability and conductance have been observed that are suggestive of mediation by I_h (Bains and Ferguson, 1997) though this has not been directly examined. Further work is required to elucidate whether HCN channels can be considered general targets for the physiological NO/cGMP pathway.

5.4. cGMP-Modulated PDEs

cGMP can potentially inhibit or enhance cAMP signalling through its actions on PDEs. In the case of PDE2, cGMP binding to a GAF domain can increase the hydrolysis of cAMP. cGMP also has a high affinity for the catalytic site of PDE3, but is inefficiently hydrolysed, and so can competitively inhibit the hydrolysis of cAMP (Beavo, 1995). Through these mechanisms, NO could cause increases or decreases in cAMP, with multiple potential consequences. For example, in cultured amacrine cells, a component of the NO-induced decrease in GABA-mediated currents, seems due to activation of PDE2 and increased hydrolysis of cAMP (Wexler et al., 1998).

The components of the NO-cGMP signalling pathway as they are currently understood are summarised in Fig. 1.

6. FUNCTION OF NO/cGMP SIGNALLING IN THE BRAIN

NO has been implicated in numerous physiological functions in the central nervous system (CNS), many of which have been extensively reviewed and so will not be discussed here. For example we refer the interested reader to reviews on the role of NO in the visual system (Cudeiro and Rivadulla, 1999), central control of autonomic function (Paton et al., 2002), monoaminergic projection systems (Kiss, 2000), neurotransmitter release (Prast and Philippu, 2001), behaviour (Nelson et al., 1997), and Ca^{2+}-signalling (Clementi and Meldolesi, 1997) as well as general roles of NO in the CNS (Garthwaite, 2000). We will focus on three areas in which recent research has clarified the role of NO as a trans-synaptic messenger: development of the CNS and the acute and long-term modulation of neuronal function (see Fig. 2).

6.1. CNS Development

NO has been implicated in three chronologically distinct steps in the development of the CNS: inhibition of proliferation and promotion of differentiation, the directionality of neurite growth, and the refinement of topographical projections.

A number of studies have found that NO inhibits the proliferation of neuronal precursors, both in developing brain and in areas of ongoing neurogenesis in the adult. In the adult, inhibition of endogenous NO production *in vivo* was found to increase the

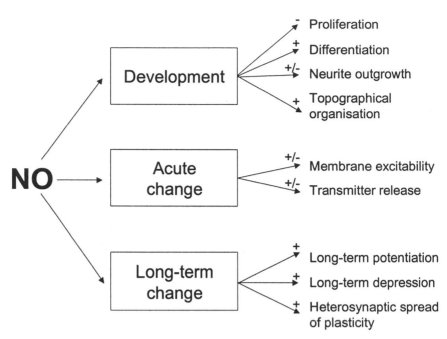

Figure 2. Functions of NO in the central nervous system.

number of proliferating cells (Moreno-Lopez et al., 2004). Similarly, proliferation of cultured immature cortical neurons was increased by both NOS inhibition and application of the NO scavenger, haemoglobin, while exogenous NO had the reverse effect (Cheng et al., 2003). nNOS is expressed in neurons adjacent to, but not within, areas of adult neurogenesis (Packer et al., 2003) and is induced by brain-derived neurotrophic factor in post-mitotic neurons that are close to proliferating cells in the developing cortex (Cheng et al., 2003). Mice lacking nNOS also show an increase in the number of proliferating cells, though the effect is less striking than after broad-spectrum NOS inhibition (Packer et al., 2003), indicating that NO derived from both eNOS and nNOS may be responsible for the control of proliferation.

NO may also accelerate the differentiation of newly-formed neurons and so, in sum, could act as a switch between proliferation and differentiation. Expression of neuronal phenotypic markers, as well as nNOS itself was delayed by those treatments that promoted proliferation (NOS inhibition, haemoglobin) and was increased by those that raised NO levels (Cheng et al., 2003; Moreno-Lopez et al., 2004). At this early stage of neuronal development, therefore, it seems that NO acts in a paracrine manner, being released from nearby differentiated neurons to stop neuronal precursors from proliferating and to induce a neuronal phenotype. The mechanism through which NO exerts these effects remains unclear. None of the above experiments tested the involvement of the NO(GC) receptor/cGMP pathway, but the low levels of NO(GC) receptor and cGMP detected immunohistochemically in neurons prior to differentiation may point to an alternative signalling process (Arnhold et al., 2002).

A number of studies have implicated NO in controlling the direction of neurite growth. The results of these studies appear contradictory, with some showing NO-induced collapse and retraction of growth cones (Gallo et al., 2002; He et al., 2002) and others finding NO/cGMP to be attractive and protective against growth cone collapse (Schmidt et al., 2002; Steinbach et al., 2002). Such opposing findings can be better understood when the concentrations of applied NO are considered. Those which showed collapse of growth cones used high micromolar to millimolar concentrations of NO donors which are likely to produce unphysiologically high NO levels and be damaging to cells. Those using lower donor concentrations or cGMP analogues, in general, demonstrate growth cone attraction and protection. For example, the semaphorin-3A-mediated attraction of apical dendrites of rat cortical pyramidal cells to the pial surface is dependent on NO(GC) receptors and cGK, while the repellent effect of semaphorin-3A on pyramidal axons is independent of this pathway (Polleux et al., 2000). Evidence therefore points to a role for NO/cGMP in directing neurites but the reliance on exogenously-applied NO and cGMP analogues means that the physiological relevance of the effects remain unclear. Only a study in PC12 cells has demonstrated that blockade of endogenous NO synthesis can block neurite outgrowth (Yamazaki et al., 2001). Further studies on primary neuronal cultures are required to help evaluate the importance on NO in this aspect of neuronal development.

The roles of NO in synaptic plasticity (see below) suggest that it is a plausible candidate for mediating the activity-dependent refinement of topographical projections that occurs after the laying down of a coarse map by chemical guidance cues. Nonetheless, NO appears to fulfil such a role only in a few specific areas. In the chick and the rat, elimination of ipsilateral retinotectal or retinocollicular projections is decreased following inhibition of NO synthesis (Campello-Costa et al., 2000; Wu et al., 2001). eNOS/nNOS double knockout mice also show a more diffuse ipsilateral retinocollicular projection, which nevertheless becomes more refined with age (Wu et al., 2000). In the ferret, the retinal representation in the dorsal lateral geniculate nucleus first becomes segregated into eye-specific columns, and later sublamination occurs to order inputs from ON- and OFF-retinal ganglion cells. The latter of these processes is dependent on NO and NO(GC) receptor activity and is correlated temporally with the presence of high cGMP levels in the dorsal lateral geniculate nucleus (Leamey et al., 2001). The organisation of other sensory projections seems to be independent of NO and the lack of gross changes in neuroanatomy and behaviour following chronic in vivo NOS inhibition would indicate that NO is not necessary to form cortical neural maps during late development (Contestabile, 2000).

Early in development, therefore, NO functions as a paracrine messenger to determine the fate of neuronal precursors and newly formed cells. Later, when its role is that of a trans-synaptic messenger, its role in the formation of topographical organisation seems to be considerably more restricted. At these later stages, and into adulthood, there is, however, ample evidence to support a role for NO in modulating the information flow between neurons, as shall be considered below.

6.2. Acute Neuronal Modulation

NO can influence neuronal function in an acute manner by altering membrane excitability and neurotransmitter release. In different neuronal systems, NO can both increase and decrease excitability, or do nothing, the effect in each case presumably being determined by which (if any) target proteins for the NO/cGMP pathway are present.

Inhibition of neuronal activity by NO has been reported in a number of cell types: exogenous NO decreases a Na^+ current in nodose ganglion neurons (Bielefeldt et al., 1999), while spinally-projecting paraventricular nucleus (PVN) neurons respond to exogenous NO or increased NO synthesis by enhanced GABA release (Li et al., 2002). In this subpopulation, blockade of NO synthesis or scavenging of basal NO has no effect, but PVN cells projecting to the rostral ventrolateral medulla (RVLM) are subject to tonic inhibition by NO and so respond to nNOS inhibition with an increase in discharge rate (Stern et al., 2003).

NO can disinhibit cells by reducing GABAergic transmission. In the cerebellum, Golgi cells are subject to a tonic NO-dependent hyperpolarisation, possibly via BK channels, and this results in decreased release of GABA from these cells (Wall, 2003). In addition, NO tonically inhibits the $GABA_A$ current in granule cells (which receive GABAergic innervation from Golgi cells) by a cGK-dependent process (Wall, 2003). The net effect of NO, therefore, is a decrease in GABAergic transmission and a subsequent disinhibition of downstream granule cell activity. NO-mediated disinhibition also occurs by a decrease in an outward current in a subpopulation of PVN cells (Bains and Ferguson, 1997) and in ventrobasal and lateral geniculate thalamic nuclei (Shaw et al., 1999). NO/cGMP stimulation in these thalamic nuclei is associated *in vivo* with increased responses to sensory stimulation, indicating that NO enhances sensory transmission through the thalamus.

NO may also have more direct excitatory effects leading to increased glutamate release in RVLM neurons (Huang et al., 2003), increased Na^+ current in striatal interneurons, increased I_h in the thalamus (Pape and Mager, 1992) and increased tonic firing in Purkinje cells (Smith and Otis, 2003).

As there are many ways in which NO could modulate neuronal excitability, the response of a given neuron will necessarily depend on the target elements expressed there. While its effects on acute neuronal function are by no means ubiquitous, its actions on both presynaptic release and postsynaptic responses support its ability to act as a signalling molecule in both anterograde and retrograde directions, but its precise role must be understood on a regional, or even cell-specific, basis.

6.3. Long-Term Modulation of Neuronal Function

The process in which neurons undergo long-term potentiation (LTP) and long-term depression (LTD) of the strength of their synaptic connections has long been studied as a likely neurophysiological correlate of learning, memory and developmental plasticity. In classical LTP, the correlated firing of the pre- and postsynaptic cell results in the synapse being strengthened, such that subsequent presynaptic activity is able to exert a stronger influence on the firing of the postsynaptic neuron. The location of the physiological change that underlies such synaptic modification has been a matter of much debate, but it is now well-accepted that both postsynaptic and presynaptic changes can occur. It is an attractive hypothesis that there is a retrograde messenger which relays information about the postsynaptic activity to the presynaptic terminal in this situation. Due to its diffusibility and association with NMDA receptors, which are known to be critical for much LTP, NO is a strong candidate for such a role.

The involvement of NO in LTP and LTD has been reviewed (Holscher, 1997; Prast and Philippu, 2001), so only the most recent advances will be considered here. Briefly, after much controversy, it has gradually become clear that NO functions in many, but not all forms of LTP and LTD. Blockade of LTP by inhibitors of NOS and/or NO scavengers has been found in several brain areas, including the hippocampus, cerebellum, amygdala and cortex. Unexpectedly, however, knockout mice for either nNOS or eNOS demonstrated normal hippocampal LTP, though double knockouts lacking both isoforms were impaired (Son et al., 1996). A subsequent study, on the other hand, reported that LTP was impaired in eNOS knockouts (Kantor et al., 1996). At the time, there was evidence that eNOS was present in CA1 pyramidal neurons but this has now been considered an artifact and both immunocytochemistry and *in situ* hybridization indicate that eNOS is present only in blood vessels (Blackshaw et al., 2003). The implication is that NO from blood vessels contributes to LTP, perhaps by providing a tonic level of NO needed for the plastic changes to take place (Bon and Garthwaite, 2003). nNOS, which is present in CA1 pyramidal cells (Burette et al., 2002), may additionally function as a stimulus-evoked signal during the triggering of LTP (Bon and Garthwaite, 2003). The results using genetic approaches remain incoherent, however. Complications may arise in the nNOS knockout animals because of the activity of the nNOSβ and γ splice variants which are upregulated in these animals (Eliasson et al., 1997). In eNOS knockouts, there may be aberrations in other second messenger pathways resulting in abnormal synaptic plasticity, as indicated by studies in the dentate gyrus (Doreulee et al., 2001).

Concerning a retrograde messenger role for NO, there is evidence consistent with postsynaptic release in CA1 pyramidal cells and in cerebellar granule cells (Arancio et al., 1996; Maffei et al., 2003; Schuman and Madison, 1991). In both cases, diffusion of NO to the presynaptic terminal then induces NO(GC) receptor- and cGK-dependent synaptic potentiation (Arancio et al., 2001; Maffei et al., 2003). Possible downstream effectors of presynaptic change include endocytotic processes by which vesicles are recycled to the readily-releasable pool (Micheva et al., 2003) and, on a longer time scale, synaptic remodelling (Nikonenko et al., 2003). NO-dependent LTP that is also sensitive to inhibition of NO(GC) receptors and cGK has been demonstrated in areas including CA1 (Kleppisch et al., 2003), the amygdala (Chien et al., 2003) and neocortex (Wakatsuki et al., 1998) but the sources and targets of NO in these synapses have not been established. Indeed, it may be erroneous to think of NO as acting solely in one direction or another across the synapse. At CA1 hippocampal synapses, for example, in addition to its presynaptic targets, NO can also act postsynaptically to induce phosphorylation of the transcription factor CREB and this pathway is also important in late stages of LTP (Lu and Hawkins, 2002).

LTD that depends on NO has also been reported in the hippocampus. The mechanism is less clear but it also appears to require cGMP/cGK-dependent processes and may involve presynaptic Ca^{2+}-release from ryanodine-sensitive stores (Reyes-Harde et al., 1999) and selective inhibition of vesicle release from the readily releasable pool (Stanton et al., 2003).

Plasticity of the parallel fibre-Purkinje cell synapse in the cerebellum is unusual as coincident presynaptic activity and postsynaptic excitation (in the form of Ca^{2+} spiking) cause LTD, rather than LTP, while parallel fibre stimulation alone induces LTP. In certain experimental paradigms, LTD can be induced by substituting NO or postsynaptic cGMP for parallel fibre stimulation, and intracellular Ca^{2+} elevation for Purkinje cell depolarisation (Lev-Ram et al., 1997). In addition, inhibitors of NOS, NO(GC) receptors and cGK block LTD induction (Casado et al., 2002; Lev-Ram et al., 1997) indicating that, at least in some situations, NO and Ca^{2+} are necessary and sufficient for LTD to occur. Other studies suggest, however, that metabotropic glutamate receptor (mGluR) and protein kinase C activation are also required (Ito, 2001). Recent work has shown that, like hippocampal plasticity, cerebellar LTD depends on NMDA receptor stimulation but, in the cerebellum, NMDA receptors are expressed presynaptically on parallel fibres (Casado et al., 2002), as is nNOS (Vincent and Kimura, 1992), while LTD is expressed postsynaptically (Ito, 2001), as are NO(GC) receptors and cGMP (Ariano et al., 1982; Honda et al., 2001). In these cerebellar synapses, therefore, NO appears to act as an anterograde messenger, being released from axon terminals and acting on postsynaptic spines. The mechanism for NO-dependent LTD expression following cGK activation is likely to involve G-substrate, which is highly expressed in Purkinje cells (Detre et al., 1984). Downstream, LTD manifests as a reduction in AMPA receptor sensitivity. Inhibition of protein phosphatase 2A (PP2A), which can occur *via* G-substrate, has been shown to increase phosphorylation of AMPA receptors and decrease synaptic currents in a manner that occludes LTD (Launey et al., 2004), suggesting a plausible mechanism for the expression of NO-mediated LTD.

NO-mediated LTP can also occur in the cerebellum following parallel fibre stimulation, an outcome favoured when the postsynaptic Ca^{2+} concentration is low. Fibre stimulation at 1Hz induces postsynaptically-expressed LTP which is dependent on NO, but not cGMP, and which can reverse LTD (Lev-Ram et al., 2003). Higher frequency stimulation (4-8 Hz) induces presynaptic LTP which does not reverse LTD but which may be similarly NO-dependent (Jacoby et al., 2001; Lev-Ram et al., 2003).

Interestingly, in the cerebellum, LTD and presynaptically-expressed LTP can spread to non-stimulated parallel fibre synapses on the same Purkinje cell (Jacoby et al., 2001; Reynolds and Hartell, 2001). These heterosynaptic changes are also dependent on NO synthesis, postsynaptic Ca^{2+} and, in the case of LTD (but not LTP), cGMP and cGK. NO is a likely candidate for mediating the spread, due to its diffusibility, a possibility supported by evidence that heterosynaptic LTP can be blocked by an NO scavenger (Jacoby et al., 2001). Other forms of NO-dependent plasticity may spread in a similar way, for example hippocampal LTP (Schuman and Madison, 1994).

A lack of synapse-specificity of NO-mediated plasticity (should it occur *in vivo*) will impinge on its ability to influence neuronal computation. It would appear to forbid Hebbian-like learning and rather require computations dependent on clusters of synapses. While such learning rules could still produce functional organisation and change (Krekelberg and Taylor, 1996), the properties of these neuronal circuits would be somewhat different from those involving synapse-specific modifications. The extent to which heterosynaptic spread of plasticity occurs physiologically is, therefore, of much relevance and requires further assessment. Stimulation of fewer neurons might be expected to produce less NO and thus less spread of NO-mediated plasticity. In the cerebellum, however, weaker stimulation, recruiting a smaller parallel fibre bundle, did not affect the degree of spread of LTD (Reynolds and Hartell, 2000). NO-dependent LTD can even occur on activation of a single parallel fibre (Casado et al., 2002), suggesting that very localised NO signals are sufficient to produce plasticity, but whether or not such spatially limited NO production permits LTD to spread to adjacent synapses has not been tested. In addition, the temporal coincidence requirements for NO and Ca^{2+} elevation in the postsynaptic cell are narrow (< 10 ms; Lev-Ram et al., 1997) which may limit the spread of LTD. If small or temporally precise NO signals cannot mediate heterosynaptic plasticity, brief, sparse activation of parallel fibres would still allow Hebbian learning, while prolonged activation of groups of fibres would engage a different learning rule. Clearly many questions remain as to the nature and relevance of this form of synaptic modulation, which can only be answered by probing the specificity of NO-mediated plasticity further and studying its dependence on various parameters of stimulation.

7. CONCLUSION

The location, identity and functional properties of the primary molecular components of the NO/cGMP pathway have now been well-characterised. These components are widespread, but not ubiquitous, across the brain, and the downstream effects of activation of this pathway are diverse and often incompletely understood. It is now important to consider the identity of these later stages of the transduction pathway and the way in which different dynamic patterns of NO and cGMP signalling induce such a variety of change in the long- and short-term function of neuronal circuitry. Future studies need to concentrate on the effects of endogenously-released NO and on the application of concentrations of exogenous NO that are physiologically relevant, in order to dissect out the downstream targets of NO signalling that relate to normal function.

8. REFERENCES

Abu-Soud, H. M., and Hazen, S. L., 2000, Nitric oxide is a physiological substrate for mammalian peroxidases, *J. Biol. Chem.* **275:** 37524.

Alderton, W. K., Cooper, C. E., and Knowles, R. G., 2001, Nitric oxide synthases: structure, function and inhibition, *Biochem. J.* **357:** 593.

Arancio, O., et al., 2001, Presynaptic role of cGMP-dependent protein kinase during long-lasting potentiation, *J. Neurosci.* **21:** 143.

Arancio, O., et al., 1996, Nitric oxide acts directly in the presynaptic neuron to produce long-term potentiation in cultured hippocampal neurons, *Cell.* **87:** 1025.

Ariano, M. A., et al., 1982, Immunohistochemical localization of guanylate cyclase within neurons of rat brain, *PNAS USA* **79:** 1316.

Arnhold, S., et al., 2002, NOS-II is involved in early differentiation of murine cortical, retinal and ES cell-derived neurons-an immunocytochemical and functional approach, *Int. J. Dev. Neurosci.* **20:** 83.

Bains, J. S., and Ferguson, A. V., 1997, Nitric oxide depolarizes type II paraventricular nucleus neurons *in vitro*, *Neuroscience* **79:** 149.

Beavo, J. A., 1995, Cyclic nucleotide phosphodiesterases: functional implications of multiple isoforms, *Physiol. Rev.* **75:** 725.

Bellamy, T. C., and Garthwaite, J., 2001a, "cAMP-specific" phosphodiesterase contributes to cGMP degradation in cerebellar cells exposed to nitric oxide, *Mol. Pharmacol.* **59:** 54.

Bellamy, T. C., and Garthwaite, J., 2001b, Sub-second kinetics of the nitric oxide receptor, soluble guanylyl cyclase, in intact cerebellar cells, *J. Biol. Chem.* **276:** 4287.

Bellamy, T. C., Griffiths, C., and Garthwaite, J., 2002, Differential sensitivity of guanylyl cyclase and mitochondrial respiration to nitric oxide measured using clamped concentrations, *J. Biol. Chem.* **277:** 31801.

Bellamy, T. C., et al., 2000, Rapid desensitization of the nitric oxide receptor, soluble guanylyl cyclase, underlies diversity of cellular cGMP responses, *PNAS USA* **97**: 2928.

Bicker, G., 2001, Sources and targets of nitric oxide signalling in insect nervous systems, *Cell Tissue Res.* **303**: 137.

Bielefeldt, K., et al., 1999, Nitric oxide enhances slow inactivation of voltage-dependent sodium currents in rat nodose neurons, *Neurosci. Lett.* **271**: 159.

Blackshaw, S., et al., 2003, Species, strain and developmental variations in hippocampal neuronal and endothelial nitric oxide synthase clarify discrepancies in nitric oxide-dependent synaptic plasticity, *Neuroscience* **119**: 979.

Bon, C. L., and Garthwaite, J., 2003, On the role of nitric oxide in hippocampal long-term potentiation, *J. Neurosci.* **23**: 1941.

Bradley, J., et al., 1997, Functional expression of the heteromeric "olfactory" cyclic nucleotide-gated channel in the hippocampus: a potential effector of synaptic plasticity in brain neurons, *J. Neurosci.* **17**: 1993.

Brenman, J. E., et al., 1996, Interaction of nitric oxide synthase with the postsynaptic density protein PSD-95 and alpha1-syntrophin mediated by PDZ domains, *Cell* **84**: 757.

Broillet, M., Randin, O., and Chatton, J., 2001, Photoactivation and calcium sensitivity of the fluorescent NO indicator 4,5-diaminofluorescein (DAF-2): implications for cellular NO imaging, *FEBS Lett.* **491**: 227.

Brown, L. A., Key, B. J., and Lovick, T. A., 1999, Bio-imaging of nitric oxide-producing neurones in slices of rat brain using 4,5-diaminofluorescein, *J. Neurosci. Meth.* **92**: 101.

Burette, A., et al., 2002, Synaptic localization of nitric oxide synthase and soluble guanylyl cyclase in the hippocampus, *J. Neurosci.* **22**: 8961.

Campello-Costa, P., et al., 2000, Acute blockade of nitric oxide synthesis induces disorganization and amplifies lesion-induced plasticity in the rat retinotectal projection, *J. Neurobiol.* **44**: 371.

Casado, M., Isope, P., and Ascher, P., 2002, Involvement of presynaptic N-methyl-D-aspartate receptors in cerebellar long-term depression, *Neuron* **33**: 123.

Cheng, A. W., et al., 2003, Nitric oxide acts in a positive feedback loop with BDNF to regulate neural progenitor cell proliferation and differentiation in the mammalian brain, *Dev. Biol.* **258**: 319.

Chien, W. L., et al., 2003, Enhancement of long-term potentiation by a potent nitric oxide-guanylyl cyclase activator, 3-(5-hydroxymethyl-2-furyl)-1-benzyl-indazole, *Mol. Pharmacol.* **63**: 1322.

Christopherson, K. S., et al., 1999, PSD-95 assembles a ternary complex with the N-methyl-D-aspartic acid receptor and a bivalent neuronal NO synthase PDZ domain, *J. Biol. Chem.* **274**: 27467.

Clementi, E., and Meldolesi, J., 1997, The cross-talk between nitric oxide and Ca2+: a story with a complex past and a promising future, *Trends Pharmacol. Sci.* **18**: 266.

Coffey, M. J., et al., 2001, Catalytic consumption of nitric oxide by 12/15- lipoxygenase: inhibition of monocyte soluble guanylate cyclase activation, *PNAS USA* **98**: 8006.

Contestabile, A., 2000, Roles of NMDA receptor activity and nitric oxide production in brain development, *Brain Res. Rev.* **32**: 476.

Cudeiro, J., and Rivadulla, C., 1999, Sight and insight--on the physiological role of nitric oxide in the visual system, *Trends Neurosci.* **22**: 109.

Cuttle, M. F., et al., 2001, Modulation of a presynaptic hyperpolarization-activated cationic current (I(h)) at an excitatory synaptic terminal in the rat auditory brainstem, *J. Physiol.* **534**: 733.

de Vente, J., et al., 2001, Localization of cGMP-dependent protein kinase type II in rat brain, *Neuroscience* **108** (1), 27-49.

de Vente, J., et al., 1998, Distribution of nitric oxide synthase and nitric oxide-receptive, cyclic GMP-producing structures in the rat brain, *Neuroscience* **87**: 207.

Detre, J. A., et al., 1984, Localization in mammalian brain of G-substrate, a specific substrate for guanosine 3',5'-cyclic monophosphate-dependent protein kinase, *J. Neurosci.* **4**: 2843.

Doreulee, N., et al., 2001, Defective hippocampal mossy fiber long-term potentiation in endothelial nitric oxide synthase knockout mice, *Synapse* **41**: 191.

El Husseini, A. E., et al., 1999, Localization of the cGMP-dependent protein kinases in relation to nitric oxide synthase in the brain, *J. Chem. Neuroanat.* **17**: 45.

Eliasson, M. J., et al., 1997, Neuronal nitric oxide synthase alternatively spliced forms: prominent functional localizations in the brain, *PNAS USA* **94**: 3396.

Endo, S., et al., 1999, Molecular identification of human G-substrate, a possible downstream component of the cGMP-dependent protein kinase cascade in cerebellar Purkinje cells, *PNAS USA* **96**: 2467.

Fawcett, L., et al., 2000, Molecular cloning and characterization of a distinct human phosphodiesterase gene family: PDE11A, *PNAS USA* **97**: 3702.

Franz, O., et al., 2000, Single-cell mRNA expression of HCN1 correlates with a fast gating phenotype of hyperpolarization-activated cyclic nucleotide-gated ion channels (Ih) in central neurons, *Eur. J. Neurosci.* **12**: 2685.

Gallo, G., et al., 2002, Transient PKA activity is required for initiation but not maintenance of BDNF-mediated protection from nitric oxide-induced growth-cone collapse, *J. Neurosci.* **22**: 5016.

Garthwaite, J., 2000, The physiological roles of nitric oxide in the central nervous system, *Nitric Oxide* **143**: 259.

Garthwaite, J., and Boulton, C. L., 1995, Nitric oxide signaling in the central nervous system, *Annu. Rev. Physiol.* **57**: 683.

Gibb, B. J., and Garthwaite, J., 2001, Subunits of the nitric oxide receptor, soluble guanylyl cyclase, expressed in rat brain, *Eur. J. Neurosci.* **13**: 539.

Gibb, B. J., Wykes, V., and Garthwaite, J., 2003, Properties of NO-activated guanylyl cyclases expressed in cells, *Br. J. Pharmacol.* **139**: 1032.

Griffiths, C., and Garthwaite, J., 2001, The shaping of nitric oxide signals by a cellular sink, *J. Physiol.* **536**: 855.

Haug, L. S., et al., 1999, Phosphorylation of the inositol 1,4,5-trisphosphate receptor by cyclic nucleotide-dependent kinases *in vitro* and in rat cerebellar slices in situ, *J. Biol. Chem.* **274**: 7467.

He, Y., Yu, W., and Baas, P. W., 2002, Microtubule reconfiguration during axonal retraction induced by nitric oxide, *J. Neurosci.* **22**: 5982.

Hobbs, A. J., 1997, Soluble guanylate cyclase: the forgotten sibling, *Trends Pharmacol. Sci.* **18**: 484.

Hofmann, F., Ammendola, A., and Schlossmann, J., 2000, Rising behind NO: cGMP-dependent protein kinases, *J. Cell Sci.* **113**: 1671.

Holscher, C., 1997, Nitric oxide, the enigmatic neuronal messenger: its role in synaptic plasticity, *Trends Neurosci.* **20**: 298.

Honda, A., et al., 2001, Spatiotemporal dynamics of guanosine 3',5'-cyclic monophosphate revealed by a genetically encoded, fluorescent indicator, *PNAS USA* **98**: 2437.

Huang, C. C., Chan, S. H., and Hsu, K. S., 2003, cGMP/protein kinase G-dependent potentiation of glutamatergic transmission induced by nitric oxide in immature rat rostral ventrolateral medulla neurons *in vitro*, *Mol. Pharmacol.* **64**: 521.

Huang, C. C., and Hsu, K. S., 2003, Reexamination of the role of hyperpolarization-activated cation channels in short- and long-term plasticity at hippocampal mossy fiber synapses, *Neuropharmacology* **44**: 968.

Huang, P. L., et al., 1993, Targeted disruption of the neuronal nitric oxide synthase gene, *Cell* **75**: 1273.

Ingram, S. L., and Williams, J. T., 1996, Modulation of the hyperpolarization-activated current (I_h) by cyclic nucleotides in guinea-pig primary afferent neurons, *J. Physiol.* **492**: 97.

Ito, M., 2001, Cerebellar long-term depression: characterization, signal transduction, and functional roles, *Physiol. Rev.* **81**: 1143.

Jacoby, S., Sims, R. E., and Hartell, N. A., 2001, Nitric oxide is required for the induction and heterosynaptic spread of long-term potentiation in rat cerebellar slices, *J. Physiol.* **535**: 825.

Kantor, D. B., et al., 1996, A role for endothelial NO synthase in LTP revealed by adenovirus-mediated inhibition and rescue, *Science.* **274**: 1744.

Kaupp, U. B., and Seifert, R., 2002, Cyclic nucleotide-gated ion channels, *Physiol. Rev.* **82**: 769.

Kingston, P. A., Zufall, F., and Barnstable, C. J., 1996, Rat hippocampal neurons express genes for both rod retinal and olfactory cyclic nucleotide-gated channels: novel targets for cAMP/cGMP function, *PNAS USA* **93**: 10440.

Kingston, P. A., Zufall, F., and Barnstable, C. J., 1999, Widespread expression of olfactory cyclic nucleotide-gated channel genes in rat brain: implications for neuronal signalling, *Synapse* **32**: 1.

Kiss, J. P., 2000, Role of nitric oxide in the regulation of monoaminergic neurotransmission, *Brain Res. Bull.* **52**: 459.

Kleppisch, T., et al., 2003, Hippocampal cGMP-dependent protein kinase I supports an age- and protein synthesis-dependent component of long-term potentiation but is not essential for spatial reference and contextual memory, *J. Neurosci.* **23**: 6005.

Kloss, S., Furneaux, H., and Mulsch, A., 2003, Post-transcriptional regulation of soluble guanylyl cyclase expression in rat aorta, *J. Biol. Chem.* **278**: 2377.

Klyachko, V. A., Ahern, G. P., and Jackson, M. B., 2001, cGMP-mediated facilitation in nerve terminals by enhancement of the spike afterhyperpolarization, *Neuron* **31**: 1015.

Koesling, D., 1999, Studying the structure and regulation of soluble guanylyl cyclase, *Methods* **19**: 485.

Kojima, H., et al., 1998, Direct evidence of NO production in rat hippocampus and cortex using a new fluorescent indicator: DAF-2 DA, *Neuroreport* **9**: 3345.

Kornau, H. C., et al., 1995, Domain interaction between NMDA receptor subunits and the postsynaptic density protein PSD-95, *Science* **269**: 1737.

Krekelberg, B., and Taylor, J. G., 1996, Nitric oxide in cortical map formation, *J. Chem. Neuroanat.* **10**: 191.

Kuzmiski, J. B., and MacVicar, B. A., 2001, Cyclic nucleotide-gated channels contribute to the cholinergic plateau potential in hippocampal CA1 pyramidal neurons, *J. Neurosci.* **21**: 8707.

Launey, T., et al., 2004, Protein phosphatase 2A inhibition induces cerebellar long-term depression and declustering of synaptic AMPA receptor, *PNAS USA* **101**: 676.

Leamey, C. A., Ho-Pao, C. L., and Sur, M., 2001, Disruption of retinogeniculate pattern formation by inhibition of soluble guanylyl cyclase, *J. Neurosci.* **21**: 3871.

Lev-Ram, V., et al., 1997, Synergies and coincidence requirements between NO, cGMP and Ca^{2+} in the induction of cerebellar long-term depression, *Neuron* **18**: 1025.

Lev-Ram, V., et al., 2003, Reversing cerebellar long-term depression, *PNAS USA* **100**: 15989.

Lev-Ram, V., et al., 2002, A new form of cerebellar long-term potentiation is postsynaptic and depends on nitric oxide but not cAMP, *PNAS USA* **99**: 8389.

Li, D. P., Chen, S. R., and Pan, H. L., 2002, Nitric oxide inhibits spinally projecting paraventricular neurons through potentiation of presynaptic GABA release, *J. Neurophysiol.* **88**: 2664.

Lohmann, S. M., et al., 1997, Distinct and specific functions of cGMP-dependent protein kinases, *Trends Biochem. Sci.* **22**: 307.

Lu, Y. F., and Hawkins, R. D., 2002, Ryanodine receptors contribute to cGMP-induced late-phase LTP and CREB phosphorylation in the hippocampus, *J. Neurophysiol.* **88**: 1270.

Lucas, K. A., et al., 2000, Guanylyl cyclases and signaling by cyclic GMP, *Pharmacol. Rev.* **52**: 375.

Maffei, A., et al., 2003, NO enhances presynaptic currents during cerebellar mossy fiber-granule cell LTP, *J. Neurophysiol.* **90**: 2478.

Malinski, T., et al., 1993, Diffusion of nitric oxide in the aorta wall monitored in situ by porphyrinic microsensors, *Biochem. Biophys. Res. Commun.* **193**: 1076.

Martinez, S. E., Beavo, J. A., and Hol, W. G., 2002, GAF Domains: Two-Billion-Year-Old Molecular Switches that Bind Cyclic Nucleotides, *Mol. Intervent.* **2**: 317.

Matyash, V., et al., 2001, Nitric oxide signals parallel fiber activity to Bergmann glial cells in the mouse cerebellar slice, *Mol. Cell Neurosci.* **18**: 664.

Micheva, K. D., et al., 2003, Retrograde regulation of synaptic vesicle endocytosis and recycling, *Nat. Neurosci.* **6**: 925.

Moncada, S., Palmer, R. M., and Higgs, E. A., 1991, Nitric oxide: physiology, pathophysiology, and pharmacology, *Pharmacol. Rev.* **43**: 109.

Moreno-Lopez, B., et al., 2004, Nitric oxide is a physiological inhibitor of neurogenesis in the adult mouse subventricular zone and olfactory bulb, *J. Neurosci.* **24**: 85.

Mullershausen, F., et al., 2003, Direct activation of PDE5 by cGMP: long-term effects within NO/cGMP signaling, *J. Cell Biol.* **160**: 719.

Nelson, R. J., et al., 1997, Effects of nitric oxide on neuroendocrine function and behavior, *Front. Neuroendocrinol.* **18**: 463.

Nikonenko, I., Jourdain, P., and Muller, D., 2003, Presynaptic remodeling contributes to activity-dependent synaptogenesis, *J. Neurosci.* **23**: 8498.

O'Donnell, V. B., et al., 1999, 15-Lipoxygenase catalytically consumes nitric oxide and impairs activation of guanylate cyclase, *J. Biol. Chem.* **274**: 20083.

Packer, M. A., et al., 2003, Nitric oxide negatively regulates mammalian adult neurogenesis, *PNAS USA* **100**: 9566.

Pape, H. C., and Mager, R., 1992, Nitric oxide controls oscillatory activity in thalamocortical neurons, *Neuron* **9**: 441.

Paton, J. F., Kasparov, S., and Paterson, D. J., 2002, Nitric oxide and autonomic control of heart rate: a question of specificity, *Trends Neurosci.* **25**: 626.

Polleux, F., Morrow, T., and Ghosh, A., 2000, Semaphorin 3A is a chemoattractant for cortical apical dendrites, *Nature* **404**: 567.

Prast, H., and Philippu, A., 2001, Nitric oxide as modulator of neuronal function, *Prog. Neurobiol.* **64**: 51.

Reyes-Harde, M., et al., 1999, Induction of hippocampal LTD requires nitric-oxide-stimulated PKG activity and Ca2+ release from cyclic ADP-ribose-sensitive stores, *J. Neurophysiol.* **82**: 1569.

Reynolds, T., and Hartell, N. A., 2000, An evaluation of the synapse specificity of long-term depression induced in rat cerebellar slices, *J. Physiol.* **527**: 563.

Reynolds, T., and Hartell, N. A., 2001, Roles for nitric oxide and arachidonic acid in the induction of heterosynaptic cerebellar LTD, *Neuroreport.* **12**: 133.

Robinson, R. B., and Siegelbaum, S. A., 2003, Hyperpolarization-activated cation currents: from molecules to physiological function, *Annu. Rev. Physiol.* **65**: 453.

Roychowdhury, S., et al., 2002, Oxidative stress in glial cultures: detection by DAF-2 fluorescence used as a tool to measure peroxynitrite rather than nitric oxide, *Glia* **38**: 103.

Russwurm, M., Wittau, N., and Koesling, D., 2001, Guanylyl cyclase/PSD-95 interaction: targeting of the nitric oxide-sensitive alpha2beta1 guanylyl cyclase to synaptic membranes, *J. Biol. Chem.* **276**: 44647.

Santoro, B., et al., 2000, Molecular and functional heterogeneity of hyperpolarization-activated pacemaker channels in the mouse CNS, *J. Neurosci.* **20:** 5264.

Sattler, R., et al., 1999, Specific coupling of NMDA receptor activation to nitric oxide neurotoxicity by PSD-95 protein, *Science* **284:** 1845.

Savchenko, A., Barnes, S., and Kramer, R. H., 1997, Cyclic-nucleotide-gated channels mediate synaptic feedback by nitric oxide, *Nature* **390:** 694.

Schmidt, H., et al., 2002, cGMP-mediated signaling via cGKIalpha is required for the guidance and connectivity of sensory axons, *J. Cell Biol.* **159:** 489.

Schuman, E. M., and Madison, D. V., 1991, A requirement for the intercellular messenger nitric oxide in long-term potentiation, *Science* **254:** 1503.

Schuman, E. M., and Madison, D. V., 1994, Locally distributed synaptic potentiation in the hippocampus, *Science* **263:** 532.

Shaw, P. J., Charles, S. L., and Salt, T. E., 1999, Actions of 8-bromo-cyclic-GMP on neurones in the rat thalamus *in vivo* and *in vitro*, *Brain Res.* **833:** 272.

Shibuki, K., and Kimura, S., 1997, Dynamic properties of nitric oxide release from parallel fibres in rat cerebellar slices, *J. Physiol.* **498:** 443.

Smith, S. L., and Otis, T. S., 2003, Persistent changes in spontaneous firing of Purkinje neurons triggered by the nitric oxide signaling cascade, *J. Neurosci.* **23:** 367.

Soderling, S. H., and Beavo, J. A., 2000, Regulation of cAMP and cGMP signaling: new phosphodiesterases and new functions, *Curr. Opin. Cell Biol.* **12:** 174.

Son, H., et al., 1996, Long-term potentiation is reduced in mice that are doubly mutant in endothelial and neuronal nitric oxide synthase, *Cell* **87:** 1015.

Southam, E., and Garthwaite, J., 1993, The nitric oxide-cyclic GMP signalling pathway in rat brain, *Neuropharmacology* **32:** 1267.

Stanton, P. K., et al., 2003, Long-term depression of presynaptic release from the readily releasable vesicle pool induced by NMDA receptor-dependent retrograde nitric oxide, *J. Neurosci.* **23:** 5936.

Stefano, G. B., and Ottaviani, E., 2002, The biochemical substrate of nitric oxide signaling is present in primitive non-cognitive organisms, *Brain Res.* **924:** 82.

Steinbach, K., Volkmer, H., and Schlosshauer, B., 2002, Semaphorin 3E/collapsin-5 inhibits growing retinal axons, *Exp. Cell Res.* **279:** 52.

Stern, J. E., Li, Y., and Zhang, W., 2003, Nitric oxide: a local signalling molecule controlling the activity of pre-autonomic neurones in the paraventricular nucleus of the hypothalamus, *Acta Physiol. Scand.* **177:** 37.

Thomas, M. K., Francis, S. H., and Corbin, J. D., 1990, Substrate- and kinase-directed regulation of phosphorylation of a cGMP-binding phosphodiesterase by cGMP, *J. Biol. Chem.* **265:** 14971.

Tsou, K., Snyder, G. L., and Greengard, P., 1993, Nitric oxide/cGMP pathway stimulates phosphorylation of DARPP-32, a dopamine- and cAMP-regulated phosphoprotein, in the substantia nigra, *PNAS USA* **90:** 3462.

Vincent, S. R., and Kimura, H., 1992, Histochemical mapping of nitric oxide synthase in the rat brain, *Neuroscience* **46:** 755.

Wakatsuki, H., et al., 1998, Layer-specific NO dependence of long-term potentiation and biased NO release in layer V in the rat auditory cortex, *J. Physiol.* **513:** 71.

Wall, M. J., 2003, Endogenous nitric oxide modulates GABAergic transmission to granule cells in adult rat cerebellum, *Eur. J. Neurosci.* **18:** 869.

Wang, X., and Robinson, P. J., 1995, Cyclic GMP-dependent protein kinase substrates in rat brain, *J. Neurochem.* **65:** 595.

Wexler, E. M., Stanton, P. K., and Nawy, S., 1998, Nitric oxide depresses GABAA receptor function via coactivation of cGMP-dependent kinase and phosphodiesterase, *J. Neurosci.* **18:** 2342.

Wood, J., and Garthwaite, J., 1994, Models of the diffusional spread of nitric oxide: implications for neural nitric oxide signalling and its pharmacological properties, *Neuropharmacology* **33:** 1235.

Wu, H. H., et al., 2000, Refinement of the ipsilateral retinocollicular projection is disrupted in double endothelial and neuronal nitric oxide synthase gene knockout mice, *Dev. Brain Res.* **120:** 105.

Wu, H. H., et al., 2001, The role of nitric oxide in development of topographic precision in the retinotectal projection of chick, *J. Neurosci.* **21:** 4318.

Wykes, V., Bellamy, T. C., and Garthwaite, J., 2002, Kinetics of nitric oxide-cyclic GMP signalling in CNS cells and its possible regulation by cyclic GMP, *J. Neurochem.* **83:** 37.

Yamazaki, M., et al., 2001, Activation of the mitogen-activated protein kinase cascade through nitric oxide synthesis as a mechanism of neuritogenic effect of genipin in PC12h cells, *J. Neurochem.* **79:** 45.

Yang, G., and Iadecola, C., 1998, Activation of cerebellar climbing fibers increases cerebellar blood flow: role of glutamate receptors, nitric oxide, and cGMP, *Stroke* **29:** 499.

Zabel, U., et al., 2002, Calcium-dependent membrane association sensitizes soluble guanylyl cyclase to nitric oxide, *Nat. Cell Biol.* **4**: 307.

Zagotta, W. N., et al., 2003, Structural basis for modulation and agonist specificity of HCN pacemaker channels, *Nature* **425**: 200.

Zhang, X., et al., 2002, Interfering with nitric oxide measurements. 4,5-diaminofluorescein reacts with dehydroascorbic acid and ascorbic acid, *J. Biol. Chem.* **277**: 48472.

SOMATODENDRITIC H$_2$O$_2$ FROM MEDIUM SPINY NEURONS INHIBITS AXONAL DOPAMINE RELEASE

Margaret E. Rice and Marat V. Avshalumov[*]

1. INTRODUCTION

Reactive oxygen species (ROS), including superoxide (O$_2$•$^-$), hydrogen peroxide (H$_2$O$_2$), and the hydroxyl radical (•OH), are often viewed only as toxic 'byproducts' of cell metabolism (Fridovich, 1979; Halliwell and Gutteridge, 1984; Halliwell, 1992; Beckman and Ames, 1998; Finkel and Holbrook, 2000). Over the past decade, however, accumulating evidence indicates that ROS are normal components of signaling pathways and can mediate a variety of physiological processes (Sundaresan et al., 1995; Aikawa et al., 1997; Finkel and Holbrook, 2000; Nishida et al., 2000; Knapp and Klann, 2002).

H$_2$O$_2$ is a particularly intriguing candidate as a signaling molecule. Because it is neutral and membrane-permeable (Ramasarma, 1983), H$_2$O$_2$ can freely diffuse away from a site of generation, a characteristic shared with the known diffusible messengers nitric oxide (NO•) and carbon monoxide (CO) (Dawson and Snyder, 1994; Kiss and Vizi, 2001). Moreover, H$_2$O$_2$ is *not* a free radical, unlike O$_2$•$^-$ and •OH, and therefore does not readily mediate oxidative damage, unless exposed to free iron or copper ions to catalyze the conversion of H$_2$O$_2$ to •OH (Cohen, 1994). Indeed, H$_2$O$_2$ has been implicated as a diffusible messenger in neuron-glia signalling (Atkins and Sweatt, 1999) and in inter-neuronal communication, including the regulation of synaptic transmission and plasticity (Samanta et al., 1998; Auerbach and Segal, 1997; Klann and Thiels, 1999; Nishida et al., 2000; Nemoto et al., 2000; Kamsler and Segal, 2003; Avshalumov et al., 2003a).

The first evidence that H$_2$O$_2$ might regulate neurotransmission was the finding that exogenous H$_2$O$_2$ could suppress the amplitude of evoked population spikes in hippocampal slices, possibly by inhibiting transmitter release (Pellmar 1986, 1987). Our laboratory confirmed that H$_2$O$_2$ can inhibit transmitter release by demonstrating reversible suppression of dopamine (DA) release during H$_2$O$_2$ exposure in striatal slices (Chen et al., 2001). More significantly, we found that *endogenous* H$_2$O$_2$ suppresses DA release. In this chapter, we review evidence that modulatory H$_2$O$_2$ in striatum is generated

---

[*] Department of Physiology and Neuroscience, NYU School of Medicine, New York, NY 10016, USA.

Dendritic Neurotransmitter Release, edited by M. Ludwig
Springer Science+Business Media, Inc., 2005

downstream from AMPA-receptor activation in the dendrites of medium spiny neurons, and that inhibition of DA release is mediated by ATP-sensitive K^+ (K_{ATP}) channels on DA axons. Thus, dendritically generated H_2O_2 must act as a diffusible messenger to inhibit DA release from presynaptic sites.

2. REGULATION OF STRIATAL DOPAMINE RELEASE BY H_2O_2

2.1. Mitochondrial Generation of H_2O_2

The major source of ROS generation in neurons and glia, as in all cells, is mitochondrial respiration (Finkel and Holbrook, 2000). The first step in this process is the one electron reduction of O_2 to form $O_2^{\bullet-}$ (Fig. 1A); H_2O_2 is generated from $O_2^{\bullet-}$ by superoxide dismutase (SOD), as well as spontaneous dismutation (Boveris and Chance, 1973; Fridovich, 1979; Halliwell, 1992). The amount of H_2O_2 produced by brain mitochondria can reach 5% of O_2 metabolized (Arnaiz et al., 1999), such that the concentration of H_2O_2 within the restricted volume of a dendrite, for example, might transiently approach mM levels (Chen et al., 2001). Absolute levels of H_2O_2 at any moment in time will depend on the balance among the rate of O_2 consumption, the activity of SOD, and the rate of metabolism of H_2O_2 back to O_2 and H_2O by the peroxidase enzymes glutathione (GSH) peroxidase and catalase (Cohen, 1994) (Fig. 1A).

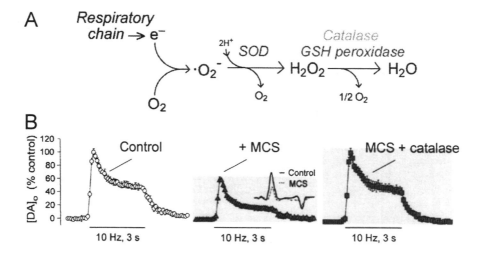

Figure 1. (A) Production of H_2O_2 by the mitochondrial respiratory chain. Levels of H_2O_2 can be altered by manipulating tissue peroxidase activity: H_2O_2 availability can be *increased* by inhibiting GSH peroxidase or *decreased* by adding exogenous catalase. (B) Striatal DA release is modulated by H_2O_2. Inhibition of GSH peroxidase by mercaptosuccinate (MCS; 1 mM) enhances H_2O_2 availability, which suppresses evoked DA release evoked by 10 Hz, 30-pulse trains in guinea-pig striatal slices; DA was monitored with carbon-fiber microelectrodes and fast-scan cyclic voltammetry (inset shows characteristic DA voltammograms in control and MCS). Catalase (500 IU mL^{-1}) in the continued presence of MCS reverses H_2O_2-dependent suppression of DA release. Data are means ± SEM (modified from Avshalumov et al. 2003a).

2.2. Endogenous H_2O_2 Inhibits Striatal Dopamine Release

We tested Pellmar's hypothesis that H_2O_2 might suppress transmitter release by examining the effect of exogenous H_2O_2 on DA release in slices of guinea-pig striatum *in vitro* (Chen et al., 2001). Release of DA was elicited by local electrical stimulation and monitored with carbon-fiber microelectrodes and fast-scan cyclic voltammetry (Bull et al., 1990; Chen et al., 2001; Avshalumov et al., 2003a). We found that transient exposure to H_2O_2 (15 min, 1.5 mM) causes a decrease in evoked extracellular DA concentration ($[DA]_o$). Release suppression is reversible upon H_2O_2 washout, with no alteration in tissue DA content or evidence of oxidative damage (Chen et al., 2001).

By manipulating tissue peroxidase activity, we discovered that *endogenous* H_2O_2 can also regulate striatal DA release (Chen et al., 2001; Chen et al., 2002; Avshalumov et al., 2003a). Exposure of striatal slices to mercaptosuccinate (MCS), a selective inhibitor of GSH peroxidase (Ying et al., 1999; Sokolova et al., 2001), causes a similar decrease in evoked $[DA]_o$ (Fig. 1B) to that seen with exogenous H_2O_2. This suppression reverses upon MCS washout (Chen et al., 2002) or when the tissue is superfused with catalase in the continued presence of MCS (Avshalumov et al., 2003a) (Fig. 1B). The inhibitory action of endogenous H_2O_2 is not species dependent: MCS causes a similar 30-40% decrease in evoked $[DA]_o$ in striatal slices from rat and non-human primate (marmoset), as well as guinea-pig (Rice et al., 2002; Avshalumov et al., 2003a). In addition, the effect of MCS persists in the presence of a DA uptake inhibitor (GBR-12909), which indicates that H_2O_2 alters DA *release* rather than DA *uptake* (Avshalumov et al., 2003a).

What cellular process generates modulatory H_2O_2 in the striatum? An important clue in understanding this was the finding that MCS has no effect on DA release evoked by a single stimulus pulse (Avshalumov et al., 2003a). This indicates that H_2O_2 must be formed after at least one pulse of a train (or action-potential burst) to activate a process that inhibits DA release elicited by subsequent pulses. Moreover, this inhibition occurs on a physiological time-scale: suppression of DA release by MCS occurs within 500 ms of a 10-Hz pulse train (Fig. 1B). Given that oxygen consumption increases during local electrical stimulation in striatal slices (Kennedy et al., 1992), the most likely source of activity-dependent H_2O_2 production is mitochondrial respiration. Recent studies from our laboratory support this hypothesis: the effect of MCS is lost when mitochondrial complex I is inhibited by rotenone (Bao and Rice, 2004). This argues against significant involvement of other $O_2^{\bullet-}/H_2O_2$ generating pathways, including NADPH oxidase (Finkel and Holbrook, 2000; Knapp and Klann, 2002). Additional experiments indicate a lack of involvement of H_2O_2 generated by monoamine oxidase (MAO), which produces one molecule of H_2O_2 for each molecule of DA metabolized (Cohen, 1994); neither evoked $[DA]_o$ nor the effect of MCS is altered by a cocktail of MAO inhibitors (Bao and Rice, 2004).

3. STRIATAL H_2O_2 GENERATION REQUIRES AMPA-RECEPTOR ACTIVATION

Glutamate-receptor activation has been shown to enhance mitochondrial H_2O_2 generation in isolated neurons in culture (Dugan et al., 1995; Reynolds and Hastings,

1995; Bindokas et al., 1996; Carriedo et al., 2000). The mechanism involves an increase in mitochondrial activity to meet increased energy demands following glutamate-induced depolarization and the re-establishment of ion gradients. Re-establishing intracellular Ca^{2+} concentration ($[Ca^{2+}]_i$), involves particularly energy demanding processes and is accompanied by enhanced mitochondrial activity and O_2 consumption (Kojima et al., 1994; Bindokas et al., 1998; Jung et al., 2000; Zenisek and Matthews, 2000). Consistent with these findings, we found that endogenously generated H_2O_2 in striatal slices is generated downstream from AMPA-receptor activation (Avshalumov et al., 2003a). In this preparation, blockade of AMPA receptors with the selective antagonist GYKI-52466 causes a ~100% increase in evoked $[DA]_o$, indicating that glutamate normally *inhibits* DA release (Fig. 2A). Blockade of NMDA receptors is without effect in this paradigm. Strikingly, the effect of AMPA-receptor blockade can be completely prevented by the application of the H_2O_2-metabolizing enzymes catalase (Fig. 2B) or GSH peroxidase (Avshalumov et al., 2003a). Is this AMPA-receptor dependent process the only source of modulatory H_2O_2? The answer appears to be yes, since the effect of MCS is abolished when AMPA receptors are blocked by GYKI-52466 (Avshalumov et al., 2003a).

The question of how, or even *if*, glutamate regulates striatal DA release has been a long-standing source of controversy (Cheremy et al., 1986; Leviel et al., 1990; Moghaddam and Gruen, 1991; Keefe et al., 1993; Westerink et al., 1992; Wu et al., 2000). The absence of ionotropic glutamate receptors on DA terminals (Bernard and

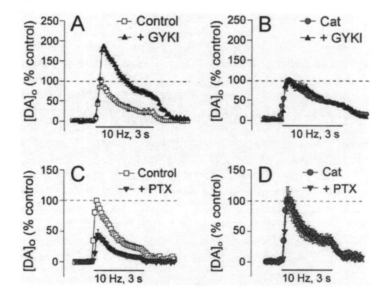

Figure 2. Regulation of striatal DA release by glutamate and GABA requires H_2O_2. A) AMPA-receptor blockade by GYKI-52466 (GYKI; 50 μM) causes a ~100% evoked $[DA]_o$ in striatum ($p < 0.001$, $n = 6$). B) The effect of AMPA-receptor blockade is completely prevented by catalase (Cat). C) $GABA_A$ receptor blockade by picrotoxin (PTX; 100 μM) causes a ~50% decrease in evoked $[DA]_o$ ($p < 0.001$, $n = 6$). D) Catalase abolishes the effect of picrotoxin. Responses in the presence of heat-inactivated catalase were the same as control. Data are means ± SEM, illustrated as percentage of same-site control (modified from Avshalumov et al., 2003a).

Bolam, 1998; Chen et al., 1998) has suggested that any glutamatergic influence must be indirect. Proof of this has been elusive, however, and the effects of glutamate receptor blockade have often seemed paradoxical. The discovery that glutamate can inhibit DA release with H_2O_2 as an inhibitory intermediate helps resolve this conundrum.

Interestingly, tests of the possible involvement of conventional GABAergic circuitry in the regulation of striatal DA release also gave surprising results: blockade of $GABA_A$ receptors with picrotoxin causes a ~50% decrease in evoked $[DA]_o$ (Fig. 2C), whereas $GABA_B$-receptor blockade with saclofen has no effect (Avshalumov et al., 2003a). These data indicate that GABA, acting through $GABA_A$ receptors, normally *enhances* DA release. The influence of GABA on DA release, like that of glutamate, must be indirect, since DA axons in striatum do not express $GABA_A$ receptors (Fujiyama et al., 2000). Indeed, GABAergic regulation also requires H_2O_2 generation; the effect of picrotoxin was completely prevented by catalase (Fig. 2D). Moreover, picrotoxin has no effect on DA release when AMPA receptors were blocked with GYKI-52466 (Avshalumov et al., 2003a). Together, these data indicate that glutamate and GABA act on the same pool of modulatory H_2O_2 that is generated downstream from AMPA-receptor activation.

4. ACTIVATION OF K_{ATP} CHANNELS UNDERLIES H_2O_2-DEPENDENT INHIBITION OF STRIATAL DOPAMINE RELEASE

The findings discussed thus far show that endogenous H_2O_2 is a messenger that mediates the effects of glutamate and GABA on striatal DA release. How does H_2O_2 inhibit DA release? The answer is that H_2O_2 generation leads to the opening of ATP-sensitive K^+ (K_{ATP}) channels (Avshalumov et al., 2003a; Avshalumov and Rice, 2003). Previous physiological studies demonstrated that *exogenous* H_2O_2 can cause membrane hyperpolarization and decreased excitation by activating a K^+ conductance in a variety of cell types, including pancreatic β-cells (Krippeit-Drews et al., 1999) and CA1 hippocampal neurons (Seutin et al., 1995). Our studies of DA release in striatal slices provided the first evidence that *endogenous* H_2O_2 can activate K_{ATP} channels. These studies showed that blockade of K_{ATP}-channels with sulfonylureas, either tolbutamide (Avshalumov et al., 2003a) or glibenclamide (Fig. 3A) (Avshalumov and Rice, 2003), causes a significant increase in evoked $[DA]_o$, which indicates that K_{ATP} channel activation inhibits DA release during normal stimulation. Blockade of K_{ATP} channels also prevents the inhibitory effect of MCS on DA release, as well as the usual effects of GYKI-52466 and picrotoxin (Fig. 3A). These data demonstrate that K_{ATP} channels are *required* for modulation of DA release by H_2O_2, glutamate, and GABA.

Which K_{ATP} channels are involved? These channels are multimeric proteins composed of an inwardly rectifying pore-forming unit, typically Kir 6.2 in neurons (Karschin et al., 1997; Aschroft and Gribble, 1998), and a sulfonylurea-binding site (SUR1 or SUR2) (Babenko et al., 1998; Aguilar-Bryan et al., 1998). SUR1- and SUR2-based channels can be distinguished by their differential sensitivity to K_{ATP}-channel openers, with preferential selectivity of diazoxide for SUR1 and selectivity of cromakalim for SUR2 (Inagaki et al., 1996; Babenko et al., 2000). In the striatum, K_{ATP} channel opening by diazoxide causes a *suppression* of DA release and prevention of the usual pattern of H_2O_2-dependent modulation by MCS, GYKI, and picrotoxin (Fig. 3B)

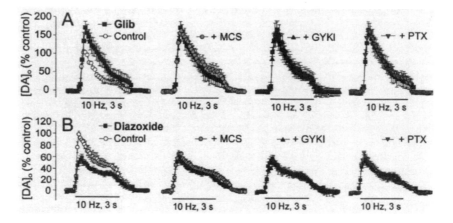

Figure 3. Inhibition of striatal DA release by endogenous H_2O_2 is mediated by K_{ATP} channels. A) K_{ATP}-channel blockade by glibenclamide (Glib; 3 μM) increased evoked $[DA]_o$ in guinea-pig striatal slices ($p < 0.01$, n = 5) and prevented the usual modulation of DA release by MCS (1 mM), GYKI-52466 (GYKI; 50 μM), and picrotoxin (PTX; 100 μM) (n = 5 for each). B) Diazoxide (30 μM), a SUR1-selective K_{ATP}-channel opener, decreased evoked $[DA]_o$ ($p < 0.01$, diazoxide vs. control; n = 5) and also abolished the effects of MCS, GYKI-52466, and PTX (n = 5). Data are means $\pm$ SEM (modified from Avshalumov and Rice, 2003).

(Avshalumov and Rice, 2003). Although SUR2-selective cromakalim also suppresses striatal DA release, it does not alter H_2O_2-dependent DA modulation by these agents, indicating selective targeting of SUR1-based K_{ATP} channels in this process (Avshalumov and Rice, 2003). The question of whether the effect of H_2O_2 is direct or mediated by additional pathways, e.g. by altering ATP levels as exogenous H_2O_2 does in pancreatic β-cells (Krippeit-Drews et al., 1999), remains open.

K_{ATP}-channels are expressed in DA cells of the substantia nigra pars compacta (SNc) and throughout the nigrostriatal pathway (Mourre et al., 1990; Xia and Haddad, 1991; Dunn-Meynell et al., 1997), as well as in striatal medium spiny neurons (Schwanstecher and Bassen, 1997). However, the pattern of effects of K_{ATP} channel blockers and openers on DA release (Fig. 3) is consistent with primary localization of H_2O_2-sensitive K_{ATP} channels on DA axons rather than on spiny neurons (Avshalumov and Rice 2003). Most DA cells in the SNc express K_{ATP} channels with either SUR1 or SUR2 subunits; intriguingly, SUR1 expression is linked to greater metabolic sensitivity in DA neurons in the SNc (Liss et al., 1999). Electrophysiological studies in SNc neurons indicate that H_2O_2 activates glibenclamide-sensitive K_{ATP} channels in one population of DA cells, but not a second population (Avshalumov et al., 2003b). Whether this pattern of 'responders' and 'non-responders' reflects SUR1 versus SUR2 expression awaits further study.

5. SOMATODENDRITIC H_2O_2 IN MEDIUM SPINY NEURONS

5.1. Which Striatal Cells Produce Modulatory H_2O_2?

To address the cellular source of H_2O_2 requires a basic understanding of striatal circuitry and receptor localization. While the overall circuitry of the basal ganglia is well known (Kemp and Powell, 1971; Albin et al., 1989; Smith and Bolam, 1990), the microchemical circuitry of individual structures, including the striatum, is only beginning to be elucidated. Motor regions of the dorsal striatum receive excitatory input from motor cortex and thalamus and provide the major inhibitory output of the basal ganglia to subcortical structures (Albin et al., 1989; Smith and Bolam, 1990). The principal striatal efferent cells are GABAergic medium spiny neurons (Kemp and Powell, 1971), which receive synaptic glutamate input to their dendrites (Smith and Bolam, 1990; Bernard and Bolam, 1998; Chen et al., 1998). These neurons also receive synaptic DA input from midbrain DA cells (Albin et al., 1989; Smith and Bolam, 1990) (Fig. 4). The absence of ionotropic glutamate receptors (Bernard and Bolam, 1998; Chen et al., 1998) and $GABA_A$ receptors (Fujiyama et al., 2000) on DA terminals suggests that glutamate-dependent H_2O_2 must be generated in non-DA cells. Prevention of the modulatory effects of endogenous glutamate and GABA by large peroxidase enzymes (Fig. 2B,D), which are likely to remain in the extracellular compartment, further suggests that H_2O_2 must diffuse via the extracellular space to reach presynaptic DA axons.

We have proposed that the most likely cells involved in generation of modulatory H_2O_2 are medium spiny neurons, which make up 90-95% of the neuronal population in striatum (Kemp and Powell, 1971). Generation in spiny neurons is supported by the pattern of sensitivity of DA release to glutamate and GABA antagonists, which mirrors the electrophysiological responsiveness of these cells (Jiang and North, 1991; Kita, 1996). Moreover, glutamate synapses can be closely apposed to DA synapses on the dendrites of medium spiny neurons (Smith and Bolam, 1990; Bernard and Bolam, 1998;

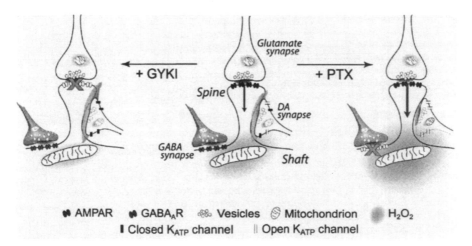

Figure 4. Triad of DA, glutamate and GABA synapses bound together *functionally* by diffusible H_2O_2. Expected H_2O_2 generation in the dendritic shaft of a medium spiny neuron is indicated for control conditions (*center*), $GABA_A$-receptor ($GABA_AR$) blockade by picrotoxin (+ PTX, *right*), and AMPA-receptor (AMPAR) blockade by GYKI-52466 (+ GYKI, *left*). In this model, AMPA-receptor-dependent excitation (red arrows) increases H_2O_2 generation, which opens presynaptic K_{ATP} channels on DA axons; activity-dependent H_2O_2 generation is opposed by $GABA_A$-receptor mediated inhibition. When AMPA-receptors are blocked (+ GYKI), local stimulation does not produce sufficient H_2O_2 to open K_{ATP} channels and alter DA release; opposition by GABA is revealed when $GABA_A$-receptors are blocked (+ PTX), which increase H_2O_2 levels, enhances K_{ATP}-channel opening and further suppresses DA release (circuitry and locations of receptors and mitochondria based on data in Smith and Bolam, 1990; Bernard and Bolam, 1998; Chen et al., 1998; Fujiyama et al., 2000). **(A color version of this figure appears in the signature between pp. 256 and 257.)**

Chen et al., 1998), placing them in an ideal position to modulate DA release via dendritically generated H_2O_2 (Fig. 4). Other striatal cells that express AMPA receptors, e.g. cholinergic interneurons (Bernard and Bolam, 1998; Chen et al., 1998) could contribute to H_2O_2 generation. However, these are sparse and do not receive DA synapses, such that it is less likely that the level and lifetime of generated H_2O_2 would be sufficient to affect DA release at distant synapses.

Medium spiny neurons also express $GABA_A$ receptors at dendritic sites near spines (Fujiyama et al., 2000), such that GABAergic input is well-positioned to oppose AMPA-receptor-mediated excitation and consequent H_2O_2 generation from mitochondria located in close proximity to dendritic spines and DA synapses (Smith and Bolam, 1990; Chen et al., 1998) (Fig. 4). Although AMPA receptors on medium spiny neurons are not Ca^{2+}-permeable, AMPA-receptor activation leads to an increase in $[Ca^{2+}]_i$ via voltage-dependent Ca^{2+} channels (Stefani et al., 1998), which would lead to an increase in mitochondrial respiration and H_2O_2 generation. We propose, therefore, that H_2O_2 generated in dendrites diffuses to adjacent DA synapses where it inhibits DA release via opening of SUR1-containing K_{ATP}-channels. By decreasing dendritic excitability, GABA lessens H_2O_2 production (Fig. 4, *center*). When $GABA_A$ receptors are blocked (+ PTX), however, H_2O_2 production increases, leading to greater suppression of DA release (Fig. 4, *right*). When AMPA receptors are blocked (+ GYKI), H_2O_2 generation is minimal, DA release increases, and $GABA_A$-dependent regulation is lost (Fig. 4, *left*). Like a brake when there is no motion, GABA has no direct influence on DA release, but rather counters the extent to which glutamate-receptor activation generates H_2O_2.

5.2. Somatodendritic Generation of H_2O_2 in Medium Spiny Neurons

Most available evidence for H_2O_2 generation in the dendrites of medium spiny neurons is indirect. However, recent studies in our laboratory indicate that medium spiny neurons do produce H_2O_2 in their cell bodies and proximal dendrites during local electrical stimulation (Fig. 5). In these studies, whole-cell recording was used to identify medium spiny neurons by their electrophysiological properties (e.g. Moriguchi, et al., 2002; Blackwell, et al., 2003). The H_2O_2-sensitive fluorescent dye 2',7'-dichlorofluorescein (DCF) (Sah and Schwartz-Bloom, 1999) was included in the backfill solution of the patch pipette. A marked increase in DCF fluorescence intensity was elicited in medium spiny neurons by the same paradigm of local electrical stimulation we typically use to evoke DA release (10 Hz, 30 pulses) (Fig. 5). This confirms that these cells indeed generate H_2O_2 during stimulus-induced cell firing (n=7). Surprisingly, detectable increases in H_2O_2 in somata and proximal dendrites were not seen until nearly a second after stimulus initiation (Fig. 5), which is later than the time course implied by our DA release studies in the presence of MCS (Section 2.2). However, more immediate H_2O_2 generation might be expected to occur in distal dendrites near spines that receive glutamatergic excitation, as suggested by a faster rate of increase in proximal dendrites than in somata (Fig. 5). The extent and time course of H_2O_2 generation in distal dendrites (i.e. confirmation of 'dendritic' rather than 'somatodendritic' generation) awaits further imaging experiments with higher spatio-temporal resolution than available at the present time. Additional studies will also be required to indicate whether medium spiny neurons are the *only* striatal cells that produce modulatory H_2O_2.

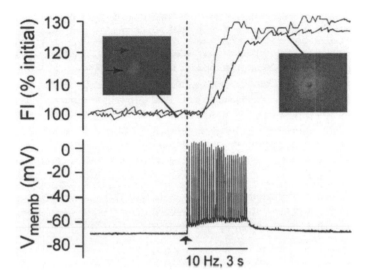

Fig. 5. Generation of H_2O_2 in a striatal medium spiny neuron during local electrical stimulation. Upper panel: fluorescence intensity (FI) in a cell body (black line) and proximal dendrite (red-orange line) in a single, DCF-filled medium spiny neuron under control conditions and during local stimulation (10 Hz, 30 pulses); the apparent plateau is a consequence of irreversible DCF activation. Inset images indicate basal and stimulated fluorescence at monitored sites in the cell body (black arrow) and proximal dendrite (red-orange arrow); DCF (7 μM) was included in the backfill solution of the patch-pipette used for whole-cell recording. DCF images were acquired using a PTI imaging system and Olympus BX-51WI microscope; excitation wavelength was 488 nm with fluorescence emission at 535 nm. Lower panel: simultaneously recorded membrane voltage (V_{memb}) during train stimulation. Dashed vertical line indicates onset of stimulus train.

(A color version of this figure appears in the signature between pp. 256 and 257.)

6. SUMMARY AND CONCLUSIONS

In our first report of DA release modulation by endogenous H_2O_2 (Chen et al., 2001), we suggested that H_2O_2 might be generated presynaptically at DA synapses to serve as an autoinhibitory signal that would limit DA release after axonal activation. We were wrong. Our subsequent studies showed that generation of modulatory H_2O_2 requires AMPA-receptor activation and can be 'fine-tuned' by GABA$_A$-receptor activation. These findings argue against generation in DA axons, since they lack AMPA and GABA$_A$ receptors (Bernard and Bolam, 1998; Chen et al., 1998; Fujiyama et al., 2000).

What we found, however, is more exciting: generation of H_2O_2 must occur in non-DA cells or processes. Our working hypothesis is that regulation of striatal DA release by glutamate and GABA involves a triad of DA, glutamate, and GABA synapses, separated by a few micrometers on the dendrites of medium spiny neurons (Smith and Bolam, 1990; Bernard and Bolam, 1998; Chen et al., 1998; Fujiyama et al., 2000), and bound together *functionally* by diffusible H_2O_2 (Fig. 4). Although we have not yet confirmed generation of H_2O_2 in distal dendrites, we have demonstrated activity dependent H_2O_2

generation in the soma and proximal dendrites of medium spiny neurons; the title of this chapter, therefore, refers to 'somatodendritic' rather than 'dendritic' H_2O_2.

Regardless of the source, endogenously generated H_2O_2 reverses conventional glutamatergic excitation by opening K_{ATP} channels to inhibit striatal DA release. These findings help clarify normal DA-glutamate interactions in striatum. Moreover, because DA-glutamate dysfunction has been implicated as a causal factor in Parkinson's disease (Olanow and Tatton, 1999; Chase and Oh, 2000; Greenamyre, 2001), schizophrenia (Deutsch et al., 2001; Sawa and Snyder, 2002), and addiction (Koob, 2000; Hyman and Malenka, 2001), exploration of this process may also suggest novel pathways through which dysfunction could occur.

One final point is that neuromodulation by H_2O_2 is a double-edged sword: an imbalance between H_2O_2 generation and regulation could result in oxidative stress, which has been implicated in nigrostriatal degeneration in Parkinson's disease (Cohen, 1994; Sonsalla et al., 1997; Olanow and Tatton, 1999; Xu et al., 2002) and, more recently, as a causal factor in schizophrenia (Do et al., 2000; Yao et al., 2001). Thus, loss of normal H_2O_2 regulation might contribute to DA-system pathology. It is relevant to note, therefore, that inhibition of GSH peroxidase by MCS leads to DA neuron hyperpolarization (Avshalumov et al., 2003b) and suppression of somatodendritic DA release in the SNc (Chen et al., 2002); however, MCS does not inhibit DA release in the adjacent ventral tegmental area (VTA) (Chen et al., 2002). This difference between SNc and VTA is potentially important, because DA neurons of the SNc degenerate in Parkinson's whereas those in the VTA are relatively spared (Yamada et al., 1990; Fearnley and Lees 1991).

7. ACKNOWLEDGEMENTS

The authors are grateful for support from NIH grants NS-36362 and NS-45325.

8. REFERENCES

Aguilar-Bryan, L., Clement, J. P. T., Gonzalez, G., Kunjilwar, K., Babenko, A., and Bryan J., 1998, Toward understanding the assembly and structure of K_{ATP} channels, *Physiol. Rev.* **78**: 227.

Aikawa, R., Komuro, I., Yamazaki, T., Zou, Y., Kudoh, S., Tanaka, M., Shiojima, I., Hiroi, Y., and Yazaki, Y., 1997, Oxidative stress activates extracellular signal-regulated kinases through Src and Ras in cultured cardiac myocytes of neonatal rats, *J. Clin. Invest.* **100**: 1813.

Albin, R. L., Young, A. B., and Penney, J. B., 1989, The functional anatomy of basal ganglia disorders, *Trends Neurosci.* **12**: 366.

Arnaiz, S. L., Coronel, M. F., and Boveris, A., 1999, Nitric oxide, superoxide, and hydrogen peroxide production in brain mitochondria after haloperidol treatment, *Nitric Oxide* **3**: 235.

Ashcroft, F. M., and Gribble, F. M., 1998, Correlating structure and function in ATP-sensitive K^+ channels, *Trends Neurosci.* **21**: 288.

Atkins, C. M., and Sweatt, J. D., 1999, Reactive oxygen species mediate activity-dependent neuron-glia signaling in output fibers of the hippocampus, *J. Neurosci.* **19**: 7241.

Auerbach, J. M., and Segal, M., 1997, Peroxide modulation of slow onset potentiation in rat hippocampus, *J. Neurosci.* **17**: 8695.

Avshalumov, M. V., and Rice, M. E., 2003, Activation of ATP-sensitive K^+ (K_{ATP}) channels by H_2O_2 underlies glutamate-dependent inhibition of striatal dopamine release. *PNAS USA* **100**:11729.

Avshalumov, M. V., Chen, B. T., Marshall, S. P., Peña, D. M., and Rice, M. E., 2003a, Glutamate-dependent inhibition of dopamine release in striatum is mediated by a new diffusible messenger, H₂O₂, *J. Neurosci.* **23**: 2744.

Avshalumov, M. V., Chen, B. T., and Rice. M. E., 2003b, Presence of different sulfonylurea receptors in nigrostriatal pathway defines sensitivity of dopamine release to H₂O₂, Program number 440.5. *2003 Abstract Viewer/Itinerary Planner,* Washington, DC: Society for Neuroscience.

Babenko, A. P., Aguilar-Bryan, L., and Bryan, J., 1998, A view of SUR/K$_{IR}$6.X, K$_{ATP}$ channels, *Annu. Rev. Physiol.* **60**: 667.

Babenko, A. P., Gonzalez, G., and Bryan J., 2000, Pharmaco-topology of sulfonylurea receptors. Separate domains of the regulatory subunits of K$_{ATP}$ channel isoforms are required for selective interaction with K$^+$ channel openers, *J. Biol. Chem.* **275**: 717.

Bao, L., and Rice, M. E., 2004, Mitochondrial H2O2 mediates glutamate-dependent inhibition of striatal dopamine release, 2004 Abstract Viewer/Itinerary Planner, Washington, DC: Society for Neuroscience (in press).

Beckman, K. B., and Ames, B. N., 1998, The free radical theory of aging matures, *Physiol. Rev.* **78**: 547.

Bernard, V., and Bolam, J. P., 1998, Subcellular and subsynaptic distribution of the NR1 subunit of the NMDA receptor in the neostriatum and globus pallidus of the rat: colocalization at synapses with the GluR2/3 subunit of the AMPA receptor, *Eur. J. Neurosci.* **10**: 3721.

Bindokas, V. P., Jordan, J., Lee, C. C., and Miller, R. J., 1996, Superoxide production in rat hippocampal neurons: selective imaging with hydroethidine, *J. Neurosci.* **16**: 1324.

Bindokas, V. P., Lee, C. C., Colmers, W. F., Miller, R. J., 1998, Changes in mitochondrial function resulting from synaptic activity in the rat hippocampal slice, *J. Neurosci.* **18**: 4570.

Blackwell, K. T., Czubayko, U., and Plenz, D., 2003, Quantitative estimate of synaptic inputs to striatal neurons during up and down states *in vitro, J. Neurosci.* **23**: 9123.

Boveris, A., and Chance, B., 1973, The mitochondrial generation of hydrogen peroxide. General properties and effect of hyperbaric oxygen, *Biochem. J.* **134**: 707.

Bull, D. R., Palij, P., Sheehan, M. J., Millar, J., Stamford, J. A., Kruk, Z. L., and Humphrey, P. P., 1990, Application of fast cyclic voltammetry to measurement of electrically evoked dopamine overflow from brain slices *in vitro, J. Neurosci. Meth.* **32**: 37.

Carriedo, S. G., Sensi, S. L., Yin, H. Z., and Weiss, J. H., 2000, AMPA exposures induce mitochondrial Ca^{2+} overload and ROS generation in spinal motor neurons *in vitro, J. Neurosci.* **20**: 240.

Chase, T. N., and Oh, J. D., 2000, Striatal dopamine- and glutamate-mediated dysregulation in experimental parkinsonism, *Trends Neurosci.* **23**: S86.

Chen, B. T., Avshalumov, M. V., and Rice, M. E., 2001, H₂O₂ is a novel, endogenous modulator of synaptic dopamine release, *J. Neurophysiol.* **85**: 2468.

Chen, B. T., Avshalumov, M. V., and Rice, M. E., 2002, Modulation of somatodendritic dopamine release by endogenous H₂O₂: susceptibility in substantia nigra but resistance in VTA, *J. Neurophysiol.* **87**: 1155.

Chen, Q., Veenman, L., Knopp, K., Yan, Z., Medina, L., Song, W. J., Surmeier, D. J., and Reiner, A., 1998, Evidence for the preferential localization of glutamate receptor-1 subunits of AMPA receptors to the dendritic spines of medium spiny neurons in rat striatum, *Neuroscience* **83**: 749.

Cheramy, A., Romo, R., Godeheu, G., Baruch, P., and Glowinski, J., 1986, *In vivo* presynaptic control of dopamine release in the cat caudate nucleus. II. Facilitatory or inhibitory influence of L-glutamate, *Neuroscience* **19**: 1081.

Cohen, G.,1994, Enzymatic/nonenzymatic sources of oxyradicals and regulation of antioxidant defenses, *Ann. N.Y. Acad. Sci.* **738**: 8.

Dawson, T. M., and Snyder, S. H., 1994, Gases as biological messengers: nitric oxide and carbon monoxide in the brain, *J. Neurosci.* **14**: 5147.

Deutsch, S. I., Rosse R. B., Schwartz B. L., and Mastropaolo, J., 2001, A revised excitotoxic hypothesis of schizophrenia: therapeutic implications, *Clin. Neuropharmacol.* **24**: 43.

Do, K. Q., Trabesinger, A. H., Kirsten-Kruger, M., Lauer, C. J., Dydak, U., Hell, D., Holsboer, F., Boesiger, P., and Cuenod, M., 2000, Schizophrenia: glutathione deficit in cerebrospinal fluid and prefrontal cortex *in vivo, Eur. J. Neurosci.* **12**: 3721.

Dugan, L. L., Sensi, S. L., Canzoniero, L. M., Handran, S. D., Rothman, S. M., Lin, T. S., Goldberg, M. P., and Choi, D. W., 1995, Mitochondrial production of reactive oxygen species in cortical neurons following exposure to N-methyl-D-aspartate, *J. Neurosci.* **15**: 6377.

Dunn-Meynell, A. A., Routh, V. H., McArdle, J. J., and Levin, B. E., 1997, Low-affinity sulfonylurea binding sites reside on neuronal cell bodies in the brain, *Brain Res.* **745**: 1.

Fearnley, J., and Lees, A. J., 1991, Aging and Parkinson's disease: substantia nigra regional selectivity, *Brain* **114**: 2283.

Finkel, T., and Holbrook, N.J., 2000, Oxidants, oxidative stress and the biology of ageing, *Nature* **408**: 239.

Fridovich, I., 1979, Hypoxia and oxygen toxicity, *Adv. Neurol.* **26**: 255.

Fujiyama, F., Fritschy, J. M., Stephenson, F. A., and Bolam, J. P., 2000, Synaptic localization of GABA$_A$ receptor subunits in the striatum of the rat, *J. Comp. Neurol.* **416**: 158.

Greenamyre, J. T., 2001, Glutamatergic influences on the basal ganglia, *Clin. Neuropharmacol.* **24**: 65.

Halliwell, B., 1992, Reactive oxygen species and the central nervous system, *J. Neurochem.* **59**: 1609.

Halliwell, B., and Gutteridge, J. M., 1984, Oxygen toxicity, oxygen radicals, transition metals and disease, *Biochem. J.* **219**: 1.

Hyman, S. E., and Malenka, R. C., 2001, Addiction and the brain: the neurobiology of compulsion and its persistence, *Nat. Rev. Neurosci.* **2**: 695.

Inagaki, N., Gonoi, T., Clement, J. P., Wang, C. Z., Aguilar-Bryan, L., Bryan, J., and Seino, S., 1996, A family of sulfonylurea receptors determines the pharmacological properties of ATP-sensitive K$^+$ channels, *Neuron* **16**: 1011.

Jiang, Z. G., and North, R.A., 1991, Membrane properties and synaptic responses of rat striatal neurones *in vitro*, *J. Physiol.* **443**: 533.

Jung, S. K., Kauri, L. M., Qian, W. J., and Kennedy, R. T., 2000, Correlated oscillations in glucose consumption, oxygen consumption, and intracellular free Ca^{2+} in single islets of Langerhans, *J. Biol. Chem.* **275**: 6642.

Kamsler, A., Segal, M., 2003, Paradoxical actions of hydrogen peroxide on long-term potentiation in transgenic superoxide dismutase-1 mice, *J. Neurosci.* **23**: 10359.

Karschin, C., Ecke, C., Ashcroft, F. M., and Karschin A., 1997, Overlapping distribution of K$_{ATP}$ channel-forming Kir6.2 subunit and the sulfonylurea receptor SUR1 in rodent brain, *FEBS Lett.* **401**: 59.

Keefe, K. A., Zigmond, M. J., and Abercrombie, E. D., 1993, *In vivo* regulation of extracellular dopamine in the neostriatum: influence of impulse activity and local excitatory amino acids, *J. Neural. Transm.* **91**: 223.

Kemp, J. M., and Powell, T. P., 1971, The structure of the caudate nucleus of the cat: light and electron microscopy, *Philos. Trans. R. Soc. Lond. B Biol. Sci.* **262**: 383.

Kennedy, R. T., Jones, S. R., and Wightman, R. M., 1992, Simultaneous measurement of oxygen and dopamine: coupling of oxygen consumption and neurotransmission, *Neuroscience* **47**: 603.

Kiss, J. P., and Vizi, E. S., 2001, Nitric oxide: a novel link between synaptic and nonsynaptic transmission, *Trends Neurosci.* **24**: 211.

Kita, H., 1996, Glutamatergic and GABAergic postsynaptic responses of striatal spiny neurons to intrastriatal and cortical stimulation recorded in slice preparations, *Neuroscience* **70**: 925.

Klann, E., and Thiels, E., 1999, Modulation of protein kinases and protein phosphatases by reactive oxygen species: implications for hippocampal synaptic plasticity, *Prog. Neuropsychopharmacol. Biol. Psychiatry* **23**: 359.

Knapp, L. T., and Klann, E., 2002, Role of reactive oxygen species in hippocampal long-term potentiation: contributory or inhibitory? *J. Neurosci. Res.* **70**: 1.

Kojima, S., Wu, S. T., Parmley, W. W., and Wikman-Coffelt, J., 1994, Relationship between intracellular calcium and oxygen consumption: effects of perfusion pressure, extracellular calcium, dobutamine, and nifedipine, *Am. Heart J.* **127**: 386.

Koob, G. F., 2000, Neurobiology of addiction. Toward the development of new therapies, *Ann. N. Y. Acad. Sci.* **909**: 170.

Krippeit-Drews, P., Kramer, C., Welker, S., Lang, F., Ammon, H. P. T., and Drews, G., 1999, Interference of H$_2$O$_2$ with stimulus-secretion coupling in mouse pancreatic β-cells, *J. Physiol.* **15**: 471.

Leviel, V., Gobert, A., and Guibert, B.,1990, The glutamate-mediated release of dopamine in the rat striatum: further characterization of the dual excitatory-inhibitory function, *Neuroscience* **39**: 305.

Liss, B., Bruns, R., and Roeper, J., 1999, Alternative sulfonylurea receptor expression defines sensitivity to K-ATP channels in dopaminergic midbrain neurons, *EMBO J.* **18**: 833.

Moghaddam, B., and Gruen, R. J., 1991, Do endogenous excitatory amino acids influence striatal dopamine release? *Brain Res.* **544**: 329.

Moghaddam, B., Gruen, R. J., Roth, R. H., Bunney, B. S., and Adams, R. N., 1990, Effect of L-glutamate on the release of striatal dopamine: *in vivo* dialysis and electrochemical studies, *Brain Res.* **518**: 55.

Moriguchi, S., Watanabe, S., Kita, H., and Nakanishi, H., 2002, Enhancement of N-methyl-D-aspartate receptor-mediated excitatory postsynaptic potentials in the neostriatum after methamphetamine sensitization. An *in vitro* slice study, *Exp Brain Res.* **144**: 238.

Mourre, C., Smith, M. L., Siesjo, B. K., and Lazdunski, M., 1990, Brain ischemia alters the density of binding sites for glibenclamide, a specific blocker of ATP-sensitive K$^+$ channels, *Brain Res.* **526**: 147.

Nemoto, S., Takeda, K., Yu, Z. X., Ferrans, V. J., and Finkel, T., 2000, Role for mitochondrial oxidants as regulators of cellular metabolism, *Mol. Cell. Biol.* **20**: 7311.

Nishida, M., Maruyama, Y., Tanaka, R., Kontani, K., Nagao, T., and Kurose, H., 2000, Gα_i and Gα_o are target proteins of reactive oxygen species, *Nature* **408**: 492.

Olanow, C. W., and Tatton, W. G., 1999, Etiology and pathogenesis of Parkinson's disease, *Annu. Rev. Neurosci.* **22**: 123.

Pellmar, T. C., 1987, Peroxide alters neuronal excitability in the CA1 region of guinea-pig hippocampus *in vitro*, *Neuroscience* **23**: 447.

Pellmar, T., 1986, Electrophysiological correlates of peroxide damage in guinea pig hippocampus *in vitro*, *Brain Res.* **364**:377.

Ramasarma, T., 1983, Generation of H₂O₂ in biomembranes, *Biochem. Biophys. Acta* **694**: 69.

Reynolds, I. J., and Hastings, T. G., 1995, Glutamate induces the production of reactive oxygen species in cultured forebrain neurons following NMDA receptor activation, *J. Neurosci.* **15**: 3318.

Rice, M. E., Forman, R. E, Chen, B. T., Avshalumov, M. V., Cragg, S.J., and Drew, K. L., 2002, Brain antioxidant regulation in mammals and anoxia-tolerant reptiles: balanced for neuroprotection and neuromodulation, *Comp. Biochem. Physiol. (Part C)* **133**: 515.

Sah, R., and Schwartz-Bloom, R. D., 1999, Optical imaging reveals elevated intracellular chloride in hippocampal pyramidal cells after oxidative stress, *J. Neurosci.* **19**: 9209.

Samanta, S., Perkinton, M.S., Morgan, M., and Williams, R.J., 1998, Hydrogen peroxide enhances signal-responsive arachidonic acid release from neurons: role of mitogen-activated protein kinase, *J. Neurochem.* **70**: 2082.

Sawa, A., and Snyder, S. H., 2002, Schizophrenia: diverse approaches to a complex disease, *Science* **296**: 692.

Schwanstecher, C., and Bassen, D., 1997, K_ATP-channel on the somata of spiny neurones in rat caudate nucleus: regulation by drugs and nucleotides, *Br. J. Pharmacol.* **121**: 193.

Seutin, V., Scuvee-Moreau, J., Masotte, L., and Dresse, A., 1995, Hydrogen peroxide hyperpolarizes rat CA1 pyramidal neurons by inducing an increase in potassium conductance, *Brain Res.* **683**: 275.

Smith, A. D., and Bolam, J. P., 1990, The neural network of the basal ganglia as revealed by the study of synaptic connections of identified neurons, *Trends Neurosci.* **13**: 259.

Sokolova, T, Gutterer, J. M., Hirrlinger, J., Hamprecht, B., and Dringen, R., 2001, Catalase in astroglia-rich primary cultures from rat brain: immunocytochemical localization and inactivation during the disposal of hydrogen peroxide, *Neurosci. Lett.* **297**: 129.

Sonsalla, P. K., Manzino, L., Sinton, C. M., Liang, C. L., German, D. C., and Zeevalk, G. D., 1997, Inhibition of striatal energy metabolism produces cell loss in the ipsilateral substantia nigra, *Brain Res.* **773**: 223.

Stefani, A, Chen, Q, Flores-Hernandez, J, Jiao, Y, Reiner, A, and Surmeier, D. J., 1998, Physiological and molecular properties of AMPA/Kainate receptors expressed by striatal medium spiny neurons, *Dev. Neurosci.* **20**: 242.

Sundaresan, M., Yu, Z. X., Ferrans, V. J., Irani, K., and Finkel, T., 1995, Requirement for generation of H₂O₂ for platelet-derived growth factor signal transduction, *Science* **270**: 296.

Westerink, B. H., Santiago, M., and De Vries, J. B., 1992, *In vivo* evidence for a concordant response of terminal and dendritic dopamine release during intranigral infusion of drugs, *Naunyn Schmied. Arch. Pharmacol.* **345**: 523.

Wu, Y., Pearl, S. M., Zigmond, M. J., and Michael, A. C., 2000, Inhibitory glutamatergic regulation of evoked dopamine release in striatum, *Neuroscience* **96**: 65.

Xia, Y., and Haddad, G. G., 1991, Major differences in CNS sulfonylurea receptor distribution between the rat (newborn, adult) and turtle, *J. Comp. Neurol.* **314**: 278.

Xu, J., Kao, S. Y., Lee, F. J., Song, W., Jin, L. W., and Yankner, B. A., 2002, Dopamine-dependent neurotoxicity of alpha-synuclein: A mechanism for selective neurodegeneration in Parkinson disease, *Nature Med.* **8**: 600.

Yamada, T., McGeer, P. L., Baimbridge, K. G., and McGeer, E. G., 1990, Relative sparing in Parkinson's disease of substantia nigra dopamine neurons containing calbindin-D28K, *Brain Res.* **526**: 303.

Yao, J. K., Reddy, R. D., and van Kammen, D. P., 2001, Oxidative damage and schizophrenia: an overview of the evidence and its therapeutic implications, *CNS Drugs* **15**: 287.

Ying, W., Han, S. K., Miller, J. W., and Swanson, R. A., 1999, Acidosis potentiates oxidative neuronal death by multiple mechanisms, *J. Neurochem.* **73**: 1549.

Zenisek, D., and Matthews, G., 2000, The role of mitochondria in presynaptic calcium handling at a ribbon synapse, *Neuron* **25**: 229.

21

HYDROGEN SULFIDE AS A SYNAPTIC MODULATOR

Hideo Kimura[1]

1. INTRODUCTION

Since the first description of the toxicity of hydrogen sulfide (H_2S) in 1713, many studies have been devoted to its toxicity. Although hydrogen sulfide (H_2S) is generally thought of in terms of a poisonous gas, relatively high endogenous levels of H_2S have been measured in the brains of rats, humans and bovine (Goodwin et al., 1989; Warenycia et al., 1989; Savage and Gould, 1990), suggesting that H_2S may have a physiological function. Physiological concentrations of H_2S specifically potentiate the activity of N-methyl-D-aspartate (NMDA) receptor and alter the induction of long-term potentiation (LTP) in the hippocampus, a synaptic model for memory (Abe and Kimura, 1996). H_2S can also regulate the release of corticotropin-releasing hormone from the hypothalamus (Russo et al., 2000). H_2S increases intracellular concentrations of Ca^{2+} ($[Ca^{2+}]_i$) in glia and induces Ca^{2+} waves, which mediate glial signal transmission (Nagai et al., 2004). Given the accumulating evidence for reciprocal interactions between glia and neurons, it has been suggested that glia modulates the synaptic transmission. H_2S may regulate the synaptic activity by modulating the activity of both neurons and glia in the brain. Based upon these observations it has been proposed that H_2S may function as a neuromodulator (Abe and Kimura, 1996).

Two other gases, nitric oxide (NO) and carbon monoxide (CO), are endogenously produced by enzymes localized in the brain (Garthwaite et al., 1988; Verma et al., 1993). Both NO and CO have been proposed as retrograde messengers in hippocampal long-term potentiation (LTP), a synaptic model of learning and memory (O'Dell et al., 1991; Schuman et al., 1991; Haley et al., 1992; Stevens and Wang, 1993; Zhuo et al., 1993; Bliss and Collingridge, 1993).

---

[1] National Institute of Neuroscience, 4-1-1 Ogawahigashi, Kodaira, Tokyo 187-8502H, Japan

Dendritic Neurotransmitter Release, edited by M. Ludwig
Springer Science+Business Media, Inc., 2005

2. H₂S PRODUCTION

Endogenous H_2S can be produced from cysteine by pyridoxal-5'-phosphate-dependent enzymes, including cystathionine β-synthase (CBS) and cystathionine γ-lyase (CSE). CBS mRNA is expressed in the brain, especially in hippocampus and cerebellum, while CSE mRNA is not detectable. The production of H_2S from brain homogenates (22.6 ± 1.6 nmol H_2S/min per gram protein) is suppressed by CBS-specific inhibitors, aminooxyacetate and hydroxylamine, while it is not suppressed by CSE-specific inhibitors, D,L-propargylglycine and β-cyano-L-alanine (Abe and Kimura, 1996). The CBS activator, S-adenosyl-L-methionine (SAM), enhances H_2S production. A model for CBS regulation has been proposed in which the C-terminal domain of CBS bends to and covers its own catalytic domain, suppressing enzymatic CBS activity. Once SAM binds to the regulatory domain of CBS, a conformational change occurs that frees the catalytic domain, and CBS becomes active (Shan et al., 2001). These observations suggest CBS as the candidate enzyme that produces H_2S in the brain. Other enzymes have also been proposed to produce H_2S in mammalian systems and their regulation may also be involved in the physiological function of H_2S (Toohey, 1989).

3. SUPPRESSION OF NEURONAL ACTIVITY BY HIGH CONCENTRATIONS OF H₂S

As H_2S is produced in the brain, we postulated that H_2S might play a role in synaptic transmission. Hippocampal field excitatory postsynaptic potentials (EPSPs) and population spikes evoked by the electrical stimulation of the Schafer collaterals in the CA1 region of rat hippocampal slices were found to have a concentration-dependent sensitivity to H_2S. Concentrations greater than 130 μM were found to suppress both field EPSPs and population spikes. The suppression by H_2S was specific to EPSPs and the population spikes as the action potentials generated by direct stimulation of presynaptic fibres were completely abolished by the Na channel blocker tetrodotoxin, but were not affected by H_2S, (Abe and Kimura, 1996).

4. H₂S FACILITATES HIPPOCAMPAL LTP

We examined the effect of physiological concentrations of H_2S on LTP. Although NaHS at concentrations less than 130 μM or a weak tetanic stimulation alone do not induce LTP, simultaneous application of both stimulators induce LTP (Abe and Kimura, 1996). This effect of H_2S is concentration-dependent in the range of 10-130μM. The simultaneous application of H_2S with a weak tetanic stimulation is required to facilitate the induction of LTP. When NaHS is applied 10 min before or after a weak tetanic stimulation, facilitation of LTP induction does not occur.

Moreover, the potentiation induced by H_2S is not additive with the LTP induced by a strong tetanic stimulation. The magnitude of LTP induced by a strong tetanic stimulation is not affected by a prior induction of LTP by a combination of weak tetanic stimulation and H_2S. In addition, after LTP had been induced by a strong tetanic stimulation, H_2S with a weak tetanic stimulation produced no further potentiation. These occlusion experiments showed that the potentiation induced by a weak tetanic stimulation in the presence of H_2S is likely to share common mechanisms with LTP induced by a strong tetanic stimulation.

The observation that NO and CO induce LTP even under the blockade of NMDA receptors (Zhuo et al., 1993) supports the idea that NO and CO act as retrograde messengers at synapses (O'Dell et al., 1991; Schuman and Madison, 1991; Stevens and Wang, 1993). In contrast, H_2S with a weak tetanic stimulation does not induce LTP in the presence of 2-amino-5-phosphonovalerate (APV), a specific blocker for NMDA receptor (Abe and Kimura, 1996), suggesting that the induction of LTP by H_2S requires the activation of NMDA receptors.

Hippocampal LTP induced by a tetanic stimulation requires the activation of NMDA receptors (Harris et al., 1984). H_2S alone does not itself induce any currents but significantly increases the NMDA-induced inward current. The enhancing effect of H_2S on the NMDA response is reversible and concentration-dependent in the same range as its LTP-facilitating effect (10-130μM) and is specific to NMDA receptors. H_2S has no effect on the currents induced by non-NMDA receptors.

Disulfide bonds play a role in modulating the function of many proteins, including NMDA receptors (Aizenman et al., 1989; Tang and Aizenman, 1993). It is therefore possible that H_2S interacts with disulfide bonds or free thiols in NMDA receptors. Application of the irreversible thiol-protecting agent dithiothreitol (DTT) with a weak tetanic stimulation significantly facilitates the induction of LTP. H_2S with a weak tetanic stimulation, however, still induces LTP even after treatment with DTT, demonstrating that DTT does not occlude the effect of H_2S (Abe and Kimura, 1996). Thus, the thiol redox sites contribute little, if any, to the potentiating effect of H_2S on the induction of LTP.

5. H₂S INCREASES INTRACELLULAR Ca²⁺ AND INDUCES Ca²⁺ WAVES IN ASTROCYTES

Glial cells have been considered to be the non-excitable and supportive elements in the nervous system, but they are now regarded as elements that respond to neuronal activity and modulate synaptic activity (Haydon, 2001). One class of glia, astrocytes, have neurotransmitter and hormone receptors and integrates neuronal input as waves of intracellular Ca^{2+} elevation. A number of conditions evoke increases in $[Ca^{2+}]_i$ in astrocytes that propagate into neighbouring astrocytes as intercellular Ca^{2+} waves (Cornell-Bell et al., 1990; Charles et al., 1991; Kang et al., 1998). These include neurotransmitters such as glutamate and ATP as well as the mechanical stimulation. Ca^{2+} waves have been well characterized in cultured astrocytes as well as acutely isolated hippocampal slices (Dani et al., 1992; Finkbeiner, 1992: Kang et al., 1998). In addition, there is accumulating evidence for reciprocal interactions between glial Ca^{2+} waves and neurons (Dani et al., 1992; Nedergaard, 1994; Parpura et al., 1994). Neuronal activity evokes glial Ca^{2+} waves (Dani et al., 1992), and glial Ca^{2+} waves drive neuronal activity (Nedergaard et al., 1994; Parpura et al., 1994). The multiple interactions between neurons and glia strongly suggest that glial cells are integral modulatory elements in synaptic transmission (Araque et al., 1999).

The observation that H_2S enhances the induction of hippocampal LTP suggests that H_2S may modulate some aspects of synaptic activity. Although H_2S enhances the NMDA receptor-mediated responses to glutamate in neurons, the effects of H_2S in the absence of glutamate on brain cells are not well understood. To investigate the effect of H_2S alone we measured changes in $[Ca^{2+}]_i$ in primary cultures of rat brain cells enriched for either neurons or glia using a Ca^{2+} imaging system with Calcium Green-1 as a Ca^{2+} sensitive fluorescent dye (Eberhard and Erne, 1991). Focal application of H_2S increased $[Ca^{2+}]_i$ in GFAP-positive astrocytes (Nagai et al., 2004). Although H_2S enhances the responses of neurons to NMDA, significant calcium responses to H_2S alone were not observed in neurons.

There is a difference in the time course of the increase in $[Ca^{2+}]_i$ between the astrocytes exposed directly to H_2S and those activated by the propagated Ca^{2+} waves. The $[Ca^{2+}]_i$ in the astrocytes exposed to H_2S sharply increased and gradually decayed, while the propagated Ca^{2+} waves showed oscillations with a faster decay (Nagai et al., 2004). The initial increase in the $[Ca^{2+}]_i$ induced by H_2S may be regulated by a different mechanism than the propagated Ca^{2+} waves.

In these primary cultures of rat brain cells enriched for glia, during first 10 days of preparation the cells are GFAP-negative (Tardy et al., 1989), but after 10 days cells become GFAP-positive and A2B5 negative astrocytes. Cells started responding to H_2S at 6 days, and the responses reached a maximum level at approximately 30 days (Nagai et al., 2004). Astrocytes started responding to NaHS at 20μM, and the amplitude of the responses increased and reached to the maximum level at 160μM. The responses then decreased at concentrations greater than 320μM, probably due to the toxic effect of H_2S observed in brain slices.

The responses to H_2S observed in cultures of astrocytes also occur in hippocampal slices. The bath application of H_2S induces increases in $[Ca^{2+}]_i$ in hippocampal brain slices (Nagai et al., 2004). It has been difficult to identify neurons and glia in brain slices during electrophysiological recording or imaging, because viable slices cannot be stained using specific cell markers. Recently, it has been found that increases in the $[Ca^{2+}]_i$ are specifically induced in astrocytes by low external concentrations of K^+ (Dallwig and Deitmer, 2002). By using this characteristic of astrocytes, we defined cells that respond to H_2S in hippocampal slices as astrocytes. Because astrocytes in acute brain slices respond to H_2S, the possibility that responses to H_2S may be the artefact caused by cell culture can be excluded. The requirement of several days in culture before astrocytes respond to H_2S may be due to the expression of proteins that are necessary for responding to H_2S or the formation of cell-cell junctions.

Glutamate and ATP are known to induce Ca^{2+} waves in astrocytes (Cornell-Bell et al., 1990; Guthrie et al., 1999). The responses to repeated applications of high concentrations of H_2S gradually decrease, while responses to repeated low concentrations increase in magnitude. The increase in intracellular Ca^{2+} induced by glutamate is dependent upon extracellular Ca^{2+}, while that induced by ATP is only dependent upon intracellular Ca^{2+} stores (Cornell-Bell et al., 1990; Kim et al., 1994; Fam et al., 2000). The increase in intracellular Ca^{2+} induced by NaHS is greatly suppressed in the Ca^{2+}-free medium, while the response to ATP was intact. H_2S increases the influx of Ca^{2+} similar to that caused by ionomycin, a Ca^{2+} ionophore (Nagai et al., 2004).

Since H_2S increases intracellular Ca^{2+}, it is possible that H_2S may activate a channel or a receptor associated with a channel that is permeable to Ca^{2+}. Trivalent cations, La^{3+} and Gd^{3+}, well known blockers of Ca^{2+} channels, potently suppress responses to H_2S (Nagai et al., 2004). Ruthenium red, which is a blocker of ryanodine receptors, which also inhibits voltage-gated Ca^{2+} channels (Cibulsky and Sather, 1999), also suppressed responses to H_2S. Three additional voltage-dependent Ca^{2+} channel blockers, flunarizine, nifedipine and ω-conotoxin GVIA were found to potently suppressed responses to H_2S.

The type of calcium channels, however, is difficult to determine, because the T-type blocker, flunarizine, L-type, nifedipine and N-type, ω-conotoxin were found to block the effect of H_2S less potently than the non-specific voltage-dependent Ca^{2+} channel blockers, La^{3+} and Gd^{3+}. Alternatively, La^{3+}, Gd^{3+} and ruthenium red are also potent inhibitors of transient receptor potential (TRP) family of channels that are permeable to Ca^{2+} and activated by the depletion of intracellular calcium stores (van Rossum et al., 2000; Clapham et al., 2001; Strubing et al., 2001). In addition, Mg^{2+} and MDL-12,330A block TRP channels (van Rossum et al., 2000; Strubing et al., 2001) and both substances suppress responses to H_2S. Interestingly, responses to H_2S were suppressed by depletion of intracellular Ca^{2+} stores by thapsigargin. So while H_2S increases intracellular concentrations of Ca^{2+}, largely by inducing Ca^{2+} influx, and to a lesser extent through the release from intracellular Ca^{2+} stores, further study is necessary to identify the specific type of Ca^{2+} channel that is activated by H_2S.

Interactions between neurons and glia may modulate synaptic transmission, as neuronal activity can evoke glial Ca^{2+} waves (Dani et al., 1992), and propagated Ca^{2+} waves in glial cells may modulate neuronal activity (Nedergaard, 1994; Parpura et al., 1994). After neurons are excited by NMDA, Ca^{2+} waves occurred in neighbouring astrocytes (Nagai et al., 2004). The Ca^{2+} waves induced by NMDA were completely suppressed by 10μM La^{3+} or 10μM Gd^{3+}. Since La^{3+} and Gd^{3+} block Ca^{2+} waves and also inhibit Ca^{2+} channels, La^{3+} and Gd^{3+} may inhibit the exocytosis of glutamate or some other factor from nerve terminals when neurons are stimulated by NMDA. H_2S released in response to neuronal excitation may increase intracellular Ca^{2+} and induce Ca^{2+} waves in neighbouring astrocytes.

6. CONCLUSION

Although H_2S enhances the induction of hippocampal LTP, the mechanism by which H_2S modulates synaptic activity is not well understood (Abe and Kimura, 1996). H_2S enhances the responses of neurons to glutamate in hippocampal slices, and H_2S alone induces the increase in intracellular Ca^{2+} in astrocytes (Nagai et al., 2004). In the presence of H_2S the induction of LTP is enhanced at the synapse and Ca^{2+} waves are induced in the surrounding astrocytes. Ca^{2+} waves propagate and reach another synapse and may modulate it. H_2S may therefore modulate synaptic activity by enhancing the responses to glutamate in neurons and inducing Ca^{2+} waves in astrocytes that propagate and modulate the neighbouring synapse.

7. REFERENCES

Abe, K., and Kimura, H., 1996, The possible role of hydrogen sulfide as an endogenous neuromodulator, *J. Neurosci.* **16**: 1066.

Aizenman, E., Lipton, D. A., and Loring, R. H., 1989, Selective modulation of NMDA responses by reduction and oxidation, *Neuron* **2**: 1257.

Araque, A., Parpura, V., Sanzgiri, R. P., and Haydon, P. G., 1999, Tripartite synapses: glia, the unacknowledged partner, *Trends Neurosci.* **22**: 208.

Bliss, T. V., and Coollingridge, G. L., 1993, A synaptic model of memory: long-term potentiation in the hippocampus, *Nature* **361**: 31.

Charles, A. C., Merrill, J. E., Dirksen, E. R., and Sanderson, M. J., 1991, Intercellular signaling in glial cells: calcium waves and oscillations in response to mechanical stimulation and glutamate, *Neuron* **6**: 983.

Cibulsky, S. M., and Sather, W. A., 1999, Block by ruthenium red of cloned neuronal voltage-gated calcium channels, *J. Pharmacol. Exp. Ther.* **289**: 1447.

Clapham, D. E., Runnels, L. W., and Strubing, C., 2001, The TRP ion channel family, *Nat. Rev. Neurosci.* **2**: 387.

Cornell-Bell, A. H., Finkbeiner, S. M., Cooper, M. S., and Smith, S. J., 1990, Glutamate induces calcium waves in cultured astrocytes: long-range glial signaling, *Science* **247**: 470.

Dallwig, R., and Deitmer, J. W., 2002, Cell-type specific calcium responses in acute rat hippocampal slices. *J. Neurosci. Methods* **116**: 77.

Dani, J. W., Chernjavsky, A., and Smith, S. J., 1992, Neuronal activity triggers calcium waves in hippocampal astrocyte networks, *Neuron* **8**: 429.

Eberhard, M., and Erne, P., 1991, Calcium binding to fluorescent calcium indicators: calcium green, calcium orange and calcium crimson, *Biochem. Biophys. Res. Commun.* **180**: 209.

Fam, S. R., Gallagher, C. J., and Salter, M. W., 2000, P2Y(1) purinoceptor-mediated Ca^{2+} signaling and Ca^{2+} wave propagation in dorsal spinal cord astrocytes, *J. Neurosci.* **20**: 2800.

Finkbeiner, S., 1992, Calcium waves in astrocytes-filling in the gaps, *Neuron* **8**: 1101.

Garthwaite, J., Charles, S. L., and Chess-Williams, R., 1988, Endothelium-derived relaxing factor release on activation of NMDA receptors suggests role as intercellular messenger in the brain, *Nature* **336**: 385.

Goodwin, L. R., Francom, D., Dieken, F. P., Taylor, J. D., Warenycia, M. W., Reiffenstein, R. J., and Dowling, G., 1989, Determination of sulfide in brain tissue by gas dialysis/ion chromatography: postmortem studies and two case reports, *J. Anal. Toxicol.* **13**: 105.

Guthrie, P. B., Knappenberger, J., Segal, M., Bennett, M. V., Charles, A. C., and Kater, S. B., 1999, ATP released from astrocytes mediates glial calcium waves, *J. Neurosci.* **19**: 520.

Harris, E. W., Ganong, A. H., and Cotman, C. W., 1984, Long-term potentiation in the hippocampus involves activation of N-methyl-D-aspartate receptors, *Brain Res.* **323**: 132.

Hayley, J. E., Wilcox, G. L., and Chapman, P. F., 1992, The role of nitric oxide in hippocampal long-term potentiation, *Neuron* **8**: 211.

Kang, J., Jiang, L., Goldman, S. A., and Nedergaard, M., 1998, Astrocyte-mediated potentiation of inhibitory synaptic transmission, *Nat. Neurosci.* **1**: 683.

Kim, W. T., Rioult, M. G., and Cornell-Bell, A. H., 1994, Glutamate-induced calcium signaling in astrocytes, *Glia* **11**: 173.

Nagai, Y., Tsugane, M., Oka, J.-I., Kimura, H., 2004, Hydrogen sulfide induces calcium waves in astrocytes, *FASEB J.* **18**: 557.

Nedergaard, M., 1994, Direct signaling from astrocytes to neurons in cultures of mammalian brain cells, *Science* **263**: 1768.

O'Dell, J. J., Hawkins, R. D., Kandel, E. R., and Arancio, O., 1991, Tests of the roles of two diffusible substances in long-term potentiation:evidence for nitric oxide as a possible early retrograde messenger, *PNAS USA* **88**: 11285.

Parpura, V., Basarsky, T. A., Liu, F., Jeftinija, K., Jeftinija, S., and Haydon, P. G., 1994, Glutamate-mediated astrocyte-neuron signaling, *Nature* **369**: 744.

Russo, C. D., Tringali, G., Ragazzoni, E., Maggiano, N., Menini, E., Vairano, M., Preziosi, P., and Navarra, P., 2000, Evidence that hydrogen sulphide can modulate hypothalamo-pituitary-adrenal axis function: in vitro and in vivo studies in the rat, *J. Neuroendocrinol.* **12**: 225.

Savage, J. C., and Gould, D. H., 1990, Determination of sulfide in brain tissue and rumen fluid by ion-interaction reversed-phase high-performance liquid chromatography. *J. Chromatogr.* **526**: 540.

Schuman, E. M., and Madison, D. V., 1991, A requirement for the intercellular messenger nitric oxide in long-term potentiation, *Science* **254**: 1503.

Shan, X., Dunbrack, R. L. J., Christopher, S. A., and Kruger, W. D., 2001, Mutation in the regulatory domain of cystathionine –synthase can functionally suppress patient-derived mutations in cis, *Human Mol. Genet.* **10**: 635.

Stevens, C. F., and Wang, Y., 1993, Reversal of long-term potentiation by inhibitors of haem oxygenase, *Nature* **364**: 147.

Strubing, C., Krapivinsky, G., Krapivinsky, L., and Clapham, D. E., 2001, TRPC1 and TRPC5 form a novel cation channel in mammalian brain, *Neuron* **29**: 645.

Tang, L.-H., Aizenman, E., 1993, The modulation of N-methyl-D-aspartate receptors by redox and alkylating reagents in rat cortical neurons in vitro, *J. Physiol.* **465**: 303.

Tardy, M., Fages, C., Riol, H., LePrince, G., Rataboul, P., Charriere-bertrand, C., and Nunez, J., 1989, Developmental expression of the glial fibrillary acidic protein mRNA in the central nervous system and in cultured astrocytes, *J. Neurochem.* **52**: 162.

Toohey, J. I., 1989, Sulphane sulphur in biological systems: a possible regulatory role, *Biochem. J.* **264**: 625.

van Rossum, D. B., Patterson, R. L., Ma, H. T., and Gill, D. L., 2000, Ca^{2+} entry mediated by store depletion, S-nitrosylation, and TRP3 channels. Comparison of coupling and function, *J. Biol. Chem.* **275**: 28562.

Verma, A., Hirsch, D. J., Glatt, C. E., Ronnett, G. V., and Snyder, S. H., 1993, Carbon monoxide: a putative neural messenger, *Science* **259**: 381.

Warenycia, M. W., Goodwin, L. R., Benishin, C. G., Reiffenstein, R. J., Francom, D. M., Taylor, J. D., and Dieken, F. P., 1989, Acute hydrogen sulfide poisoning. Demonstration of selective uptake of sulfide by the brainstem by measurement of brain sulfide levels, *Biochem. Pharmacol.* **38**: 973.

Zhuo, M., Small, S. A., Kandel, E. R., and Hawkins, R. D., 1993, Nitric oxide and carbon monoxide produce activity-dependent long-term synaptic enhancement in hippocampus, *Science* **260**: 1946.

INDEX